CULTURE OF ANIMAL CELLS

A Manual of Basic Technique

CULTURE OF ANIMAL CELLS
A Manual of Basic Technique

R. Ian Freshney

Department of Clinical Oncology
Cancer Research Campaign Laboratories
University of Glasgow

Alan R. Liss, Inc., New York

Address all Inquiries to the Publisher
Alan R. Liss, Inc., 150 Fifth Avenue, New York, NY 10011

Printed in the United States of America.

Second Printing, April 1984
Third Printing, November 1984

Library of Congress Cataloging in Publication Data

Freshney, R. Ian.
Culture of animal cells.

Bibliography: p.
Includes index.
1. Tissue culture. 2. Cell culture. I. Title.
QH585.F74 1983 591'.07'24 82-24960
ISBN 0-8451-0223-0

To my dear wife, Mary

Contents

10 Preparation and Sterilization

11 Disaggregation of the Tissue and Primary Culture

21 Specialized Techniques. I

22 Specialized Techniques. II

Preface

Tissue culture is not a new technique. It has been in existence since the beginning of this century and has passed through its simple exploratory phase, a later expansive phase in the 1950's, and is now in a phase of specialization concerned with control mechanisms and differentiated function. Matching the current trends, recent additions to the range of available tissue culture books have been concerned with specialized techniques and the result of this is that the basic procedures have become a little neglected.

It has been my objective in preparing this book to provide the novice to tissue culture with sufficient information to perform the basic techniques. It is anticipated that the reader will have a fundamental grasp of elementary anatomy, histology, cell physiology, and the basic principles of biochemistry, but will have had little or no experience in tissue culture. This book should prove useful at the advanced undergraduate level for technicians in training, for graduate studies, and at the post-doctoral level. It is intended as an introduction to the theory of the technique, and biology of cultured cells as well as a practical, step-by-step guide to procedures, and should be of value to anyone without any, or with little, prior experience in tissue culture. Of necessity, some of the more exciting developments in recent years, e.g., production of monoclonal antibodies by hybridoma cultures, can only be described briefly and references provided to further reading.

A list of reagents and commercial suppliers is located at the end of the book. Occasionally, a supplier's name is incorporated in the text but in most cases reference should be made to the trade index. Other reference materials included at the rear of the book are a glossary, a list of cell banks, a subject index, and the literature references cited in the text.

It is inevitable when preparing a text such as this that, in addition to my own experience, I have called upon the help and advice of many others both during the preparation of the book and in the twenty years or so since I was first introduced to the field. As with many other similar techniques, there is much of tissue culture that is never documented, but passed on by word of mouth at meetings, or, more often, in moments of conviviality after meetings. Hence there may be occasions when I have reproduced advice or information as if it were my own, without due acknowledgment to published work, because I have been unable to trace a reference, or none exists. In all such cases I would like to thank those who have contributed consciously or unconsciously to my own accumulated experience in the field.

While it would be impossible to recall all of those with whom contact over the past two decades has influenced my current understanding of the field, there are those of whom I must make special mention. First among these is Dr. John Paul, who introduced me to the field and whose sound common sense and practicality were a good introduction to what can, in the correct hands, be a very precise discipline. I owe him my sincere gratitude, as his one-time student and now associate and friend.

In my years with the Beatson Institute I have had the privilege to work with many people, both resident and visitors, and share in their experience in the development of techniques to which I would otherwise not have been exposed. In some cases they are acknowledged in the text or figure legends, but I hope any who are not mentioned by name will still recognize my gratitude.

Among others who should be named are those who have worked most closely with me in recent years, helped in my own research activities, and generated some of the data that appear on these pages. They include Ms Diana Morgan, Mrs. Elaine Hart, Mrs. Margaret Frame, Mr. Alistair McNab, Mrs. Irene Osprey, and Miss Sheila Brown. Although my wife and I do not work together usually, I have had the benefit of her skilled assistance at times, and, in addition, her experience in the field has added greatly to my own. Others who have worked with me for shorter periods, and elements of whose work may be reported here in part, are Mohammad Hassanzadah, Peter Crilly, Fadik Akturk, Metyn Guner, Fahri Celik, Aileen Sherry, Bob Shaw, and Carolyn MacDonald.

I have also been indebted to many people in Glasgow and elsewhere for helpful advice and collaboration. Among many others, these include David G.T. Thomas, David I. Graham, Michael Stack-Dunne, Peter Vaughan, Brian McNamee, David Doyle, Rona

MacKie, Kenneth C. Calman, and the late John Maxwell Anderson, with whom I had my first introduction to clinical collaboration.

I must also record my good fortune to have been able to spend time in other laboratories and learn from the approaches of others such as Robert Auerbach, Richard Ham, and Wally McKeehan.

I am also grateful to Flow Laboratories for their help and collaboration in running basic tissue culture courses and the resultant opportunity to broaden my knowledge of the field.

I would like to express my gratitude to Paul Chapple who first persuaded me that I should write a basic techniques book on tissue culture, and to numerous others, including Don Dougall, Wally and Kerstin McKeehan, Peter del Vecchio, John Ryan, Jim Smith, Rob Hay, Charity Waymouth, Sergey Federoff, Mike Gabridge, and Dan Lundin for help and advice during the preparation of the manuscript.

I would also like to thank Miss Donna Madore for converting my often illegible manuscript into typescript, Mrs. Marina LaDuke for expert photography, Miss Diane Leifheit for further help with the illustrations, and Ms Jane Gillies for preparing the line drawings. These four ladies spent many hours on my behalf and their patience and skill is greatly appreciated. My thanks are also due to Mrs. Norma Wallace for completing the final retype quickly and efficiently and at very short notice.

It would not be fitting for me to conclude this preface without further major acknowledgment to my wife, Mary, my daughter, Gillian, and son, Norman. Not only did I enjoy their sympathy and understanding at home, when I am sure, at times, I did not deserve it, but I also benefitted from the fruits of their labors during the day: drawing graphs, collecting references, researching and tabulating methods and information. My wife's experience in the field, plus countless hours in reading, revising, and collecting information, made her share in this work indispensable.

Chapter 1 Introduction

BACKGROUND

Tissue culture was first devised at the beginning of this century [Harrison 1907, Carrel, 1912] as a method for studying the behavior of animal cells free of systemic variations that might arise in the animal both during normal homeostasis and under the stress of an experiment. As the name implies, the technique was elaborated first with undisaggregated fragments of tissue, and growth was restricted to the migration of cells from the tissue fragment, with occasional mitoses in the outgrowth. Since culture of cells from such primary explants of tissue dominated the field for more than 50 yr, it is not surprising that the name "tissue culture" has stuck in spite of the fact that most of the explosive expansion in this area since the 1950s has utilized dispersed cell cultures.

Throughout this book the term "tissue culture" will be used as the generic term to include organ culture and cell culture. The term "organ culture" will always imply a three-dimensional culture of undisaggregated tissue retaining some or all of the histological features of the tissue *in vivo*. "Cell culture" will refer to cultures derived from dispersed cells taken from the original tissue, from a primary culture, or from a cell line or cell strain, by enzymatic, mechanical, or chemical disaggregation. The term "histotypic culture" will imply that cells have been reassociated in some way to recreate a three-dimensional tissue-like structure, e.g., by perfusion and overgrowth of a monolayer, reaggregation in suspension, or infiltration of a three-dimensional matrix such as collagen gel.

Harrison chose the frog as his source of tissue presumably because it was a cold-blooded animal, and consequently incubation was not required. Furthermore, since tissue regeneration is more common in lower vertebrates, he perhaps felt that growth was more likely to occur than with mammalian tissue. Although his technique may have sparked off a new wave of interest in cultivation of tissue *in vitro*, few later workers were to follow his example in the selection of species. The stimulus from medical science carried future interest into warm-blooded animals where normal and pathological development are closer to human. The accessibility of different tissues, many of which grew well in culture, made the embryonated hen's egg a favorite choice; but the development of experimental animal husbandry, particularly with genetically pure strains of rodents, brought mammals to the forefront as favorite material. While chick embryo tissue could provide a diversity of cell types in primary culture, rodent tissue had the advantage of producing continuous cell lines [Earle et al., 1943].

The demonstration that human tumors could also give rise to continuous cell lines [e.g., HeLa: Gey et al., 1952], encouraged interest in human tissue, helped later by Hayflick and Moorhead's classical studies with normal cells of a finite life-span [1961].

For many years the lower vertebrates and the invertebrates have been largely ignored though unique aspects of their development (tissue regeneration in amphibia, metamorphosis in insects) make them attractive systems for the study of the molecular basis of development. More recently agriculture has encouraged toxicity and virological studies in insects, and fish farming has required new exploration of fish diseases.

In spite of this resurgence of interest, tissue culture of lower vertebrates and the invertebrates remains a very specialized area, and the bulk of interest remains in avian and mammalian tissue. This has naturally influenced the development of the art and science of tissue culture, and much of what will be described in the ensuing chapters of this book reflect this as well as my own personal experience. Hence advice on incubation and the physical and biochemical properties of media refers to homiotherms and guidance on the appropriate modification for poikilothermic animals will require recourse to the literature. This will be discussed in a little more detail in a later chapter. Many of the basic techniques of asepsis, preparation and sterilization, primary culture, selection and cell separation, quantitation, and so on, apply equally to poikilotherms and will require only minor modification; on the whole the principles remain the same.

The types of investigation which lend themselves particularly to tissue culture are summarized in Figure 1.1.: (1) Intracellular activity, e.g., the replication and transcription of deoxyribonucleic acid (DNA), protein synthesis, energy metabolism; (2) intracellular flux, e.g., movement of ribonucleic acid (RNA) from the nucleus to the cytoplasm, translocation of hormone receptor complexes, fluctuations in metabolite pools; (3) "ecology," e.g., nutrition, infection, virally or chemically induced transformation, drug action; and (4) cell-cell interaction, e.g., embryonic induction, cell population kinetics, cell-cell adhesion.

The development of tissue culture as a modern sophisticated technique owes much to the needs of two major branches of medical research: the production of antiviral vaccines and the understanding of neoplasia. The standardization of conditions and cell lines for the production and assay of viruses undoubtedly provided much impetus to the development of modern tissue culture technology, particularly the production of large numbers of cells suitable for biochemical analysis. This and other technical improvements made possible by the commercial supply of reliable media and sera, and by the greater control of contamination with antibiotics and clean air equipment, has made tissue culture accessible to a wide range of interests.

In addition to cancer research and virology, other areas of research have come to depend heavily on tissue culture techniques. The introduction of cell fusion techniques [Barski, et al., 1960; Soreuil and Ephrussi, 1961; Littlefield, 1964; Harris and Watkins, 1965] established somatic cell genetics as a major component in the genetic analysis of higher animals including man, and contributed greatly, via the monoclonal antibody technique, to the study of immunology, already dependent on cell culture for assay techniques and production of hemopoietic cell lines.

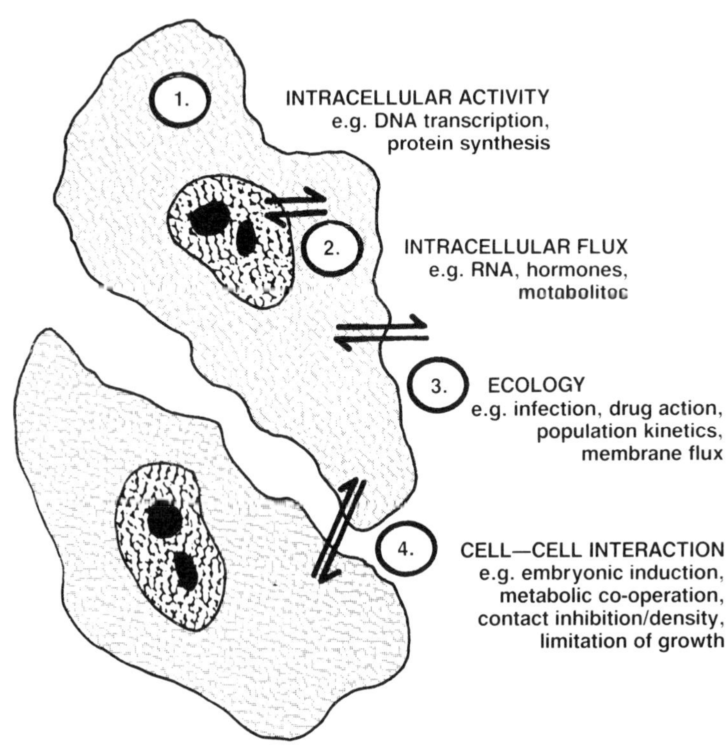

Fig. 1.1. *Areas of interest in tissue culture.*

Other areas of major interest include the study of cell interactions and intracellular control mechanisms in cell differentiation and development [Auerbach and Grobstein, 1958; Cox, 1974; Finbow and Pitts, 1981] and attempts to analyze nervous function [Bornstein and Murray, 1958; Minna et al., 1972]. Progress in neurological research has, however, not had the benefit of working with propagated cell lines as propagation of neurons has not so far been possible *in vitro* without resorting to the use of transformed cells (see Chapter 20).

Tissue culture technology has also been adopted into many routine applications in medicine and industry. Chromosomal analysis of cells derived from the womb by amniocentesis can reveal genetic disorders in the unborn child, viral infections may be assayed qualitatively and quantitatively on monolayers of appropriate host cells, and the toxic effects of pharmaceutical compounds and potential environmental pollutants can be measured in colony-forming assays.

Further developments in the application of tissue culture to medical problems may follow from the demonstration that cultures of epidermal cells form functionally differentiated sheets in culture [Green et al., 1979], and endothelial cells may form capillaries [Folkman and Haudenschild, 1980], suggesting possibilities in homografting and reconstructive surgery using an individual's own cells. The introduction of heterologous genetic material into mammalian cells [Willecke et al., 1979; Wigler et al., 1979], although somewhat overshadowed by current propagation in bacteria, may yet prove a desirable means for producing biologically significant compounds such as growth hormone and insulin. Similarly, the production of monoclonal antibodies [Kohler and Milstein, 1975] in hybrids between human plasma cells and human myeloma cells may prove a valuable technique for the production of specific antibodies.

It is clear that the study of cellular activity in tissue culture may have many advantages; but in summarizing these, below, considerable emphasis must also be placed on its limitations, in order to maintain some sense of perspective.

ADVANTAGES OF TISSUE CULTURE

Control of the Environment

The two major advantages, as implied above, are the control of the physicochemical environment (pH, temperature, osmotic pressure, O_2, CO_2 tension), which may be controlled very precisely, and the physiological conditions, which may be kept relatively constant but cannot always be defined. Most media still require supplementation with serum which is highly variable [Olmsted, 1967; Honn et al., 1975], and contains undefined elements such as hormones and other regulatory substances. Gradually, however, the functions of serum are being understood; and as a result, it is being replaced by defined constituents [Birch and Pirt, 1971; Ham and McKeehan, 1978; Barnes and Sato, 1980].

Characterization and Homogeneity of Sample

Tissue samples are invariably heterogeneous. Replicates even from one tissue vary in their constituent cell types. After one or two passages, cultured cell lines assume a homogeneous, or at least uniform, constitution as the cells are randomly mixed at each transfer and the selective pressure of the culture conditions tends to produce a homogeneous culture of the most vigorous cell type. Hence, at each subculture each replicate sample will be identical, and the characteristics of the line may be perpetuated over several generations. Since experimental replicates are virtually identical, the need for statistical analysis of variance is seldom required.

Economy

Cultures may be exposed directly to a reagent at a lower and defined concentration, and with direct access to the cell. Consequently, less is required than for injection *in vivo* where >90% is lost by excretion and distribution to tissues other than those under study.

DISADVANTAGES

Expertise

Culture techniques must be carried out under strict aseptic conditions, because animal cells grow much less rapidly than many of the common contaminants such as bacteria, molds, and yeasts. Furthermore, unlike microorganisms, cells from multicellular animals do not exist in isolation, and consequently, are not able to sustain independent existence without the provision of a complex environment, simulating blood plasma or interstitial fluid. This implies a level of skill and understanding to appreciate the requirements of the system and to diagnose problems as they arise. Tissue culture should not be undertaken casually to run one or two experiments.

Quantity

A major limitation of cell culture is the expenditure of effort and materials that goes into the production of relatively little tissue. A realistic maximum per batch for most small laboratories (2 or 3 people doing tissue culture) might be 1–10 g of cells. With a little more effort and the facilities of a larger laboratory, 10–100 g is possible; above 100 g implies industrial pilot plant scale, beyond the reach of most laboratories, but not impossible if special facilities are provided.

The cost of producing cells in culture is about ten times that of using animal tissue. Consequently, if large amounts of tissue (>10 g) are required, the reasons for providing them by tissue culture must be very compelling. For smaller amounts of tissue ($\leqslant 10$ g), the costs are more readily absorbed into routine expenditure; but it is always worth considering whether assays or preparative procedures can be scaled down. Semimicro- or micro-scale assays can often be quicker due to reduced manipulation times, volumes, centrifuge times, etc. and are often more readily automated (see under Microtitration, Chapter 19).

Instability

This is a major problem with many continuous cell lines resulting from their unstable aneuploid chromosomal constitution. Even with short-term cultures, although they may be genetically stable, the heterogeneity of the cell population, with regard to cell growth rate, can produce variability from one passage to the next. This will be dealt with in more detail in Chapters 12 and 18.

MAJOR DIFFERENCES *IN VITRO*

Many of the differences in cell behavior between cultured cells and their counterparts *in vivo* stem from the dissociation of cells from a three-dimensional geometry and their propagation on a two-dimensional substrate. Specific cell interactions characteristic of the histology of the tissue are lost, and, as the cells spread out, become mobile and, in many cases, start to proliferate, the growth fraction of the cell population increases. When a cell line forms it may represent only one or two cell types and many heterotypic interactions are lost.

The culture environment also lacks the several systemic components involved in homeostatic regulation *in vivo*, principally those of the nervous and endocrine systems. Without this control, cellular metabolism may be more constant *in vitro* than *in vivo*, but may not be truly representative of the tissue from which the cells were derived. Recognition of this fact has led to the inclusion of a number of different hormones in culture media (see Chapter 9) and it seems likely that this trend will continue.

Energy metabolism *in vitro* occurs largely by glycolysis, and although the citric acid cycle is still functional it plays a lesser role.

It is not difficult to find many more differences between the environmental conditions of a cell *in vitro* and *in vivo* and this has often led to tissue culture being regarded in a rather skeptical light. Although the existence of such differences cannot be denied, it must be emphasized that many specialized functions are expressed in culture and as long as the limits of the model are appreciated, it can become a very valuable tool.

Origin of Cells

If differentiated properties are lost, for whatever reason, it is difficult to relate the cultured cells to functional cells in the tissue from which they were derived. Stable markers are required for characterization (see Chapter 15); and in addition, the culture conditions may need to be modified so that these markers are expressed (see next chapter).

DEFINITIONS

There are three main methods of initiating a culture [Schaeffer, 1979] (see Glossary and Fig. 1.2): (1) *Organ culture* implies that the architecture characteristic of the tissue *in vivo* is retained, at least in part, in the culture. Toward this end, the tissue is cultured at the liquid/gas interface (on a raft, grid, or gel) which favors retention of a spherical or three-dimensional shape. (2) In *primary explant culture* a fragment of tissue is placed at a glass (or plastic)/liquid interface where, following attachment, migration is promoted in the plane of the solid substrate. (3) *Cell culture* implies that the tissue or outgrowth from the primary explant is dispersed (mechanically or enzymatically) into a cell suspension which may then be cultured as an adherent monolayer on a solid substrate, or as a suspension in the culture medium.

Organ cultures, because of the retention of cell interactions as found in the tissue from which the culture was derived, tend to retain the differentiated properties of that tissue. They do not grow rapidly (cell proliferation is limited to the periphery of the explant and is restricted mainly to embryonic tissue) and hence cannot be propagated; each experiment requires fresh explantations and this implies greater effort and poorer sample reproducibility than with cell culture. Quantitation is, therefore, more difficult and the amount of material that may be cultured is limited by the dimensions of the explant ($\leqslant 1$ mm^3) and the effort required for dissection and setting up the culture.

However, it must be emphasized that organ cultures do retain specific histological cell interactions without which it may be difficult to reproduce the characteristic cell behavior of the tissue.

Cell cultures may be derived from primary explants or dispersed cell suspensions. Because cell proliferation is often found in such cultures, propagation of cell

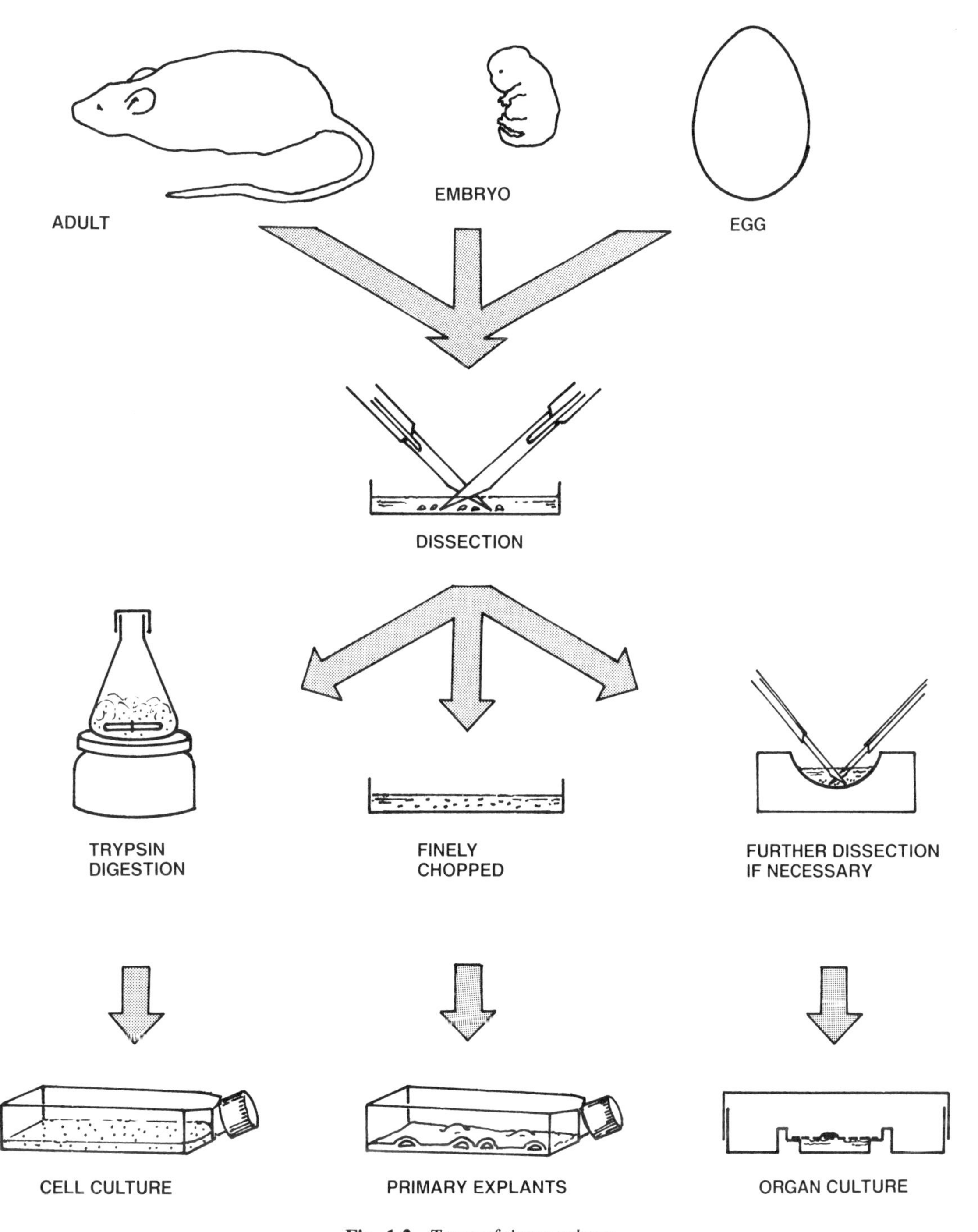

Fig. 1.2. *Types of tissue culture.*

lines becomes feasible. A monolayer or cell suspension, with a significant growth fraction, may be dispersed by enzymatic treatment or simple dilution and reseeded, or subcultured, into fresh vessels. This constitutes a "passage" and the daughter cultures so formed are the beginnings of a "cell line."

The formation of a cell line from a primary culture implies (1) an increase in total cell number over several generations, (2) that cells or cell lineages with similar high growth capacity will predominate, resulting in (3) a degree of uniformity in the cell population. The line may be characterized, and those characteristics will apply for most of its finite lifespan. The derivation of "continuous" (or "established" as they were once known) cell lines usually implies a phenotypic change or "transformation" and will be dealt with in Chapters 2 and 16.

When cells are selected from a culture, by cloning or some other method, the subline is known as a "cell strain." Detailed characterization is then implied. Cell lines may be propagated as an adherent monolayer or in suspension. *Monolayer* culture implies that adherence to the substrate is an integral part of survival and subsequent cell proliferation and is the mode of culture common to most normal cells with the exception of mature hemopoietic cells. *Suspension* cultures are derived from cells which can survive and proliferate without attachment; and this property seems to be reserved for hemopoietic cells, transformed cell lines, or cells from malignant tumors. It can be shown, however, that a small proportion of cells exists in many normal tissues, capable of proliferation in suspension (see Chapter 16). The identity of these cells remains unclear, but a relationship to the stem cell or uncommitted precursor cell compartment has been postulated. This concept implies that some cultured cells represent precursor pools within the tissue of origin, and the generality of this observation will be discussed more fully in the next chapter. Are cultured cell lines more representative of precursor cell compartments *in vivo* than of fully differentiated cells, bearing in mind that most differentiated cells do not normally divide?

Because they may be propagated as a uniform cell suspension or monolayer, cell cultures have many advantages in quantitation, characterization, and replicate sampling, but lack the potential for cell-cell interaction and cell-matrix interaction afforded by organ cultures. For this reason many workers have attempted to reconstitute three-dimensional cellular structures using aggregated cell suspension ("spheroids") or perfused high-density cultures (such as Vitafiber, Amicon) (see Chapter 21). In many ways some of the more exciting developments in tissue culture arise from recognizing the necessity of specific cell interaction in homogeneous or heterogeneous cell populations in culture. This may mark the transition from an era of fundamental molecular biology, where the regulatory processes have been worked out at the cellular level, to an era of cell or tissue biology where this understanding is applied to integrated populations of cells.

Chapter 2
Biology of the Cultured Cell

THE CULTURE ENVIRONMENT

The validity of the cultured cell as a model of physiological function *in vivo* has frequently been criticized. There are problems of characterization due to the alteration of the cellular environment; cells proliferate *in vitro* which would not normally *in vivo*, cell-cell and cell-matrix interactions are reduced because purified cell lines lack the heterogeneity and three-dimensional architecture found *in vivo*, and the hormonal and nutritional milieu is altered. This creates an environment which favors the spreading, migration, and the proliferation of unspecialized cells rather than the expression of differentiated functions. The provision of the appropriate environment, nutrients, hormones, and substrate is fundamental to the expression of specialized functions. Before considering such specialized conditions, let us examine the events accompanying the formation of a primary cell culture and a cell line derived from it (Fig. 2.1).

INITIATION OF THE CULTURE

Primary culture techniques are described in detail in Chapter 11. Briefly, a culture is derived either by outgrowth of migrating cells from a fragment of tissue, or by enzymatic or mechanical dispersal of the tissue. Regardless of the method employed, this is the first in a series of selective processes (Table 2.1) which may ultimately give rise to a relatively uniform cell line. In primary explantation (see Chapter 11) selection occurs by virtue of the cells' capacity to migrate from the explant, while with dispersed cells, only those cells which (1) survive the disaggregation technique and (2) adhere to the monolayer or survive in suspension will form the basis of a primary culture.

If the primary culture is maintained for more than a few hours, a further selection step will occur. Cells capable of proliferation will increase, some cell types will survive but not increase, and yet others will be unable to survive under the particular conditions used. Hence, the distribution of cell types will change and continue to do so until, in the case of monolayer cultures, all the available culture substrate is occupied. After confluence is reached the proportion of density-limited cells gradually decreases, and the proportion of cells which are less sensitive to density limitation of growth (see Chapters 12 and 16) increases. Virally or spontaneously transformed cells will overgrow their normal counterparts. Keeping the cell density low, e.g., by frequent subculture, helps to preserve the normal phenotype in cultures such as mouse fibroblasts, where spontaneous transformants tend to overgrow at high cell densities [Todaro and Green, 1963; Brouty-Boyé et al., 1979, 1980].

Some aspects of specialized function are expressed more strongly in primary culture, particularly when the culture becomes confluent (i.e., all the available growth area is utilized and the cells make close contact with one another). At this stage the culture will show its closest morphological resemblance to the parent tissue.

EVOLUTION OF CELL LINES

After the first subculture—or passage (see Fig. 2.1.)—the primary culture becomes a cell line (Chapter 1) and may be propagated and subcultured several times. With each successive subculture, the component of the population with the ability to proliferate most rapidly will gradually predominate, and nonproliferating or slowly proliferating cells will be diluted out. This is most strikingly apparent after the first subculture, where differences in proliferative capacity are compounded with varying abilities to withstand the trauma of trypsinization and transfer (see Chapter 12).

Although some selection and phenotypic drift will continue, by the third passage the culture becomes more stable, typified by a rather hardy, rapidly prolif-

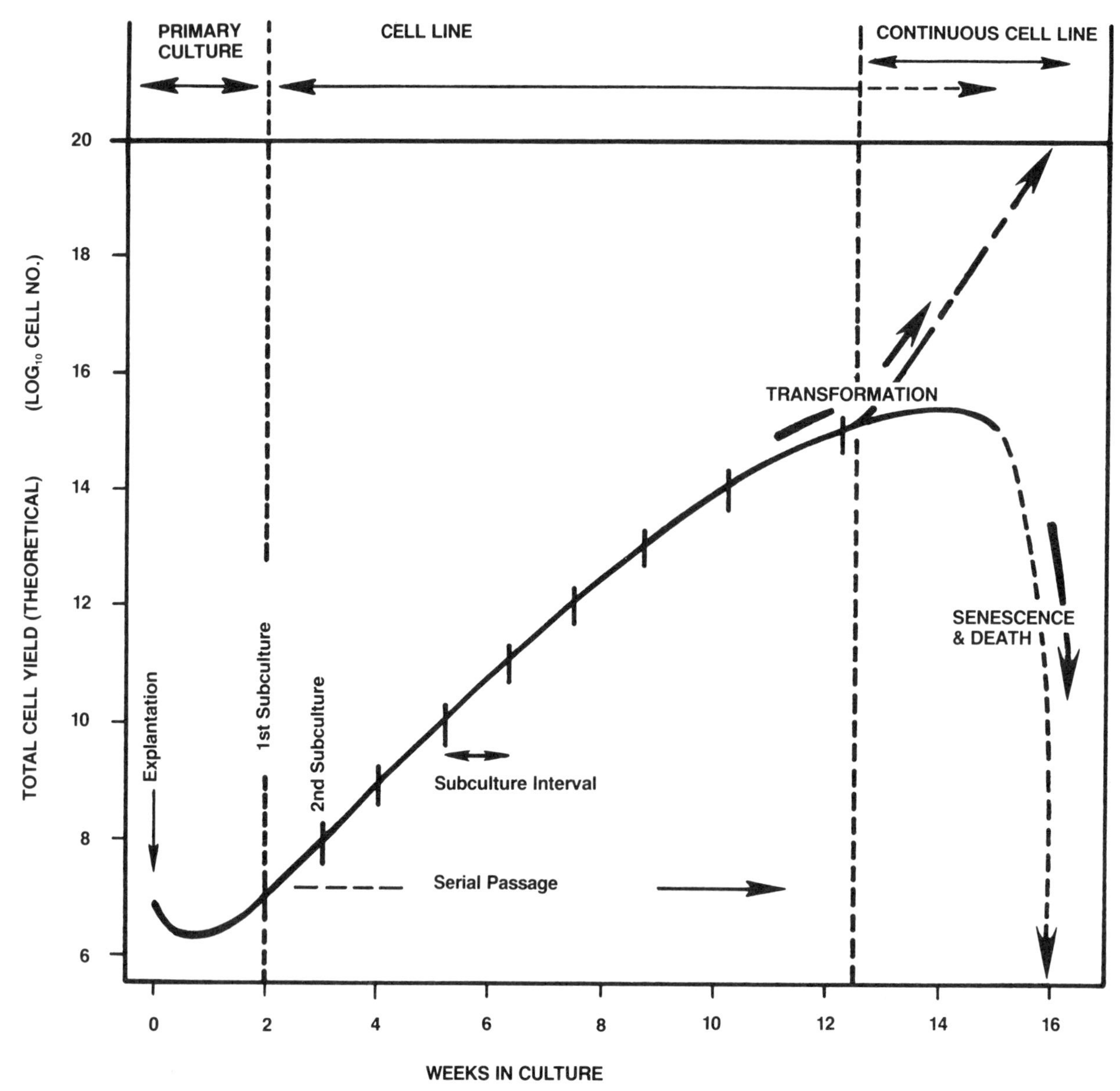

Fig. 2.1. *Evolution of a cell line. The vertical axis represents total cell growth (assuming no reduction at passage) on a log scale, the horizontal axis, time in culture on a linear scale, for a hypothetical cell culture. Although a continuous cell line is depicted as arising at 12½ wk, it could, with different cultures, arise at any time. Likewise, senescence may arise at any time, but for human diploid fibroblasts, it is most likely betwen 30 and 60 cell doublings or ten to 20 wk, depending on the doubling time.*

TABLE 2.1. Elements of Selection in the Evolution of Cell Lines

Stage	Factors influencing selection: Primary explant	Factors influencing selection: Enzymatic disaggregation
Isolation	Mechanical damage	Enzymatic damage
Primary culture	Outgrowth (migration)	Attachment and spreading
Subculture	Trypsin sensitivity	
	Nutrient, hormone, and substrate limitations	
Propagation as a cell line	Relative growth rates of different cells	
	Effect of cell density on predominance of normal or transformed phenotype (see text)	
	Nutrient, hormone, and substrate limitations	
Senescence, transformation	Normal cells die out	
	Transformed cells overgrow	

erating cell. This cell is frequently mesenchymal in origin, derived from connective tissue fibroblasts or vascular elements. The phenomenon is known as fibroblastic overgrowth, and while it has given rise to some very useful cell lines—e.g., WI38 human embryonic lung fibroblasts [Hayflick and Moorhead, 1961], BHK21 baby hamster kidney fibroblasts [MacPherson and Stoker, 1962], and perhaps the most famous of all, the L-cell, a subcutaneous fibroblast treated with methylcholanthrene [Earle et al., 1943; Sanford et al., 1948]—it has presented one of the major challenges of tissue culture since its inception: namely, how to prevent the overgrowth of the more fragile or slower-growing specialized cells such as hepatic parenchyma or epidermal keratinocytes. Inadequacy of the culture conditions is largely to blame for this problem and considerable progress has now been made in the use of selective media and substrates for the maintenance of many specialized cell lines (see Chapter 20).

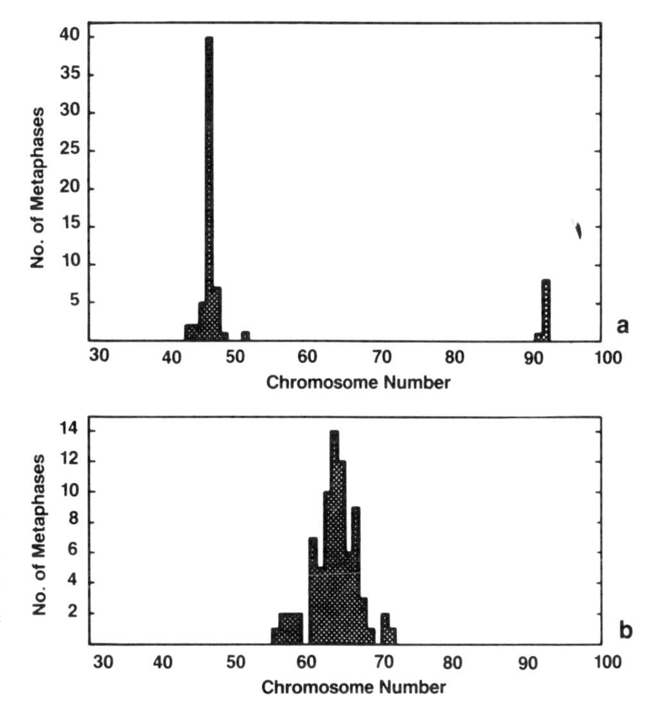

Fig. 2.2. *Chromosome numbers of finite and continuous cell lines. a. A normal human glial cell line. b. A continuous cell line from human metastatic melanoma.*

"CRISIS" AND THE DEVELOPMENT OF CONTINUOUS CELL LINES

Most cell lines may be propagated in an unaltered form for a limited number of cell generations, beyond which they may either die out or give rise to continuous cell lines (Fig. 2.1). The ability of a cell line to grow continuously probably reflects its capacity for genetic variation allowing subsequent selection. Human fibroblasts remain predominantly euploid throughout their culture life-span and never give rise to continuous cell lines [Hayflick and Moorhead, 1961], while mouse fibroblasts and cell cultures from a variety of human and animal tumors often become aneuploid in culture and give rise to continuous cultures with fairly high frequency. The alteration in a culture giving rise to a continuous cell line is commonly called "in vitro transformation" (see Chapter 16) and may occur spontaneously or be chemically or virally induced.

Continuous cell lines are usually aneuploid and often have a chromosome complement between the diploid and tetraploid value (Fig. 2.2). It is not clear whether the cells that give rise to continuous lines are present at explantation in very small numbers or arise later as a result of transformation of one or more cells. The second would seem to be more probable on cell kinetic grounds as continuous cell lines can appear quite late in a culture's life history, long after the time it would have taken for even one preexisting cell to overgrow. The possibility remains, however, that there is a subpopulation in such cultures with a predisposition to transform not shared by the rest of the cells.

The term "transformation" has been applied to the process of formation of a continuous cell line partly because the culture undergoes morphological and kinetic alterations, but also because the formation of a continuous cell line is often accompanied by an increase in tumorigenicity. A number of the properties of continuous cell lines are also associated with malignant transformations such as reduced serum requirement, reduced density limitation of growth, growth in semisolid media, and aneuploidy (see also Table 18.1), etc. These will be reviewed in more detail in Chapter 16. Similar morphological and behavioral changes can also be observed in cells which have undergone virally or chemically induced transformation.

Many (if not most) normal cells do not give rise to continuous cell lines. In the classic example [Hayflick and Moorhead, 1961] normal human fibroblasts remain euploid throughout their life-span and at crisis (usually around 50 generations) will stop dividing, though they may remain viable for up to 18 months thereafter. Human glia [Pontén and Westermark, 1980] and chick fibroblasts [Hay and Strehler, 1967] behave similarly. Epidermal cells, on the other hand, have shown gradually increasing life-spans with improvements in culture techniques [Green et al., 1979] and may yet be shown capable of giving rise to continuous growth. This may be related to the self-renewal capacity of the tissue *in vivo* (see below, this chapter). Continuous culture of lymphoblastoid cells is also possible [Moore et al., 1967], although in this case, transformation with Epstein Barr virus may be implicated.

It is possible that the condition that predisposes to the development of a continuous cell line is inherent genetic variation, so it is not surprising to find genetic instability perpetuated in continuous cell lines. A common feature of many human continuous cell lines is the development of a subtetraploid chromosome number (see Fig. 2.2).

For a further discussion of variation and instability see Chapter 18.

DEDIFFERENTIATION

This implies that differentiated cells lose their specialized properties *in vitro*, but it is often unclear whether (1) undifferentiated cells of the same lineage (Fig. 2.3) overgrow terminally differentiated cells of reduced proliferative capacity or (2) the absence of the appropriate inducers (hormones: cell or matrix interaction) causes deadaptation. In practice both may occur. Continuous proliferation may select undifferentiated precursors, which, in the absence of the correct inductive environment, do not differentiate.

An important distinction should be made between dedifferentiation, deadaptation, and selection. Dedifferentiation implies that the specialized properties of the cell are lost irreversibly, e.g., a hepatocyte would lose its characteristic enzymes (arginase, aminotrans-

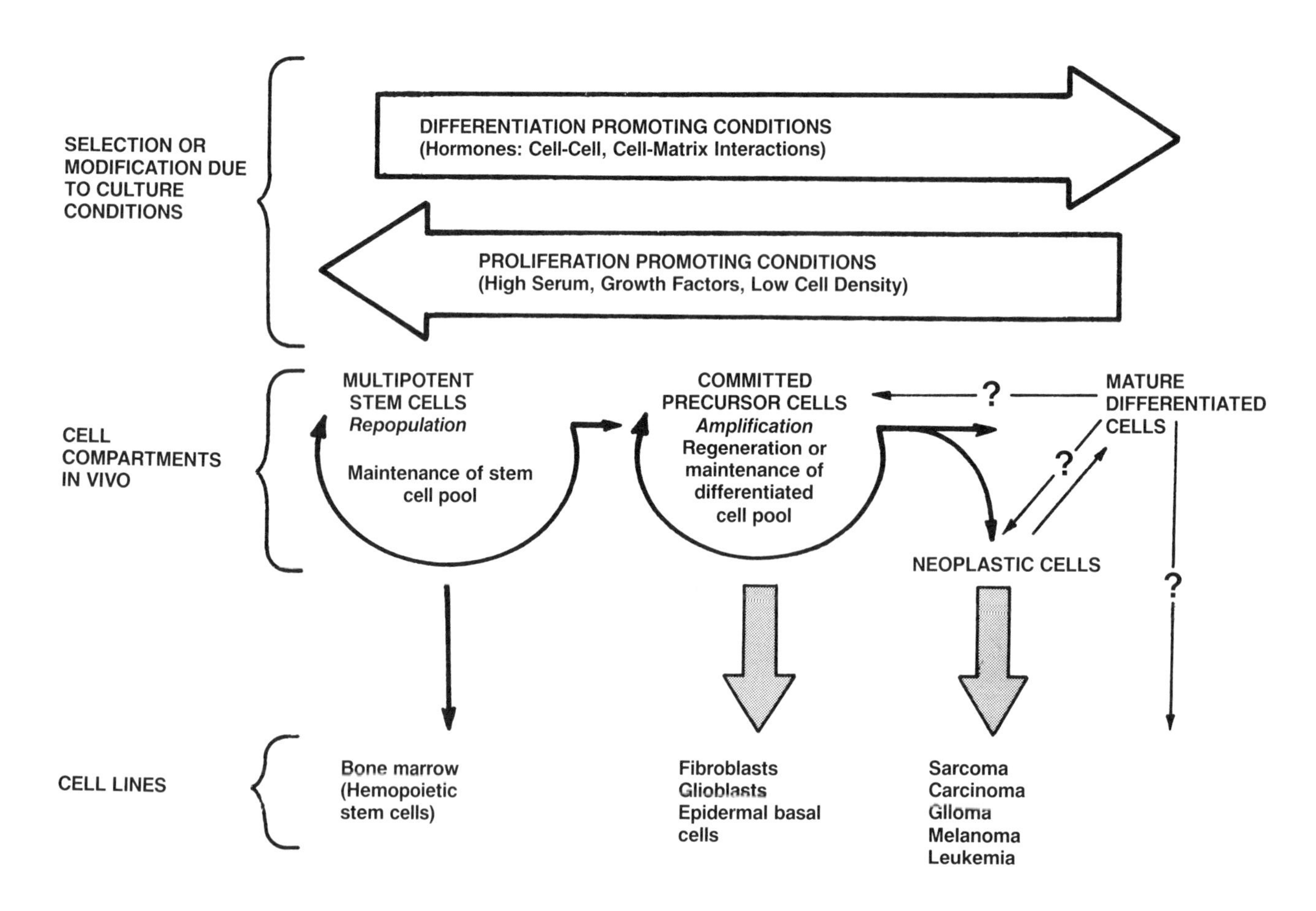

Fig. 2.3. *Origin of cell lines. With a few exceptions (e.g., differentiated tumor cells) culture conditions select for the proliferating cell compartment of the tissue or induce cells which are partially differentiated to revert to a precursor status. While neoplastic cells, and cell lines, may be derived from differentiated cells, it seems more likely that they arise from malignant precursor cells, some of which retain the capability to divide while continuing to differentiate.*

ferases, etc.), could not store glycogen or secrete serum proteins, and these properties could not be reinduced once lost. Deadaptation, on the other hand, implies that synthesis of specific products, or other aspects of specialized function, are under regulatory control by hormones—cell-cell interaction, cell-matrix interaction, etc.—and can be reinduced, given that the correct conditions can be recreated. Taking the hepatocyte again as an example, Michalopoulos and Pitot [1975] and Sattler et al. [1978] have shown that induction of tyrosine aminotransferase in normal rat liver cells requires hormones (insulin, hydrocortisone) and the correct matrix interaction (collagen). It is gradually becoming apparent that, given the correct culture conditions, differentiated functions can be expressed by a number of different cell types (Table 2.2) [Breen and DeVellis, 1974; Benda et al., 1968; Okazaki and Holtzer, 1966; Friend et al., 1971; Scher et al., 1971; Moscona and Piddington, 1966; Linser and Moscona, 1980], and the concept of dedifferentiation is now regarded as an unlikely explanation for the loss of specialized functions.

For correct inducing conditions to act, the appropriate cells must be present. In early attempts at liver cell culture, lack of expression of hepatocyte properties was due partly to overgrowth of the culture by connective tissue fibroblasts or endothelium from blood vessels or sinusoids. By using the correct disaggregation technique [Berry and Friend, 1969] and the correct culture conditions [Michalopoulos and Pitot, 1975; Malan-Shibley and Iype, 1981], hepatocytes can be selected preferentially. Similarly, epidermal cells can be grown either by using a confluent feeder layer [Rheinwald and Green, 1975] or selective medium [Peehl and Ham, 1980; Tsao et al., 1982]. The appearance of other examples, e.g., feeder selection for breast and colonic epithelium [Freshney et al., 1981], D-valine for the isolation of kidney epithelium [Gilbert and Migeon, 1975], and the use of cytotoxic antibodies [Edwards et al., 1980] (selection procedures reviewed in Chapters 13, 14, and 20), clearly demonstrate that the selective culture of specialized cells is not the insuperable problem that it once appeared.

WHAT IS A CULTURED CELL?

The question remains open, however, as to the exact nature of the cells that grow in each case. Expression of differentiated markers under the influence of inducing conditions may either mean that the cells being cultured are mature and only require induction to maintain synthesis of specialized proteins, or that the culture is composed of precursor or stem cells which are capable of proliferation but remain undifferentiated until the correct inducing conditions are applied, whereupon some or all of the cells mature to differentiated cells. It may be useful to think of a cell culture as being in an equilibrium betwen multipotent stem cells, undifferentiated but committed precursor cells, and mature differentiated cells (see Fig. 2.3) and that the equilibrium may shift according to the environmental conditions. Routine serial passage at relatively low cell densities would promote cell proliferation and little differentiation while high cell densities, low serum, and the appropriate hormones would promote differentiation and inhibit cell proliferation.

The source of the culture will also determine which cellular components may be present. Hence cell lines derived from the embryo may contain more stem cells and precursor cells and be capable of greater self-renewal than cultures from adults. In addition, cultures from tissues which are undergoing continuous renewal *in vivo* (epidermis, intestinal epithelium, hemopoietic cells) will still contain stem cells which, under the appropriate culture conditions, may survive indefinitely, while cultures from tissue which renew only under stress (fibroblasts, muscle, glia) may only contain committed precursor cells with a limited culture life-span.

Thus, the identity of the cultured cell is not only defined by its lineage *in vivo* (hemopoietic, hepatocyte, glial, etc.) but also by its position in that lineage (stem cell, committed precursor cell, or mature differentiated cell). With the exception of mouse teratomas and one or two other examples from lower vertebrates, it seems unlikely that cells will change lineage (transdifferentiate), but they may well change position in the lineage, and may even do so reversibly in some cases.

When cells are cultured from a neoplasm, they need not adhere to these rules. Thus a hepatoma from rat may proliferate *in vitro* and still express some differentiated features, but the closer they are to the normal phenotype, the more induction of differentiation may inhibit proliferation. Although the relationship between position in the lineage and cell proliferation may become relaxed (though not lost; B16 melanoma cells still produce more pigment at high cell density and at a low rate of cell proliferation than at a low cell density and a high rate of cell proliferation), transfer between lineages has not been clearly established.

TABLE 2.2. Examples of Cultured Cell Lines and Strains Which Express Differentiated Properties *In Vitro*

Origin	Cell line	Status	Species	Property	Reference
Spleen	Friend	c*	Mouse	Hemoglobin synthesis	[Scher et al., 1971]
Pigmented retina	Pigmented retina		Chick	Pigmentation	[Coon and Cahn, 1966]
	Cartilage		Chick	Cartilage synthesis	[Coon and Cahn, 1966]
Reuber hepatoma	H 4-II-E-C3	c	Rat	Tyrosine aminotransferase induction	[Pitot et al., 1964]
Morris hepatoma	HTC	c	Rat	Tyrosine aminotransferase induction	[Granner et al., 1968]
Myeloid leukemia	K562	c	Human	Hemoglobin synthesis	[Andersson et al., 1979a, b]
Glioma	C_6	c	Rat	Glial fibrillary acidic protein Glycerol phosphate dehydrogenase	[Benda et al., 1968]
Pituitary tumor	GH_2, GH_3	c	Rat	Growth hormone production	[Buonassisi et al., 1962; Tashjian, 1979]
Adrenal cortex tumor	Adrenal cortex	c	Rat	Steroid synthesis	"
Melanoma	B16	c	Mouse	Pigmentation	[Nilos and Makarski, 1978]
Liver	Hepatocytes		Rat	Tyrosine aminotransferase	[Malan-Shibley and Iype, 1981]
Glioma	CCM	c	Human	Glial fibrillary acidic protein	[Freshney et al., unpublished observations]
Brain (fetal)	Various		Human	(same as above)	(same as above)
Skeletal muscle			Chick	Myogenesis Creatinine phosphokinase induction	[Richler and Yaffe, 1970]
Epidermis	Keratinocytes		Mouse	Cornification	[Fusening et al., 1972; Lillie et al., 1980]
Epidermis	Keratinocytes		Human	Cornification	[Rheinwald and Green, 1975b]
Teratoma	Keratinizing epithelium	c	Mouse	Cornification	[Rheinwald and Green, 1975a]
Neuroblastoma	C1300	c	Rat	Neurite outgrowth	[Lieberman and Sachs, 1978]
Hypothalamus	C7		Mouse	Synthesis of neurophysin and vasopressin	[DeVitry et al., 1974]
Myeloid leukemia	HL60	c	Human	Globulin synthesis	[Olsson and Ologsson, 1981]
Kidney	MDCK	c	Dog		[Rindler et al., 1979]
Placenta			Human	Choriogonadotropin	[Yang, 1978]
Teratocarcinoma			Mouse	Various	[Martin, 1975]
Myeloma			Mouse	Immunoglobulin production	[Horibata and Harris, 1970]

*c, continuous.

FUNCTIONAL ENVIRONMENT

Since the inception of tissue culture as a viable technique, culture conditions have been adapted to suit two major requirements: (1) production of cells by continuous proliferation and (2) preservation of specialized functions. The upsurge of interest in cellular and molecular biology and virology in the 1950s and 1960s concentrated mainly on fundamental intracellular processes such as the regulation of protein synthesis, often requiring large numbers of cells. Later, the development of such techniques as molecular hybridization and gene transfer allowed the emphasis to shift to the study of the regulation of specialized functions.

While the need for bulk cultures remains, more attention has been directed to the creation of an environment which will permit the controlled expression of differentiation.

It has been recognized for many years that specific functions are retained for longer where the three-dimensional structure of the tissue is retained, as in organ culture (see Chapter 21). Unfortunately, organ cultures cannot be propagated, must be prepared *de novo* for each experiment, and are more difficult to quantify than cell cultures. For this reason there have been numerous attempts to recreate three-dimensional structures by perfusing monolayer cultures [Kruse et al., 1970; Whittle and Kruse, 1973; Knazek et al., 1972; Knazek, 1974; Gullino and Knazek, 1979] and to reproduce elements of the environment *in vivo* by culturing cells [Michalopoulos and Pitot, 1975] on or in special matrices like collagen gel [Tang et al., 1981; Burwen and Pitelka, 1980], cellulose [Leighton, 1951] or gelatin sponge [Douglas et al., 1976], or matrices from other natural tissue matrix glycoproteins such as fibronectin, chondronectin, and laminin [Gospodarowicz et al., 1980; Kleinman et al., 1981; Reid and Rojkind, 1979] (see Chapter 8). These techniques present some limitations, but with their provision of homotypic cell interactions, cell matrix interactions, and the possibility of introducing heterotypic cell interactions, they may hold considerable promise for the examination of tissue-specific functions.

The development of normal tissue functions in culture would facilitate investigation of pathological behavior such as demyelination and malignant invasion. But, from a fundamental viewpoint, it is only when cells *in vitro* express their normal functions that any attempt can be made to relate them to their tissue of origin. Expression of the differentiated phenotype need not be complete, since the demonstration of a single cell type-specific cell surface antigen may be sufficient to place a cell in the correct lineage. More complete functional expression may be required, however, to place a cell in its correct position in the lineage, and to reproduce a valid model of its function *in vivo*.

Chapter 3
Design and Layout of the Laboratory

The major requirement that distinguishes tissue culture from other laboratory techniques is the need to maintain asepsis. This is accentuated by the much slower growth of cultured animal cells relative to most of the major potential contaminants. The introduction of laminar flow cabinets has greatly simplified the problem and allows the utilization of unspecialized laboratory accommodation (see below and Chapter 5).

There are six main functions to be accommodated: sterile handling, incubation, preparation, wash-up, sterilization, and storage (Table 3.1). The clean area for sterile handling should be located at one end of the room and wash-up and sterilization at the other, with preparation, storage, and incubation in between. The preparation area should be adjacent to the wash-up and sterilization areas, and storage and incubators should be readily accessible to the sterile working area.

STERILE HANDLING AREA

This should be located in a quiet part of the laboratory, its use should be restricted to tissue culture, and there should be no through traffic or other disturbance likely to cause dust or draughts. Use a separate room or cubicle if laminar flow cabinets are not available. The work area, in its simplest form, should be a plastic laminate-topped bench, preferably plain white or neutral gray, to facilitate observation of cultures, dissection, etc., and enable accurate reading of pH when using phenol red as an indicator. Nothing should be stored on this bench and any shelving above should only be used in conjuction with sterile work, e.g., for holding pipette cans and instruments. The bench should either be freestanding (away from the wall) or sealed to the wall with a plastic sealing strip.

Laminar Flow

The introduction of laminar flow cabinets with sterile air blown over the work surface (Fig. 3.1) (see Chapter 4) affords greater control of sterility at a lower cost than providing a separate sterile room. Individual cabinets are preferable as they separate operators and can be moved around, but laminar flow wall or ceiling units in batteries can be used (Fig. 3.2). With cabinets, only the operator's arms enter the sterile area, while with laminar flow wall or ceiling units, there is no cabinet and the operator is part of the work area. While this may give more freedom of movement, particularly with large pieces of apparatus (roller bottles, fermentors), greater care must be taken by the operator not to disrupt the laminar flow, and it may prove necessary to wear caps and gowns to avoid contamination.

Make sure you select hoods that suit your accommodation—freestanding or bench top—and allow plenty of leg room underneath with space for pumps, aspirators, etc. Select chairs which are of a suitable height; preferably with adjustable seat height and back angle.

It is a good principle, made easier by the introduction of laminar flow, to create a "sterility gradient" in the tissue culture laboratory. Hence, a single room housing all the necessary functions of a tissue culture laboratory should have its sterile cabinets located at one end, furthest from the door, while wash-up, preparation of glassware or reagents, centrifugation, etc., would be best performed at the opposite end of the room (Fig. 3.3,3.4). This principle still applies where laminar flow is not available, in fact, even more so; but the introduction of laminar flow, particularly horizontal laminar flow, makes the gradient easier to maintain.

In addition, you may wish to provide a small room or cubicle for use as a containment area (Fig. 3.5). This must be separated by a door or air lock from the rest of your suite, and will need its own incubators, freezer, refrigerator, centrifuge, etc. It will also re-

TABLE 3.1. Tissue Culture Facilities

Minimum requirements (essential)	Desirable features (beneficial)	Useful additions
Sterile area: clean and quiet area, no through traffic Separate from animal house and microbiological labs	Filtered air (air conditioning)	Piped CO_2 and compressed air
Preparation area	Hot room with temperature recorder	Storeroom for bulk plastics
Storage areas: liquids—ambient, 4° C, −20° C; glassware (shelving); plastics (shelving); small items (drawers); specialized equipment (slow turnover), cupboard(s); chemicals—ambient, 4° C, −20° C (share with liquids but keep chemicals in sealed container over desiccant)	Microscope room Dark-room Service bench adjacent to culture area Separate prep room Separate sterilizing room	Containment room for biohazard work Liquid N_2 storage tank (~500 L)
Space for incubator(s)		
Space for liquid N_2 freezer(s)		
Wash-up area (not necessarily within tissue culture laboratory, but adjacent)		

quire a special laminar flow cabinet or pathogen hood with separate extract and pathogen trap (for fuller description of containment facilities, see Chapter 7).

INCUBATION

The requirement for cleanliness is not as stringent as with sterile handling, but clean air, low disturbance level, and no through traffic will endow your incubation area with a better chance of avoiding dust, spores, and drafts that carry them.

Incubation may be carried out in separate incubators or in a thermostatically controlled hot room. Incubators, bought singly, are inexpensive and economic in space; but as soon as you require more than two, their cost is more than a simple hot room and their use less convenient. As a rough guide, you will need 0.2 m^3 (200 l, 6 ft^3) of incubation space (0.5 m^2, 6 ft^2 shelf space) per person. Extra provision may need to be made for a humid incubator(s) with a controlled CO_2 level in the atmosphere.

Hot Room

If you have the space within the laboratory area or have an adjacent room readily available and accessible, it may be possible to convert this into a hot room (Fig. 3.6). It need not be specifically constructed but should be sufficiently well insulated not to allow "cold spots" on the walls. If insulation is required, line with plastic laminate-veneered board, separated from the wall by about 5 cm (2 in) fiberglass, mineral wool, or fire-retardant plastic foam. Mark the location of the straps or studs carrying the lining panel to identify anchorage points for shelving. Shelf supports should be spaced at 500–600 mm (21 in) to support shelving without sagging.

Do not underestimate the space that you will require in the lifetime of the hot room. It costs very little more to equip a large hot room than a small one. Calculate on the basis of the amount of shelf space you will require; if you have just started, multiply by five or ten; if you have been working for some time, by two or four. Allow 200–300 mm (9 in) between shelves and use wider shelves (450 mm, 18 in) at the bottom, and narrower (250–300 mm, 12 in) above eye level. Slatted shelving (mounted on adjustable brackets) should be used to allow for air circulation. It must be flat and perfectly horizontal with no bumps or irregularities.

Wooden furnishings should be avoided as much as possible as they warp in the heat and can harbor infestations.

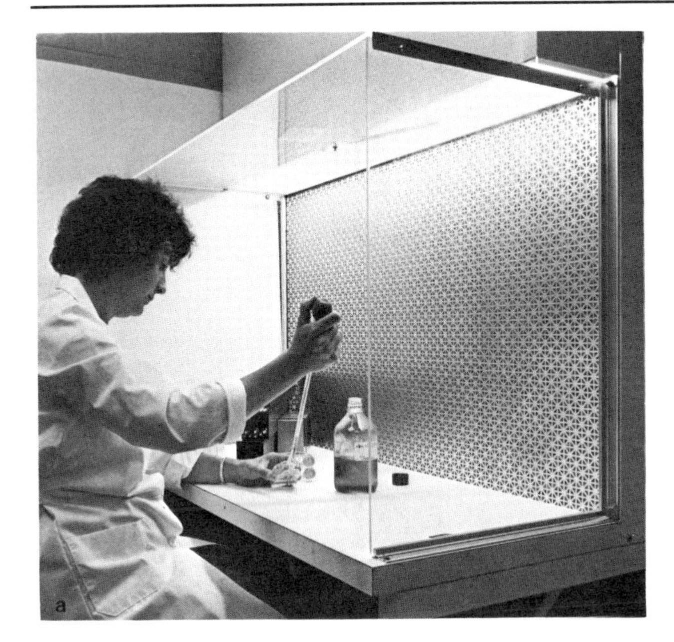

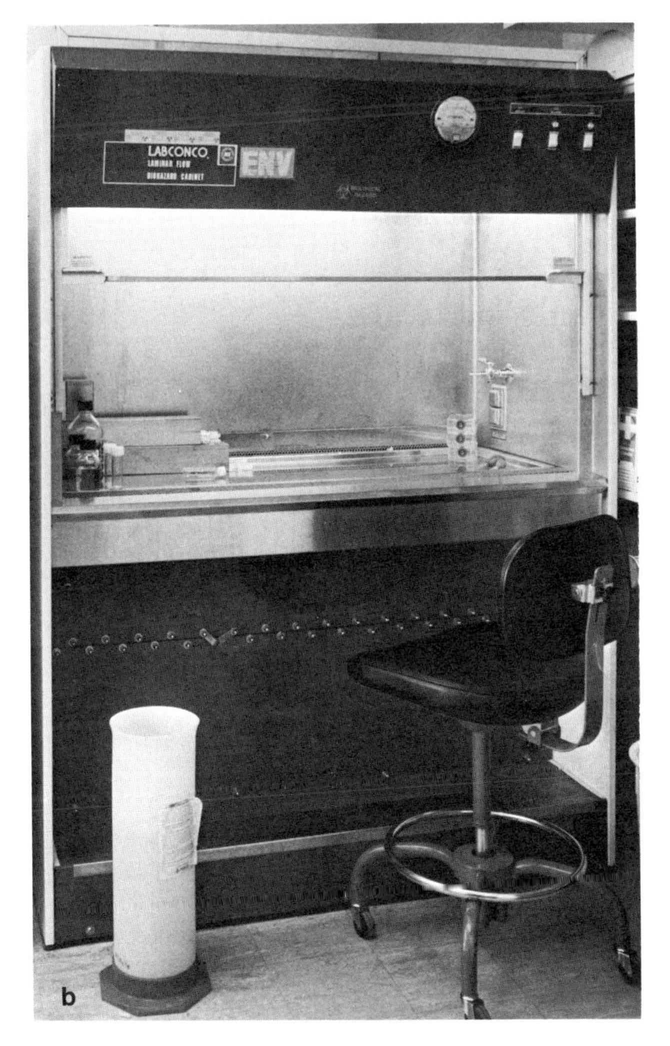

Fig. 3.1. *Laminar flow hoods. a. horizontal flow. b. vertical flow (Biohazard type).*

Fig. 3.2. *Aseptic room. Filtered air supplied from the ceiling, and the whole room is regarded as a sterile working area.*

A small bench, preferably stainless steel or solid plastic laminate, should be provided at some part of the hot room. This should accommodate a microscope, its transformer, and the flasks that you wish to examine. If you contemplate doing cell synchrony experiments or having to make any sterile manipulations at 36.5° C, you should also allow space for a small laminar flow unit (300 × 300 or 450 × 450 mm, 12–18 in, square filter size) either wall mounted or on a stand over part of the bench. Alternatively a small laminar flow hood (not more than 1,000 mm, 3 ft, wide) could be located in the room. The fan motor of the hood should be tropically wound and should not run continuously. Apart from wear of the motor, it will generate heat.

Incandescent lighting is preferable to fluorescent which can cause degradation of constituents of the medium. Furthermore, some fluorescent tubes have difficulty in striking in a hot room.

The temperature should be controlled within ± 0.5° C at any point and at any time. This depends on (1) the sensitivity and accuracy of the control gear, (2) the

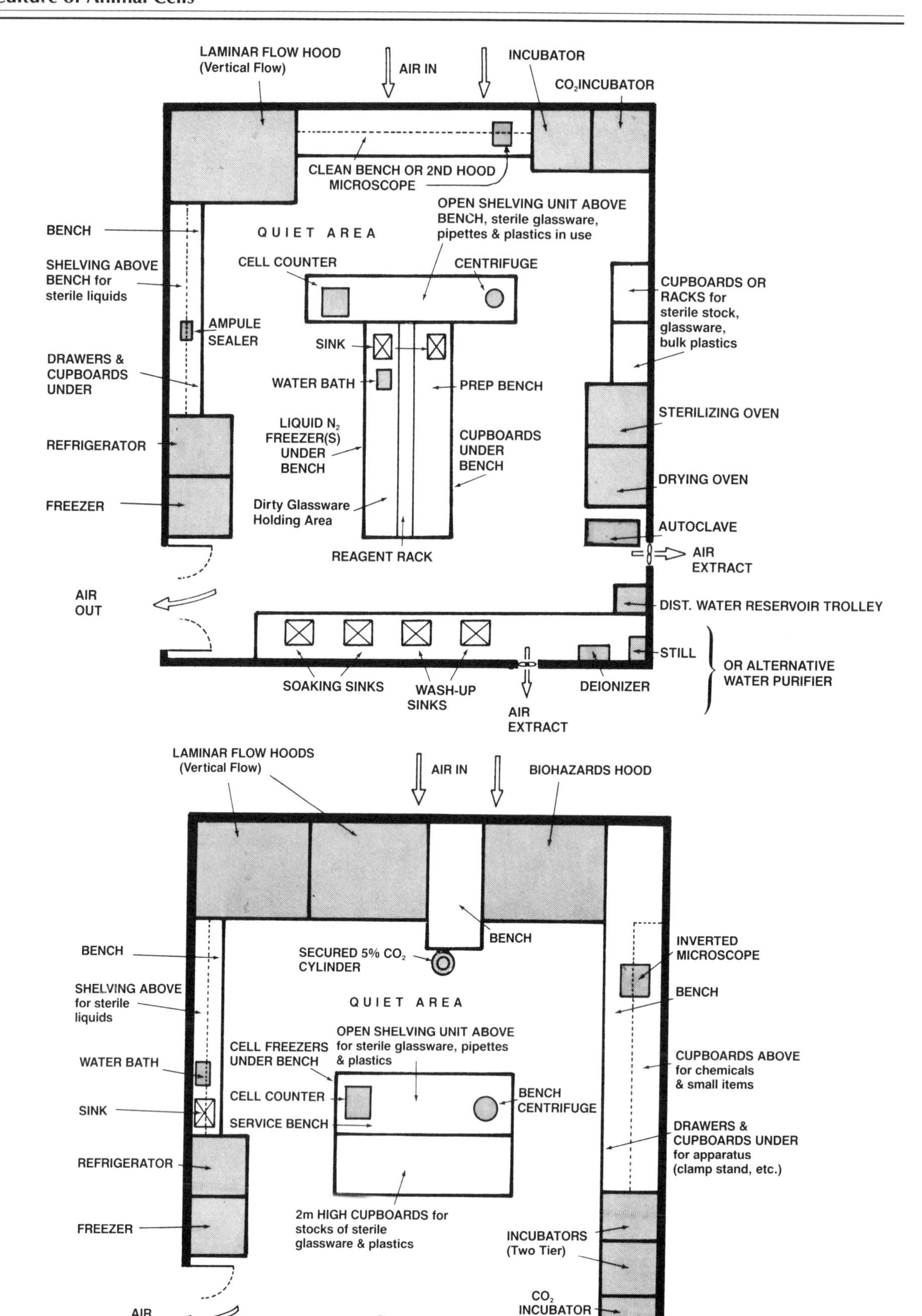

LAMINAR FLOW HOOD (Vertical Flow)
AIR IN
INCUBATOR
CO_2 INCUBATOR
CLEAN BENCH OR 2ND HOOD
MICROSCOPE
OPEN SHELVING UNIT ABOVE BENCH, sterile glassware, pipettes & plastics in use
BENCH
QUIET AREA
SHELVING ABOVE BENCH for sterile liquids
CELL COUNTER
CENTRIFUGE
CUPBOARDS OR RACKS for sterile stock, glassware, bulk plastics
AMPULE SEALER
SINK
DRAWERS & CUPBOARDS UNDER
WATER BATH
PREP BENCH
STERILIZING OVEN
LIQUID N_2 FREEZER(S) UNDER BENCH
CUPBOARDS UNDER BENCH
REFRIGERATOR
DRYING OVEN
FREEZER
Dirty Glassware Holding Area
AUTOCLAVE
REAGENT RACK
AIR EXTRACT
AIR OUT
DIST. WATER RESERVOIR TROLLEY
STILL
OR ALTERNATIVE WATER PURIFIER
SOAKING SINKS
WASH-UP SINKS
DEIONIZER
AIR EXTRACT
3.3
LAMINAR FLOW HOODS (Vertical Flow)
AIR IN
BIOHAZARDS HOOD
BENCH
BENCH
SECURED 5% CO_2 CYLINDER
INVERTED MICROSCOPE
SHELVING ABOVE for sterile liquids
QUIET AREA
BENCH
OPEN SHELVING UNIT ABOVE for sterile glassware, pipettes & plastics
CELL FREEZERS UNDER BENCH
CUPBOARDS ABOVE for chemicals & small items
WATER BATH
CELL COUNTER
BENCH CENTRIFUGE
SINK
SERVICE BENCH
DRAWERS & CUPBOARDS UNDER for apparatus (clamp stand, etc.)
REFRIGERATOR
FREEZER
2m HIGH CUPBOARDS for stocks of sterile glassware & plastics
INCUBATORS (Two Tier)
CO_2 INCUBATOR
AIR OUT
WASH-UP TROLLEY
SINK UNIT
CO_2 CYLINDERS
3.4

3.5

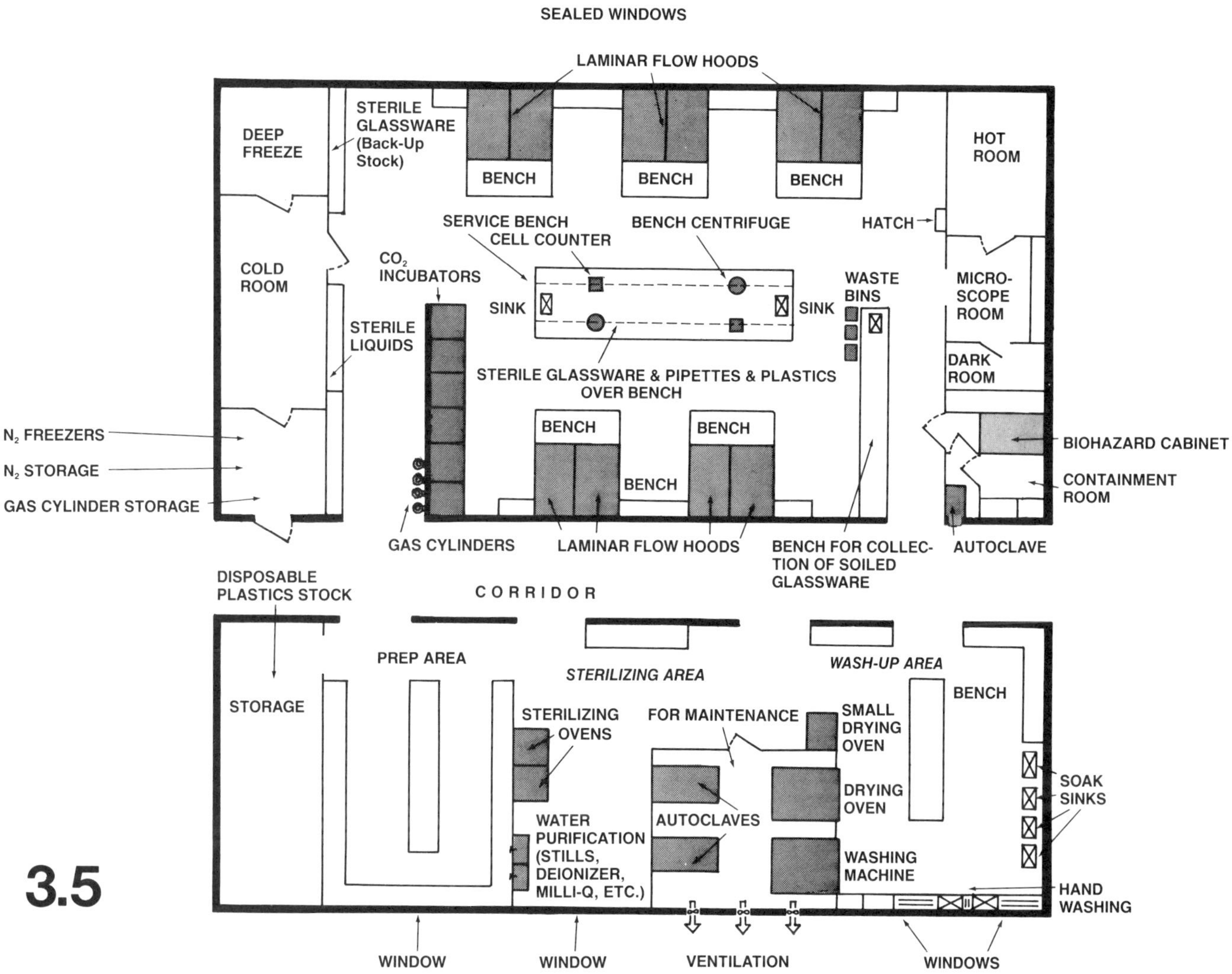

Fig. 3.3. *Suggested layout for simple, self-contained tissue culture laboratory for use by two or three persons. Shaded areas represent movable equipment.*

Fig. 3.4. *Tissue culture laboratory suitable for five or six persons with washing-up and preparation facility located elsewhere. Shaded areas represent movable equipment.*

Fig. 3.5. *Large scale tissue culture laboratory with adjacent washing up, sterilization, and preparation area. Suitable for 20 to 30 persons. Shaded areas represent equipment as distinct from furniture.*

siting of the thermostat sensor, (3) the circulation of air in the room, (4) correct insulation, and (5) the evolution of heat by other apparatus (stirrers, etc.) in the room.

Heaters. Heat is best supplied via a fan heater, domestic or industrial, depending on the size of the room. Approximately 2–3 kW per 20 m^3 (700 ft^3) will be required (or two 1.0–1.5 kW) depending on the insulation. The fan on the fan heater should run continuously, and the power to the heating element should come from a proportional controller (see below).

Air circulation. A second fan, positioned on the opposite side of the room, and with the air flow opposing that of the fan heater, will ensure maximum circulation. If the room is more than 2 m × 2 m (6 ft × 6 ft) some form of ducting may be necessary. Blocking off the corners as in Figure 3.6 is often easiest and most economical in space in a square room. In a long rectangular room, a false wall may be built at either end but be sure to insulate it from the room and make it strong enough to carry shelving.

Thermostats. Thermostats should be of the "proportional controller" type acting via a relay to supply heat at a rate proportional to the difference between the room temperature and the set point. So, when the door opens and the room temperature falls, recovery is rapid; but the temperature does not overshoot, as

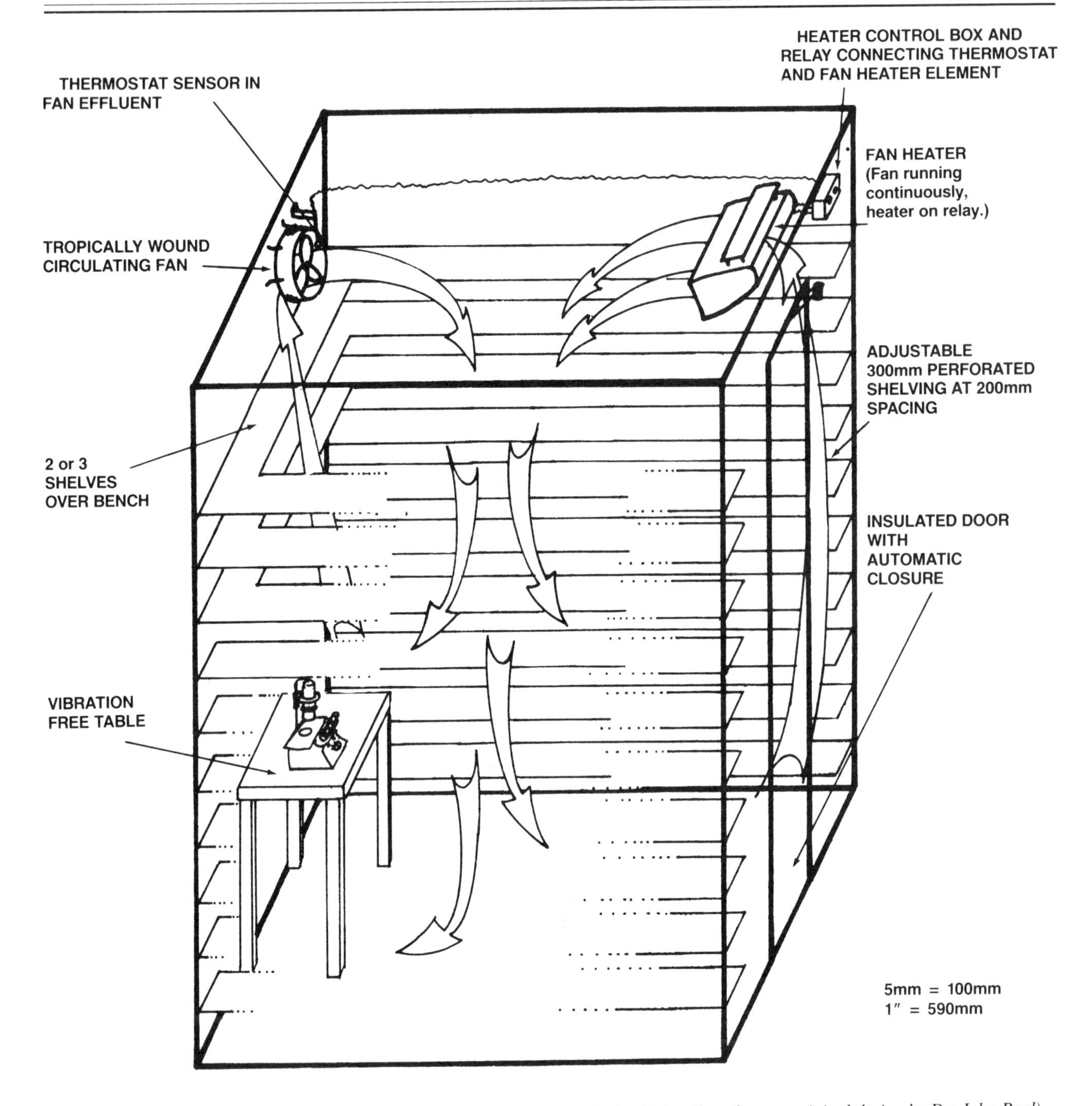

Fig. 3.6. *Suggested design for a simple hot room. Arrows represent air circulation (based on an original design by Dr. John Paul).*

the closer the room temperature approaches the set point, the less heat is supplied.

If possible, dual thermostats in parallel, but preferably controlling separate heaters, should be installed so that if one fails it is overridden by the other (Fig. 3.7). One thermostat ("regulating") is set at the required temperature, 36.5° C, with a narrow flunctuation range, say 0.4° C (± 0.2° C). The second, or safety, thermostat is set slightly below the first so that if the temperature falls below the range controlled by the first, the second will be activated. Finally, there should be an overriding thermostat in series (on both heaters if two are installed) set at 38.5° C, so that if a thermostat locks on, the override thermostat will cut out at 38.5° C and illuminate a warning light (Fig. 3.8).

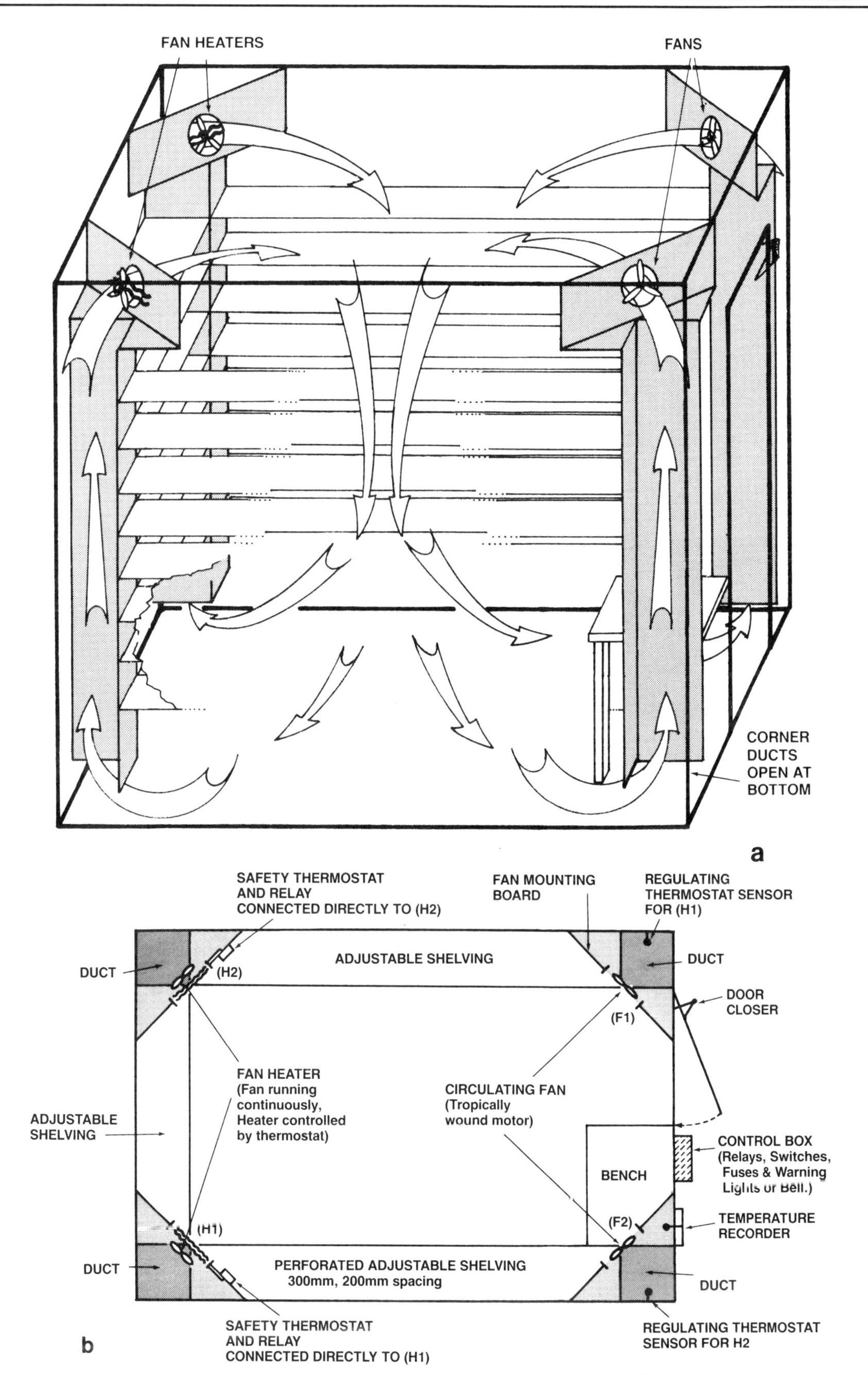

Fig. 3.7. *Hot room with dual heating circuits and safety thermostats. a. Oblique view. b. Plan view. Arrows represent air circulation.*

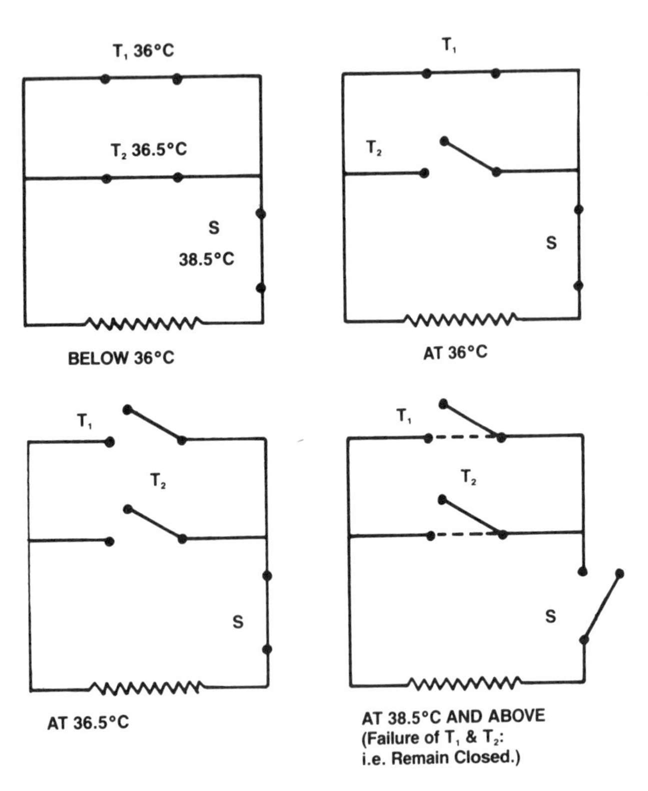

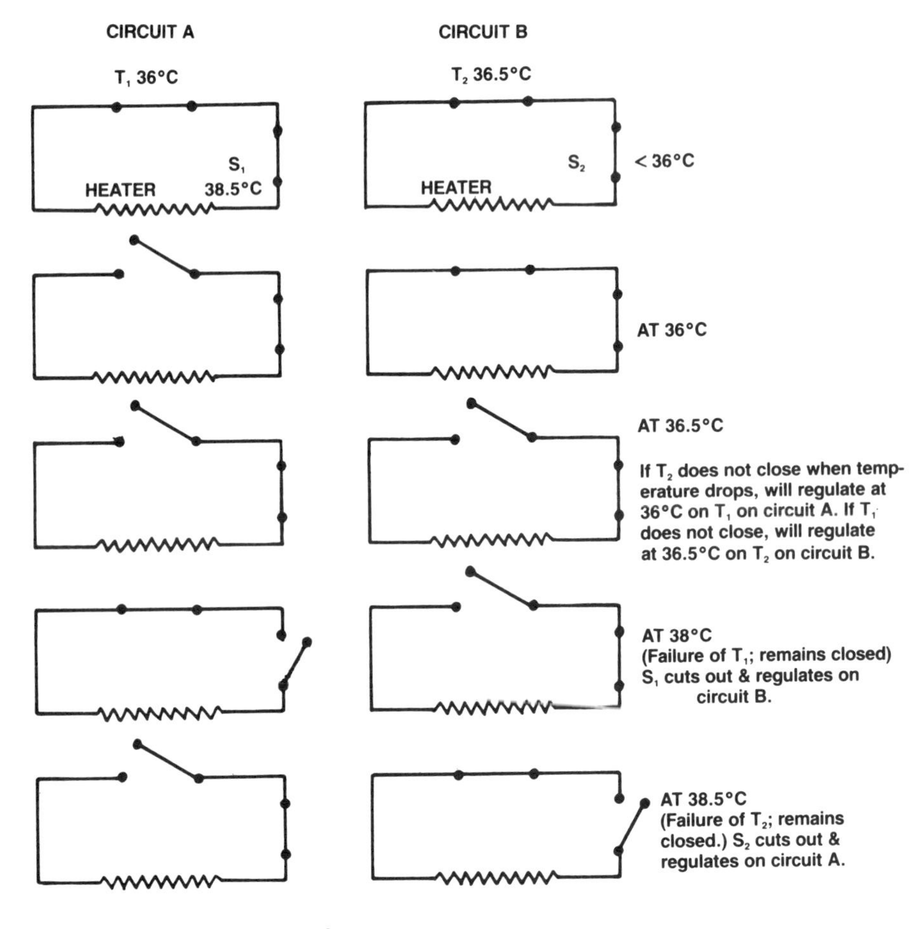

Fig. 3.8. *Circuit diagrams for regulating and safety thermostats for hot room illustrated in Figures 3.6, 3.7. The switches indicated are electronic relays activated by thermostats or thermistors. T_1 and T_2, regulating thermostats, S_1 and S_2, safety thermostats. a. Single heater circuit. b. Dual heater circuit.*

The thermostat sensors should be located in an area of rapid air flow close to the effluent from the second, circulating, fan for greatest sensitivity. A rapid response, high thermal conductivity sensor (thermistor or thermocouple) should be used in preference to a pressure bulb type.

Overheating. Since so much care is taken to provide heat and replenish its loss rapidly, another problem is often forgotten—namely, unwanted heat gain. This can arise (1) because of a rise in ambient temperature in the laboratory in hot weather or (2) due to heat produced from within the hot room by apparatus such as stirrer motors, roller racks, laminar flow units, etc.

Try to avoid heat-producing equipment in the hot-room, and arrange for heat dissipation either by a thermostatically controlled fan extract (and inlet) or an air conditioner. In either case, set the thermostat well below (>2° C) the heater thermostats so that the latter will regulate the temperature.

Access. If a proportional controller, good circulation, and adequate heating are provided, an air lock will not be required. The door should still be well insulated (foam plastic or fiberglass filled), light, and easily closed, preferably self-closing. It is also useful to have a hatch leading into the tissue culture area, with a shelf on both sides, so that cultures may be transferred easily into the room. The hatch door should also have an insulated core. If the hatch is located above the bench, this will avoid any risk of a "cold spot" on the shelving.

A temperature recorder should be installed such that the chart is located in easy and obvious view of the people working in the tissue culture room. A weekly change of chart is convenient and still has sufficient resolution. If possible, one high-level and one low-level warning light should be placed beside the chart or at a different, but equally obvious, location.

SERVICE BENCH

It may be convenient to position a bench to carry cell counter, microscope, etc., close to the sterile handling area, either dividing the area, or separating it from the other end of the lab (see Figs. 3.3,3.4,3.5). The service bench should also have provisions for storage of sterile glassware, plastics, pipettes, screw caps, syringes, etc., in drawer units below and shelves above. This bench may also house other accessory equipment such as an ampule sealing device and a small bench centrifuge. The bench should provide a close supply of all the immediate requirements.

PREPARATION

The need for extensive media preparation in small laboratories can be avoided if there is a proven source of reliable commercial culture media. While a large laboratory (~50 people doing tissue culture) may still find it more economic to prepare their own media, most smaller enterprises may prefer to purchase ready-made media. This reduces preparation to reagents, such as salt solutions, EDTA, etc., bottling these and water, and packaging screw caps and other small items for sterilization. While this area should still be clean and quiet, sterile handling is not necessary as all the items will be sterilized.

If there is difficulty in obtaining reliable commercial media, a larger area should be allocated for preparation to accommodate a coarse and fine balance, pH meter, and, preferably, an osmometer. Bench space will be required for dissolving and stirring solutions, and for bottling and packaging. If possible, an extra horizontal laminar flow hood should be provided in the sterile area for filtering and bottling sterile liquids, and incubator space must be allocated for quality control of sterility, i.e., incubation of samples of media in broth and after plating out.

Heat-stable solutions and equipment can be autoclaved or dry-heat sterilized at the nonsterile end of the area. Both streams then converge on the storage areas (see below).

WASH-UP

If possible wash-up and sterilization facilities should be provided outside the tissue culture lab as the humidity and heat that they produce may be difficult to dissipate without increasing air flow above desirable limits. Autoclaves, ovens, and distillation apparatus should be located in a separate room if possible (see Fig. 3.5), with an efficient extraction fan. The wash-up area should have plenty of space for soaking glassware, deep sinks for manual washing of glassware, and space for an automatic washing machine, should you require one. There should also be plenty of bench space for handling baskets of glassware, sorting pipettes, packaging and sealing sterile packs, and a pipette washer and drier. If the sterilization facilities must be located in the tissue culture lab, site them where greatest ventilation is possible and furthest from the sterile handling area.

Trolleys are often useful for collecting dirty glassware and redistributing fresh sterile stocks, but remember to allocate parking space for them.

STORAGE

Storage must be provided for: (1) sterile liquids (a) at room temperature (salt solutions, water, etc.), (b) at 4° C (media), and (c) at −20° C or −70° C (serum, trypsin, glutamine, etc.), (2) sterile glassware: (a) media bottles, glass culture flasks and (b) pipettes, (3) sterile disposable plastics: (a) culture flasks and petri dishes, (b) centrifuge tubes and vials, and (c) syringes, (4) screw caps, filter tubes, stoppers, etc., (5) apparatus: filters and large receiver flasks, and (6) gloves, disposal bags, etc. All should be within easy reach of the sterile working area. Refrigerators and freezers should be located toward the nonsterile end of the lab as the doors and compressor fans create dust and drafts and they may harbor fungal spores. Also, they require maintenance and periodic defrosting which creates a level and kind of activity best separated from your sterile working area.

The keynote of storage areas is ready access both for withdrawal and replenishment. Double-sided units are useful because they may be restocked from one side and used from the other. Storage boxes or trays which can be taken away and filled and replaced when full are also useful.

Remember to allocate sufficient space for storage as this will allow you to make bulk purchases, and thereby save money, and at the same time reduce the risk of running out of valuable stocks at times when they cannot be replaced. As a rough guide, you will need 200 l (~8 ft^3) of 4° C storage and 100 l (~4 ft^3) of −20° C storage per person. The volume per person increases with fewer people. Thus, one person may need a 250-l (10-ft^3) fridge and a 150-l (6-ft^3) freezer. This refers to storage space only; and allowance must be made for access for working in the cold-room where walk-in cold-rooms and deep freezers are planned.

CONSTRUCTION AND LAYOUT

The room should be supplied with air filtered to usual industrial or office standards and be designed for easy cleaning. Furniture should fit tight to the floor or be suspended from the bench allowing space to clean underneath. Cover the floor with vinyl or other dust-proof finish and allow a slight fall in the level toward a floor drain located in the center of the room. This allows liberal use of water if the floor has to be washed, but, more important, it protects equipment from damaging floods if stills, autoclaves, or sinks overflow.

If the tissue culture lab and preparation, wash-up, and sterilization areas can be separated, so much the better. Adequate floor drainage should still be provided in both areas although clearly the wash-up and sterilization area will be most important. If you have a separate wash-up and sterilization facility, it will be convenient to have this on the same floor and adjacent, with no steps to negotiate, so that trolleys may be used. Across a corridor is probably ideal (see Fig. 3.5), (see next chapter for sinks, soaking baths, etc.)

Try to imagine the flow of traffic—people, reagents, wash-up, trolleys, etc.—and arrange for minimum conflict, easy and close access to stores, and easy withdrawal of soiled items. Make sure your doors are wide enough to allow entry of all the equipment you want, particularly laminar flow units, and allow space for maintenance.

Inevitably space will be the first problem and some compromising will be inevitable, but a little thought ahead of time can save much space and ultimately people's tempers.

Chapter 4
Equipping the Laboratory

The specific needs of a tissue culture laboratory, like most labs, can be divided into three categories: (1) essential—you cannot perform a job without them; (2) beneficial—the work would be done better, more efficiently, quicker, or with less labor; (3) useful—it would make life easier, improve working conditions, reduce fatigue, enable more sophisticated analyses to be made, or generally make your working environment more attractive.

Equipment that might be used in tissue culture is listed in Table 4.1 in the grouping suggested above. Remember two main points: Assuming you need it and can afford it, you must be able to get it into the room (access) and you must have space for it (accommodation).

ESSENTIAL EQUIPMENT

Incubator

This should be large enough, probably 200 l (6 ft^3) per person, have forced air circulation, temperature control $\pm$ 0.5°C, and a safety thermostat which cuts out if the incubator overheats or, better, which regulates it if the first fails. It should be corrosion resistant, e.g., stainless steel (anodized aluminum is acceptable for a dry incubator), and easily cleaned. A double cabinet, one above the other, independently regulated, gives you more accommodation with the added protection that if one-half fails the other can still be used. This is also useful when you need to clean out one compartment.

Sterilizer

The simplest of these is a domestic pressure cooker which will generate 1 atm (15 lb/in^2) above ambient; but this will only accommodate the short variety of pipette, which is not suitable for flasks larger than 75 cm^2. More complex autoclaves exist, but the main consideration is the capacity; will it accommodate all you want to do? A simple bench top autoclave (Fig. 4.1a) may be sufficient but a larger model with a timer and a choice of pre- and poststerilization evacuation (Fig. 4.1b) will give more capacity and greater flexibility in use. A "wet" cycle (water, salt solutions, etc.) is performed without evacuation before or after sterilization. Dry items (instruments, swabs, screw caps, etc.) require the chamber to be evacuated before sterilization, to allow efficient access of hot steam, and should be evacuated after sterilization to remove steam and promote subsequent drying; otherwise the articles will emerge wet, leaving a trace of contamination from the condensate on drying. To minimize this risk always use deionized or reverse osmosis water to supply the autoclave.

If you require a high sterilization capacity (300 l, 9 ft^3, or more), buy two smaller autoclaves rather than one large one, so that during routine maintenance and accidental breakdowns you still have one functioning machine. Furthermore, a smaller machine will heat up and cool more quickly and can be used more economically for small loads. Leave sufficient space around them for maintenance and ventilation and provide adequate air extraction to remove heat and steam.

Refrigerators and Freezers

Usually a domestic item will be found to be quite efficient and cheaper than special laboratory equipment. However, if you require a lot of accommodation (400 l, 12 ft^3, or more; see Chapter 3), a large hospital (blood bank) or catering freezer may be better. While autodefrost freezers may be bad for some reagents (enzymes, antibiotics, etc.), they are very useful for most tissue culture stocks where their bulk and nature

TABLE 4.1. Tissue Culture Equipment

Minimum requirements (essential)	Desirable features (beneficial)	Useful additions
Incubator	Laminar flow hood(s), vertical, horizontal, biohazard	−70°C freezer
Sterilizer (autoclave, pressure cooker, oven)	Cell counter	Glassware washing machine
Refrigerator	Vacuum pump	Closed-circuit TV for inverted microscope(s)
Freezer (for −20°C storage)	CO_2 incubator	Colony counter
Inverted microscope	Coarse and fine balance	High-capacity centrifuge (6 × 1 liter)
Soaking bath or sink	pH meter	Cell sizer (e.g., Coulter ZB series)
Deep washing sink	Osmometer	Time-lapse cinemicrographic equipment
Pipette cylinder(s)	Phase-contrast and fluorescence microscope(s)	Interference-contrast microscope
Pipette washer	Portable temperature recorder	Polythene bag sealer (for packaging sterile items for long-term storage)
Still or water purifier	Permanent temperature recorders on sterilizing oven and autoclave	Controlled-rate cooler (for cell freezing)
Bench centrifuge	Roller racks for roller bottle culture	Filing for freezer records and catalogues
Liquid N_2 freezer (~ 35 l, 1,500–3,000 ampules)	Magnetic stirrer racks for suspension cultures	Centrifugal elutriator centrifuge and rotor
Liquid N_2 storage flask (~ 25 l)	Pipette drier	Fluorescence-activated cell sorter
	Trolleys for collecting soiled glassware and redistributing fresh supplies	Pipette plugger
	Autopipette or other form of automatic dispensor, dilutor	Densitometer
	Separate sterilizing oven and drying oven	Density meter (for density gradient cell separation)

precludes severe cryogenic damage. Conceivably, serum could deteriorate during oscillations in the temperature of autodefrost freezer, but in practice it does not seem to. Many of the essential constituents of serum are small proteins, polypeptides, and simpler organic and inorganic compounds which may be insensitive to cryogenic damage.

Microscope

It cannot be overstressed that, in spite of considerable and highly desirable progress toward quantitative analysis of cultured cells, it is still vital to look at them regularly. A morphological change is often the first sign of deterioration in a culture (see Chapter 15) and the characteristic pattern of microbiological infection (see Chapter 17) is easily recognized.

A simple inverted microscope is essential (Fig. 4.2). Make certain that the stage is large enough to accommodate large roller bottles (see Chapter 21) in case you should require them. There are many simple and inexpensive inverted microscopes on the market; but if you foresee the need for photography of living cultures, then you should invest in one with high-quality optics, a long working distance phase-contrast condenser and objectives, with provision to take a camera. The Leitz Diavert is good value and very serviceable.

Washing-up Equipment

Soaking baths or sinks. Soaking baths or sinks should be deep enough so that all your glassware (except pipettes and large aspirators) can be totally immersed in detergent during soaking, but not so deep that the weight of glass is sufficient to break smaller items at the bottom, e.g., 400 mm (15 in) wide × 600 mm (24 in) long × 300 mm (12 in) deep.

If you are designing a lab from scratch, then you can get sinks built in of the size that you want. Stainless steel or polypropylene are best, the former if you plan to use radioisotopes and the latter for hypochlorite disinfectants.

Washing sinks should be deep enough (450 mm, 18 in) to allow manual washing and rinsing of your largest

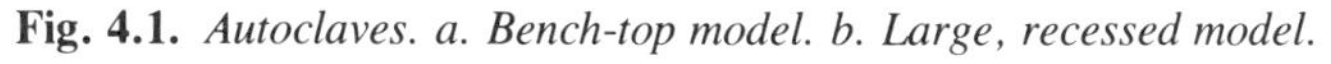

Fig. 4.1. *Autoclaves. a. Bench-top model. b. Large, recessed model.*

items without having to stoop too far to reach into them, and about 900 mm (3 ft) from floor to rim (Fig. 4.3). It is better to be too high than too low; a short person can always stand on a raised step to reach a high sink but a tall person will always have to bend down if the sink is too low. There should be a raised edge around the top of the sink to contain a spillage and prevent the operator getting wet when bending over the sink. The raised edge should go around behind the taps at the back.

Each washing sink will require four taps: a single cold water, combined hot/cold mixer, a cold hose connection for a rinsing device, and a nonmetallic tap for deionized water, from a reservoir above the sink (see Fig. 4.3). Piped systems for deionized water should be avoided as they can build up dirt and algae and are difficult to clean.

Pipette cylinders. These should be made from polypropylene and freestanding, distributed around the lab as required.

Pipette washer. Following an overnight soak in detergent, reusable pipettes are easily washed in a standard siphon-type washer (see Chapter 10 and Fig. 10.4). This should be placed at floor rather than bench level to avoid awkward lifting of the pipettes and connected to the deionized water supply so that the final rinse can be done in deionized water. If possible a simple changeover valve should be incorporated into the deionized water feed line (see Fig. 4.3).

Pipette drier. If a stainless steel basket is used in the washer, this may then be transferred directly to an electric drier. Alternatively, pipettes can be dried on a rack or in a regular drying oven.

Sterilizing and Drying Oven

Although all sterilizing can be done in an autoclave, it is preferable to sterilize pipettes and other glassware by dry heat, avoiding the possibility of chemical contamination from steam condensate or corrosion of pipette cans. This will require a high-temperature (160–

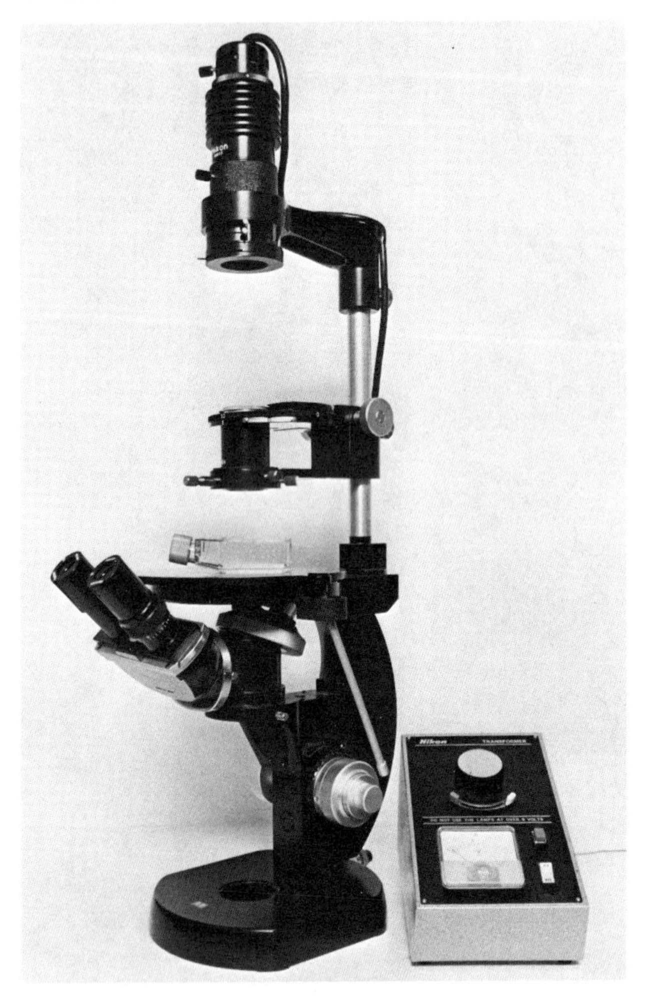

Fig. 4.2. *Nikon inverted microscope.*

180°C) fan-powered oven to ensure even heating throughout the load. As with autoclaves, do not get an oven that is too big for the size of glassware that you use. It is better to use two small ovens than one big one; heating is easier, more uniform, quicker, and more economical when only a little glassware is being used. You are also better protected during breakdowns.

Water Purification

Water is purified for two main purposes: (1) rinsing glassware and (2) making up media and reagents. For the first, a simple deionizer or reverse osmosis unit is adequate. For the second, higher-purity water is required. The traditional method was to use double-glass distilled water (glass- or silica-sheathed elements), but many laboratories now replace the first distillation stage with deionization (Fig. 4.4). The deionizer should have a conductivity meter monitoring the effluent to signify when the cartridge must be changed. Deionization alone is insufficient, as many organic contaminants are not retained by the ion-exchange resin. Distillation is usually recommended as a necessary second step, although in some areas it may be replaced by carbon filtration.

Recently, high-grade deionization coupled with charcoal filtration has been used to replace distillation entirely (see Fig. 10.7) with reverse osmosis recommended as a preliminary purification stage (Millipore, Fisons, Elga). Water prepared by this method is cheaper to produce as the power consumption is less than an electric still, although replacement of deionizer and charcoal filter cartridges is expensive.

Centrifuge

Periodically, cell suspensions require centrifugation to increase the concentration of cells or to wash off a reagent. A small bench-top centrifuge is sufficient for most purposes and has the advantage of rapid acceleration and deceleration without braking, which tends to disrupt cell pellets. Cells sediment satisfactorily at 80-100 *g*; higher *g* may cause damage and promote agglutination of the pellet. A simple unrefrigerated centrifuge of 6 × 50 ml maximum capacity should suffice, unless you intend to use large-scale suspension cultures, when a large-capacity refrigerated centrifuge, 4 × 1 liters or 6 × 1 liters, will be required.

Cell Freezing

The procedures for cell freezing will be dealt with in detail elsewhere (Chapter 18), but the basic facilities should be considered here. The freezing process can be carried out satisfactorily without sophisticated equipment, but storage requires a properly constructed liquid N_2 freezer and storage dewar (see Fig. 18.2). Freezers range in size from around 25 l to 500 l, i.e., 250 to 15,000 one-milliliter ampules. It is best to freeze a minimum of five ampules for each cell strain, 20 for a commonly used strain, and 100 for one in continuous use. A capacity of 1,200–1,500 is appropriate for most small laboratories.

The choice of freezer is determined by three factors: (1) capacity (No. of ampules); (2) economy and static holding time (the time taken for all the liquid N_2 to evaporate)—both of which are governed by the evaporation rate; and (3) convenience of access. Generally speaking, 2 is inversely proportional to 1 and 3. There

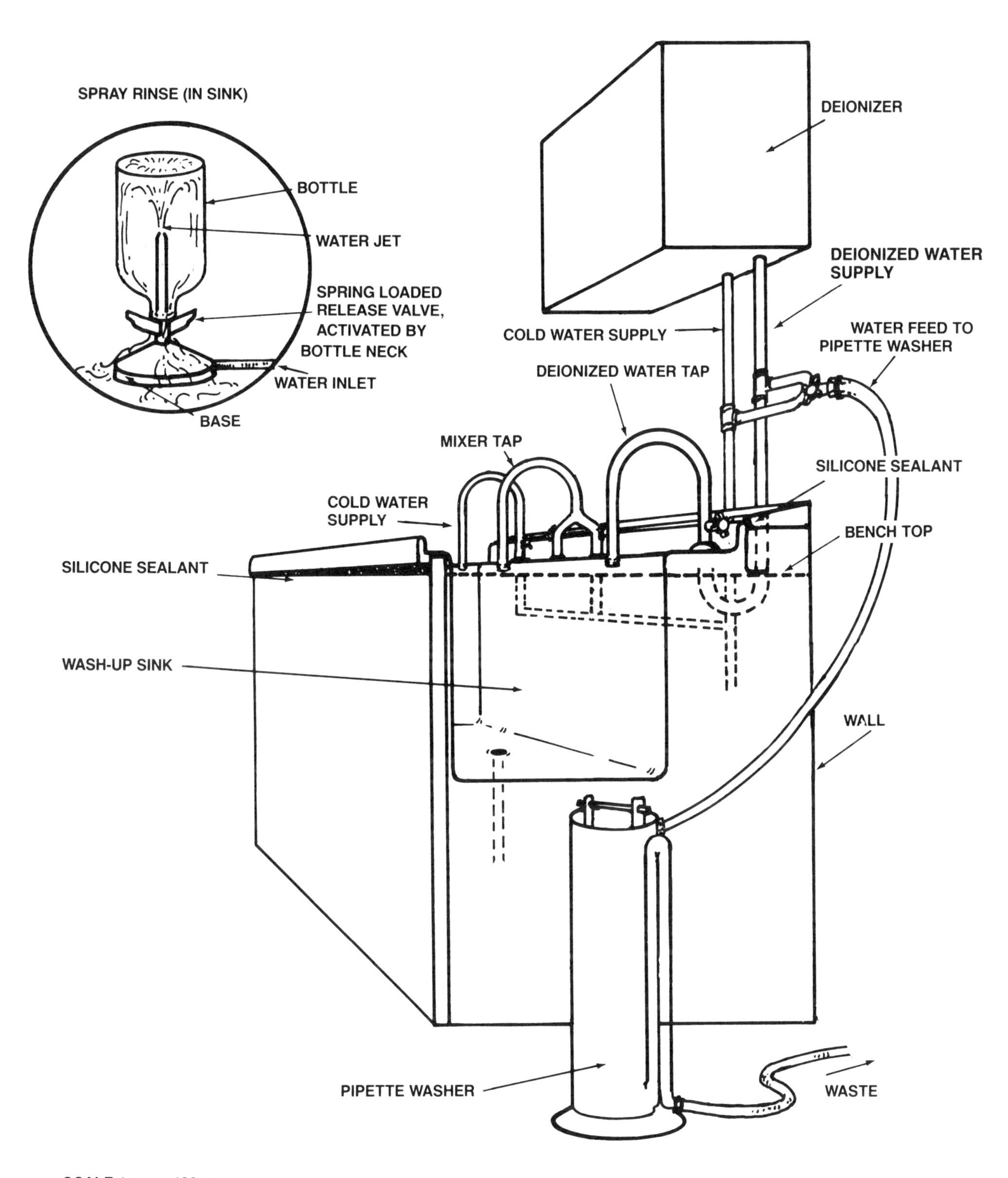

Fig. 4.3. *Washing-up sink and pipette washer, drawn to scale (bench height, 900 mm). Inset: bottle rinsing device, located in sink.*

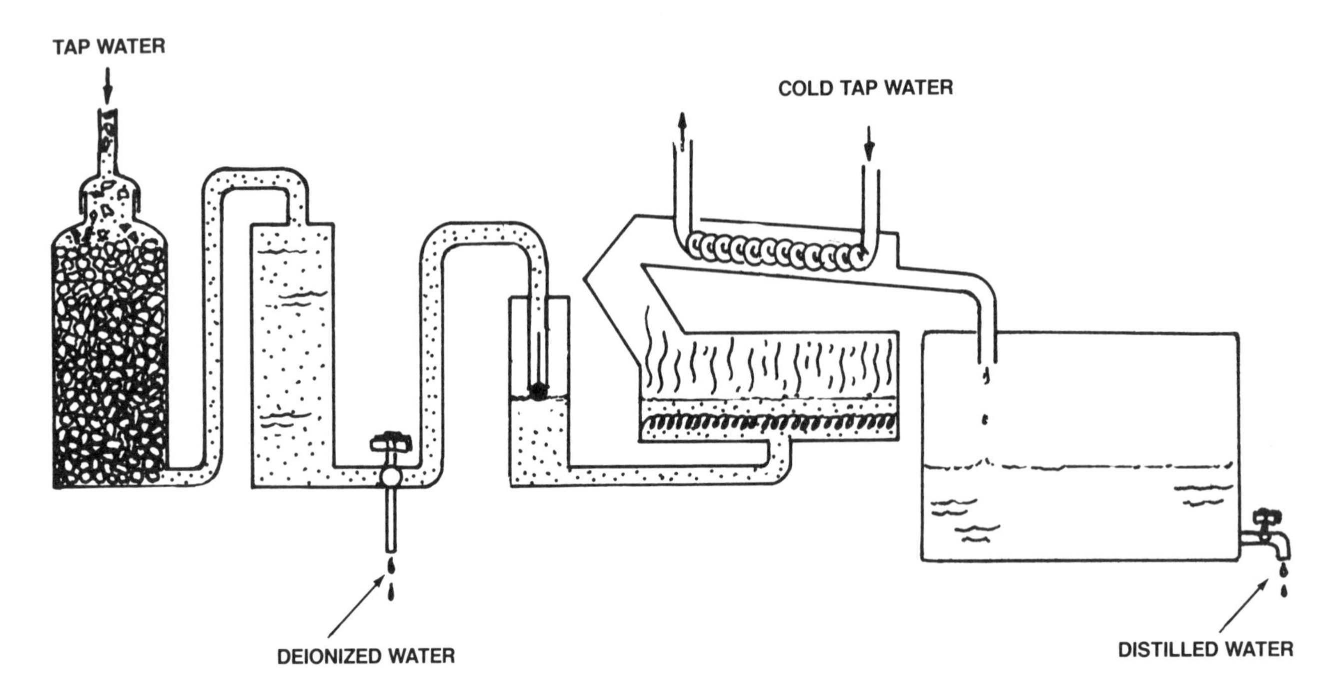

Fig. 4.4. *Deionized distilled water production. Deionizer coupled to still with automatic liquid level control and cutout switches.*

are two main types of freezer: narrow-necked with slow evaporation but with more difficult access, and wide-necked with easier access but three times the evaporation rate (see Fig. 18.6). If the cost of liquid nitrogen and its supply presents no problem, then a wide-necked freezer may be more convenient (e.g., Union Carbide LR40, 3,000-ampules capacity, 3–5 l/d evaporation) although the holding time will only be about 1 wk to 10 d. A narrow-necked freezer, on the other hand, will be more economical and last up to 2 months if N_2 supplies run out (e.g., L'aire Liquide 35-l, 1,500-ampules capacity, 0.5 l/d evaporation).

If you require bulk storage (~10,000 ampules), then you will need to consider a vessel of around 300 l capacity. Wide-necked freezers are most common in this size because of the mechanical difficulties in operating narrow-necked freezers of high capacity, but the latter are available and will save a considerable amount in expenditure on liquid N_2. At 300 l the evaporation rate is approximately 10 l/d in a wide necked freezer.

The advantages of gas-phase and liquid-phase storage will be discussed elsewhere (Chapter 18), but one major implication of storing in the gas phase is that the liquid phase is necessarily reduced to the space below your ampule storage area, usually 20–30% of the full volume. Hence, the static holding time is reduced to one-third or one-fifth of that of the filled freezer, filling must be carried out more regularly, and the chances of accidental thawing are increased. Where the investment is higher (many ampules or rare cell strains) automatic alarm systems should be fitted and, for the high-capacity freezers, an automatic filling system is recommended. However, automatic systems can fail and a twice-weekly check of liquid levels with a dipstick should be maintained and a record kept.

An appropriate storage vessel should also be purchased to enable a backup supply of liquid N_2 to be held. The size of this depends (1) on the size of the freezer, (2) the frequency and reliability of delivery of liquid N_2, and (3) the rate of evaporation. A 40 l, wide-necked freezer will require about 20–30 l/wk, so a 50 l dewar flask (or two 25 l flasks, which are easier to handle) is advisable. A 35 l, narrow-necked freezer, on the other hand, using 5–10 l/wk will only require a 25 l dewar. Larger freezers are best supplied on line from a dedicated storage tank; e.g., 160 l storage vessel linked to a 320 l freezer with automatic filling and alarm.

BENEFICIAL EQUIPMENT

The above describes the essential equipment for a modest tissue culture facility; but there are several items of equipment, which, if your budget will stretch to them, will make your laboratory easier to use and more efficient.

Laminar Flow Hood

Usually one hood is sufficient for two to three people (see Chapter 5 and Figs. 3.1, 5.3). A horizontal flow hood is cheaper and gives best sterile protection to your cultures; but for potentially hazardous materials (radioisotopes, carcinogenic or toxic drugs, virus-producing cultures, or any primate (including human) cell lines), a vertical laminar flow, preferably a Class II Biohazard, cabinet should be used.

Choose a hood that is (1) large enough (usually 1,200 mm (4 ft) wide × 600 mm (2 ft) deep), (2) quiet (noisy hoods are more fatiguing), (3) easily cleaned both inside the working area and below the work surface in the event of spillage, and (4) comfortable to sit at (some cabinets have awkward ducting below the work surface which leaves no room for your knees, or have screens which obscure your vision). The front screen should be able to be raised or removed completely to facilitate cleaning and handling bulky culture apparatus. Remember, however, that a Biohazard cabinet will not give you, the operator, the required protection if you lift the front screen.

Cell Counter

A cell counter (see Fig. 19.2) is a great advantage when more than two or three cell lines are carried and is essential for precise quantitative growth kinetics. Several companies now market models ranging in sophistication from simple particle counting up to automated cell size analysis. For routine counting, the Coulter "D Industrial" is more than adequate and much less expensive than equipment with cell sizing facilities (see also Cell Counting, Chapter 19).

Vacuum Pump

A vacuum pump or simple tap siphon saves a lot of time and effort when handling large numbers of cultures or large fluid volumes. Tap siphons require a minimum of 6 m (20 ft) head of water to create sufficient suction, but are by far the cheapest, simplest, and most efficient way to dispose of nonhazardous tissue culture effluent. If you do not have sufficient water pressure, or are handling potentially hazardous material, use a vacuum pump similar to that supplied for sterile filtration. If necessary, the same pump will serve both duties. The effluent should be collected in a reservoir into which a sterilizing agent such as glutaraldehyde or hypochlorite may be added when work is finished and at least 30 min before the reservoir is emptied (Fig. 4.5). A drying agent, hydrophobic filter (Gelman), or second trap placed in the line to the pump prevents fluid being carried over. Avoid vacuum lines; if they become contaminated with fluids, they can be very difficult to clean out.

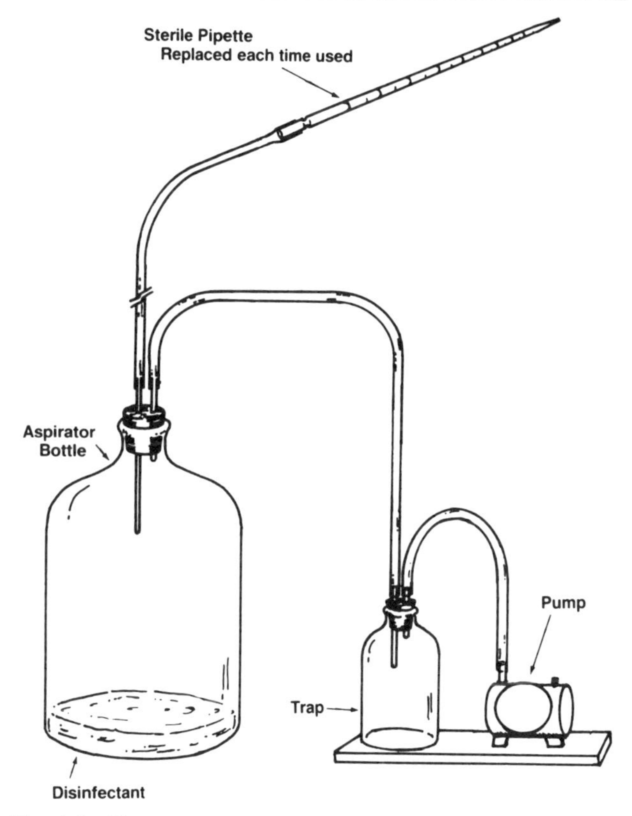

Fig. 4.5. *Vacuum pump assembly for withdrawing spent medium, etc.*

CO_2 Incubator

Although incubations can be performed in sealed flasks in a regular dry incubator or hot-room, some vessels, e.g., petri dishes or multiwell plates, require a controlled atmosphere with high humidity and elevated CO_2 tension (Fig. 4.6). The cheapest way of controlling the gas phase is to place the cultures in a plastic box, a desiccator, or an anaerobic jar (Fig. 4.7). Gas the container with the correct CO_2 mixture and then seal. If the container is not full, include an open dish of water to increase the humidity. Making the culture medium about 10% hypotonic will also help to counteract evaporation.

CO_2 incubators are rather expensive, but their ease of use and the superior control of CO_2 tension and temperature (anaerobic jars and desiccators take longer

Fig. 4.6. *Automatic CO_2 incubator (Napco). This is a dual-chamber model although only the top chamber is shown. Dual controls are located in the top panel, temperature regulation on the left, and a CO_2 controller on the right (see also Fig. 4.7).*

Fig. 4.7. *Becton Dickinson Anaerobic Jar. This type of jar or a desiccator (preferably plastic) can be used to maintain a regulated atmosphere in the absence of a CO_2 incubator.*

to warm up) justify the expenditure. A controlled atmosphere is achieved by blowing air over a humidifying tray (Fig. 4.8) and controlling the CO_2 tension with a CO_2 monitoring device. Alternatively, CO_2 tension may be controlled by mixing air and CO_2 in the correct ratio; but CO_2 controllers, although they add to the capital cost of the incubator, reduce CO_2 consumption considerably and give better control and recovery after opening the incubator. They function by drawing air from the incubator into the sample chamber, determining the concentration of CO_2, and injecting pure CO_2 into the incubator to make up any deficiency. Air is circulated around the incubator to keep both the CO_2 level and the temperature uniform.

Since humid incubators require regular cleaning, the interior should dismantle readily without leaving inaccessible crevices or corners.

Preparation and Quality Control

A coarse and a fine balance and a simple pH meter are useful additions to the tissue culture area for the preparation of media and special reagents. Although a phenol red indicator is sufficient for monitoring pH in most solutions, a pH meter will be required when phenol red cannot be used, e.g., in preparation of cultures for fluorescence assays.

One of the most important physical properties of culture medium, and one that is often difficult to predict, is the osmolality. An osmometer (see Fig. 9.2) is, therefore, a useful accessory to check solutions as they are made up, to adjust new formulations, or to compensate for the addition of reagents to the medium. They usually work by freezing-point depression or elevation of vapor pressure. Choose one with a low sample volume ($\leqslant$ 1 ml) since on occasion you may want to measure a valuable or scarce reagent.

Upright Microscope

An upright microscope may be required, in addition to an inverted microscope, for chromosome analysis, mycoplasma detection, and autoradiography. Select a high-grade research microscope, such as the Reichert Polyvar, with regular brightfield optics up to $\times$ 100 objective magnification, phase-contrast up to at least

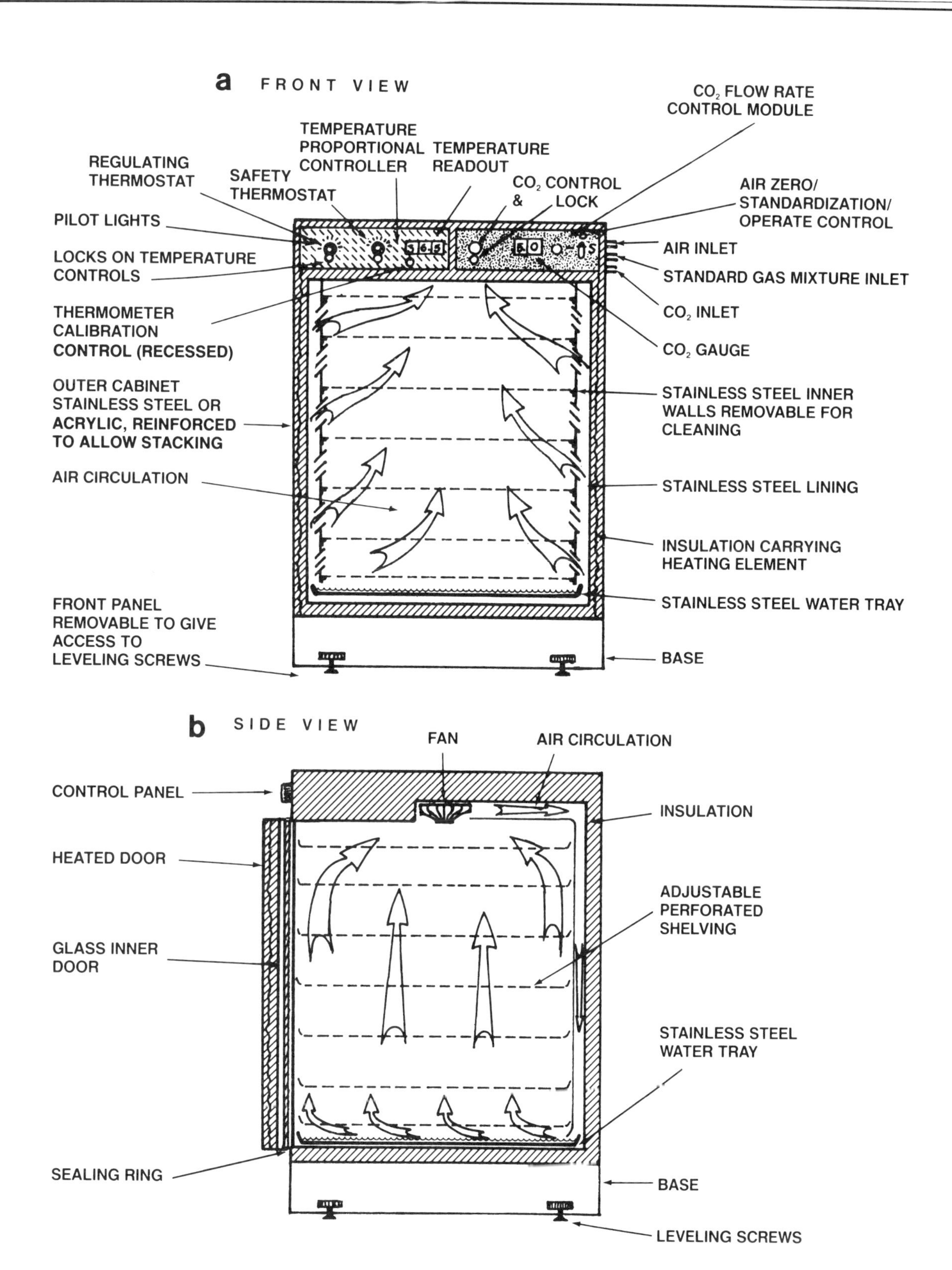

Fig. 4.8. *Components of a typical CO_2 incubator. a. Front view. b. Side view.*

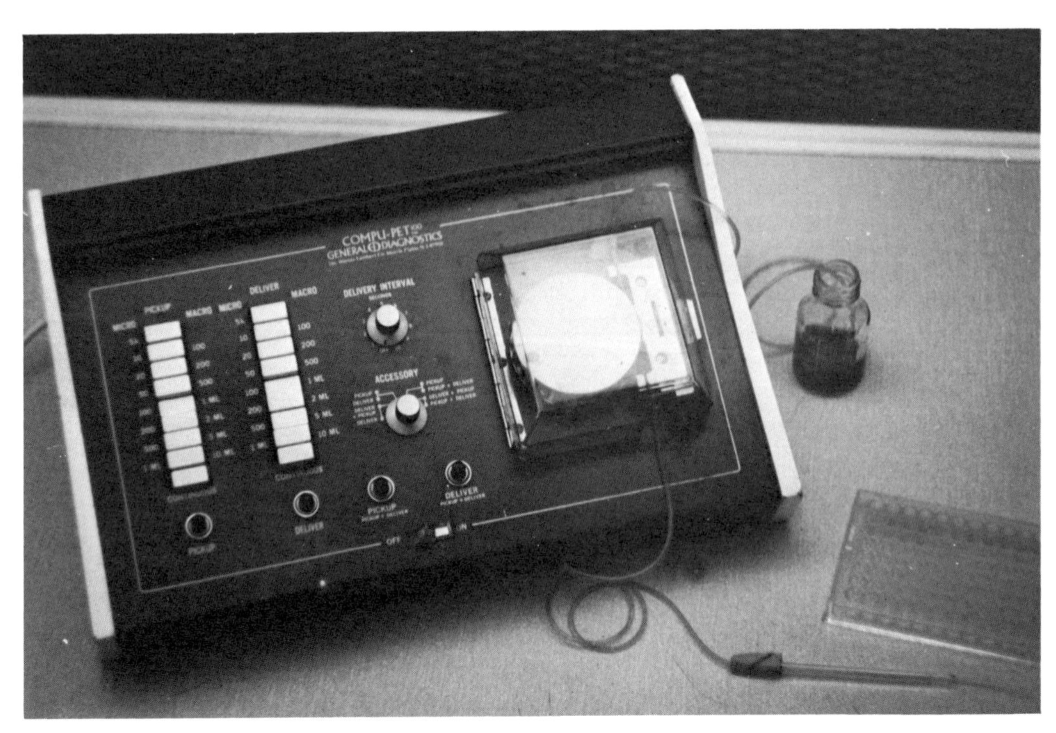

Fig. 4.9. *Automated pipetting device. The Compu-pet, suitable for repetitive dispensing and dilution in the 5 μl–10 ml range. Only the delivery tube requires sterilization.*

× 40 objective magnification, and preferably × 100, and fluorescence optics with epi-illumination and × 40 and × 100 objectives. Leitz supplies a × 50 water immersion objective which is particularly useful for observation of routine mycoplasma preparations with Hoechst stain (see Chapter 17). An automatic camera should also be fitted.

Temperature Recording

Ovens, incubators, and hot-rooms should be monitored regularly for uniformity and stability of temperature control. A recording thermometer with ranges from below −50°C to about +200°C will enable you to monitor cell freezing, incubators, and sterilizing ovens with one instrument fitted with a resistance thermometer or thermocouple with a long Teflon-coated lead.

Ideally, in addition to the above, recording thermometers should be permanently fixed into your hot-room sterilizing oven and autoclave, and a regular check kept for abnormal behavior.

Bulk Culture

Increasing the scale of your cultures beyond normal static cultures (see Chapter 21) may require that you provide roller bottle racks for monolayer cultures (see Fig. 21.8) or magnetic stirrer racks for suspension cultures (Figs. 21.4, 21.13).

Pipette Aids and Automatic Pipetting

If a large number of cultures is to be made, an automatic pipette such as the Compu-pet or Watson-Marlowe (Fig. 4.9) will be found to be an advantage. (Pipette aids and automatic pipetting are reviewed in Chapter 6.) For smaller numbers, Gilson-type pipettors are good for small volumes; and Bellco supplies an automated pipette aid which takes regular pipettes (Fig. 4.10). Only the disposable pipette tips of micropipettes need be sterile; the Bellco pipette aid uses regular sterile plugged pipettes.

There are also a number of automated devices designed specifically for microtitration plates which are useful, and these will be dealt with under automated techniques and microtitration in Chapter 6.

USEFUL ADDITIONAL EQUIPMENT

Low-Temperature Freezer

Most tissue culture reagents can be stored at 4°C or −20°C, but occasionally some drugs, reagents, or derivatives from cultures may require a temperature of −70°C where most, if not all the water is frozen and most chemical and radiolytic reactions are se-

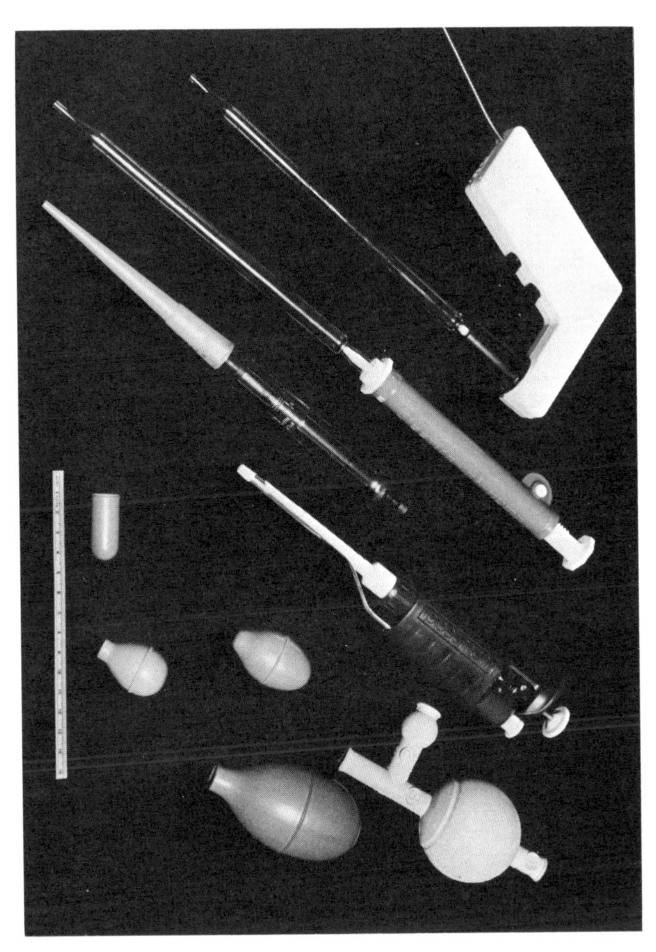

Fig. 4.10. *Pipetting aids. Top to bottom: Bellco, Pi-pump (standard pipettes); Finn-pipette, Gilson (micropipettors, take special plastic tips); various plain bulbs and bulb with inlet and outlet valves (see also Figs. 5.2, 5.5, 6.2).*

verely limited. Such a freezer is also a useful accessory for cell freezing (see Chapter 18). The chest type is more efficient at maintaining a low temperature for minimum power consumption, but vertical cabinets are much less extravagant in floor space. If you do choose a cabinet type, make sure that it has individual compartments (six to eight in a 400 l (15 ft^3) freezer) with separate close-fitting doors, and expect to pay 20% more than for a chest type.

Glassware Washing Machine

A reliable person doing your washing-up is probably the best way of producing clean glassware; but when the amount gets to be too great or reliable help is not readily available, it may be worth considering an automatic washing machine (Fig. 4.11). There are several of these currently available which are quite satisfactory. You should look for the following principles of operation:

(1) Choice of racks with individual spigots over which you can place bottles, flasks, etc. Open vessels such as petri dishes and beakers will wash satisfactorily in a whirling arm spray, but narrow-neck vessels need individual jets. Each jet should have a cushion at its base to protect the neck of the bottle from chipping.

(2) The water pump which pumps the water through the jets should have a high delivering pressure, requiring around 5 hp.

(3) Washing water should be heated to 90°C.

(4) There should be a facility for a deionized water rinse at the end of the cycle. This should be heated to 50–60°C; otherwise the glassware may crack after the hot wash and rinse, and should be delivered as a continuous flush and not recycled. If recycling is unavoidable, a minimum of three separate deionized rinses will be required.

(5) Preferably, rinse water from the end of the previous wash cycle should be discarded and not retained for the prerinse of your next wash. This reduces the risk of cross contamination when the machine is used for chemical and radioisotope wash-up.

(6) The machine should be lined with stainless steel and plumbed in stainless steel or nylon pipework.

(7) If possible, a glassware drier should be chosen that will accept the same racks (see Fig. 4.11), so that they may be transferred directly via a suitably designed trolley without unloading.

Betterbuilt makes such machines of different sizes with compatible drying ovens.

Closed-Circuit TV

Since the advent of cheap microcircuits, television cameras and monitors have become a valuable aid to the discussion of cultures and the training of new staff or students (Fig. 4.12). Choose a high-resolution, but not high-sensitivity, camera, as the standard camera sensitivity is usually sufficient, and high sensitivity may lead to problems of over-illumination. Black and white usually gives better resolution and is quite adequate for phase-contrast observation of living cultures. Color is preferable for fixed and stained specimens. If you will be discussing cultures with a technician or one or two associates, a 12- or 15-in monitor is adequate and gives better definition, but if you are teaching a group of ten or more students then go for a 19- or 21-in monitor. Addition of a video recorder will enable time-lapse films to be made.

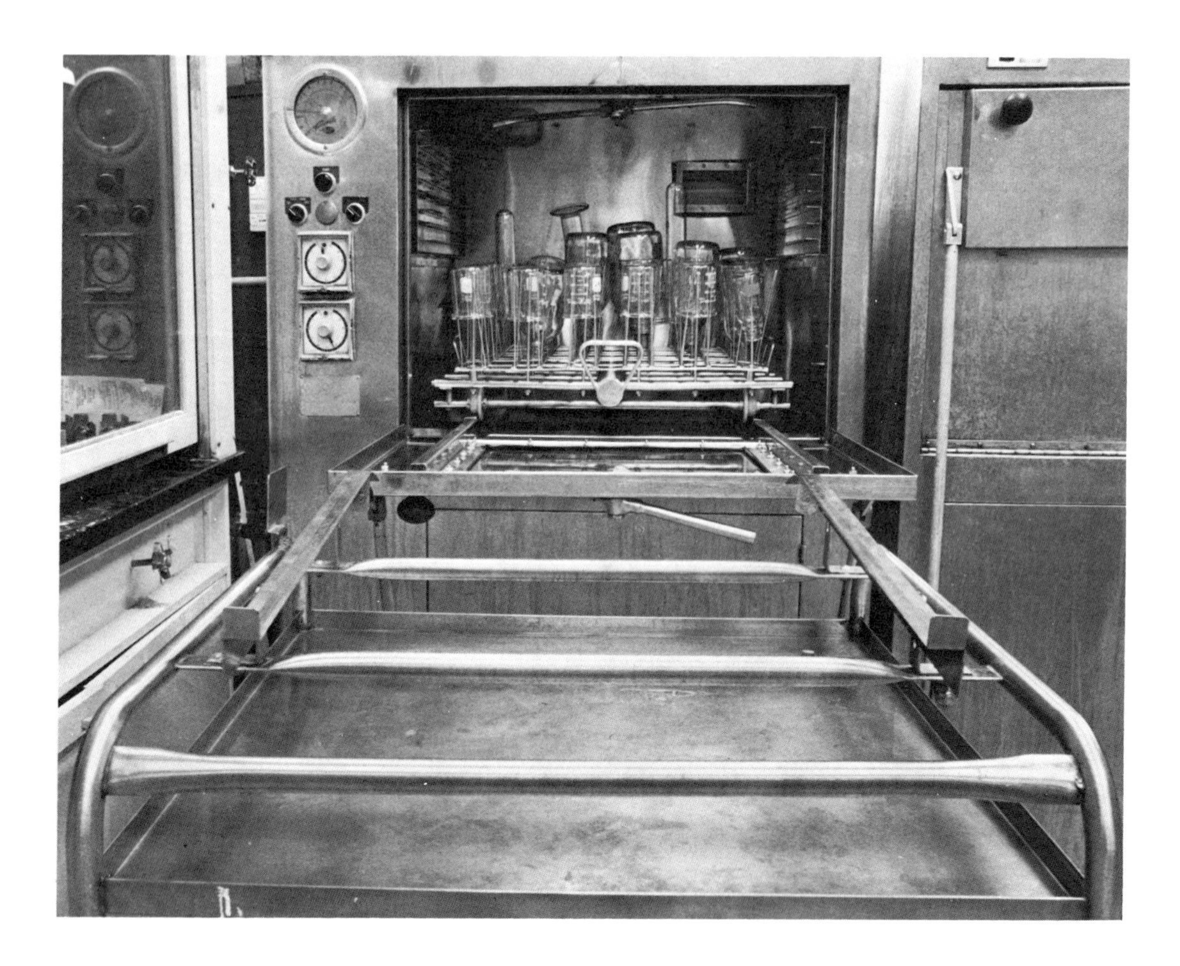

Fig. 4.11. *Automatic glassware washing machine. Glassware is placed on individual jets which ensures thorough washing and rinsing. After washing, glassware is withdrawn on the rack onto the trolley (front) and transferred to drier (right) fitted with same rails as washing machine and drier (Betterbuilt).*

Colony Counters

Monolayer colonies are easily counted by eye or on a dissecting microscope with a felt-tip pen to mark off the colonies, but if a lot of plates are to be counted, then an automated counter will help.

There are three levels of sophistication in colony counters. The simplest use an electrode-tipped marker pen which counts when you touch down on a colony. They often have a magnifying lens to help visualize the colonies (see Fig. 19.5). From there, a large increase in sophistication and cost takes you to an electronic counter employing a fixed program which counts colonies using a hard-wired program. These counters are very rapid and can discriminate between colonies of different diameters (though this is not necessarily proportional to cell number per colony; see also Chapter 19).

At the highest level of sophistication, image-analysis equipment may be used for colony counting but will need more skill and experience in programming. Because of this programmable feature, however, image analysis will cope with almost any size or shape of colony and will perform other complex tasks such as measuring area of outgrowth round an explant.

Cell Sizing

A dual-threshold cell counter (e.g., the Coulter ZB) with the facility for pulse-height analysis scans a cell population at a range of threshold settings simultaneously and prints out cell size distributions automatically.

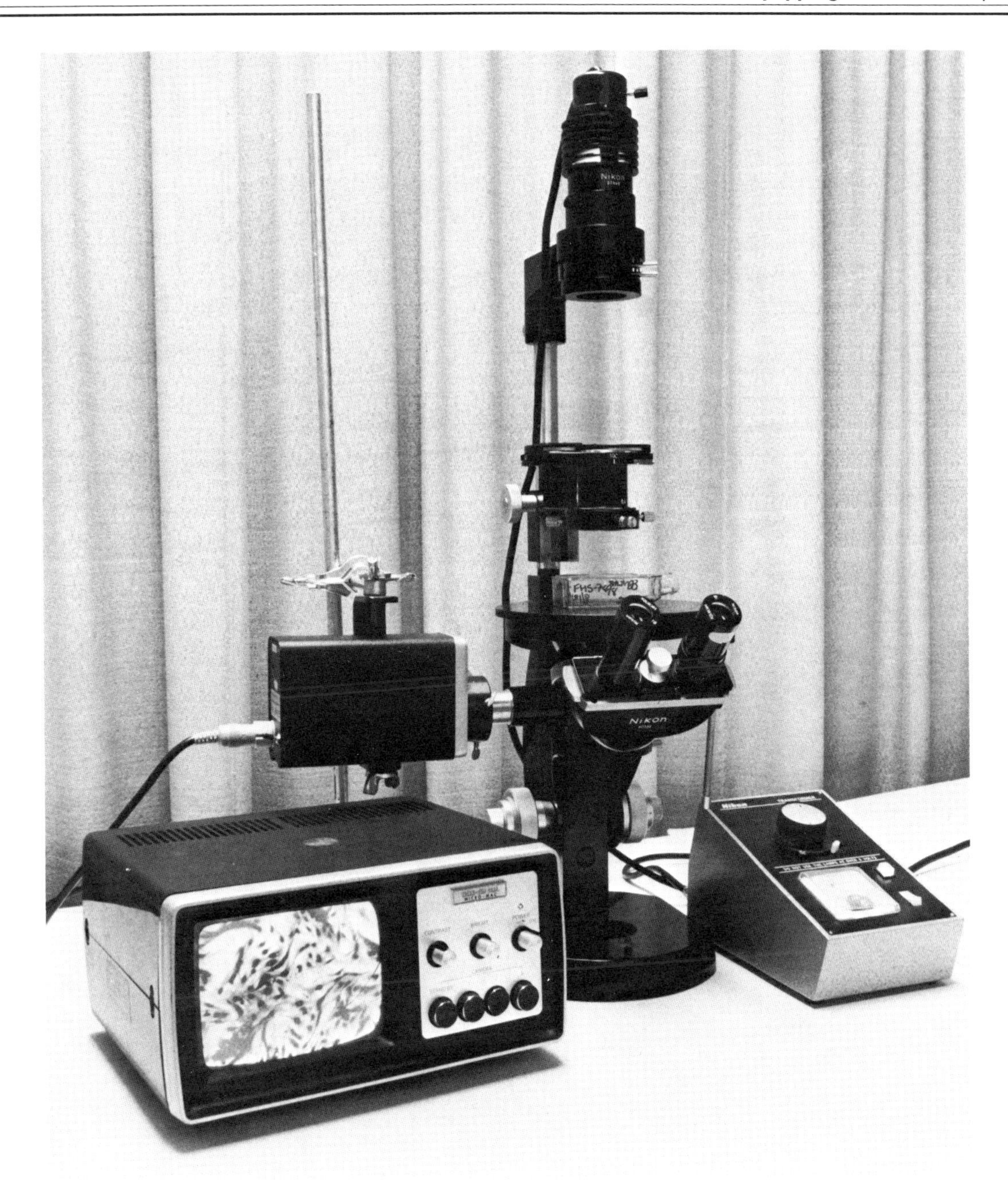

Fig. 4.12. *Closed-circuit television attached to Nikon inverted microscope.*

Time-lapse Cinemicrography

This technique is discussed in more detail in Chapter 21. The apparatus may be added to most good-quality inverted microscopes.

Controlled-Rate Cooler

While cells may be frozen by simply placing them in an insulated box at −70°C, some cells may require different cooling rates or differently shaped cooling curves [Mazur et al., 1970; Leibo and Mazur, 1971]. A programmable freezer enables the cooling rate to be varied by blowing liquid nitrogen into the freezing chamber, under the control of a preset program (see Fig. 18.6).

Centrifugal Elutriator

This is an especially adapted centrifuge suitable for separating cells of different sizes (see under Cell Separation, Chapter 14). They are costly but very effective.

Fluorescence-Activated Cell Sorter

The fluorescence-activated cell sorter (impulse cytophotometer, flow cytometer, cytofluorimeter) can analyze cell populations and separate them according to a variety of criteria (see Chapters 14 and 19). It has almost unlimited potential but is too expensive to come within most tissue culture laboratory equipment budgets.

It is always very tempting to purchase new pieces of equipment as they appear on the market, but weigh the advantages that they may offer against the space that they will occupy and what they will cost. Try to be sure also that (1) they will be of lasting benefit and (2) you and others will want to use them.

Chapter 5 Aseptic Technique

In spite of the introduction of antibiotics, contamination by microorganisms remains a major problem in tissue culture. Bacteria, yeasts, and fungal spores may be introduced via the operator, the atmosphere, work surfaces, solutions, and many other sources (see Table 17.1). Contaminations may be minor and confined to one or two cultures, can spread between several and infect a whole experiment, or can be widespread and wipe out your, or even the whole laboratory's, entire stock. Catastrophes can be minimized if (1) cultures are checked on the microscope, preferably by phase contrast, every time that they are handled, (2) they are kept antibiotic-free for at least part of the time to reveal cryptic contaminations (see Chapters 12 and 17), (3) reagents are checked for sterility before use (by yourself or the supplier), (4) bottles of media, etc., are not shared or used for different cell lines, and (5) the standard of sterile technique is kept high at all times.

Mycoplasmal infection, invisible under regular microscopy, presents one of the major threats. Undetected, it can spread to other cultures around the laboratory. It is, therefore, essential to back up visual checks with a mycoplasma test, particularly if cell growth appears abnormal. (For a more detailed account of contamination see Chapter 17.)

OBJECTIVES OF ASEPTIC TECHNIQUE

Correct aseptic technique should provide a barrier between microorganisms in the environment outside the culture and the pure uncontaminated culture within its flask or dish. Hence, all materials which will come into direct contact with the culture must be sterile and manipulations designed such that there is no direct link between the culture and its nonsterile surroundings.

It is recognized that the sterility barrier cannot be absolute without working under conditions which would severely hamper most routine manipulations. Since testing the need for individual precautions would be an extensive and lengthy controlled trial, procedures are adopted largely on the basis of common sense and experience. Aseptic technique is a combination of procedures designed to reduce the probability of infection, and the correlation between the omission of a step and subsequent contamination is not always clear. The operator may abandon several precautions before the probability rises sufficiently that a contamination occurs. By then, the cause becomes multifactorial and consequently no simple solution is obvious. If, once established, all precautions are maintained consistently, breakdown will be rarer and more easily detected.

Although laboratory conditions have improved in some respects (air conditioning and filtration, laminar flow facilities, etc), the modern laboratory is often more crowded and accommodation may have to be shared. However, with reasonable precautions, maintenance of sterility is not difficult.

QUIET AREA

In the absence of laminar flow, a separate sterile room should be used if possible (see Fig. 3.2). If not, pick a quiet corner of the laboratory with little or no traffic and no other activity (see Chapter 3). With laminar flow, an area should be selected which is free from through drafts and traffic should still be kept to a minimum. Animals and microbiological culture should be excluded from the tissue culture area. It should be kept clean and dust free and should not contain equipment other than that connected with tissue culture.

WORK SURFACE

One of the most frequent examples of bad technique is the failure to keep the work surface clean and tidy.

The following rules should be observed:

(1) Start with a completely clear surface.

(2) Swab down liberally with 70% alcohol.

(3) Bring on to it only those items you require for a particular procedure and swab bottles, cans, etc., with 70% alcohol beforehand.

(4) Arrange your apparatus (a) to have easy access to all of it without having to reach over one item to get at another; (b) to leave a wide, clear space in the center of the bench (not just the front edge!) to work on (Fig. 5.1). If you have too much equipment too close to you, you will inevitably brush the tip of a sterile pipette against a nonsterile surface.

(5) Work within your range of vision, e.g., insert a pipette in a bulb with the tip of the pipette pointing away from you so that it is in your line of sight continuously and not hidden by your arm.

(6) Mop up any spillage immediately and swab with 70% alcohol.

(7) Remove everything when you have finished and swab down again.

PERSONAL HYGIENE

There has been much discussion about whether handwashing encourages or reduces the bacterial count on the skin. Regardless of this debate, washing will moisten the hands and remove dry skin likely to blow onto your culture and reduce loosely adherent microorganisms which are the greatest risk to your cultures. Surgical gloves may be worn and swabbed frequently, but it may be preferable to work without (where no hazard is involved) and retain the extra sensitivity that this allows.

Caps, gowns, and face masks are often worn but are not always strictly necessary, particularly when working with laminar flow. However, if you have long hair, tie it back. When working on the open bench, do not talk while working asceptically; and if you have a cold, wear a face mask, or, better still, do not do any tissue culture during the height of the infection. Talking is permissible when working in vertical laminar flow with a barrier between you and the culture but should still be kept to a minimum.

PIPETTING

Standard glass or disposable plastic pipettes are still the easiest form of manipulating liquids. Syringes are often used, but regular needles are too short to reach into most bottles. Syringing may produce high shearing forces when dispensing cells and increase the risk of self-inoculation.

Pipettes of a convenient size range should be selected—1 ml, 2 ml, 5 ml, 10 ml, and 25 ml cover most requirements. If you only require a few of each, make up mixed cans and save space. Mouth pipetting, even with plugged pipettes or a filter tube/mouthpiece should be avoided, as it has been shown to be a contributory factor in mycoplasmal infection and may introduce an element of hazard to the operator, e.g., with virus-infected cell lines and human biopsy or autopsy specimens and biohazards (see Chapter 7). Inexpensive bulbs and pipetting devices are available; try a selection of these to find one that suits you (see Fig. 4.10). They should accept securely all the sizes of pipette that you use without forcing them and without the pipette falling out. The regulation of flow should be easy and rapid, but at the same time, capable of fine adjustment. You should be able to draw liquid up and down repeatedly (e.g., to disperse cells) and there should be no fear of carry-over. The device should fit comfortably in your hand and should be easy to operate with one hand.

The Marburg-type pipette (see Fig. 4.10) (Gilson, Oxford, Eppendorf, etc.) is particularly useful for small volumes (1 ml and less) though there can be some difficulty in reaching down into larger vessels with most of them. They are best used in conjunction with a shallow vial or bottle and are particularly useful when dealing with microtitration assays and other multiwell dishes. Multipoint pipettors (4, 6, or 12 point) are available for microtitration dishes (see Fig. 4.10).

It is necessary to insert a cotton plug in the top of a glass pipette before sterilization to maintain sterility in the pipette during use. If this becomes wet in use, discard the pipette into the wash-up. Plugging pipettes for sterile use is a very tedious job, as is the removal of plugs before washing. Automatic pipette pluggers are available, and although expensive, speed up the process and reduce the tedium (see Fig. 10.5). Alternatively, a sterile filter tube (Fig. 5.2) may be attached to the bulb, eliminating the need to plug pipettes. It is important that the filter tube is changed between handling of different cell lines to avoid the risk of cross contamination. Automatic pipetting devices and repeating dispensers are available; these will be discussed in Chapter 6.

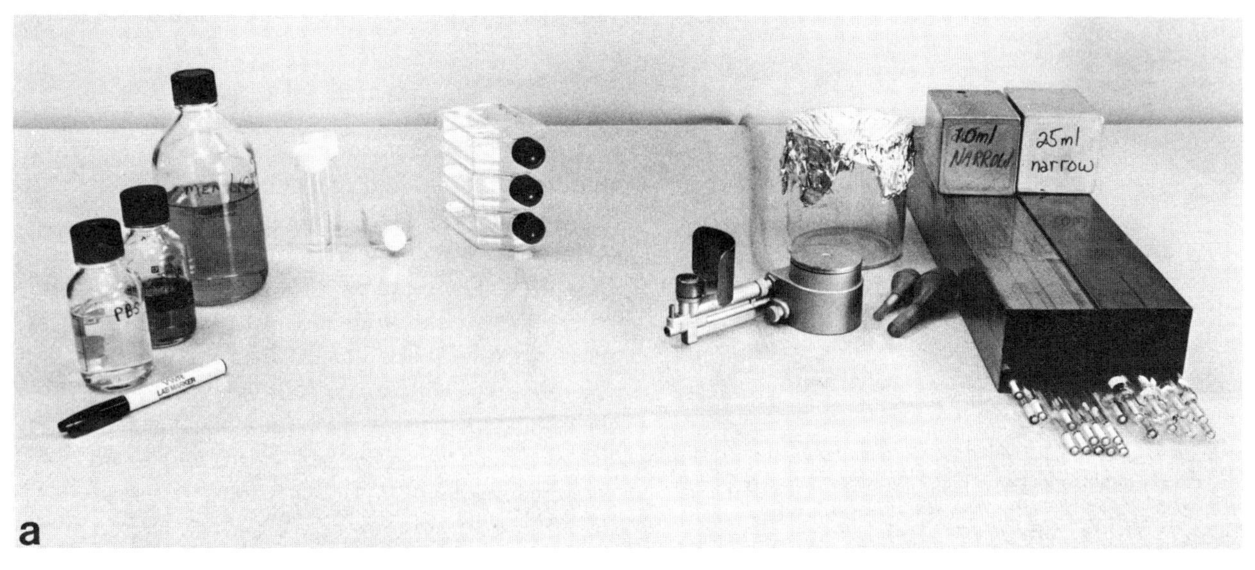

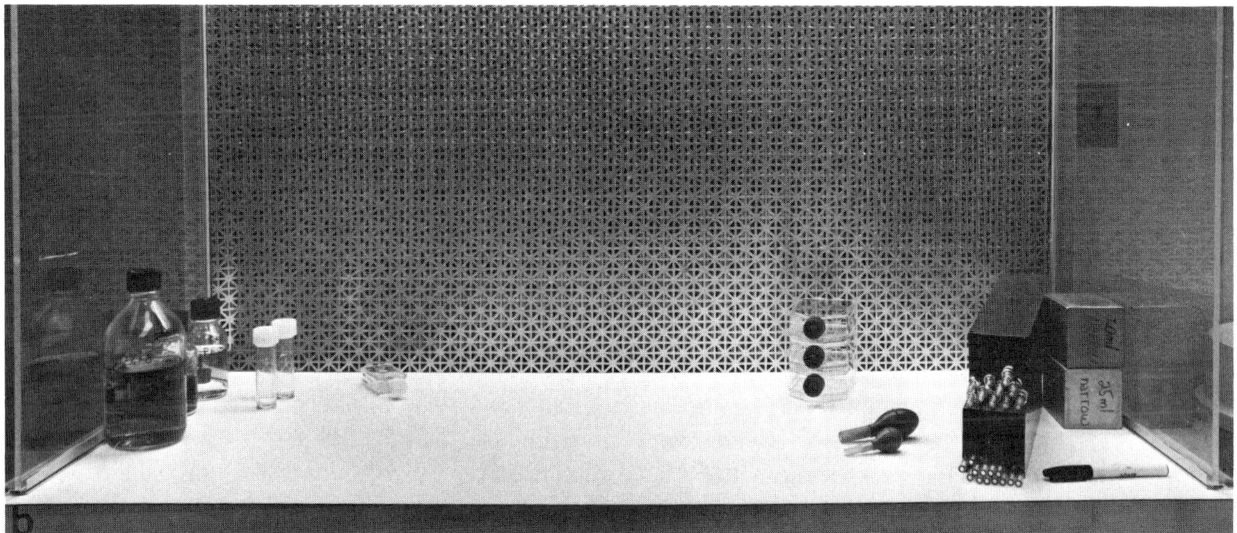

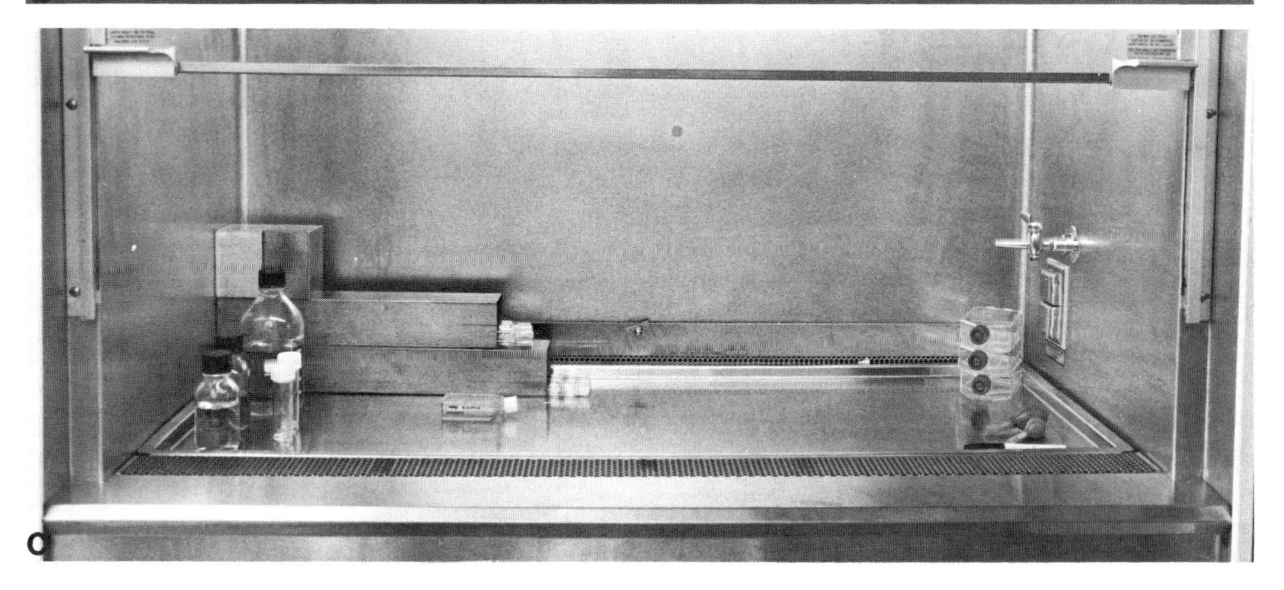

Fig. 5.1. *Suggested layout of work area. a. Open bench. b. Horizontal laminar flow. c. Vertical laminar flow.*

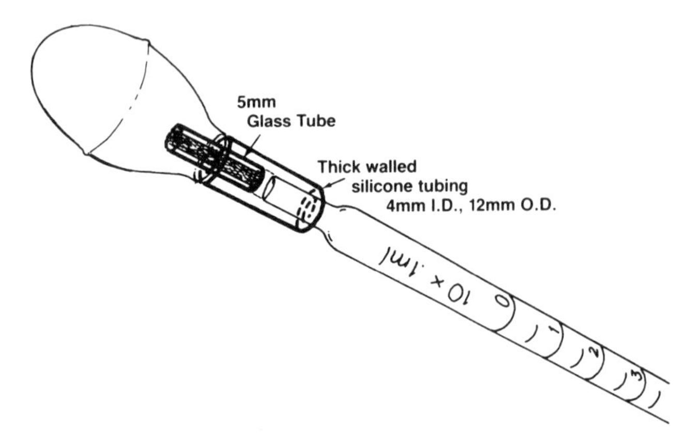

Fig. 5.2. *Filter tube. Interposed between bulb and pipettes, avoids the necessity to plug pipettes. Filter tube must be renewed between cell lines or if wetted (developed at the Beatson Institute from an original idea by Dr. John Paul).*

STERILE HANDLING

Swabbing

Swab bottles, particularly from the cold-room, before using for the first time each day.

Capping

Deep screw caps should be used in preference to stoppers although care must be taken when washing caps to ensure that all detergent is rinsed from behind rubber liners. The screw cap should be covered with aluminum foil to protect the neck of the bottle from sedimentary dust.

Flaming

When working on the open bench, the necks of bottles and screw caps should be flamed before and after opening a bottle and before and after closing. Pipettes should be flamed before use. Work close to the flame where there is an up-current due to convection, and do not leave bottles open. Screw caps should be placed open side down on a clean surface and flamed before replacing on the bottle.

Pouring

Whenever possible, do not pour from one sterile container into another unless the bottle you are pouring from is to be used once only and, preferably, is to deliver all its contents (premeasured) in one single delivery. The major risk in pouring lies in the generation of a bridge of liquid between the outside of the bottle and the inside which may permit infection to enter the bottle.

LAMINAR FLOW

The major advantage of working in laminar flow is that the working environment is protected from dust and contamination by a constant stable flow of filtered air passing over the work surface (Fig. 5.3) (see also Fig. 3.1). There are two main types: (1) horizontal, where the air flow blows from the side facing you, parallel to the work surface, and is not recirculated; (2) vertical, where the air blows down from the top of the cabinet on to the work surface and is drawn through the work surface and either recirculated or vented. In recirculating hoods, 20% is vented and made up by drawing in air at the front of the work surface. This is designed to minimize overspill from the work area of the cabinet. Horizontal flow hoods give the most stable airflow and best sterile protection to the culture and reagents; vertical flow gives more protection to the operator. If potentially hazardous material (radioisotopes, mutagens, human- or primate-derived cultures, virally infected cultures, etc.) is being handled, a Class II vertical flow biohazard hood should be used (see Fig. 7.3a).

If known human pathogens are handled, a Class III pathogen cabinet with a pathogen trap on the vent is obligatory (see Fig. 7.3b).

Laminar flow hoods depend, for their efficiency, on a minimum pressure drop across the filter. When filter resistance builds up, the pressure drop increases and the flow rate of air in the cabinet falls. Below 0.4 m/s (80 ft/min), the stability of the laminar air flow is lost and sterility can no longer be maintained. The pressure drop can be monitored with a manometer fitted to the cabinet, but direct measurement of airflow with an anemometer is preferable.

Routine maintenance checks are required (every 3–6 months) of the primary filters, which may be removed (after switching off the fan) and washed in soap and water, as they are usually made of polyurethane foam. Every 6 months the main filter should be checked for air flow and holes (detectable by locally increased air flow and an increased particulate count). This is best done on a contract basis.

Regular weekly checks should be made below the work surface and any spillage mopped up and the area sterilized. Spillages should, of course, be mopped up when they occur; but occasionally they go unnoticed, so a regular check is imperative.

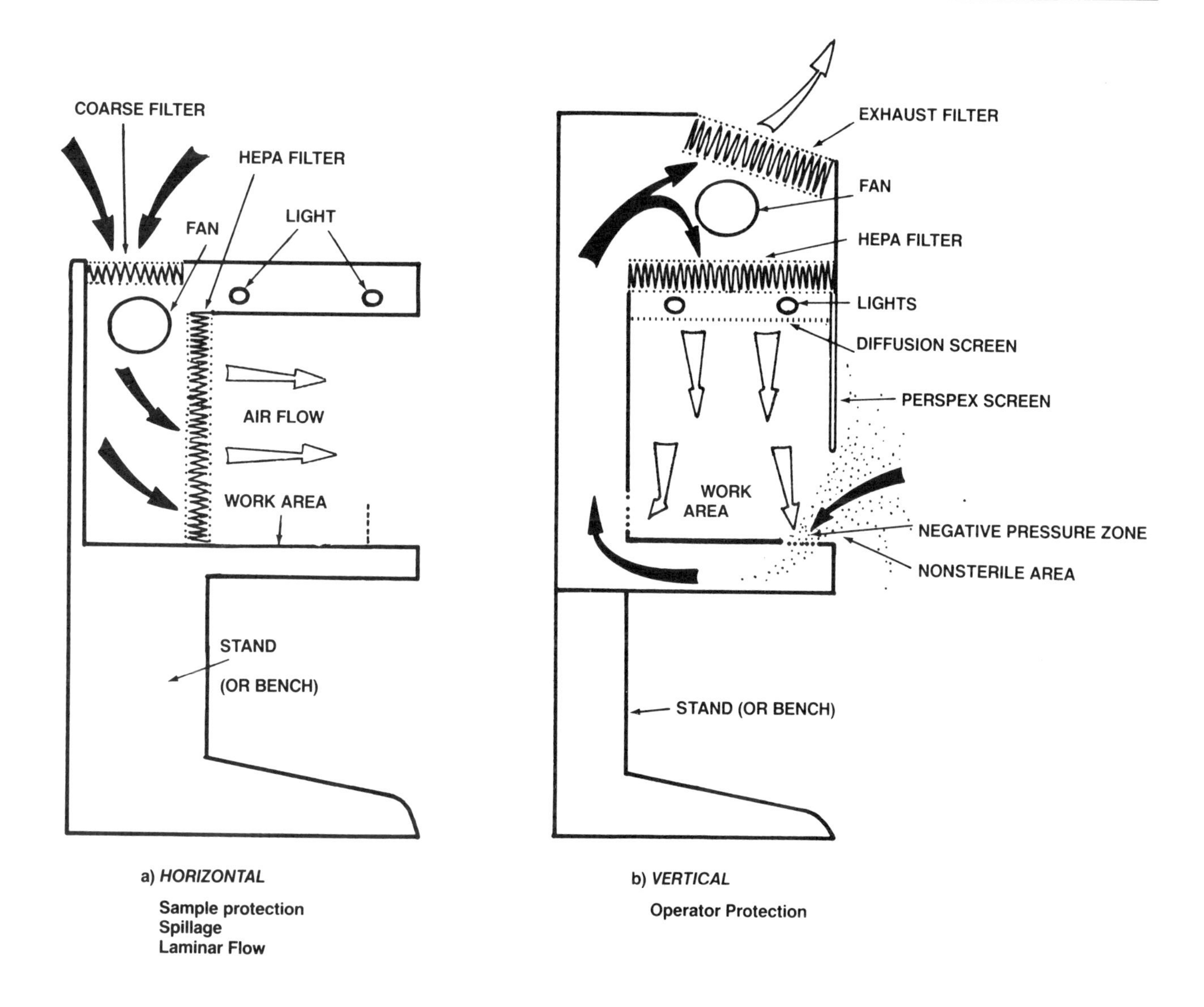

Fig. 5.3. *Horizontal (a) and vertical (b) laminar flow hoods. Filled arrows, nonsterile air; open arrows, sterile air.*

Laminar flow hoods are best left running continuously because this keeps the working area clean. Should any spillage occur, either on the filter or below the work surface, it dries fairly rapidly in sterile air, reducing the chance of growth of microorganisms.

Ultraviolet (uv) lights are used to sterilize the air and exposed work surfaces in laminar flow cabinets between use. The effectiveness of this is doubtful because crevices are not reached, and these are treated more effectively with alcohol or other sterilizing agents, which will run in by capillarity. Ultraviolet irradiation will also lead to crazing of some clear plastic panels (e.g., Perspex) after 6 months to 1 yr.

STANDARD PROCEDURE

Emphasis is being placed here on aseptic technique: Media preparation and other manipulations will be discussed under the appropriate headings.

Outline

Clean and prepare work area, with bottles, pipettes, etc. Carry out preparative procedures first before culture work. Flame articles as necessary and keep the work surface clean and clear. Finally, tidy up and wipe over surface with 70% alcohol.

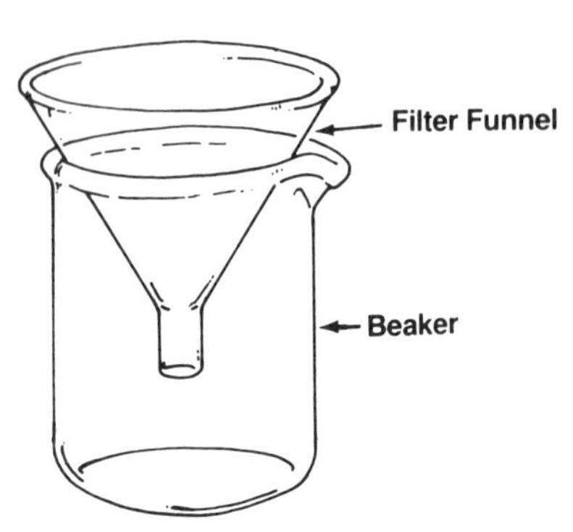

Fig. 5.4. *Waste beaker. Filter funnel prevents splashback from beaker.*

Materials

70% alcohol
swabs
bunsen (not in laminar flow) gas lighter
pipette-aid or bulb (see Fig. 4.10)
waste beaker (Fig. 5.4) or aspiration pump (see Fig. 4.5)
scissors
marker pen

For media preparation from concentrates: (all sterile)

water
stock media × 10
serum
glutamine
antibiotics
sodium bicarbonate
cans of pipettes, e.g., 25 ml, 10 ml, 5 ml, 1 ml, filter tubes(if pipettes are not already plugged with cotton wool)
glass bottles for media
glass or plastic flasks for cultures
notebook
diary or record cards
pen

Protocol

1.
Swab down bench surface or all inside surfaces of laminar flow hood with 70% alcohol
2.
Bring media, etc., from cold store and freezer, swab bottles with alcohol and place those that you will need first on the bench or in the hood
3.
Collect pipettes and place at the rear or side of the work surface in an accessible position (see Fig. 5.1). Open pipette cans and place lids out of the way but still in your sterile work area (within the hood)
4.
Collect any other glassware, plastics, intruments, etc. that you will need and place them close by
5.
Prepare medium as required (dilute from concentrate, add serum, etc.) as follows:

(a) Flame necks of bottles, rotating neck in flame, and slacken caps—they may be flamed outside the hood, if one is used—and flame again after slackening.

(b) On open bench (i) take pipette from can, touching other pipettes as little as possible, particularly at the tops; (ii) flame top to burn off any cotton protruding from the pipette; (iii) insert in bulb, pointing pipette away from you and holding it well above the graduations. Take care not to exert too much pressure as pipettes can break when being forced into a bulb; (iv) flame pipette by pushing lengthwise through flame, rotate 180°, and pull back through flame. This should only take 2–3 s or the pipette will get too hot. You are not attempting to sterilize the pipette, merely to fix any dust which may have settled on it. If you have touched anything or contaminated the pipette in any other way, discard it into the wash-up; do not attempt to resterilize it by flaming; (v) holding the pipette still pointing away from you, remove the cap of your first bottle into the crook formed between your little finger and the heel of your hand (Fig. 5.5); (vi) flame the neck of the bottle; (vii) withdraw the requisite amount of fluid and hold; (viii) flame bottle neck and recap; (ix) remove caps of receiving bottle, flame neck, insert fluid, reflame, and replace cap; (x) when finished, tighten caps, flame thoroughly, and replace foil. Work with the bottles tilted so that your hand does not come over the open neck.

If you have difficulty in holding the cap in your hand while you pipette, leave the foil in place and place the cap on the bench resting on the skirt of the foil. If bottles are to be left open, they should be sloped as close to horizontal as possible, laying them on the bench or on a bottle rest (Fig. 5.6).

(c) In laminar flow, proceed as for open bench but omit flaming during manipulations. Bottles may be left open more safely but should still be closed if you leave the hood for more than a few minutes.

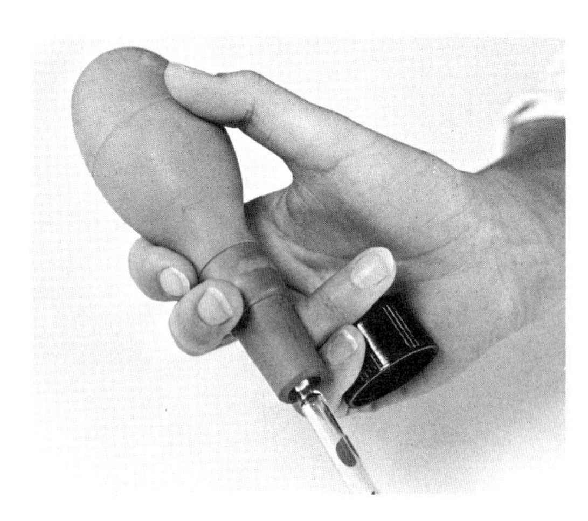

Fig. 5.5. *Uncapping bottle and holding cap. Hand may need to be moved up or down bulb between uncapping and pipetting. With this particular bulb (Aspirette), the forefinger is used to seal the top of the bulb when pipetting.*

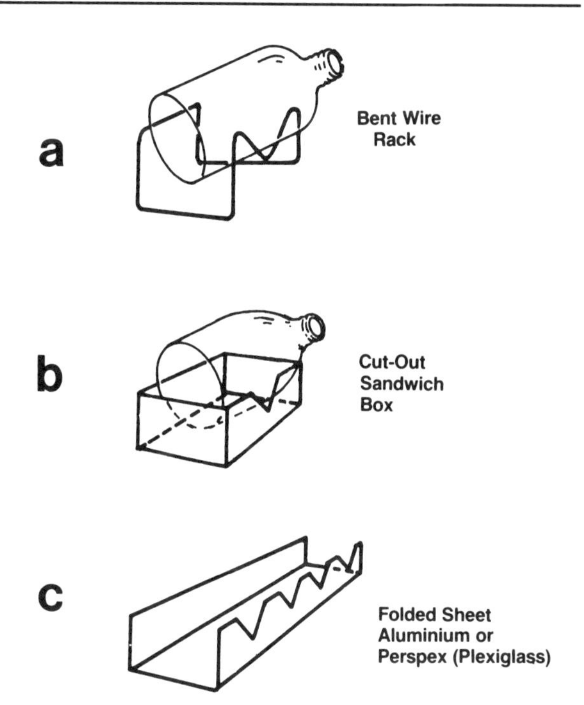

Fig. 5.6. *Suggested designs for bottle rest for use during pipetting. a. Wire rack, suggested by M. Stack-Dunne. b. V-cut in a plastic storage box. c. Folded aluminium or plexiglas suggested by A.C. McKirdy.*

In vertical laminar flow, do not work immediately above an open bottle or dish. In horizontal laminar flow, do not work behind an open bottle or dish

6. On completion of media preparation, etc., remove stock solutions from work surface keeping only the bottles that you will require

7. Check cultures, decide what they require, and bring to sterile work area

8. Swab bottles, flame necks (plastics very briefly), and place on work surface, preferably one cell strain at a time with its own bottle of medium and other solutions

9. For fluid change, proceed as follows:

(a) Take sterile pipette, flame (if not in hood), and insert bulb.

(b) Flame neck of culture bottle, open, withdraw medium and discard into waste beaker (see Fig. 5.4). A suction pump with a collection trap, or a tap siphon, with a suction line into the hood is the quickest and easiest method of withdrawing fluid for disposal (see Fig. 4.5).

(c) Transfer fresh medium to culture flask as in (5) above.

(d) Tighten caps, flame necks, replace foil

10. Return flasks to incubator and media to cold-room

11. Clear away all pipettes, glassware, etc., and swab down the work surface

The essence of good sterile technique is similar to good laboratory technique anywhere. Keep a clean clear space to work and have on it only what you require at one time. Prepare as much as possible in advance so that cultures are out of the incubator for the shortest possible time and the various manipulations can be carried out quickly, easily, and smoothly. Keep everything in direct line of sight and develop an awareness of accidental contacts between sterile and nonsterile surfaces. Leave the area clean and tidy when you finish.

Chapter 6
Mechanical Aids and Automation

There is at present insufficient standardization of routine tissue culture for the many techniques to lend themselves readily to automation and, in most cases, the scale of the operations does not warrant it. Furthermore, there is always a danger when culture techniques are automated that the operator becomes less aware of minor fluctuations in culture behavior. Consequently, this chapter will concentrate on semiautomated or mechanized dispensing and dilution.

AUTOPIPETTES

Repetitive pipettes have been designed in a variety of patterns and those most suited to tissue culture are the syringe types. They operate either by alternately

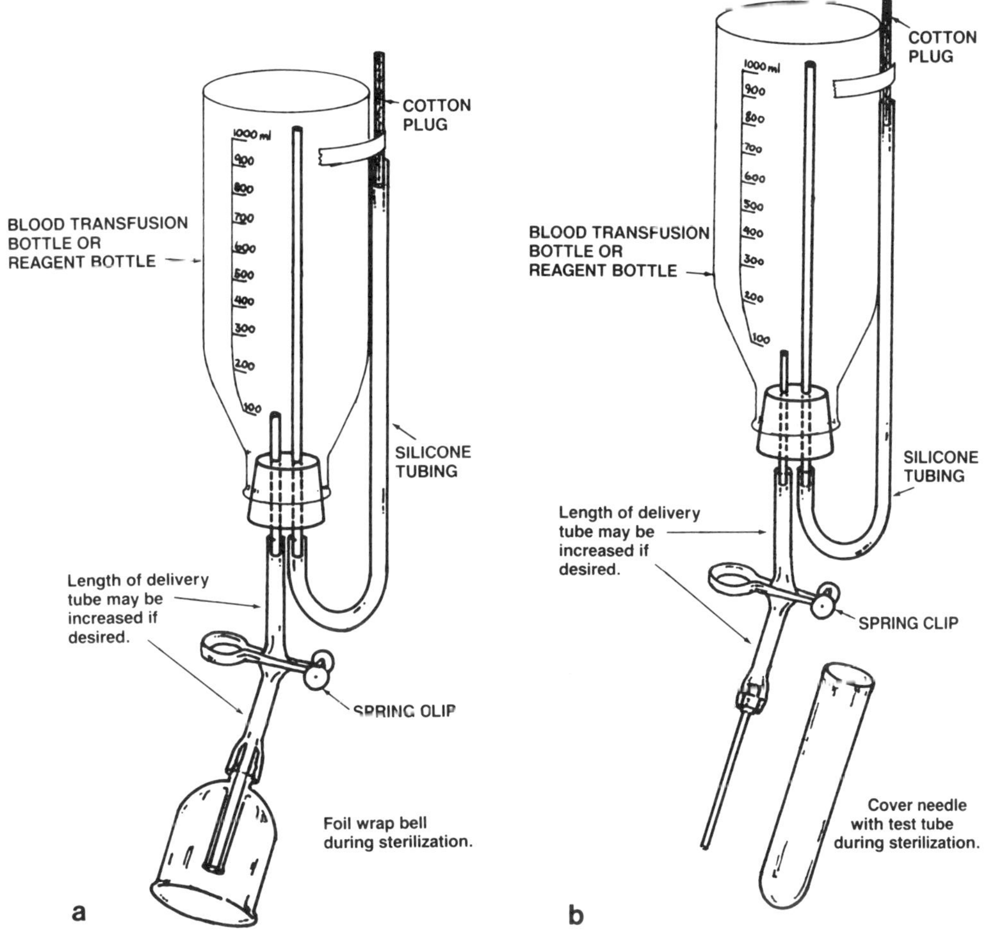

Fig. 6.1. *Simple dispensing devices for use with a graduated bottle. a. With bell used in conjunction with open bottle. b. With needle for slower delivery via a skirted cap or membrane type closure. (From a design by Dr. John Paul.)*

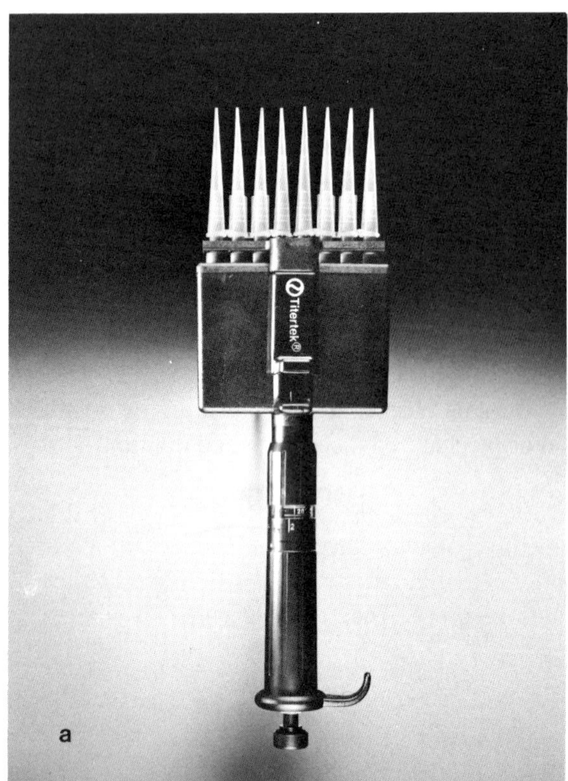

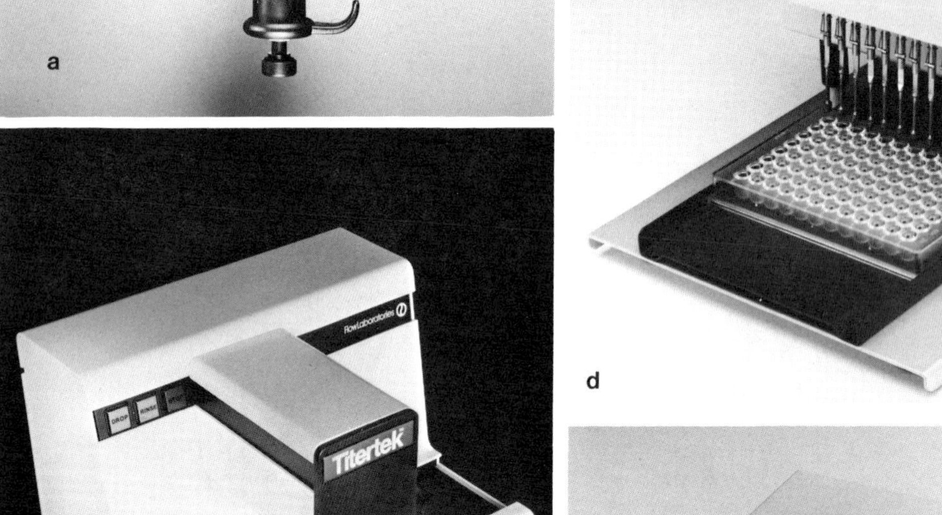

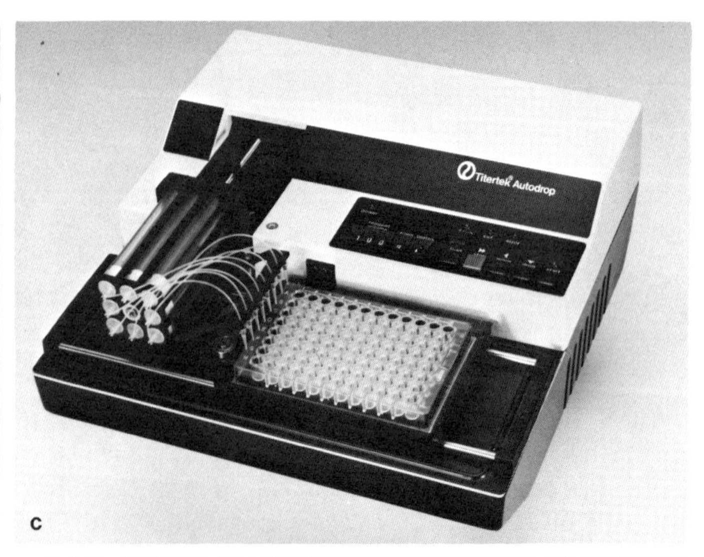

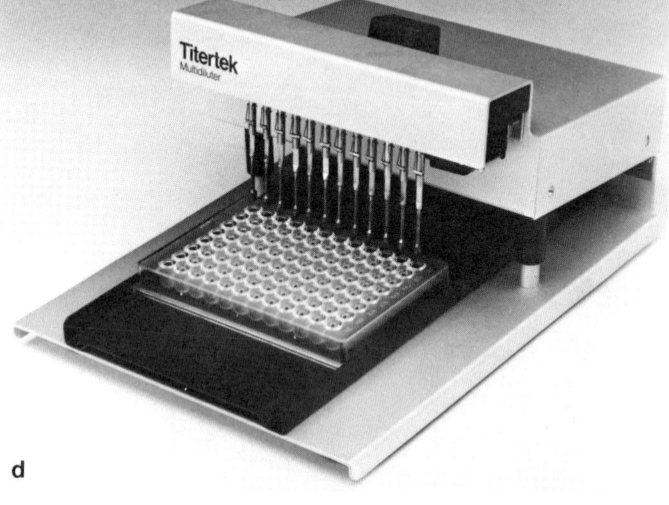

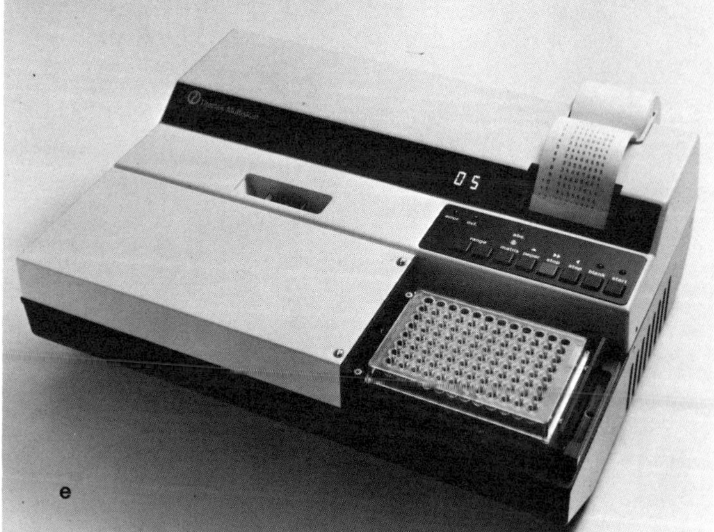

Fig. 6.2. *Microtitration instruments. a. Multipoint micropipette. b, c. Automatic multipoint pipettor. d. Multidiluter. e. Densitometer. Photographs reproduced by permission of Flow Laboratories Ltd.*

drawing up and expressing liquid through a two-way valve (Cornwall Syringe) or by incremental movement of the syringe piston (Hamilton, Flow).

Repetitive dispensers can also be mounted on reagent bottles (Boehringer/Oxford), in which case the culture flask is taken to the pipette rather than vice versa.

All of these repeating pipettes have problems in use resulting from the necessity to autoclave glass syringes, two-way valves, etc. The valves tend to stick (though making these of Teflon helps), syringe pistons deform, or, if Teflon, may contract due to compression during autoclaving. It is preferable to have a nonsterile metering and repeating mechanism so that only the dispensing element need be sterile. The Bellco automated pipette handle, though it has no facility for repetitive pipetting, conforms to this requirement, as does the Tridak Stopper (Bellco) syringe dispenser.

Although more expensive, the best method of automated pipetting is provided by a peristaltic pump controlled in small increments (e.g., Compu-pet) (see Fig. 4.9). In these, only the delivery tube is autoclaved, and accuracy and reproducibility can be maintained to high levels over ranges from 10 μl up to 10 ml. In addition, a number of delivery tubes may be sterilized and held in stock, allowing a quick changeover in the event of accidental contamination or change in cell type or reagent. Larger pumps (10–50 ml) are also available (e.g., Watson-Marlow). In this volume range, it is also possible to use a simple transfusion device with a graduated reservoir (Fig. 6.1). Graduated reservoirs are less convenient where smaller volumes or greater accuracy is required, although a burette, preferably with a two-way valve, can be used.

The introduction of microtitration trays (see Fig. 8.4), has brought with it many automated dispensers, diluters (Fig. 6.2), and other accessories. Transfer devices using perforated trays or multipoint pipettes make it easier to seed from one plate to another, and there are also plate mixers and centrifuge carriers available. The range of equipment is so extensive that it cannot be covered here and the appropriate trade catalogues should be consulted (Flow, Microbiological Associates, Dynatech, Gibco.).

Chapter 7
Laboratory Safety and Biohazards

A major problem which arises constantly in establishing safe practices in a biology laboratory is the disproportionate concern given to the more esoteric and poorly understood risks, such as those arising from genetic manipulation, relative to the known proven hazards of chemicals, toxins, fire, ionizing radiation, electrical shock, and broken glass. No one should ignore potential biohazards [Barkley, 1979], but they should not displace the recognition of everyday safety problems.

The following typical examples should not be interpreted as a code of practice, but rather as advice which might help in compiling safety regulations.

GENERAL SAFETY

Glassware and Sharp Items

The most common form of injury in tissue culture results from accidental handling of broken glass and syringes, e.g., broken pipettes in a wash-up cylinder when too many pipettes, particularly Pasteur pipettes, are forced into too small a container. Pasteur pipettes should be discarded. Avoid syringes unless they are needed for loading ampules or withdrawing fluid from a vial. When disposable needles are discarded, the point should be bent over and trapped inside the sheath. Provide separate receptacles for the disposal of sharp items and broken glass and do not use them for general waste.

Take care when fitting a bulb or pipetting device onto a pipette. Choose the correct size to guard against the risk of the pipette breaking at the neck and lacerating your hand.

Chemical Toxicity

Relatively few major toxic substances are used in tissue culture, but when they are, the conventional precautions should be taken paying particular attention to the distribution of aerosols by laminar flow cabinets (see Biohazards, this chapter). Detergents, particularly those used in automatic machines, are usually caustic; even when they are not they can cause irritation to the skin, eyes, and lungs. Use dosing devices where possible, wear gloves, and avoid procedures which cause the detergent to spread as dust. Liquid detergent concentrates are more easily handled but are often more expensive.

Chemical disinfectants such as hypochlorite should also be used cautiously and with a dispenser. Hypochlorite disinfectants will bleach clothing and cause skin irritations and will even corrode stainless steel.

Specific chemicals used in tissue culture requiring special attention are (1) Dimethyl sulphoxide (DMSO), which is a powerful solvent and skin penetrant and can, therefore, carry many substances through the skin (Horita and Weber, 1964) and even through protective gloves; and (2) mutagens and carcinogens, which should be handled only in a safety cabinet (see below) and are sometimes dissolved in DMSO.

Gases

Most gases used in tissue culture (CO_2, O_2, N_2) are not harmful in small amounts but are, nevertheless, dangerous if handled improperly. They are contained in pressurized cylinders which must be properly secured (Fig. 7.1). When a major leak occurs, there is a risk of asphyxiation from CO_2 and N_2 and of fire from O_2. Evacuation and maximum ventilation are necessary in each case and for O_2, call the fire department.

Ampule sealing is usually performed in a gas-oxygen flame, so great care must be taken both to guard the flame and to prevent unscheduled mixing of the gas and oxygen. A one-way valve should be incorporated in the gas line so that oxygen cannot blow back.

Liquid N_2

There are three major risks associated with liquid N_2: frostbite, asphyxiation, and explosion. Since the temperature of liquid N_2 is $-196°$ C, direct contact

Fig. 7.1. *Cylinder clamp. Clamps onto edge of bench and secures gas cylinder with fabric strap. Fits different sizes of cylinder and can be moved from one position to another if necessary. Available from most laboratory suppliers.*

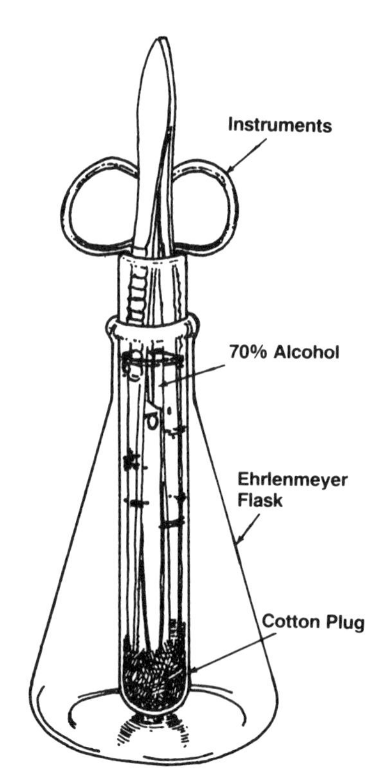

Fig. 7.2. *Flask for alcohol sterilization of instruments. Wide base prevents tipping and center tube reduces the amount of alcohol required so that spillage, if it occurs, is minimized (from an original idea by Mrs. M.G. Freshney).*

with it (splashes, etc.), or with anything, particularly metallic, submerged in it, presents a serious hazard. Gloves thick enough to act as insulation but flexible enough to allow manipulation of ampules should be worn. When liquid N_2 boils off during routine use of the freezer, regular ventilation is sufficient to remove excess nitrogen; but when nitrogen is being dispensed, or a lot of material is being inserted in the freezer, extra ventilation will be necessary.

When ampules are submerged in liquid N_2, a high-pressure difference results between the outside and the inside of the ampule. If it is not perfectly sealed, this results in inspiration of liquid N_2 which will cause the ampule to explode violently when thawed. This can be avoided by storing in the gas phase (see Chapter 18) or by ensuring that the ampules are perfectly sealed. Thawing from storage under liquid N_2 should always be performed in a container with a lid, such as a plastic bucket (see Chapter 18).

Fire

Particular fire risks associated with tissue culture stem from the use of bunsen burners for flaming together with alcohol for swabbing or sterilization. Keep the two separate; always ensure that alcohol for sterilizing instruments is kept in a narrow-necked bottle or flask which is not easily upset, and with the minimum volume of alcohol (Fig. 7.2). Alcohol for swabbing should be kept in a plastic wash bottle. When instruments are sterilized in alcohol and the alcohol subsequently burnt off, care must be taken not to return the instruments to the alcohol while still alight.

Radiation

Hazards from radiation are not exceptional in tissue culture and are usually covered by existing codes of practice for the use of radioisotopes. Periodically, however, uv light is used as a sterilizing agent or mutagen, and care should be taken in these cases to protect the skin and eyes. Ultraviolet irradiation should be performed in a protective glass or plastic cabinet.

BIOHAZARDS

The need for protection against biological hazards [see also Barkley, 1979] is defined (1) by the source of the material and (2) by the nature of the operation being carried out. It is also governed by the conditions under which culture is performed. Standard microbio-

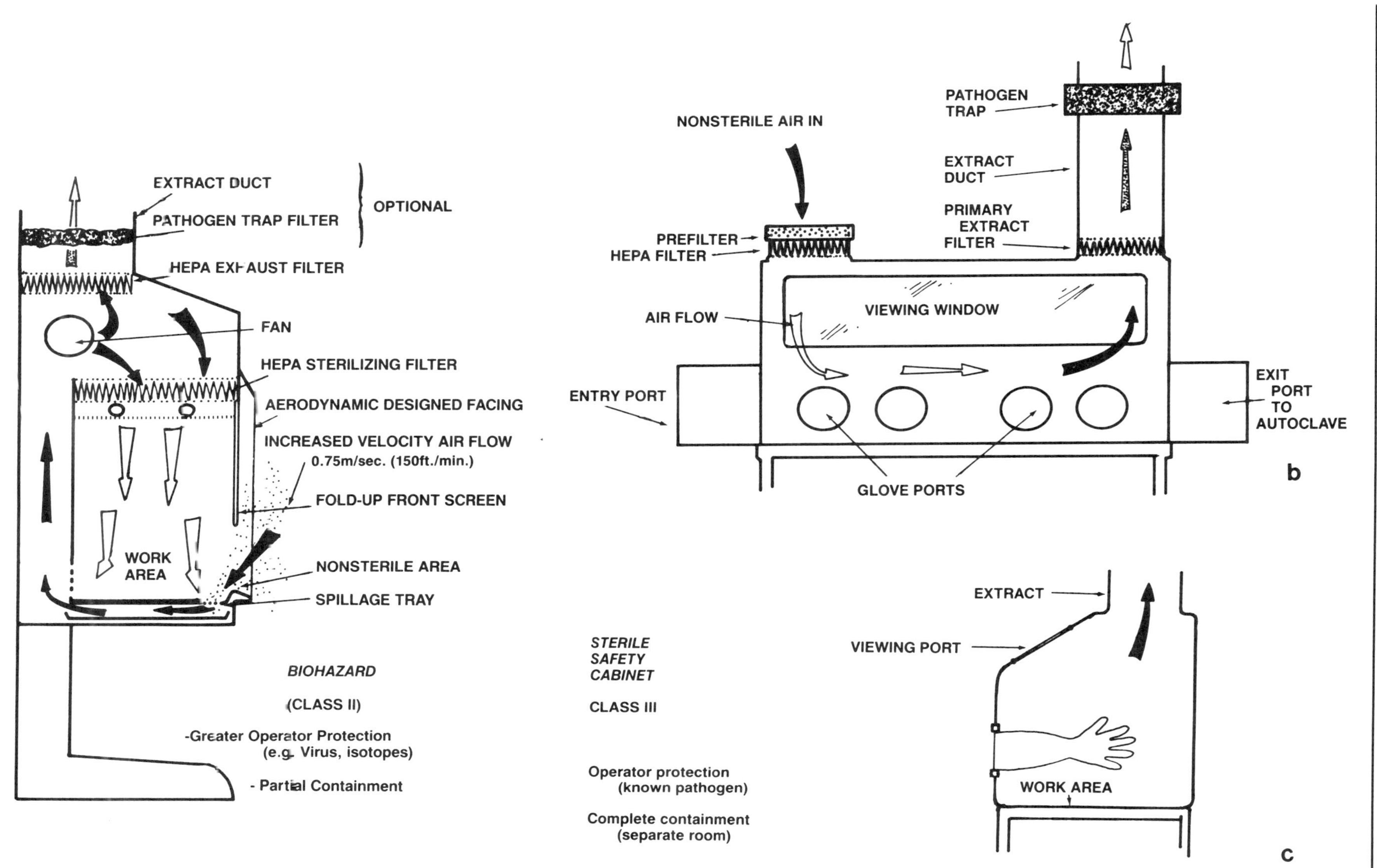

Fig. 7.3. *Biohazard cabinets. a. Class II. Vertical laminar flow, recirculating 70% to 80% of the air. Air (20–30%) exhausted via a filter and discharged into the room, or ducted out of the room through an optional pathogen trap. Air taken in at front of cabinet to make up recirculating volume and prevent overspill from work area. b. Class III. Nonrecirculating, sealed cabinet with glove pockets. Works at negative pressure and with air lock for entry of equipment and direct access to autoclave either connected or adjacent. c. Side view of Class III cabinet.*

logical technique on the open bench has the advantage that the techniques in current use have been established as a result of many years of accumulated experience. Problems arise when new techniques are introduced or when the number of people sharing the same area increases. With the introduction of horizontal laminar flow cabinets, the sterility of the culture was protected more effectively, but the exposure of the operator to aerosols was increased. This led to the development of vertical laminar hoods with an air curtain at the front (see Chapter 5) to minimize overspill from within the cabinet.

We can define three levels of handling: (1) a sealed pathogen cabinet with filtered air entering and leaving via a pathogen trap filter (Class III); (2) a vertical laminar flow cabinet with front protection in the form of an air curtain (Class II); (Fig. 7.3, National Sanitation Foundation Standard 49, NIH specification NIH 03-112, British Standard BS5726); and (3) open bench, depending on good microbiological technique. Table 7.1 lists common procedures with suggested levels of containment. These are suggestions only, however, and you should seek the advice of your local safety committee if in any doubt.

There is often less doubt when known classified pathogens are being used since the regulations are laid down by the Howie Report (U.K.) [1978] and Center for Disease Control (U.S.A.) or when harmless, sterile solutions are being prepared. It is the "gray area" in the middle that causes concern, as development of new techniques such as interspecific cell hybridization and new facilities such as laminar flow, introduces putative risks for which there are no epidemiological data available for assessment. Transforming viruses, transformed human cell lines, and human-mouse hybrids, for example, should be treated cautiously until data accumulates that they carry no risk.

Risks which are more easily recognized are those associated with biopsy and autopsy specimens from human and primate tissue. Where infection has been confirmed, the type of organism will determine the degree of containment, but where there is no known infection, the possibility remains that the sample may yet carry hepatitis B, tuberculosis, or other pathogens, as yet undiagnosed. Such samples should be handled with caution (Class II Biohazard cabinet, no sharp instruments used in handling, discard into disinfectant for autoclave) until they can be shown to be uninfected by the appropriate clinical diagnostic tests.

TABLE 7.1. Biohazard Procedures and Suggested Levels of Containment

Procedure	Level of protection
Media preparation	Open bench, standard microbiological practice, or horizontal laminar flow
Cell lines other than human and other primates	Open bench, standard microbiological practice, or horizontal or vertical laminar flow
Primary culture and serial passage of human and other primate cells	Vertical laminar flow cabinet with air curtain protection at front and filtered extract (Class II)
Interspecific hybrids or other recombinants between human cells and animal tumor cells	Vertical laminar flow cabinet with air curtain protection at front and filtered extract (Class II).
Virus-producing human cell lines	Pathogen cabinets with glove pockets, filtered air entering and pathogen trap on vented air (Class III). Located in a separate room with separate provision for incubation, cetrifugation, cell counting, etc. No access except to designated personnel. All waste, soiled glassware, etc., to be sterilized as it leaves the room and extracted air to be filtered.
Tissue samples and cultures carrying known human pathogens	Pathogen cabinet with glove pockets, filtered air entering and pathogen trap on vented air (Class III). Located in a separate room with separate provision for incubation, centrifugation, cell counting, etc. No access except to designated personnel. All waste, soiled glassware, etc., to be sterilized as it leaves the room and extracted air to be filtered.

The Culture Environment: I. Substrate, Gas Phase, and Temperature

Chapter 8

The regulation of the environment of the culture is expressed via four routes: (1) the nature of the substrate or phase on or in which the cells grow. This may be solid, as in monolayer growth on plastic, semisolid as in a gel such as collagen or agar, or liquid as in suspension culture; (2) the physicochemical and physiological constitution of the medium; (3) the constitution of the gas phase; and (4) the incubation temperature. It is, perhaps, useful to think of the four elements of the ancient alchemists in remembering these routes: "air"—the gas phase, "earth"—the substrate, "fire"—the temperature, and "water"—the medium. Since the constitution of culture medium is such a major component of the culture environment it will be dealt with separately in the next chapter.

THE SUBSTRATE

The majority of vertebrate cells cultured in vitro have been grown as monolayers on an artificial substrate. Spontaneous growth in suspension is restricted to hemopoietic cell lines, rodent ascites tumors, and a few other selected cell lines. From the earliest attempts, glass has been used as the substrate, initially because of its optical properties, but subsequently because it appears to carry the correct charge for cells to attach and grow. With the exception of the abovementioned cells and other transformed cell lines, most cells require to spread out on a substrate in order to proliferate [Fisher and Solursh, 1979; Folkman and Moscona, 1978]. Inadequate spreading due to poor adhesion or overcrowding will inhibit cell proliferation. Cells shown to require attachment for growth are said to be "anchorage dependent." Cells which have undergone transformation frequently become anchorage independent and can grow in suspension when stirred or held in suspension with semisolid media such as agar.

This assumes, however, that cell proliferation is the principal objective. It may not be; cells which are anchored only to each other as spheroids in suspension or which are growing as a secondary layer on top of a confluent monolayer may proliferate more slowly but may still reflect more accurately behavior *in vivo*.

Artificial Substrates

Glass. Glass is most commonly used, and has been since tissue culture first began. It is cheap, easily washed without losing its growth-supporting properties, can be sterilized readily by dry or moist heat, and is optically clear. Treatment with strong alkali (e.g., NaOH or caustic detergents) renders glass unsatisfactory for culture until it is neutralized by an acid wash.

Disposable plastic. Single-use polystyrene flasks provide a simple, reproducible substrate for culture. They are usually good optical quality, and the growth surface is flat, providing uniform and reproducible cultures. Polystyrene, as manufactured, is hydrophobic, and does not provide a suitable surface for cell growth, so tissue culture plastics are treated by γ-radiation, chemically, or with an electric arc to produce a charged surface which is then wettable. As the resulting product varies in quality from one manufacturer to another, samples from a number of sources should be tested by determining the plating efficiency and growth rate of your cells (see Chapters 13 and 19), in medium containing the normal and half-normal concentration of serum (high serum concentrations may mask imperfections in the plastic).

While polystyrene is by far the most common and cheapest plastic substrate, cells may also be grown on polyvinylchloride, polycarbonate, polytetrafluorethylene (P.T.F.E), TPX, and a number of other plastics. If you need to use a different plastic, it is worth trying to grow a regular monolayer and then attempting to clone cells on it (see Chapters 13 and 19), with and without pretreatment of the surface (see below).

Polytetrafluorethylene is available in a charged (hydrophilic) and uncharged (hydrophobic) form; the charged form can be used for regular monolayer cells and the uncharged for macrophages and some transformed cell lines. Polytetrafluorethylene films are available as disposable petri dishes ("Petriperm," Heraeus), or as membranes to be incorporated in an autoclave reusable culture vessel ("Chamber/Dish," Bionique).These dishes have two other advantages: (1) The substrate is permeable to O_2 and CO_2 and (2) the plastic is thin and, therefore, well suited to histological sectioning for light or electron microscopy.

Permeable substrates have been in use for many years. In 1965, Sandström suggested that hepatocytes survived better in the higher oxygen tension provided by growth in a cellophane sandwich. Growth of cells on floating collagen [Michalopoulos and Pitot, 1975; Lillie et al., 1980] and cellulose nitrate membranes [Savage and Bonney, 1978] have been used to improve the survival of epithelial cells and promote terminal differentiation (see Chapter 20).

It is possible that growth of cells on a permeable substrate contributes more than increased diffusion of oxygen, CO_2, and nutrients. Attachment to a natural substrate such as collagen may exert some biological control of phenotypic expression due to the interaction of receptor sites on the cell surface with specific sites in the extracellular matrix. Permeability of the surface to which the cell is anchored may, in itself, signify polarity to the cell by simulating the basement membrane underlying an epithelial cell layer or between tissue cells and endothelium surrounding the vascular space. Such polarity may be vital to full functional expression in secretory epithelia and many other cell types. This prompted Reid and Rojkind [1979], Gospodarowicz [Vlodavsky et al., 1980], and others to explore the growth of cells on natural substrates related to basement membrane (see below).

Microcarriers. Polystyrene (Nunclon, GIBCO), Sephadex (Flow Laboratories and Pharmacia) and polyacrylamide (Biorad) are available in bead form for propagation of anchorage-dependent cells in suspension (see Chapter 21).

Sterilization of plastics. Disposable plasticware is usually supplied sterile and cannot be reused as washing in detergents renders the surface unsuitable for monolayer culture. For the sterilization of other plastics, see Chapter 10.

Alternative artificial substrates. Although glass and plastic are employed for more than 90% of all cell propagation, there are alternative substrates which can be used for specialized applications. Westermark [1978] developed a method for the growth of fibroblasts and glia on palladium. Using electron microscopy shadowing equipment, he produced islands of palladium on agarose, which does not allow cell attachment in fluid media. The size and shape of the islands was determined by masks made in the manner of electronic printed circuits, and the palladium was applied by "shadowing" under vacuum, as used in electron microscopy.

Cells grown on stainless steel discs and labeled with radioactive isotopes can be counted directly by end-window counting [Birnie and Simons, 1967], and there are other reports of growing cells on metallic surfaces [Litwin, 1973]. Observation of the cells on an opaque substrate requires surface interference microscopy, unless very thin metallic films are used, as with Westermark's palladium islands.

Treated surfaces. Cell attachment and growth can be improved by pretreating the substrate in a variety of ways. It is a well-established piece of tissue culture lore that used glassware supports growth better than new. This may be due to etching of the surface or minute traces of residue left after culture. Growth of cells in a flask also improves the surface for a second seeding and this type of conditioning may be due to collagen [Hauschka and Konigsberg, 1966] or fibronectin [Thom et al., 1979] released by the cells. The substrate can be conditioned by treatment with spent medium from another culture [Stampfer et al., 1980] or by purified fibronectin [Gilchrest et al., 1980] or collagen [Elsdale and Bard, 1972; Kleinman et al., 1979, 1981]. Treatment with denatured collagen improves the attachment of many cells such as epithelial cells [Lillie et al., 1980; Freeman et al., 1976] and muscle cells [Hauschka and Konigsberg, 1966], and it may be necessary for the expression of differentiated functions by these cells (see Chapter 20). Collagen may also be applied as an undenatured gel, and this type of substrate has been shown to support neurite outgrowth from chick spinal ganglia [Ebendal, 1976], morphological differentiation of breast [Yang et al., 1981], and other epithelia [Sattler et al., 1978] and to promote expression of tissue-specific functions of a number of other cells in vitro [Meier and Hay, 1974, 1975; Kosher and Church, 1975].

Evidence is gradually accumulating that specific

treatment of the substrate with biologically significant compounds can induce specific alterations in attachment or behavior of specific cell types. For example, chondronectin enhances chondrocyte adherence and laminin epithelial cells [Kleinman et al., 1981]. Reid and Rojkind [1979] described methods for preparing reconstituted "basement membrane rafts" from tissue extracts for optimization of culture conditions for cell differentiation.

Gelatin coating has been found to be beneficial for the culture of muscle [Richler and Yaffe, 1970], endothelial cells [Folkman et al., 1979], and it is necessary for some mouse teratomas. McKeehan and Ham [1976] found that it was necessary to coat the surface of plastic dishes with 1 mg/ml poly-D-lysine before cloning in the absence of serum (see Chapter 13).

This raises the interesting question of whether the cell requires at least two components of interaction with the substrate: (1) adhesion to allow the attachment and spreading necessary for cell proliferation [Folkman and Moscona, 1978] and (2) specific interactions, reminiscent of the interaction of an epithelial cell with basement membrane, with other extracellular matrix constituents, or with adjacent tissue cells [Auerbach and Grobstein, 1958]. The second type of interaction may be less critical to sustained proliferation of undifferentiated cells but may be required for the expression of some specialized functions (see Chapter 2).

While inert coating of the surface may suffice, it may yet prove necessary to provide a monolayer of an appropriate cell type as an underlay for maintenance of some specialized cells. Gospodarowicz et al. [1980] were able to grow endothelium on confluent monolayers of 3T3 cells which had been extracted with Triton × 100, leaving cell coat residue on the surface of the substrate. This so-called extracellular matrix (ECM) has also been used to promote differentiation in ovarian granulosa cells [Gospodarowicz, 1980] and in studying tumor cell behavior [Vlodarsky et al., 1980].

Feeder layers. Cultures of mouse embryo fibroblasts, or other cells, have been used for many years to enhance growth particularly at low cell densities (see Chapter 13) [Puck and Marcus, 1955]. This action is due partly to supplementation of the medium but may also be due to conditioning of the substrate by cell products. Feeder layers grown as a confluent monolayer may make the surface suitable for attachment for other cells. We have shown selective growth of breast and colonic epithelium, and of glioma, on confluent feeder layers of normal fetal intestine [Freshney et al., 1982].

The survival and neurite extension by central and peripheral neurons can be enhanced by culturing the neurons on a monolayer of glial cells, although in this case the effect is due to a diffusible factor rather than direct cell contact [Lindsay, 1979].

After a monolayer culture reaches confluence subsequent proliferation causes cells to detach from the artificial substrate and migrate over the surface of the monolayers. Their morphology may change (Fig. 8.1), and the cells are less well spread, more densely staining, and may be more highly differentiated. Apparently, the interaction of a cell with a cellular underlay is different from the interaction with a synthetic substrate. This can cause change in morphology and reduce proliferative potential.

Three-dimensional matrices. It has long been realized that while growth in two dimensions is a convenient way of preparing and observing a culture and allows a high rate of cell proliferation, it lacks the cell-cell and cell-matrix interaction characteristic of whole tissue *in vivo*. The very first attempts to culture animal tissues [Harrison, 1907; Carrel, 1912] were performed with gels formed of clotted lymph or plasma on glass. In these cases, however, the cells migrated along the glass/clot interface rather than within the gel and tissue architecture and cell-cell interaction was gradually lost. Migration was often accompanied by proliferation of cells in the outgrowth, leading, in later studies, to the development of propagated cell lines. It gradually became apparent that many functional and morphological characteristics were lost during serial subculture, as discussed in Chapter 2.

These deficiencies encouraged the exploration of three-dimensional matrices such as collagen gel [Douglas et al., 1980]; cellulose sponge, alone [Leighton et al., 1951] or collagen-coated [Leighton et al., 1968]; or Gelfoam (see Chapter 2). Many different cell types can be shown to penetrate such matrices and establish a tissuelike histology. Breast epithelium, seeded within collagen gel, displays a tubular morphology, while breast carcinoma grows in a more disorganized fashion, confirming the correlation between this mode of growth and the condition *in vivo* [Yang et al., 1981].

Neurite outgrowth from sympathetic ganglia neurons growing on collagen gels follows the orientation of the collagen fibers in the gel [Ebendal, 1976] (see further discussion of three-dimensional cultures, Chapter 21).

Nonadhesive substrates. There are situations where attachment of the cell is undesirable. The selection of virally transformed colonies, for example, can be

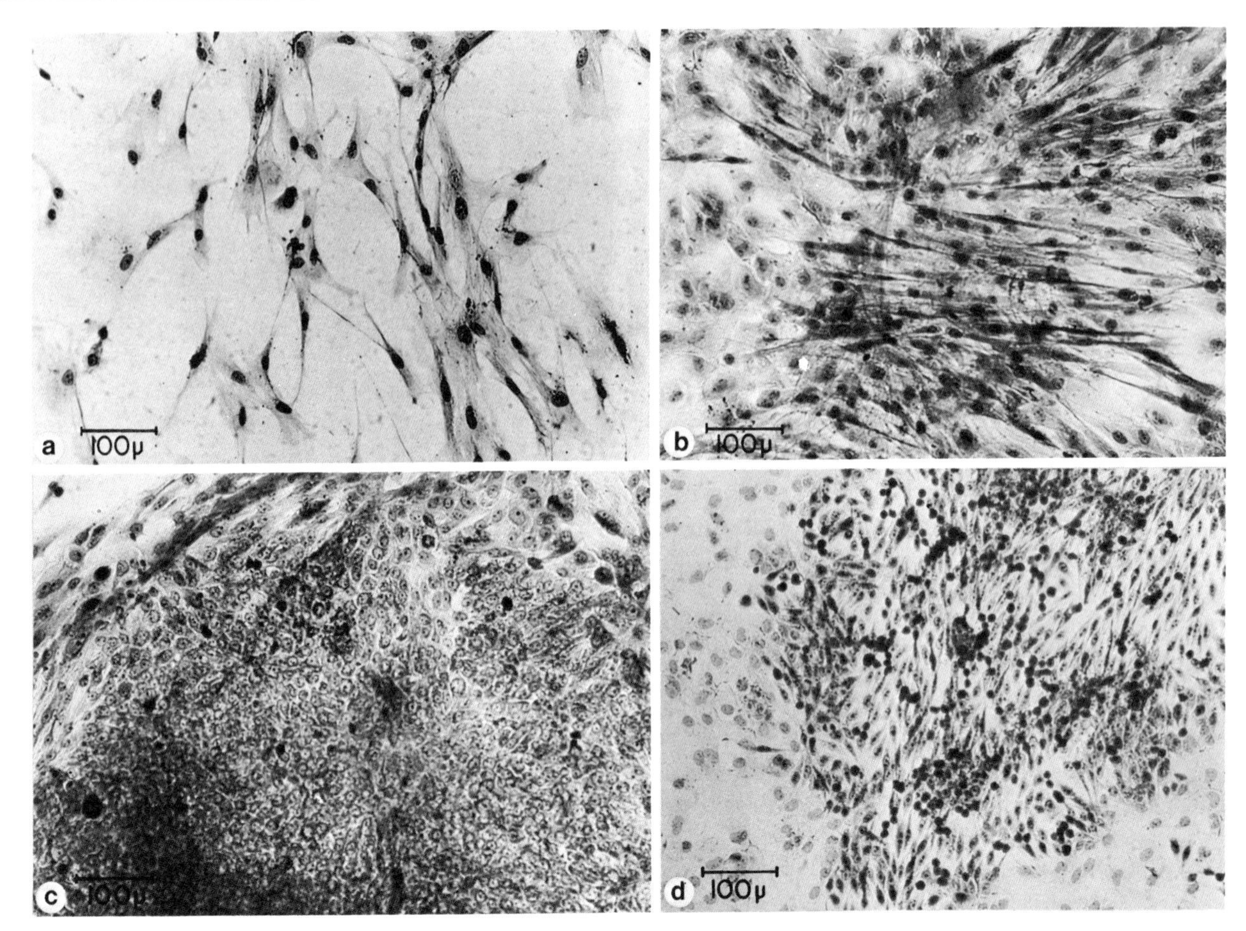

Fig. 8.1. *Morphological alteration in cells growing on feeder layers. a. Fibroblasts from human breast carcinoma growing on plastic and (b) growing on a confluent feeder layer of fetal human intestinal cells (FHI). c. Epithelial cells from human breast carcinoma growing on plastic and (d) on same confluent feeder layer as in b.*

achieved by plating cells in agar [Macpherson and Montagnier, 1964], as the untransformed cells do not form colonies readily in this matrix.

There are two principles involved in this system: (1) prevention of attachment at the base of the dish where spreading would occur and (2) immobilization of the cells such that daughter cells remain associated with the colony even if nonadhesive. The usual agents employed are agar, agarose, or Methocel (Methylcellulose viscosity 4,000 cps). The first two are gels and the third is a high-viscosity sol. Because Methocel is a sol, cells will sediment slowly through it. It is, therefore, commonly used with an underlay of agar (see Chapter 13). Non-tissue culture grade dishes can be used without an agar underlay, but some attachment and spreading may occur.

Liquid-gel or liquid-liquid interfaces. While the Methocel-over-agar system usually gives rise to discrete colonies at the interface of the agar and the Methocel, some cells can migrate across the gel surface and form monolayers or cords of cells (Fig. 8.2). The reason for this remains obscure. Rosenberg [1965] observed cell spreading and monolayer formation with HeLa cells at the liquid-liquid interface between various fluorinated hydrocarbons (FC43, FC73) and aqueous culture media. The occurrence of spreading and locomotion on nonrigid substrates conflicts somewhat with current concepts of cell adhesion and locomotion unless denatured serum protein or some other substance forms a layer at the interface sufficient to permit anchorage. Methocel, particularly, often contains particulate debris which may help to promote this.

Perfused microcapillary bundles. Knazek et al. [1972] developed a technique for the growth of cells on the outer surface of bundles of plastic microcapillaries (Fig. 8.3) (see Chapter 21). The plastic allows the diffusion of nutrients and dissolved gases from medium perfused through the capillaries. Cells will grow up to several cells deep on the outside of the capillaries and an analogy with whole tissue is suggested.

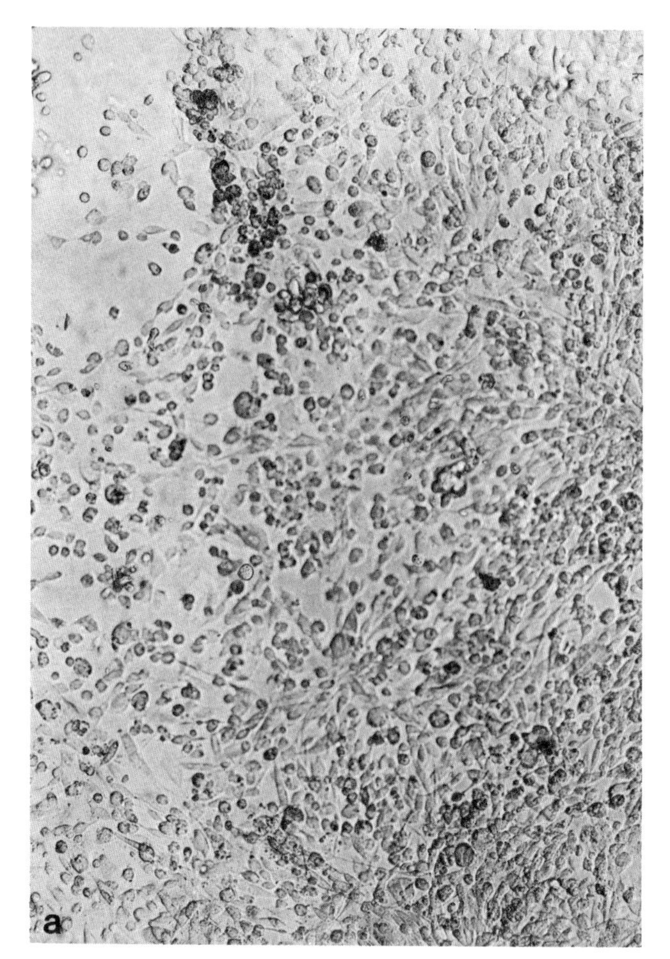

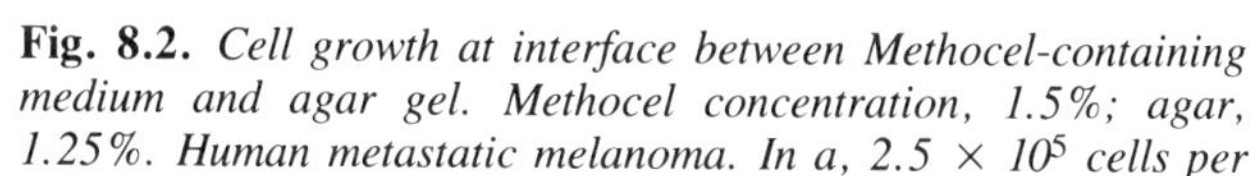

Fig. 8.2. *Cell growth at interface between Methocel-containing medium and agar gel. Methocel concentration, 1.5%; agar, 1.25%. Human metastatic melanoma. In a, 2.5 × 10^5 cells per ml, cloned alone. In b, 5 × 10^4 cell cloned with 2 × 10^5 homologous feeder cells per ml.*

Culture Vessels

Some typical culture vessels are listed in Table 8.1. The anticipated yield of HeLa cells is quoted for each vessel; the yield from a finite cell line, e.g., diploid fibroblasts, would be about one-fifth of the HeLa figure. Several factors govern the choice of culture vessel including (1) the cell yield, (2) whether the cells grow in suspension or as a monolayer, (3) whether the culture should be vented to the atmosphere or sealed, (4) what form of sampling and analysis is to be performed, and (5) the anticipated cost.

Cell yield. For monolayer cultures, the cell yield is proportional to the available surface area of the flask. Small volumes and multiple replicates are best performed in multiwell dishes (Fig. 8.4) which range from Teresaki plates (60–72 wells, 10-μl culture volume) up to four wells, 50 mm in diameter. The most popular are microtitration dishes (96 or 144 wells, 0.1–0.2 ml, 0.25 cm^2 growth area) and 24-well "cluster dishes" (1–2 ml each well, 1.75 cm^2 (see Table 8.1). The middle of the size range embraces both petri dishes (Fig. 8.5) and flasks ranging 20 cm^2–150 cm^2. Flasks are usually designated by their surface area, e.g., No. 25 or No. 120. Plastic culture flasks come in the range 25, 75, 120, 150, and 175 cm^2 (Fig. 8.6); glass bottles are more variable since they are usually drawn from standard pharmaceutical supplies (Fig. 8.7). They should have (1) one reasonably flat surface, (2) a deep screw cap with a good seal and nontoxic liner, and (3) shallow sloping shoulders to facilitate harvesting monolayer cells after trypsinization and to improve the efficiency of washing.

If you require large cell yields (e.g., ~ 10^9 HeLa cervical carcinoma cells or 2 × 10^8 MCR-5 diploid human fibroblast), then increasing the size and number of conventional bottles becomes cumbersome and special vessels are required. These are described in Chap-

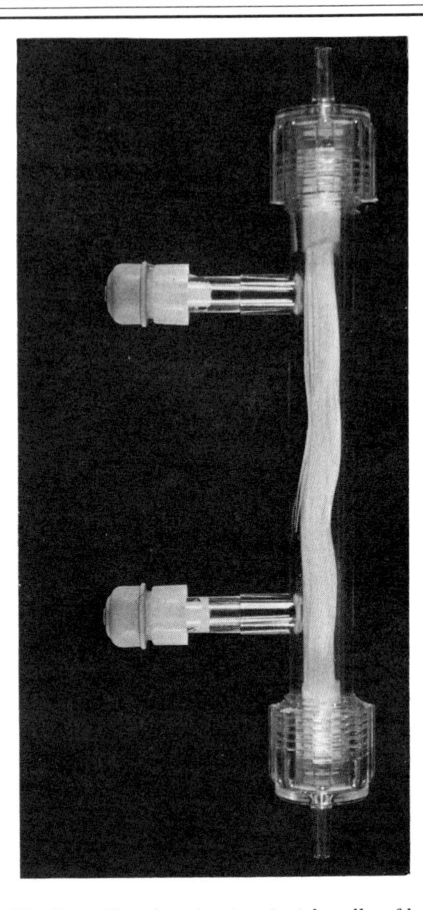

Fig. 8.3. *Vitafiber Chamber (Amicon). A bundle of hollow fibers of permeable plastic is enclosed in a transparent plastic outer chamber, accesible via either of the two side arms for seeding cells. During culture, the chamber is perfused down the center of the hollow fibers through connections attached to either end of the chamber (see also Fig. 21.3).*

ter 21. Increasing the yield of cells growing in suspension requires only that the medium volume be increased, as long as cells in deep culture are kept agitated and sparged with 5% CO_2 in air (see Chapter 21).

Venting. Multiwell dishes and petri dishes chosen for replicate sampling or cloning have loose-fitting lids to give easy access to the dish. Consequently, they are not sealed and will require a humid atmosphere with control of the CO_2 tension (see above and Chapter 4). Because a thin film of liquid may form around the inside of the lid, partially sealing some dishes, loose-fitting lids should be provided with supports to raise them off the rim of the dish (Fig. 8.8). If a perfect seal is required, some multiwell dishes can be sealed with self-adhesive Mylar film (Flow). Flasks may be vented by slackening the caps. Again, because of variable sealing due to liquid inside the cap, the cap must be slackened one full turn. Flasks are vented in this way to allow CO_2 to enter (in a CO_2 incubator) or to allow excess CO_2 to escape in excessive acid-producing cell lines (see above).

Sampling and analysis. Multiwell plates are ideal for replicate culture if all samples are to be removed simultaneously and processed in the same way. If, on the other hand, samples need to be withdrawn at different intervals and processed immediately, it may be preferable to use separate vessels (flasks, test tubes, etc.) (Fig. 8.9). Individual wells in microtitration plates can be sampled by cutting and removing only that part of the adhesive plate sealer overlying the wells to be sampled. Alternatively, microtitration plates are available with removable wells for individual processing.

If processing of the sample involves extraction in acetone, toluene, ethyl acetate or certain other organic solvents, then a problem will arise with polystyrene. Since this problem is often associated with histological procedures, Lux supplies Thermanox (TPX) plastic coverslips, suitable for histology, to fit into regular multiwell dishes (which need not be tissue culture grade). However, they are of poor optical quality and should be mounted cells uppermost with a conventional glass coverslip on top.

Glass vessels are required for procedures such as hot perchloric acid extractions of DNA. Plain-sided test tubes or Erlenmeyer flasks (no lip) used in conjunction with sealing tape or Oxoid caps are quick to use and are best kept in a humid CO_2-controlled atmosphere. Regular glass scintillation vials, or "minivials," are also good culture vessels as they are flat-bottomed and have a screw closure. Once used with scintillant, however, they should not be reused for culture.

Cost. Cost always has to be balanced against convenience—e.g., petri dishes are always cheaper than flasks of an equivalent surface area but require humid CO_2-controlled conditions and are more prone to infection.

Cheap soda glass bottles, though not always of good optical quality, are often better for culture than higher grade Pyrex or optically clear glass, which usually contains lead.

A major disadvantage of glass is that it is labor intensive in preparation as it must be carefully washed and resterilized before it can be reused. The cost of this will depend on your existing staff and the number of flasks used. To employ a new member of staff to wash and sterilize glassware will cost about half the amount per flask, relative to disposable plastic, for an

TABLE 8.1. Culture Vessel Characteristics*

Culture vessel	Plastic or glass	No. of replicates	Vol.	Surface area	Approx. cell yield	Supplier
Microtest (Terasaki)	P	60, 72	0.01 ml	0.78 mm^2	2.5×10^3	F, N
Microtitration plate	P	96, 144	0.1 ml	32.00 mm^2	10^5	C, F, N, L
Multiwell plate†	P	4 round	1.0 ml	2.00 cm^2	5×10^5	C, F, N
Multiwell plate	P	12 round	2.0 ml	4.5 cm^2	10^6	F
Multiwell plate	P	24 round	1.0 ml	2.00 cm^2	5×10^5	C, F, L
Multiwell plate	P	8 rectangular	2.0 ml	7.8 cm^2	2×10^6	Lu
Multiwell plate	P	4 rectangular	3.0 ml	16.08 cm^2	4×10^6	Lu
Multiwell plate	P	6 round	2.5 ml	9.62 cm^2	2.5×10^6	C, L
Multiwell plate	P	4 round	5.0 ml	28.27 cm^2	7×10^6	F, N, L
Petri dishes†						
30 mm	P		2.0 ml	6.85 cm^2	1.7×10^6	S
35 mm	P		3.0 ml	8.00 cm^2	2.0×10^6	C, Cg
50 mm	P		4.0 ml	17.50 cm^2	4.4×10^6	S, F, N
60 mm	P		5.0 ml	21.00 cm^2	5.2×10^6	C, Cg, S, N
90 mm	P		10.0 ml	49.00 cm^2	12.2×10^6	S, F
100 mm	P		10.0 ml	55.00 cm^2	13.7×10^6	C, Cg, F, N
100 mm^2	P		15.0 ml	100.00 cm^2	20×10^6	S
Tissue culture tubes						
Leighton†	P & G		1.0 ml	4.00 cm^2	10^6	Be
One side flattened	P		2.0 ml	5.50 cm^2	10^6	N
Round with screw cap	P		2.0 ml	100 mm × 14 mm		N
Flasks						
25	P		5.0 ml	25.00 cm^2	5×10^6	C, Cg, F, N
50	G		10–20 ml	50.00 cm^2	10^7	
75	P & G		15–30 ml	75.00 cm^2	2×10^7	C, Cg, F, N
120	G		40–100 ml	120.00 cm^2	5×10^7	
150	P		75 ml	150.00 cm^2	6×10^7	C, Cg
175	P		50–100 ml	175.00 cm^2	7×10^7	F, N
Roller bottles						
2 1/2 1	G		100–250 ml	700.00 cm^2	2.5×10^8 (~ 1 g)	N.B
Roller disposable	P		100–250 ml	850.00 cm^2	3.0×10^8	F, Cg
Large	G		100–500 ml	1,585.00 cm^2	6.0×10^8	N.B
Nunc cell factory	P		1,800 ml	6,000.00 cm^2	2.0×10^9	N
Spiral	P		1,600 ml	8,500.00 cm^2	2.5×10^9 (~ 10 g)	S
Microcarriers		see Chapter 21	See Stirrer bottles, below			Ph, B, N, F
Stirrer bottles						
Reagent bottle, round (500 ml)	G		200 ml		3×10^8	Cg, Be
Reagent bottle, round (1,000 ml)	G		400 ml		8×10^8	Cg, Be
Aspirator (2,000 ml)	G		600 ml		10^9	P
Aspirator (5,000 ml)	G		4,000 ml	Gas with 5% CO_2	6×10^9	P
Aspirator (10,000 ml)	G		8,000 ml	Gas with 5% CO_2	8×10^9	P

*Abbreviations: B, Bio Rad; Be, Bellco; Cg, Corning; C, Costar; F, Falcon; L, Linbro (Flow); Lu, Lux; N, Nunc (GIBCO); N.B, New Brunswick; P, Pyrex; Ph, Pharmacia; S, Sigma.

†Dishes can be used on their own or with a coverslip, e.g., glass, TPX (Lux), Polystyrene (Lux), Melinex (I.C.I.). Non-tissue culture grade dishes may be used with coverslips. Petri dish sizes often refer to the outside diameter of the base or lid. Surface area must be calculated from the inside diameter of the base.

annual output of 10,000 flasks. If your usage is substantially less than this, it will be better to use disposable plastic, particularly for smaller flasks (25 cm^2). If you do not need to meet the cost of employing washing-up staff, glass will be found much cheaper than plastic.

THE GAS PHASE

Oxygen

The significant constituents of the gas phase are oxygen and carbon dioxide. Cultures vary in their oxygen requirement; the major distinction lying between organ and cell cultures. While atmospheric, or lower, oxygen tensions [Cooper et al., 1958; Balin et al., 1976] are preferable for most cell cultures, some organ cultures, particularly from late stage embryo, newborn or adult, require up to 95% O_2 in the gas phase [Trowell, 1959; DeRidder and Mareel, 1978]. This may be a problem of diffusion related to the geometry of organ cultures (see Chapter 21) rather than a distinct cellular requirement, since most dispersed cells prefer lower oxygen tensions, and some systems, e.g., human tumor cells in clonogenic assay

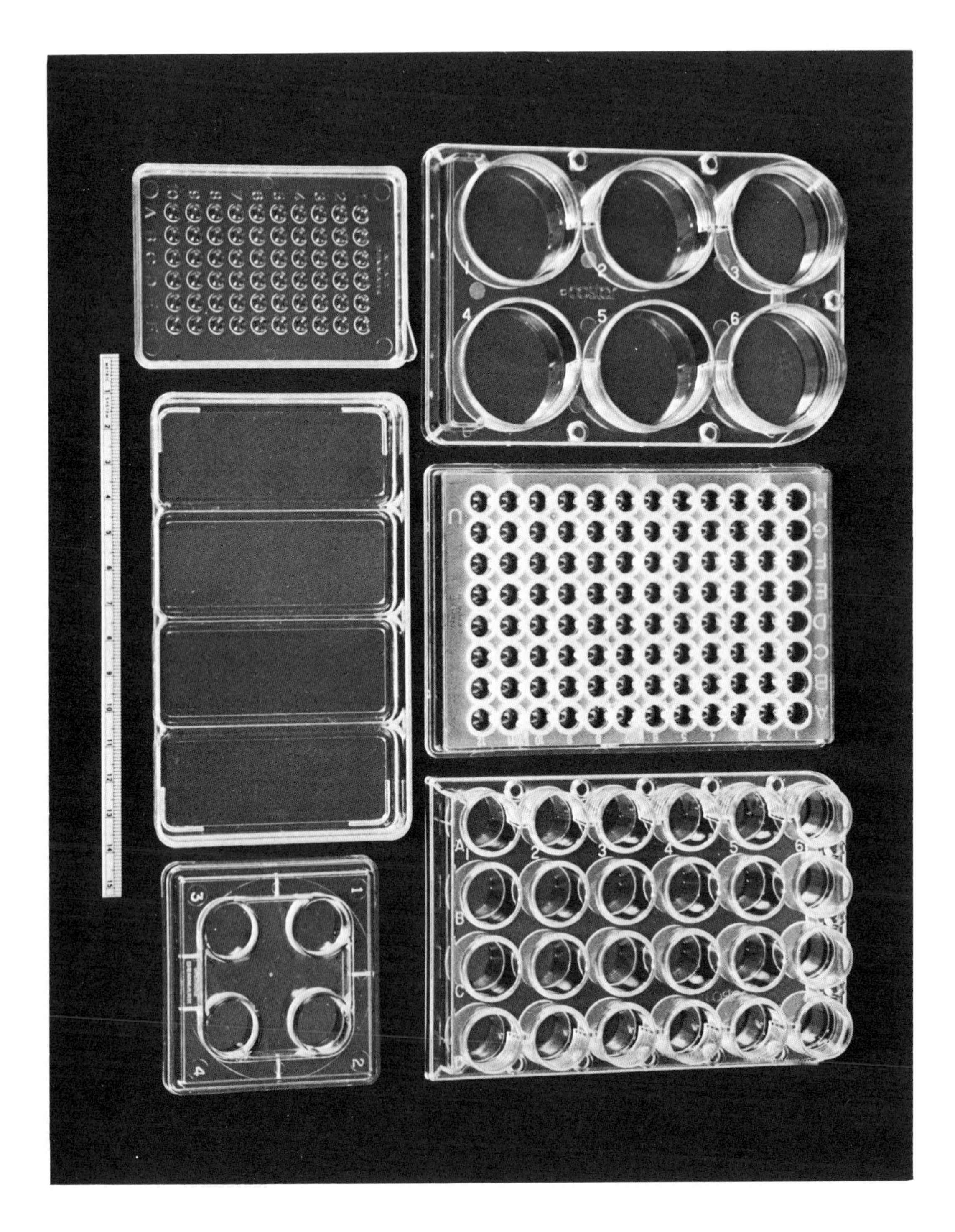

Fig. 8.4. *Multiwell plates (see Table 8.1 for sizes and capacities).*

Fig. 8.5. *Some common sizes of disposable plastic petri dishes. Sizes range from 35 mm to 90 mm diameter, circular, and 9 cm × 9 cm, square. Larger dishes are available but are seldom used for cell culture. A grid pattern can be provided to help in scanning the dish—to count colonies, for example.*

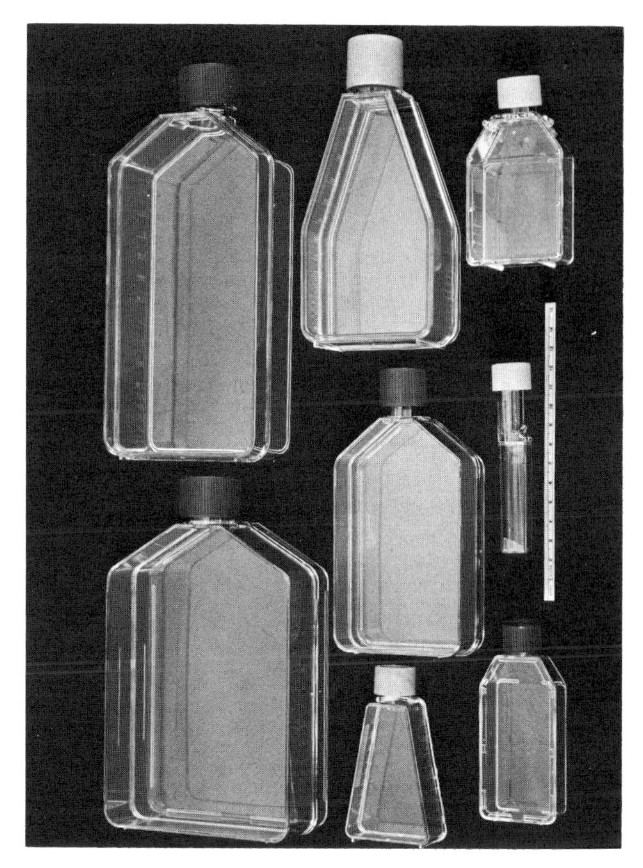

Fig. 8.6. *Disposable plastic culture vessels (Falcon, Costar and Corning). The triangular bottles (Costar) are designed to improve access to all of the growth surface when dispersing a monolayer (see Table 8.2 for sizes and capacities).*

[Courtney et al., 1978], and human embryonic lung fibroblasts [Balin et al., 1976], do better in less than the normal atmospheric oxygen tension. It has been suggested [McKeehan et al., 1976] that the requirement for selenium in medium is related to oxygen tension and that this element detoxifies dissolved oxygen. This requirement may only arise in the absence of serum proteins and, as there is a trend toward serum-free media, the role of dissolved O_2 may become more important in the future, requiring, perhaps, controlled O_2 tension during incubation. As the depth of the culture medium can influence the rate of oxygen diffusion to the cells it is advisable to keep the depth of medium within the range 2–5 mm (0.2–0.5 ml/cm^2) in static culture.

Carbon Dioxide

Carbon dioxide has a rather complex role to play, and because many of its actions are interrelated, e.g., dissolved CO_2, pH, and HCO_3^- concentration, it is difficult to determine its major direct effect. The atmospheric CO_2 tension will regulate the concentration of dissolved CO_2 directly, as a function of temperature. This in turn produces H_2CO_3 which dissociates:

$$H_2O + CO_2 \rightleftharpoons H_2CO_3 \rightleftharpoons H^+ + HCO_3^- \qquad (1)$$

As HCO_3^- has a fairly low dissociation constant with most of the available cations, it tends to reassociate, leaving the medium acid. The net result of increasing atmospheric CO_2 is to depress the pH, so the effect of elevated CO_2 tension is neutralized by increasing the bicarbonate concentration:

$$NaHCO_3 \rightleftharpoons Na^+ + HCO_3^- \qquad (2)$$

The increased HCO_3^- concentration pushes equation 1 to the left until equilibrium is reached at pH 7.4. If another alkali, e.g., NaOH, is used instead, the net result is the same.

$$\mathrm{NaOH + H_2CO_3 \rightleftharpoons NaHCO_3 + H_2O \rightleftharpoons Na^+ + HCO_3^- + H_2O} \quad (3)$$

The equivalent $NaHCO_3$ concentrations commonly used with different CO_2 tensions are listed in Table 8.2.

Intermediate values of CO_2 and HCO_3^- may be used provided the concentration of both are varied simultaneously. As many media are made up in acid solution and may incorporate a buffer, it is difficult to predict

Fig. 8.7. *Examples of standard glass bottles which may be used as culture flasks.*

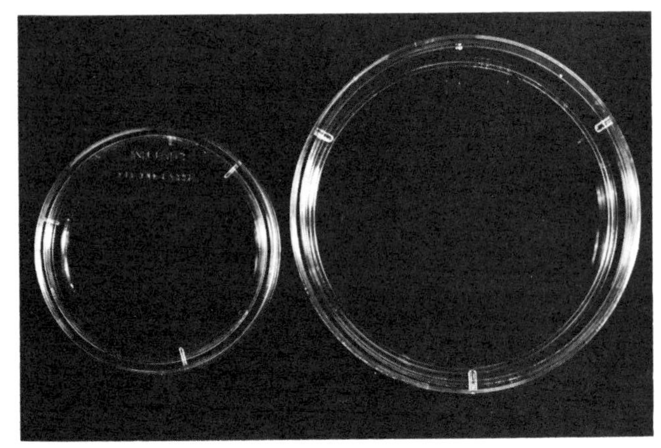

Fig. 8.8. *"Vented" dishes (9 cm and 6 cm diameter).*

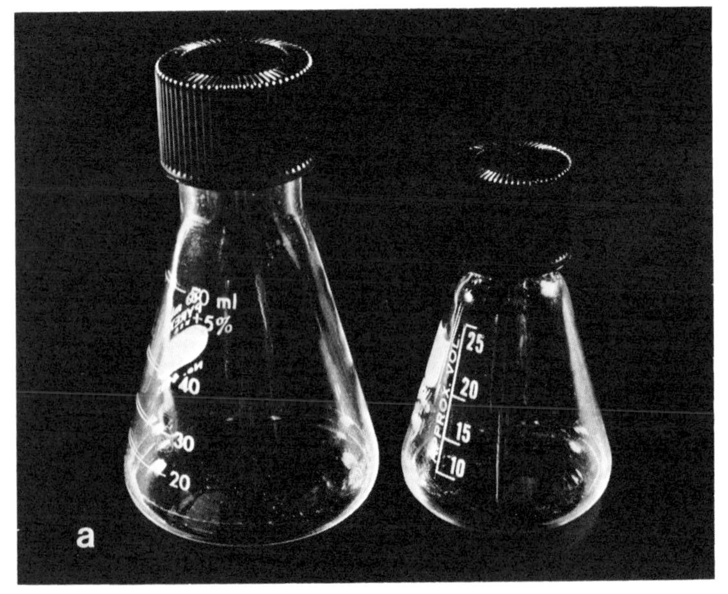

Fig. 8.9. *Screw-cap vials and conical flasks suitable for replicate cultures or sample storage. a. Screw caps are preferable to stoppers as they are less likely to leak and protect the neck of the flask from contamination. b. Scintillation vials are particularly useful for isotope incorporation studies but should not be reused for culture after containing scintillation fluid.*

TABLE 8.2. Variations in HCO_3^- and CO_2 Concentrations in Variants of Eagle's Minimum Essential Medium (MEM)

	Eagles's MEM Hank's salts	Eagle's MEM Earle's salts	Dulbecco's modification of Eagle's MEM
$NaHCO_3$	4 m*M*	26 m*M*	44 m*M*
CO_2	Atmospheric & evolved CO_2 from culture	5%	10%

TABLE.8.3. Relationship Between HCO_3^-, CO_2 and HEPES Concentrations

Gas phase CO_2	Liquid Phase	
	HCO_3^-	HEPES
Atmospheric	4 m*M*	10 m*M*
2%	8 m*M*	20 m*M*
5%	24 m*M*	50 m*M*

how much bicarbonate to use when other alkali may also end up as bicarbonate, as in equation 3 above. It is best to adopt the procedure outlined in Chapter 10—i.e., add the specified amount of bicarbonate and then sufficient 1 N NaOH such that the medium equilibrates to the desired pH after incubation at 36.5° C overnight. When dealing with media already at working strength, vary the amount of HCO_3^- to suit the gas phase (Table 8.2) and leave overnight to equilibrate at 36.5° C. Each medium has a recommended bicarbonate concentration and CO_2 tension to achieve the correct pH and osmolality but minor variations will occur in different methods of preparation.

With the introduction of Good's buffers [Good et al., 1966] into tissue culture, there was some speculation that since CO_2 was no longer necessary to stabilize the pH, it could be omitted. This has since proved to be untrue [Itagaki and Kimura, 1974], at least for a large number of cell types, particularly at low cell concentrations. Although 20 mM HEPES buffer can control pH within the physiological range, the absence of atmospheric CO_2 allows equation 1 to move to the left, eventually eliminating dissolved CO_2, and ultimately HCO_3^-, from the medium. This appears to limit cell growth, although whether the cells require the dissolved CO_2 or the HCO_3^- (or both) is not clear. Recommended HCO_3^-, CO_2, and HEPES concentrations are given in Table 8.3.

The inclusion of pyruvate in the medium enables cells to increase their endogenous production of CO_2, making them independent of exogenous CO_2 and HCO_3^-. Leibovitz L15 medium [Leibovitz, 1963] contains a higher concentration of sodium pyruvate (550 mg/l) but no $NaHCO_3$ and does not require CO_2 in the gas phase. Sodium-glycerophosphate can also be used to buffer autoclavable media lacking CO_2 and HCO_3^- [Waymouth, 1979].

In summary, cultures at low cell concentration in an open vessel need to be incubated in an atmosphere of CO_2, the concentration of which is in equilibrium with the sodium bicarbonate in the medium. At very low cell concentrations (e.g., during cloning), it is necessary to add CO_2 to the gas phase of sealed flasks for most cultures. At high cell concentrations, it will not be necessary to add CO_2 to the gas phase in sealed flasks, but it may yet be necessary in open dishes. Where the culture produces a lot of acid, and the endogenous production of CO_2 is high, it may be desirable to slacken the cap of a culture flask and allow excess CO_2 to escape. In these cases it is advisable to incorporate HEPES (20 mM) in the medium to stabilize the pH.

INCUBATION TEMPERATURE

The optimal temperature for cell culture is dependent on (1) the body temperature of the animal from which the cells were obtained, (2) any regional variation in temperature (e.g., skin may be lower), and (3) the incorporation of a safety factor to allow for minor errors in incubator regulation. Thus, the temperature recommended for most human and warm-blooded animal cell lines is 36.5° C, close to body heat but set a little lower for safety.

Avian cells, because of the higher body temperature in birds, should be maintained at 38.5° C for maximum growth but will grow quite satisfactorily, if more slowly, at 36.5° C.

Cultured cells will tolerate considerable drops in temperature, can survive several days at 4° C and can be frozen and cooled to −196° C (see Chapter 18), but they cannot tolerate more than about 2° C above normal (39.5° C) for more than a few hours, and will die quite rapidly at 40° C and over.

Epidermal cells from homiothermic animals may grow better at a slightly lower temperature—33° C.

Much of the preceding discussion has been based on observations with warm-blooded animals but is, nevertheless, applicable in principle to lower vertebrates, and perhaps to a lesser extent, to invertebrates. Temperature must be considered separately, however. In general the cells of poikilothermic animals have a wide temperature tolerance but should be maintained at a constant level within the normal range of the donor species. This requires incubators with cooling as well as heating as the incubator temperature may need to be below ambient (e.g., for fish). As for a hot-room, cooling capacity should be sufficient to lower the temperature about 2° C, or more, below ambient so that regulation is performed by the heater circuit which is more sensitive.

If necessary, poikilothermic animal cells can be maintained at room temperature, but the variability of the ambient temperature in laboratories makes this undesirable.

Regulation of temperature should be kept within ±0.5° C; consistency is more important than accuracy. Cells will grow quite well between 33° and 39° C but will naturally vary in growth rate and metabolism. The incubation temperature should be kept constant both in time and at different parts of the incubator. Water baths give the most accurate control of temperature, but present problems of contamination, particularly since the flasks need to be immersed for proper temperature control. They are, therefore, seldom used and incubators are preferable. The air should be circulated by a fan to give even temperature distribution, and cultures should be placed on perforated shelves and not on the floor or touching the sides of the incubator. A further discussion of incubator and hot-room design is given in Chapters 3 and 4.

Chapter 9
The Culture Environment: II. Media and Supplements

The discovery that cells from explants could be subcultured and propagated *in vitro* led to attempts to provide more defined media to sustain continuous cell growth, from the simple basal media of Eagle [1955, 1959] to the more complex medium 199 of Morgan et al. [1950] and CMRL 1066 of Parker et al. [1957]. Although most media in current use are now defined, they are still usually supplemented with 5–20% serum, and it was the desire to eliminate this remaining undefined constituent that led to the evolution of such complex media as NCTC 109, Evans et al. [1956] and 135, Evans and Bryant [1965], Waymouth's MB 572/1 [1959], Ham's F10 [1963], and F12 [1965], Birch and Pirt [1971], the MCDB series [Ham and McKeehan, 1978], and Sato's hormone-supplemented media [Barnes and Sato, 1980].

One approach to developing a medium is to start with a rich medium such as Ham's F12 [1965] or medium 199 supplemented with a high concentration of serum (say 20%) and gradually attempt to reduce the serum concentration by manipulating the concentrations of existing constitutents and by adding new ones. This is a very laborious procedure but it has resulted in a number of different formulations for the culture of human fibroblasts and other cell types either in low serum concentrations or in its complete absence (see below under Serum-Free Media.)

Even after many years of exhaustive research into matching particular media to specific cell types and culture conditions, the choice of medium is not obvious and is often empirical. However, before considering this further, we should first consider the constitution of media.

PHYSICAL PROPERTIES

pH

Most cell lines will grow well at pH 7.4. Although the optimum pH for cell growth varies relatively little among different cell strains, some normal fibroblast lines perform best at pH 7.4–7.7, and transformed cells may do better at pH 7.0–7.4 [Eagle, 1973]. There have been reports that epidermal cells may be maintained at pH 5.5 [Eisinger et al., 1979].

Phenol red is commonly used as an indicator. It is red at pH 7.4 becoming orange at pH 7.0, yellow at pH 6.5, slightly bluish red at pH 7.6, and purple at pH 7.8. Since the assessment of color is highly subjective, it is essential to make up a set of standards using sterile balanced salt solution (BSS) and phenol red at the correct concentration and in the same type of bottle that you normally use for preparing medium.

Preparations of pH Standards

Materials

- Hank's Balanced Salt Solution, × 10 concentrate or powder, without bicarbonate and glucose
- nine bottles of a size closest to your standard medium bottles or culture flasks
- distilled deionized water
- sterile 0.1 N NaOH (make up in distilled deionized water and filter sterilize; see Chapter 10)
- pH meter

Protocol

1. Make up the BSS at pH 6.5, dispense into nine bottles (2 extra to allow for breakage) of the appropriate size and autoclave
2. Adjust the pH to 6.5, 6.8, 7.0, 7.2, 7.4, 7.6, 7.8, with sterile 0.1 *N* NaOH, checking the pH of a sample of each bottle on a pH meter after allowing each bottle to equilibrate with the atmosphere
3. Keep sterile and sealed

Buffering

Culture media require to be buffered under two sets of conditions: (1) open dishes, where evolution of CO_2 causes the pH to rise, and (2) overproduction of CO_2 and lactic acid in transformed cell lines at high cell concentrations, when the pH will fall. A buffer may be incorporated in the medium to stabilize the pH but in (1) exogenous CO_2 may still be required by some cell lines, particularly at low cell concentrations, to prevent total loss of dissolved CO_2 and bicarbonate from the medium (see Chapter 8). In (2) it is usually preferable to leave the cap slack (shrouded in aluminum foil) or to use a CO_2-permeable cap (Cam Lab) to promote the release of CO_2.

A bicarbonate buffer is still used more frequently than any other, in spite of its poor buffering capacity at physiological pH, because of its low toxicity, low cost, and nutritional benefit to the culture. HEPES is a much stronger buffer in the pH 7.2–7.6 range and is now used extensively at 10 or 20 m*M*. When HEPES is used with exogenous CO_2, it has been found that the HEPES concentration must be more than double that of the bicarbonate for adequate buffering (see Table 8.3). A variation of Ham's F12 with 20 m*M* HEPES, 8 m*M* bicarbonate, and 2% CO_2 has been used successfully in the author's laboratory for the culture of a number of different cell lines.

Osmolality

Most cultured cells have a fairly wide tolerance for osmotic pressure [see also Waymouth, 1970]. Since the osmolality of human plasma is about 290 mOsm/kg, it is reasonable to assume that this is the optimum for human cells *in vitro,* although it may be different for other species (e.g., around 310 mOsm/kg for mice [Waymouth, 1970]). In practice, osmolalities between 260 mOsm/kg and 320 mOsm/kg are quite acceptable for most cells. Slightly hypotonic medium may be better for petri dish culture to compensate for evaporation during incubation.

Osmolality is usually measured by freezing-point depression or elevation of vapor pressure (Fig. 9.1). The measurement of osmolality is a useful quality control step if you are making up medium yourself as it helps to guard against errors in weighing and dilution, etc. It is particularly important to monitor osmolality if alterations are made in the constitution of the medium. Addition of HEPES, drugs dissolved in strong acids and bases, and subsequent neutralization, can all markedly affect the osmolality.

Temperature

Apart from the direct effect of temperature on cell growth, it will also influence pH due to the increased solubility of CO_2 at lower temperatures, and possibly, due to changes in ionization and the pK_a of the buffer. The pH should be adjusted to 0.2 units lower at room temperature than at 36.5°C. It is best to make up the medium complete with serum and incubate a sample overnight at 36.5°C under the correct gas tension to check the pH when making up a medium for the first time.

Viscosity

The viscosity of culture medium is influenced mainly by the serum content, and in most cases will have little effect on cell growth. It becomes important, however, whenever a cell suspension is agitated, e.g. when a suspension culture is stirred, or when cells are dissociated after trypsinization. If there is cell damage under these conditions, then this may be reduced by increasing the viscosity of the medium with carboxy methyl cellulose or polyvinyl pyrrolidone [Birch and Pirt, 1971]. This becomes particularly important in low serum concentrations and in the absence of serum.

Surface Tension and Foaming

The surface tension of medium may be used to promote adherence of primary explants to the substrate (see Chapter 11) but is seldom controlled in any way. In suspension cultures, where 5% CO_2 in air is bubbled through medium containing serum, foaming may result. The addition of a silicone antifoam (Dow Chemical, Co.) helps to prevent this by reducing surface tension.

The effects of foaming have not been clearly defined. The rate of protein denaturation may increase, and the risk of contamination increases if the foam reaches the neck of the culture vessel.

CONSTITUENTS OF MEDIA

Balanced Salt Solutions

It is useful to distinguish balanced salt solutions from media for the present discussion. A balanced salt solution (BSS) is composed of inorganic salts, usually including sodium bicarbonate, and supplemented with glucose, although glucose and bicarbonate are often omitted. The compositions of some common examples are given in Table 9.1. HEPES buffer (5–20 m*M*) may be added to these if necessary and the equivalent weight of NaCl omitted to maintain the correct osmolality.

BSS is used as a diluent for more complete media,

Fig. 9.1. *Osmometer (Advanced Instruments).*

as a washing or dissection medium, and for a variety of other purposes which require an isotonic solution which is not necessarily nutritionally complete.

The choice of BSS is dependent on (1) the CO_2 tension (see Chapter 8 and Table 9.2). The bicarbonate concentration must be such that equilibrium is reached at pH 7.4 at 36.5° C, e.g., Earle's BSS (EBSS) is commonly used as a diluent for Eagle's media for equilibration with 5% CO_2 while Hanks' BSS [Hanks and Wallace, 1949] (HBSS) (see Reagent Appendix) is used with air; (2) its use for tissue disaggregation, or monolayer dispersal. In these cases Ca^{2+} and Mg^{2+} are usually omitted as in Moscona's [1952] calcium- and magnesium-free saline (CMF), or Dulbecco and Vogt's [1954] phosphate-buffered saline, solution A (PBSA); and (3) its use for suspension culture of adherent cells. MEM(S) is a variant of Eagle's [1959] minimum essential medium, deficient in Ca^{2+}, to reduce cell aggregation and attachment.

HBSS, EBSS, and PBS rely on the relatively weak buffering of phosphate, which is not at its most effective at physiological pH. Paul [1975] constructed a tris-buffered BSS which is more effective, but for which the cells sometimes require a period of adaptation. HEPES is currently the most effective buffer in the pH 7.2–7.8 range and TRICINE in the pH 7.4–8.0 range, although these tend to be expensive if used in large quantities. Sodium β-glycerophosphate (pK_a 6.6) also works well [Waymouth, personal communication].

TABLE 9.1. Balanced Salt Solutions

Component	Earle's balanced salt solution gm/l	Dulbecco's phosphate buffered saline (solution A) (PBSA) gm/l	Hanks's balanced salt solution gm/l	Spinner salt solution (Eagle) gm/1
Inorganic salts				
$CaCl_2$ (anhyd.)	0.02	—	0.14	—
$CaCl_2$•$2H_2O$	—	—	—	—
KCl	0.04	0.20	0.40	0.40
KH_2PO_4	—	0.20	0.06	—
MgCl (anhyd.)	—	—	—	—
$MgCl_2$•$6H_2O$	—	—	0.10*	—
$MgSO_4$ (anhyd.)	—	—	—	—
$MgSo_4$•$7H_2O$	0.20	—	0.10	0.20
NaCl	6.68	8.00	8.00	6.80
$NaHCO_3$	2.20	—	0.35	2.20
Na_2HPO_4	—	—	—	—
Na_2HPO_4•$7H_2O$	—	2.16	0.09**	—
NaH_2PO_4•H_2O	0.14†	—	—	1.40
Other components				
D-glucose	1.00	—	1.00	1.00
Phenol red	0.01	—	0.01‡	0.01

*$MgCl_2$•$6H_2O$ added to original formula.
†Original formulation calls for 150.0 mg/l NaH_2PO_4•$2H_2O$, In Vitro, 9, #6 (1974).
**Original formulation calls for 0.06 gm/l Na_2HPO_4•$2H_2O$, Proc. Soc. Exp. Biol. and Med., 71 (1949).
‡Original formulation calls for 0.02 gm/l, Proc. Soc. Exp. Biol. and Med., 71 (1949).

TABLE 9.2. Selection of Media*

Cells or cell line			Media†									
			MEM	α-MEM	DME	GMEM	F10	F12	RPMI 1640	Iscove's MEM	Fischer's	AMDB 647 (autoclavable)
1	C	HeLa Hu cervical Ca	S		S		S	S				
2	C	L929 Ad. mouse fibr.	S		S							
3	C	3T3 mouse emb. fibr.	S		S							
4	PL	Specialized cells from mouse										H
5	C	NRK normal rat kidney	S		S							
6	C	BHK baby hamster kidney				S						
7	C	MCDK dog kidney			S							
8	C	Lymphoblastoid cell line							S			
9	C	Mouse myelomas			S				S	P		
10	C	Mouse leukemias							S		F	
11	C	Mouse erythroleukemia			S			S	S			
12	PL	Chick embryo fibr.	S					S				
13	PL	Mouse emb. fibr.	S									
14	PL	Human diploid fibroblasts	S									
15	C	CHO Chinese hamster ovary	S				F	F				
16	PL	Human tumor	S		S			S				
17	PL	Specialized cultures from chick embryo						S				
18	PL	Hemopoietic cells		F					S	H		

*PL, primary culture or finite cell line; C, continuous cell line or cell strain; S, with serum supplement; P, with undefined protein supplement; H, defined protein, hormone, and/or growth factor supplement; F, serum free, sometimes with adaptation. These are suggestions for media which can be used for the cells specified; a blank space in this table does not imply that these cells will not grow in that medium.
†MEM, Eagle [1959]; α-MEM, Stanners et al. [1971]; DME and GMEM, Dulbecco and Freeman [1959]; Morton [1970]; F10, Ham [1963]; F12, Ham [1965]; RPMI 1640, Moore et al, [1967]; Iscove's MEM, Iscove and Melchers [1978]; Iscove et al, [1980]; Fischer's, Fischer and Sartorelli [1964]; AMDB 647 (autoclavable), Waymouth [1978]; MB 752/1, Waymouth [1959]; MB 706/1, Kitos et al. [1962]; MAB 87/3, Gorham and Waymouth [1965]; Waymouth [1977]; MCDB 104, McKeehan et al. [1977]; MCD3 301, Hamilton and Ham [1977]; MCDB 202, Ham and McKeehan [1978]; MCDB 401, Ham and McKeehan [1978]; DME F12, Barnes and Sato [1980]; 5A, McCoy et al, [1959]; L15 (no CO_2), Leibovitz [1963]; NCTC 135, Evans [1965]; 199, Morgan et al. [1950]; CMRL 1066, Parker et al. [1957].

DEFINED MEDIA

Defined media vary in complexity from Eagle's minimum essential medium (MEM) [Eagle, 1959], which contains essential amino acids, vitamins, and salts, to complex media such as 199 [Morgan et al., 1950], CMRL 1066 [Parker et al., 1957], RPMI 1640 [Moore et al., 1967], and F12 [Ham, 1965] (Table 9.3). The complex media contain a larger number of different amino acids and vitamins and are often supplemented with extra metabolites (e.g., nulceosides) and minerals. Nutrient concentrations are, on the whole, low in F12 and high in Dulbecco's modification of Eagle's MEM (DMEM) [Dulbecco and Freeman, 1959; Morton, 1970], although the latter has fewer constitutents. Barnes and Sato [1980] employed a 1:1 mixture of DMEM and F12 as the basis for their serum-free formulations to combine the richness of F12 and the higher nutrient concentration of DMEM.

The common constituents of medium may be grouped as follows (Table 9.3).

Amino Acids

The essential amino acids, i.e., those which are not synthesized in the body, are required by cultured cells with, in addition, cysteine and tyrosine, although individual requirements for amino acids will vary from one cell line to another. Other nonessential amino acids are often added to compensate either for a particular cell type's incapacity to make them or because they are made but lost into the medium. The concentration of amino acids usually limits the maximum cell concentration attainable, and the balance may influence cell survival and growth rate. Glutamine is required by most cells although some cell lines will utilize glutamate. Recent evidence suggests that glutamine is used in cultured cells as an energy and carbon source [Reitzer et al., 1979].

TABLE 9.2 (continued)

	Media†												
	MB 752/1	MB 705/1	MAB 87/3	MCDB 104	MCD3 301	MCDB 202	MCDB 401	DME/ F12	5A	L15 (no CO_2)	NCTC 135	199	CMRL 1066
1													
2	F	F										F	F
3			F				P						
4			F										
5													
6													
7													
8													
9													S
10													
11													
12						P							
13			F				P						
14				P					S	S		S	S
15					F								
16									S	S		S	S
17								H					
18												S	S

*PL, primary culture or finite cell line; C, continuous cell line or cell strain; S, with serum supplement; P, with undefined protein supplement; H, defined protein, hormone, and/or growth factor supplement; F, serum free, sometimes with adaptation. These are suggestions for media which can be used for the cells specified; a blank space in this table does not imply that these cells will not grow in that medium.

†MEM, Eagle [1959]; α-MEM, Stanners et al. [1971]; DME and GMEM, Dulbecco and Freeman [1959]; Morton [1970]; F10, Ham [1963]; F12, Ham [1965]; RPMI 1640, Moore et al, [1967]; Iscove's MEM, Iscove and Melchers [1978]; Iscove et al, [1980]; Fischer's, Fischer and Sartorelli [1964]; AMDB 647 (autoclavable), Waymouth [1978]; MB 752/1, Waymouth [1959]; MB 706/1, Kitos et al. [1962]; MAB 87/3, Gorham and Waymouth [1965]; Waymouth [1977]; MCDB 104, McKeehan et al. [1977]; MCD3 301, Hamilton and Ham [1977]; MCDB 202, Ham and McKeehan [1978]; MCDB 401, Ham and McKeehan [1978]; DME F12, Barnes and Sato [1980]; 5A, McCoy et al, [1959]; L15 (no CO_2), Leibovitz [1963]; NCTC 135, Evans [1965]; 199, Morgan et al. [1950]; CMRL 1066, Parker et al. [1957].

TABLE 9.3.

Component	Eagle's MEM mg/l	Dulbecco's modification mg/l	Ham's F12 mg/l	CMRL 1066 mg/l	RPMI 1640 mg/l	Waymouth's MB 752/1 mg/l	McCoy's 5a mg/l	Iscove's modified Dulbecco's medium* mg/l
Inorganic salts								
$CaCl_2$ (anhyd.)	200.00	200.00	—	200.00	—	—	100.00	165.00
$CaCl_2 \cdot 2H_2O$	—	—	44.00	—	—	120.00	—	—
$Fe(NO_3)_3 \cdot 9H_2O$	—	0.10	—	—	—	—	—	—
KCl	400.00	400.00	223.60	400.00	400.00	150.00	400.00	330.00
KH_2PO_4	—	—	—	—	—	80.00	—	—
$MgCl_2$ (anhyd.)	—	—	—	—	—	—	—	—
$MgCl_2 \cdot 6H_2O$	—	—	122.00	—	—	240.00	—	—
$MgSO_4 \cdot$ (anhyd.)	—	—	—	—	—	—	—	97.67
$MgSO_4 \cdot 7H_2O$	200.00	200.00	—	200.00	100.00	200.00	200.00	—
NaCl	6,800.00	6,400.00	7,599.00	6,799.00	6,000.00	6,000.00	6,400.00	4,505.00
$NaHCO_3$	2,200.00	3,700.00	1,176.00	2,200.00	2,200.00	2,240.00	2,200.00	3,024.00
$NaH_2PO_4 \cdot H_2O$	140.00	125.00	—	140.00	—	—	530.00	125.00
$Na_2HPO_4 \cdot H_2O$	—	—	—	—	—	—	—	—
Na_2HPO_4 (anhyd.)	—	—	—	—	—	—	—	—
$Na_2HPO_4 \cdot 7H_2O$	—	—	268.00	—	1,512.00	566.00	—	—
KNO_3	—	—	—	—	—	—	—	0.076
$Na_2SeO_3 \cdot 5H_2O$	—	—	—	—	—	—	—	0.0173
$CuSO_4 \cdot 5H_2O$	—	—	0.00249	—	—	—	—	—
$FeSO_4 \cdot 7H_2O$	—	—	0.834	—	—	—	—	—
KH_2PO_4	—	—	—	—	—	—	—	—
$ZnSO_4 \cdot 7H_2O$	—	—	0.863	—	—	—	—	—
$CaNO_3 \cdot 4H_2O$	—	—	—	—	100.00	—	—	—
$CoCl_2 \cdot 6H_2O$	—	—	—	—	—	—	—	—
$MnSO_4 \cdot H_2O$	—	—	—	—	—	—	—	—
$(NH_4)_6Mo_7O_{24} \cdot 4H_2O$	—	—	—	—	—	—	—	—
Other components								
D-glucose	1,000.00	4,500.00	1,802.00	1,000.00	2,000.00	5,000.00	3,000.00	4,500.00
Lipoic acid	—	—	0.21	—	—	—	—	—
Phenol red	10.00	15.00	12.0	20.0	5.00	10.00	10.00	15.00
Potassium penicillin G	—	—	—	—	—	—	—	—
Sodium pyruvate	—	110.00	110.00	—	—	—	—	110.00
Sodium succinate	—	—	—	—	—	—	—	—
Streptomycin sulfate	—	—	—	—	—	—	—	—
Succinic acid	—	—	—	—	—	—	—	—
HEPES	—	—	—	—	—	—	—	5,958.00
Hypoxanthine	—	—	4.10	—	—	25.00	—	—
Hypoxanthine (Na Salt)	—	—	—	—	—	—	—	—
Linoleic acid	—	—	0.084	—	—	—	—	—
Putrescine 2HCl	—	—	0.161	—	—	—	—	—
Thymidine	—	—	0.73	10.0	—	—	—	—
Cocarboxylase	—	—	—	1.0	—	—	—	—
Coenzyme A	—	—	—	2.5	—	—	—	—
Deoxyadenosine	—	—	—	10.0	—	—	—	—
Deoxycytidine HCl	—	—	—	10.0	—	—	—	—
Deoxyguanosine	—	—	—	10.0	—	—	—	—
Diphyosphopyridine nucleotide $\cdot 4H_2O$	—	—	—	7.0	—	—	—	—
Ethanol for solubilizing lipid components	—	—	—	16.0	—	—	—	—
Flavin adenine dinucleotide	—	—	—	1.0	—	—	—	—
Glutathione (reduced)	—	—	—	10.0	1.00	15.00	0.50	—
5-methyl-deoxycytidine	—	—	—	0.1	—	—	—	—
Sodium acetate $\cdot 3H_2O$	—	—	—	83.0	—	—	—	—
Sodium glucuronate $\cdot H_2O$	—	—	—	4.2	—	—	—	—

Table 9.3 continued on following page.

TABLE 9.3. (continued)

Component	Eagle's MEM mg/l	Dulbecco's modification mg/l	Ham's F12 mg/l	CMRL 1066 mg/l	RPMI 1640 mg/l	Waymouth's MB 752/1 mg/l	McCoy's 5a mg/l	Iscove's modified Dulbecco's medium* mg/l
Triphosphopyridine nucleotide	—	—	—	1.0	—	—	—	—
Tween 80	—	—	—	5.0	—	—	—	—
Uridine triphosphate·$4H_2O$	—	—	—	1.0	—	—	—	—
Insulin (crystalline) (bovine)	—	—	—	—	—	—	—	—
Bacto-peptone	—	—	—	—	—	—	600.00	—
Fetal bovine serum	—	—	—	—	—	—	—	—
Amino acids								
L-alanine	—	—	8.90	25.0	—	—	13.90	25.00
L-arginine (free base)	—	—	—	—	200.00	—	—	—
L-arginine·HCl	126.00	84.00	211.00	70.0	—	75.00	42.10	84.00
L-asparagine	—	—	—	—	50.00	—	45.00	—
L-asparagine·H_2O	—	—	15.01	—	—	—	—	28.40
L-aspartic acid	—	—	13.30	30.0	20.00	60.00	19.97	30.00
L-cysteine (free base)	—	—	—	—	—	61.00	31.50	—
L-cystine	24.00	48.00	—	20.0	50.00	15.00	—	—
L-cystine-2HCl	—	—	—	—	—	—	—	91.24
L-cysteine·HCl·H_2O	—	—	35.12	260.0	—	—	—	—
L-glutamic acid	—	—	14.70	75.0	20.00	150.00	22.10	75.00
L-glutamine	292.00	584.00	146.00	100.0	300.00	350.00	219.20	584.00
Glycine	—	30.00	7.50	50.0	10.00	50.00	7.50	30.00
L-histidine	—	—	—	—	—	—	—	—
L-histidine (free base)	—	—	—	—	15.00	128.00	—	—
L-histidine HCl·H_2O	42.00	42.00	20.96	20.0	—	—	20.96	42.00
L-hydroxy-proline	—	—	—	10.0	20.00	—	19.67	—
L-isoleucine	52.00	105.00	3.94	20.0	50.00	25.00	39.36	105.00
L-leucine	52.00	105.00	13.10	60.0	50.00	50.00	39.36	105.00
L-lysine	—	—	—	—	—	—	—	—
L-lysine (free base)	—	—	—	—	—	—	—	—
L-lysine HCl	73.10	146.00	36.50	70.0	40.00	240.00	36.50	146.00
L-methionine	15.00	30.00	4.48	15.0	15.00	50.00	14.90	30.00
L-phenylalanine	33.00	66.00	4.96	25.0	15.00	50.00	16.50	66.00
L-proline	—	—	34.50	40.0	20.00	50.00	17.30	40.00
L-serine	—	42.00	10.50	25.0	30.00	—	26.30	42.00
L-threonine	48.00	95.00	11.90	30.0	20.00	75.00	17.90	95.00
L-tryptophane	10.00	16.00	2.04	10.0	5.00	40.00	3.10	16.00
L-tyrosine	36.00	72.00	5.40	40.0	20.00	40.00	18.10	—
L-tyrosine (disodium salt)	—	—	—	—	—	—	—	104.20
L-valine	47.00	94.00	11.70	25.0	20.00	65.00	17.60	94.00
Vitamins								
L-ascorbic acid	—	—	—	50.000	—	17.50	0.50	—
Biotin	—	—	0.0073	0.010	0.20	0.02	0.20	0.013
D-Ca pantothenate	1.00	4.00	0.4800	0.010	0.25	1.00	0.20	4.00
Choline bitartrate	—	—	—	—	—	—	—	—
Choline chloride	1.00	4.00	13.9600	0.500	3.00	250.00	5.00	4.00
Folic acid	1.00	4.00	1.300	0.010	1.00	0.40	10.00	4.00
i-inositol	2.00	7.20	18.000	0.050	35.00	1.00	36.00	7.20
Nicotinamide	1.00	4.00	0.04	0.025	1.00	1.00	0.50	4.00
Pyridoxal HCl	1.00	4.00	0.062	0.025	—	—	0.50	4.00
Riboflavin	0.10	0.40	0.0380	0.010	0.20	1.00	0.20	0.40
Thiamine HCl	1.00	4.00	0.3400	0.010	1.00	10.00	0.20	4.00
Vitamin B_{12}	—	—	1.3600	—	0.005	0.20	2.00	0.013
Pyridoxine HCl	—	—	0.0620	0.025	1.00	1.00	0.50	—
Cholesterol	—	—	—	0.200	—	—	—	0.02
Para-aminobenzoic acid	—	—	—	0.050	1.00	—	1.00	—
Nicotinic acid	—	—	—	—	—	—	0.50	—
								—

*Supplemented with 400 µg/ml purified bovine serum albumin, 50–100 µg/ml soybean lipid, and 1 µg/ml transferrin [Iscove and Melchers, 1978].

Vitamins

Eagle's MEM contains only the B group vitamins (see Table 9.3), other requirements presumably being derived from the serum. With increasing medium complexity (usually designed to reduce the serum requirement), the list of vitamins increases. The requirement for extra vitamins is most apparent where the serum concentration is reduced, but there are other cases (e.g., low cell densities for cloning) where they may be essential even in the presence of serum. Vitamin limitation is usually expressed in terms of cell survival and growth rate rather than maximum cell density.

Salts

The salts are chiefly those of Na^+, K^+, Mg^{2+}, Ca^{2+}, Cl^-, SO_2^{2-}, PO_4^{3-}, and HCO_3^- and are the major components contributing to the osmolality of the medium. Calcium is reduced for suspension cultures to minimize cell aggregation and attachment (see above). The sodium bicarbonate concentration is determined by the concentration of CO_2 in the gas phase (see above and Chapter 8).

Glucose

Glucose is included in most media as an energy source. It is metabolized principally by glycolysis to form lactic acid and via the citric acid cycle to form CO_2. The accumulation of lactic acid in the medium, particularly evident in embryonic and transformed cells, implies that the citric acid cycle may not function entirely as *in vivo* and recent data have shown that much of its carbon is derived from glutamine rather than glucose. This may explain the exceptionally high requirement of some cultured cells for glutamine or glutamate.

Buffer

As described above under balanced salt solutions, a limited buffering capacity is present in most media due to phosphate. Where extra buffering capacity is required, HCO_3^- in equilibrium with CO_2 in the gas phase is commonly employed for open vessels; HEPES is also widely used, usually at 20 m*M* (see above). The contribution of the amino acids to buffering is slight as their concentration is normally too low, but Leibovitz [1963] constructed a medium (L15) with elevated amino acid concentrations sufficient to buffer the medium without bicarbonate or exogenous CO_2.

Minerals

As with some of the vitamins, most minerals required by the cells are provided by the serum and supplementation of the medium is generally not required unless the serum is reduced. In low serum, or in its complete absence, requirements for iron, copper, zinc, selenium, and other trace elements become apparent.

Organic Supplements

A variety of other compounds including nucleosides, citric acid cycle intermediates, pyruvate, and lipids appear in complex media. Again these constituents have been found to be necessary when the serum concentration is reduced and may help in cloning and in maintaining certain specialized cells.

Hormones and Growth Factors

See below under Serum and Serum-free Media.

SERUM

The sera used most in tissue culture are calf serum, fetal bovine serum, horse serum, and human serum. The choice of serum will be discussed below in more detail. Calf serum is the most widely used, fetal bovine second, usually for more demanding cell lines, and human serum in conjunction with some human cell lines. Horse serum is preferred to calf serum by some workers as it can be obtained from a closed herd and is often more consistent from batch to batch.

Although most cell lines still require the supplementation of the medium with serum, there are now many instances where cultures may be maintained and may proliferate serum free. Continuous cell strains such as the L929 and HeLa were among the first to be grown serum free [Evans et al., 1956; Waymouth, 1959; Birch and Pirt, 1971; Higuchi, 1977] and a degree of selection may have been involved. However, results from the laboratories of Ham [Ham and McKeehan, 1978], Sato [Barnes and Sato, 1980], and others [Carney et al., 1981] have demonstrated that serum may be reduced or omitted without cellular adaptation if nutritional and hormonal modifications are made to the media appropriate to the cell line being studied [Sato et al., 1982] (see below). This has provided indirect evidence for the constitution of serum which will be considered before further discussion of serum-free media.

Protein

Although proteins are a major component of serum, the functions of many of these *in vitro* remain obscure; and it may be that relatively few proteins are required other than as carriers for minerals, fatty acids, and hormones, or as hormones themselves. Those proteins which have been found beneficial are albumin [Iscove and Melchers, 1978; Barnes and Sato, 1980] and globulins [Tozer and Pirt, 1964]. Fibronectin (Chapter 8; cold-insoluble globulin) promotes attachment and α_2-macroglobulin inhibits trypsin [DeVonne and Mouray, 1978]. Fetuin enhances cell attachment in fetal serum [Fisher et al., 1958], and transferrin [Guilbert and Iscove, 1976] binds iron making it less toxic but available to the cell. There may be other proteins as yet uncharacterized, essential for cell attachment and growth.

Polypeptides

Natural clot serum stimulates cell proliferation more than serum from which the cells have been removed physically. This appears to be due to the release of a polypeptide from the platelets during clotting. This polypeptide, platelet-derived growth factor (PDGF) [Heldin et al., 1979; Antoniades et al., 1979] is one of a family of polypeptides with mitogenic activity and is probably the major growth factor in serum. Others, such as fibroblast growth factor (FGF) [Gospodarowicz, 1974], epidermal growth factor (EGF) [Cohen, 1962; Carpenter and Cohen, 1977; Gospodarowicz et al., 1978], endothelial growth factor [Folkman et al., 1979; Maciag et al., 1979], and multiplication-stimulating activity (MSA) [Dulak and Temin, 1973a, b] which have been isolated from whole tissue or released into the medium by cells in culture, have varying degrees of specificity [Hollenberg and Cuatrecasas, 1973] and are probably present in serum in small amounts [Gospodarowicz and Moran, 1974]. Many of these growth factors are available commercially (see "Growth Factors", Trade Index) in pure form.

Hormones

Hormones may exhibit a variety of different effects on cells, and it is often difficult to recognize the key pathway. Insulin promotes the uptake of glucose and amino acids [Hokin and Hokin, 1963; Segal, 1964; Fritz and Knobil, 1964; Kelly et al., 1978] and may owe its mitogenic effect to this property. Some growth factors bind to the insulin receptor on the cell surface and may act similarly. Growth hormone may be present in sera, particularly fetal sera, and in conjunction with the somatomedins, may have a mitogenic effect. Hydrocortisone is also present in serum in varying amounts. It can promote cell attachment [Ballard and Tomkins, 1969; Fredin et al., 1979] and cell proliferation [Guner et al., 1977] but under certain conditions (e.g., high cell density) may be cytostatic [Freshney et al., 1980a, b] and can induce cell differentiation [Moscona and Piddington, 1966; Ballard, 1979].

Serum replacement experiments suggest that other hormones required for culture may be present in serum (see below), making it necessary at present to retain serum for culture but underlining the desirability for its replacement.

Metabolites and Nutrients

Serum also contains amino acids, glucose, ketoacids, and a number of other nutrients and intermediary metabolites. These may be important in simple media but less so in complex media, particularly those with higher amino acid concentrations and other supplements.

Minerals

Serum replacement experiments have also suggested that trace elements and iron, copper, and zinc may be provided by serum [Ham and McKeehan, 1978], bound to serum protein. A requirement for selenium has also been demonstrated in the same way [McKeehan et al., 1976].

Inhibitors

Serum may also contain substances inhibiting cell proliferation [Harrington and Godman, 1980]. Some of these may be artifacts of preparation, e. g., bacterial toxins from contamination prior to filtration; the γ-globulin fraction may contain antibodies cross-reacting with the culture. Heat inactivation removes complement from the serum and reduces the cytotoxic action of immunoglobulins without damaging polypeptide growth factors, but it may remove some more labile constitutents, and is not always as satisfactory as untreated serum.

The presence of tissue-specific inhibitors of cell proliferation, or chalones, in serum has not been verified although some hormones may be inhibitory in a nonspecific fashion.

SERUM-FREE MEDIA

Ever since the observation in the 1950s that natural media could be replaced in part by synthetic media, attempts have been made to culture cells without serum. NCTC 109 [Evans et al., 1956], 135 [Evans and Bryant, 1965], Waymouth's MB752/1 [Waymouth, 1959], MB705/1 [Kitos et al., 1962], and the media of Birch and Pirt [1970, 1971] and Higuchi [1977] were all able to sustain the growth of L-cells without serum. Pirt and co-workers further modified their formulation by adding insulin and, with other minor modifications, were able to culture HeLa cells without serum [Blaker et al., 1971]. Ham [1963, 1965] was able to clone CHO cells serum free and, more recently, specific formulations, (e.g., MCDB104) have been derived to culture human fibroblasts [Ham and McKeehan, 1978], many normal and neoplastic murine and human cells [Barnes and Sato, 1980], lymphoblasts [Iscove and Melchers, 1978], and several different primary cultures [Mather and Sato, 1979a, b] in the presence of serum with, in some cases, traces of serum proteins added. Two approaches have been followed in the development of serum-free media for specific cell types: (1) the development of new formulations based on the replacement of serum by specific nutrients [e.g., Ham and McKeehan, 1978] and (2) supplementation of existing media (e.g., Ham's F12) with hormones or other characterized macromolecular fractions [e.g., Barnes and Sato, 1980]. Sato and co-workers use a 1:1 mixture of Ham's F12 and Dulbecco's modification of Eagle's MEM.

Hormones which have been used include insulin at 1–10 units/ml, which improves plating efficiency in a number of different cell types, and hydrocortisone, which improves the cloning efficiency of glia and fibroblasts (see Chapter 13) and which has been found necessary for the maintenance of epidermal keratinocytes and some other epithelial cells (see Chapter 20). Barnes and Sato [1980] describe the use of 5-tri-iodo tyrosine (T_3) as a necessary supplement for MDCK (dog kidney) cells, and various combinations of estrogen, androgen, or progesterone with hydrocortisone and prolactin can be shown to be necessary for the maintenance of mammary epithelium (see Chapter 20.)

Other hormones with functions not usually associated with the cells they were tested on were found to be effective in replacing serum, e.g., FSH (follicle-stimulating hormone) with B16 murine melanoma [Barnes and Sato, 1980]. It is possible that sequence homologies exist between some growth-stimulating polypeptides and well-established peptide hormones. Alternatively, processing of some of the large proteins or polypeptides may release active peptide sequences with quite different functions.

The family of polypeptides which have been found to be mitogenic *in vitro* is now quite extensive and includes FGF, EGF, PDGF, and MSA (see above). In some cases they may act synergistically or additively with prostaglandin $F_2\alpha$, somatomedin C, and hydrocortisone. The use of these supplements together with binding proteins such as transferrin, adhesion factors such as fibronectin and fetuin, and low molecular weight nutrients such as selenium, retinol, and linoleic acid have been reviewed comprehensively by Barnes and Sato [1980]. The use of hormonal and other defined supplements by Sato and co-workers has produced an ever-increasing list of continuous cell lines and primary cultures which can be maintained serum free.

So far there seem to be no clear guidelines to indicate which supplements may be required. Some may be fairly universal, like insulin, transferrin, and selenium, while others such as estrogens, androgens, and T_3 may be more specific for individual cell types (though not necessarily those cell types that would be traditionally associated with those hormones). As with medium and serum selection, trial and error may be the only method to select the correct supplements. Some may be grouped together as being most likely to be effective and least likely to be toxic, e.g., insulin, 1 mg/ml, FGF, 10 ng/ml, growth hormone, 50 ng/ml, transferrin, 10 μg/ml, fatty acid-free bovine serum albumin, 1 mg/ml, fibronectin, 1 ng/ml, H_2SeO_3, 20 n*M*, putrescine, 100 μM, and 3 μg/ml linoleic acid. If such a cocktail is found effective in reducing serum supplementation, the active constitutents may be identified by systematic omission of single components, and then their concentrations optimized. If this is insufficient, in conjunction with substrate modification by polylysine or collagen (see Chapter 8), then other supplements such as EGF (for epithelial cells), 10 ng/ml, T_3, 10 p*M*, hydrocortisone, 10 n*M*, or Sato's "Gimmel factor," 1 μg/ml, may be tried.

The optimization of serum-free medium is a long and time-consuming business with few short cuts and little obvious direction other than prior examples. It is advisable therefore to consult the literature before embarking on such a task, starting, perhaps, with such reviews as Barnes and Sato [1980], Ham and McKeehan [1979], Sato et al. [1982], and then the more specific literature as listed in Table 9.2.

Although many of the above requirements have been discovered in serum replacement experiments, supplementing serum-containing media with growth factors and/or hormones can influence plating efficiency and the rate of cell proliferation. Insulin and hydrocortisone will stimulate the plating efficiency of a number of different cell types (see Chapter 13), and EGF is used with cultures of epidermal keratinocytes, and other epithelial cells, in the presence of serum.

Plating hemopoietic cells requires colony-stimulating factor (CSF) and propagation of normal T-lymphocytes requires T-cell growth factor (TCGF) (see Chapter 20) and the identification of other factors regulating the proliferation, mobility, and differentiation of hemopoietic cells *in vitro* suggest that this is a very complex area [Burgess and Metcalf, 1980; Schnook et al., 1981; Gillis and Watson, 1981].

Special care may be required when trypsinizing cells from serum-free media as they are more fragile and may need to be chilled to reduce damage [McKeehan, 1977]. A trypsin inhibitor may be required in the medium after transfer [Rockwell et al., 1980] and special treatment of the growth substrate with fibronectin or polylysine [McKeehan and Ham, 1976] may be necessary for cell attachment and spreading (see Chapter 8).

It seems unlikely that a single medium will ever be devised suitable for all cell types, but it may be possible to develop a common basal medium with variable hormonal and other defined supplements to suit a variety of cell types. This may pave the way for the long-awaited standardization of culture technique which has so far eluded tissue culturists.

SELECTION OF MEDIUM AND SERUM

Unfortunately, there are few good guidelines for the selection of the appropriate medium for a given cell type. Information is usually available in the literature, or from the source of the cells, for cell lines currently available; failing this, the choice is either empirical or by comparative testing of several media (see below). Many continuous cell lines (e.g., HeLa, L-cells, BHK-21), primary cultures of human, rodent, and avian fibroblasts, and cell lines derived from them can be maintained on a relatively simple medium such as Eagle's MEM, supplemented with calf serum. More complex media are required where a specialized function is being expressed (e.g., aminotransferase activity in rat hepatoma [Pitot et al., 1964], cartilage and pigment secretion in chick embryo cells [Coon and Cahn, 1966], or when cells are passaged at low seeding density ($<10^3$/ml), as for cloning (see Chapter 13). Frequently the more demanding culture conditions that require complex media also require fetal bovine serum rather than calf or horse serum.

Some examples are given in Table 9.2 of cell types and the media used for them, but this list is neither exhaustive nor binding. For a more complete list, see Ham and McKeehan [1979] and Morton [1970]. If a clear indication of the correct culture conditions is not available, a simple cell growth experiment with commercially available media and multiwell plates (see Chapter 19) can be carried out in about 2 wks. Assaying for clonal growth (see Chapters 13 and 19) and measuring the expression of specialized functions may narrow the choice further. You may be surprised to find that your best conditions do not agree with the literature, and you will have to decide between the optimal growth and behavior of the cells as you find them, or reproducing the conditions found in another laboratory, which may be difficult due to variations in the impurities present in reagents and water and in batches of serum. It is to be hoped that as serum requirements are reduced and reagent purity increases, medium standardization will improve and the need for such a choice will not arise.

A reasonable range of media to test would be (1) Eagle's MEM, (2) Dulbecco's DMEM, (3) Ham's F12, (4) CMRL 1066, (5) RPMI 1640, (6) Waymouth's MB752/1, (7) McCoy's 5A, and (8) a 1:1 mixture of DMEM and F12 (for references, see Table 9.2).

Finally, you may have to compromise in your choice of medium or serum because of cost. Autoclavable media are available from commercial suppliers (Flow, GIBCO). They are simple to prepare from powder and are suitable for many continuous cell strains. They may need to be supplemented with glutamine for some cells and will usually require serum. The cost of serum should be calculated on the basis of medium volume where cell yield is not important, but where the objective is to produce large quantities of cells, calculate serum costs on a per cell basis. If a culture grows to 10^6/ml in serum A and 2×10^6/ml in serum B, serum B becomes the less expensive by a factor of two.

If fetal bovine serum seems essential, try mixing it with calf serum. This may allow you to reduce the concentration of the more expensive fetal serum. If you can, leave out serum altogether, or reduce the concentration and use one of the serum-free formulations suggested above.

Batch Reservation

Serum standardization is difficult as batches vary considerably and one batch will only last about 1 yr, stored at −20° C. Select the type of serum that is most appropriate and request batches to test from a number of suppliers. Most serum suppliers will normally reserve a batch until a customer can select the most suitable one (provided this does not take longer than 3 wks or so). When a suitable batch has been selected, the supplier is requested to hold it for up to 1 yr for regular dispatch. Other suppliers should also be informed so that they may return theirs to stock.

Testing Serum

The quality of a given serum is assured by the supplier, but their quality control is usually performed with one of a number of continuous cell lines. If your requirements are more discriminating, then you will need to do your own testing. There are four main parameters for testing serum:

Cloning efficiency. During cloning the cells are at a low density and hence are at their most sensitive, making this a very stringent test. Plate the cells out at ten to 100 cells/ml and look for colonies after 10 d to 2 wks. Stain and count the colonies (see Chapters 13 and 19) and look for differences in cloning efficency. Each serum should be tested at a range of concentrations from 2% up to 20%. This will reveal whether one serum is equally effective at a lower concentration, thereby saving money and prolonging the life of the batch, and will show up any toxicity at high serum concentration.

Growth curve. A growth curve should be performed in each serum (see Chapter 19) determining the lag period, doubling time, and saturation density (density at "plateau"). A long lag implies that the culture is having to adapt; short doubling times are preferable if you want a lot of cells quickly; and a high saturation density will provide more cells for a given amount of serum and will be more economical.

Preservation of cell culture characteristics. Clearly the cells must do what you require of them in the new serum, whether they are acting as host to a given virus, producing a certain differentiated cell product, or expressing a characteristic sensitivity to a given drug.

Sterility. Serum from a reputable supplier will have been tested and shown to be free of microorganisms, but occasionally odd bottles or parts of a batch may slip through even the most stringent sampling procedures. To be certain, you should grow cells in antibotic-free medium supplemented with the serum, look for any microbiological contamination, and stain the cells for mycoplasmas (see Chapter 17).

The previous precautions and tests are advisable in selecting a new batch of serum partly because of the investment both in terms of cash and cell lines, but also because of the need for stringent control of your culture conditions. However, it is not always within the capacity of a small laboratory to cover all these requirements, and short cuts may have to be taken. If this is the case, you may be obliged to accept the assurances of the supplier regarding sterility and growth-promoting capacity, but you must still test the serum for any special requirements of your own.

OTHER SUPPLEMENTS

Tissue extracts and digests have been used for many years as supplements to tissue culture media in addition to serum. Many are derived from microbiological culture techniques and autoclavable broths, e.g., bactopeptone, tryptose, and lactalbumin hydrolysate, which are proteolytic digests of beef heart, or lactalbumin, and contain mainly amino acids and small peptides. Bactopeptone and tryptose may also contain nucleosides and other heat-stable tissue constitutents such as fatty acids and carbohydrates.

Embryo extract. Embryo extract (or embryo juice) is a crude homogenate of 10-day chick embryo clarified by centrifugation (see Appendix). The crude extract was fractionated by Coon and Cahn [1966] to give high and low molecular weight fractions. The low molecular weight fraction promoted cell proliferation while the high molecular weight fraction promoted pigment and cartilage cell differentiation.

Conditioned medium. Puck and Marcus [1955] found that the survival of low-density cultures could be improved by growing the cells in the presence of feeder layers (see Chapter 13, Cloning). While part of this effect may have been due to conditioning of the substrate, the main effect was presumed to be conditioning of the medium by release of small molecular metabolites and macromolecules into the medium [Takahashi and Okada, 1970]. Hauschka and Konigsberg [1966] showed that the conditioning of culture medium necessary for the growth and differentiation of myoblasts was due to collagen released by the feeder cells. Feeder layers and conditioning of the medium by embryonic fibroblasts or other cell lines remains a useful method of culturing difficult cells [e.g., Stampfer et al., 1980] (see Chapter 13).

Chapter 10
Preparation and Sterilization

A good general principle to follow is that all stocks of chemicals and glassware used in tissue culture should be reserved for that purpose alone. Traces of heavy metals or other toxic substances can be difficult to detect other than by a gradual deterioration of your cultures. It also follows that separate stocks imply separate glassware washing. The requirements of tissue culture washing are higher than for general glassware; a special detergent may be necessary (see below) and cross-contamination from chemical glassware must be avoided.

All apparatus and liquids which come in contact with cultures or other reagents must be sterile. A summary of the procedures used is given in Tables 10.1 and 10.2.

PROCEDURES FOR THE PREPARATION AND STERILIZATION OF APPARATUS

Glassware

Items of glassware used for dispensing and storage of media, and for cell culture, must be cleaned very carefully to avoid traces of toxic materials, contaminating the inner surfaces, becoming incorporated into the medium (Fig. 10.1). Where the glass surface is to be used for cell propagation it must not only be clean but also carry the correct charge. Caustic alkaline detergents render the surface of the glass unsuitable for cell attachment and require subsequent neutralization with HCl or H_2SO_4, but many modern detergents can be used which do not alter the glass surface and which can be removed completely.

For the most effective washing procedure: (1) Do not let soiled glassware dry out. A sterilizing agent, such as sodium hypochlorite, should be included in the water used to collect soiled glassware (a) to remove any potential biohazard and (b) to prevent microbial contamination growing up in the water. (2) Select a detergent which is effective in the water of your area, rinses off easily, and is nontoxic (see below). (3) Ensure that the glassware is thoroughly rinsed in tap water and deionized or distilled water, before drying. (4) Dry inverted. (5) Sterilize by dry heat to minimize the risk of depositing toxic residues from steam sterilization.

Plastic culture flasks are, on the whole, meant for single use, as washing in detergent renders them unsuitable for cell propagation (in monolayer) and resterilization is difficult. Cells may be reseeded back into the same flask after subculture but this tends to increase the risk of contamination.

Sterilization procedures are designed not just to kill the bulk of microorganisms but to eliminate spores which may be particularly resistant. Moist heat is more effective than dry heat but does carry a risk of leaving a residue. Dry heat is preferable but at a minimum of 160° C for 1 hr. Moist heat (for fluids and perishable items) need only be maintained at 121° C for 15–20 min. For moist heat to be effective, steam penetration must be assured and for this the sterilization chamber must be evacuated prior to steam injection.

Materials

- Pipette cylinders (to collect used pipettes)
- disinfectant (if required)
- detergent
- soaking baths
- bottle brushes
- stainless steel baskets (to collect washed and rinsed glassware for drying)
- aluminum foil
- sterility indicators (Browne's tubes, Steriletest Strips)
- glass petri dishes (for screw caps)

TABLE 10.1. Sterilization of Equipment and Apparatus

Item	Sterilization
Apparatus containing glass & silicone tubing	Autoclave*
Disposable tips for micropipettes	Autoclave
Dispenser tubing for Compu-pet	Autoclave
Filters—Millipore, Sartorius	Autoclave—do not pre- or postvac.
Glassware	Dry heat†
Glass bottles with screw caps	Autoclave
Glass coverslips	Dry heat
Glass slides	Dry heat
Instruments	Dry heat
Magnetic stirrer bars	Autoclave
Pasteur pipettes—glass	Dry heat
Pipettes—glass	Dry heat
Screw caps	Autoclave
Silicone tubing	Autoclave
Stoppers—rubber, silicone	Autoclave
Test tubes	Dry heat

*Autoclave—100 kPa (15 lb/in^2) 121° C for 20 min.
†Dry heat—160° C/l hr.

TABLE 10.2. Sterilization of Liquids

Solution	Sterilization	Storage
Agar	Autoclave*	Room temperature
Amino acids	Filter†	4° C
Antibiotics	Filter	−20° C
Bacto-peptone	Autoclave	Room temperature
Bovine serum albumin	Filter—use stacked filters (see text)	4° C
Carboxylmethyl cellulose	Steam—30 min**	4° C
Collagenase	Filter	−20° C
DMSO	Self-sterilizing, aliquot into sterile tubes	Room temperature—keep dark, avoid contact with rubber or plastics
EDTA	Autoclave	Room temperature
Glucose—20%	Autoclave	Room temperature
Glucose—1-2%	Filter (low concentrations caramelize if autoclaved)	Room temperature
Glutamine	Filter	−20° C
Glycerol	Autoclave	Room temperature
HEPES	Autoclave	Room temperature
HCl 1 N	Filter	Room temperature
Lactalbumin hydrolysate	Autoclave	Room temperature
Methocel	Autoclave	4° C
$NaHCO_3$	Filter	Room temperature
NaOH 1 N	Filter	Room temperature
Phenol red	Autoclave	Room temperature
Salt solutions (without glucose)	Autoclave	Room temperature
Serum	Filter—use stacked filters (see text)	−20° C
Sodium pyruvate 100 m*M*	Filter	−20° C
Transferrin	Filter	−20° C
Tryptose	Autoclave	Room temperature
Trypsin	Filter	−20° C
Vitamins	Filter	−20° C
Water	Autoclave	Room temperature

*Autoclave—100 kPa (15 lb/in^2) 121° C for 20 min.
†Filter—0.2-μm pore size.
**Steam—100° C for 30 min.

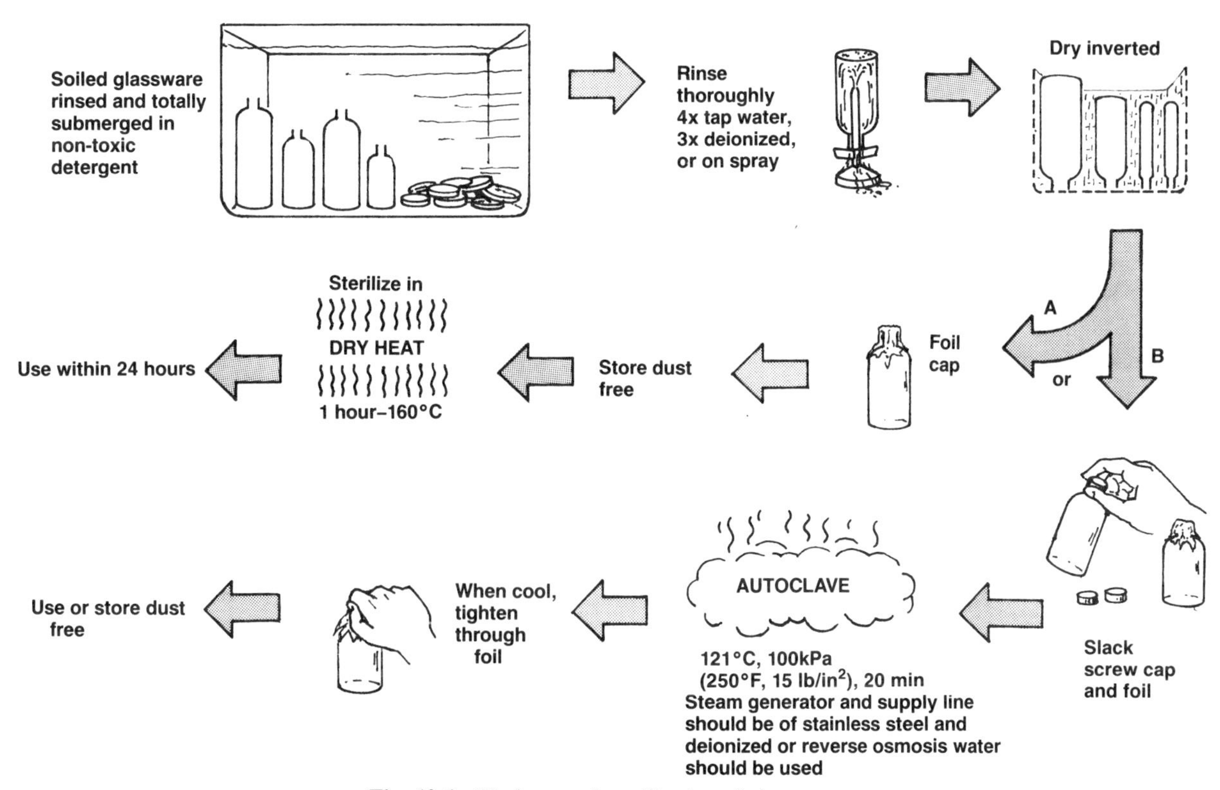

Fig. 10.1. *Wash-up and sterilization of glassware.*

autoclavable plastic film (Portex, Cedanco) or paper sterilization bags
sterile-indicating autoclave tape

Collection and washing

1.
Collect immediately after use into water which may contain a disinfectant such as sodium hypochlorite. It is important that apparatus should not dry before soaking, or cleaning will be much more difficult

2.
Soak overnight in a suitable detergent (see below)

3.
Machine wash or brush by hand or machine (Fig. 10.2) the following morning and rinse thoroughly in three complete changes of tap water followed by two changes of deionized water. If rinsing is done by hand, a sink spray (Fig. 4.3) is a useful accessory; otherwise bottles must be emptied and filled completely each time. Clipping bottles in a basket will help to speed up this stage

4.
After rinsing thoroughly, invert bottles, etc., in stainless steel wire baskets and dry upside down

Fig. 10.2. *Motorized bottle brushes. Care must be taken not to press down too hard on the brush lest the bottle break.*

5.
Cap with aluminum foil when cool and store

Sterilization

1.
Place in an oven with fan-circulated air at 160° C

2.
Check that temperature has returned to 160° C, seal the oven with a strip of tape with the time recorded on it, and leave for 1 hr; ensure that the center of the load achieves 160° C by using a sterility indicator or recording thermometer with the sensor in the middle of the load. Do not pack the load too tightly; leave room for hot air circulation
3.
After 1 hr, switch off the oven and allow to cool with the door closed. It is convenient to put the oven on an automatic timer so that it can be left to switch off on its own
4.
Use within 24–48 hr

Alternatively, bottles may be loosely capped with screw caps, autoclaved for 20 min at 120° C with pre- and postvacuum cycle (see Chapter 4), and the caps tightened when cool. Unfortunately, misting often occurs when bottles are autoclaved and a slight residue may be left when this evaporates. There is also a risk of the bottles becoming contaminated as they cool by drawing in nonsterile air before they are sealed. Dry heat sterilization is better, allowing the bottles to cool down within the oven before removal

Pipettes

1.
Place in water or a sterilizing agent (hypochlorite or glutaraldehyde) immediately after use (Fig. 10.3). Do not put pipettes which have been used with agar or silicones (water repellents, antifoams, etc.) in the same cylinder as regular pipettes. Use disposable pipettes for silicones and either rinse agar pipettes after use in hot tap water or use disposable pipettes
2.
If plugged, blow out plugs with compressed air,

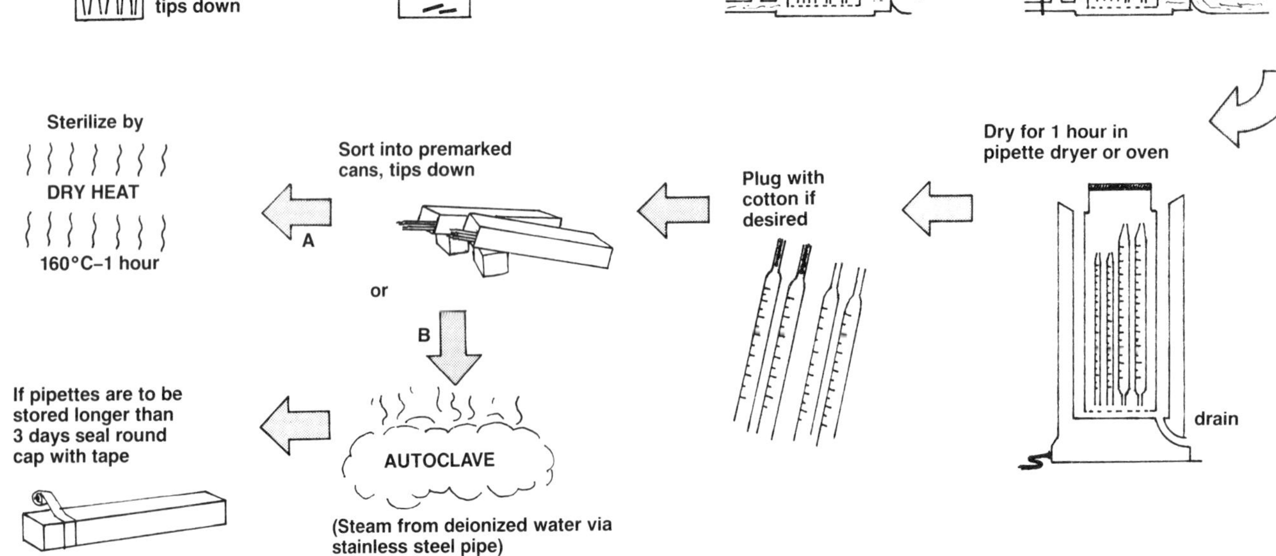

Fig. 10.3. *Wash-up and sterilization of pipettes.*

transfer to detergent (allow to drain first), and soak overnight

3. Allow pipettes to drain and transfer to pipette washer (Fig. 10.4, see also Fig. 4.3), points uppermost

4. Rinse by siphoning action of pipette washer for a minimum of 2 hr

5. Turn over valve to deionized water (see Fig. 4.3), or wait until last tap water finally runs out, turn off tap water, and empty and fill three times with deionized water

6. Transfer to pipette drier or drying oven and dry with points uppermost

7. Plug with cotton (Fig. 10.5). Alternatively, pipette plugs may be dispensed with and a short sterile filter tube placed on the pipette bulb during use (see Fig. 5.2). To avoid cross-contamination, this must be replaced before starting work and every time that you change to a new cell strain

8. Sort pipettes by size and store in drawers

Fig. 10.5 *Semiautomatic pipette plugger.*

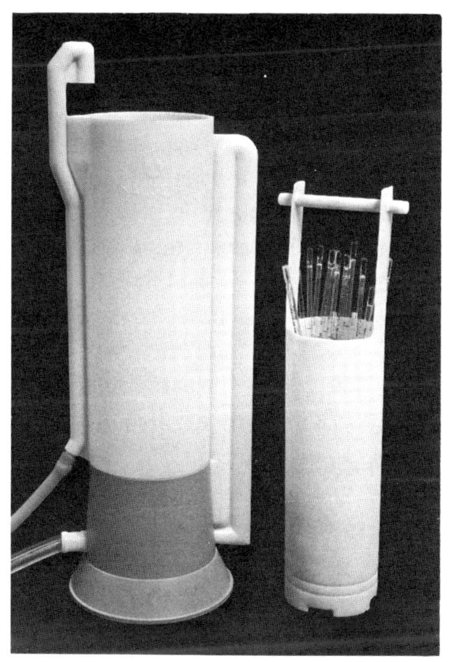

Fig. 10.4. *Siphon pipette washer. The model illustrated is made of polypropylene. Versions are available made of stainless steel, allowing the pipette carrier to be transferred directly to a drier.*

Sterilization

1. Place pipettes in pipette cans (square aluminum or stainless steel with silicone cushions at either end; square cans do not roll on the bench) and label both ends of the cans. Put one pipette size per can with a few cans containing an assortment of 1-ml, 2-ml, 10-ml, and 25-ml pipettes in the ratio 1:1:3:2

2. Seal with sterile-indicating tape and sterilize by dry heat for 1 hr at 160° C. The temperature should be measured in the center of the load, to ensure that this, the most difficult part to reach, attains the miminum sterilizing conditions. Leave spaces between cans when loading the oven, to allow for circulation of hot air

3. Remove from oven, allow to cool, and transfer cans to tissue culture laboratory. If you anticipate that pipettes will lie for more than 48 hr before use, seal cans around the cap with adhesive tape

Screw Caps

There are two main types of caps for glass bottles in common use: (1) aluminum caps with a synthetic rubber or silicone lining and (2) disposable polypropylene caps which are self-sealing and have no liner. Reusable aluminum caps seal better (important for CO_2 retention and pH control) but are more difficult to clean. Disposable polypropylene caps need not be cleaned (but can be easily) but will only seal if screwed down tightly on a bottle with no chips or imperfections on the lip of the opening.

Do not leave aluminum caps, or any other aluminum items in alkaline detergents for more than 30 min, as they will corrode. Do not have glassware in the same detergent bath or the aluminum may contaminate the glass. Avoid machine washing detergents as they are very caustic.

Aluminum caps. Soak 30 min in detergent and rinse thoroughly for 2 hr (make sure all caps are submerged). Rinsing may be done in a beaker (or pail) with running tap water led by a tube to the bottom of the beaker. Stir the caps by hand every 15 min. Alternatively, place in a basket, or better in a pipette washing attachment, and rinse in an automatic washing machine, but do not use detergent in machine.

Disposable caps. These should not need to be washed unless reused. They may be washed and rinsed by hand as above (extending the detergent soak if necessary). Because these caps may float, they must be weighted down during soaking and rinsing. For automatic washers, use pipette washing attachment and normal cycle with machine detergent.

Stoppers. Use silicone or heavy metal-free white rubber in preference to natural rubber. Wash and sterilize as for disposable caps (there will be no problem with floatation in washing and rinsing).

Sterilization. Place caps in a glass petri dish with the open side down. Wrap in cartridge paper or steam permeable nylon film (Portex), and seal with autoclave tape (Fig. 10.6). Autoclave for 20 min at 120° C, 100 kPa (15 lb/in^2) (see Fig. 10.8).

Keep organic matter out of the oven. Do not use paper tape or packaging unless you are sure that it will not release volatile products on heating. Such products will eventually build up on the inside of the oven, making it smell when hot, and some deposition may occur inside the glassware being sterilized.

Fig. 10.6. *Packaging screw caps for sterilization.*

Selection of Detergent

Solicit samples from local suppliers and test them (1) for their ability to wash heavily soiled glassware, (2) for the quality of the growth surface afterwards, and (3) for the toxicity of the detergent (see also Chapters 13 and 19).

1. Washing Efficiency:

Materials

Standard 75-cm^2 or 120-cm^2 glass culture vessels
samples of detergents made up to working strength
HBSS

Protocol

(i) Autoclave flasks carrying cell monolayers, standing vertically, for 20 min at 120° C

(ii) Soak overnight in detergent, three flasks per detergent

(iii) Rinse out detergent with water, note flasks with residue and brush as necessary to get them clean

(iv) Rinse four times in water and three times in deionized water

(v) Dry and sterilize by dry heat as above

(vi) Check flasks again for apparent cleanliness and any sign of residue. Add a little BSS containing phenol red to each flask (approximately 1 ml/100 cm^2), rinse over the inside of the bottle, and look for any color change. If it becomes pink (alkaline), detergent has not been completely rinsed out of the bottle. When you are satisfied that the flasks are clean, sterilize them by dry heat see above

2. Quality of Growth Surface:

Materials

Monolayer cells with good cloning efficiency (10% or more)
medium
serum
(for materials for fixing and staining cells, see Chapter 15)

Protocol

(i)
Taking flasks from 1, (vi above), add cell suspension to each flask and to three control flasks (disposable plastic of same surface area), using a cell concentration suitable for cloning (e.g., 20 cells/ml, 5 cells/cm^2, CHO-Kl; 100 cells/ml, 25/cm^2, MRC-5; see Chapter 13). Use the minimum concentration of serum which will allow the cells to clone (see under Testing Serum—Chapter 9)
(ii)
Incubate for 10–20 d, depending on the growth rate of the cells (see Chapters 13 and 19), fix stain, and count colonies (see Chapters 13 and 15)
(iii)
Determine relative plating efficiency (see Chapter 19) using disposable plastic flask as control, and record

3. Cytotoxicity: (See Chapter 19.)

Materials

Microtitration plate
medium
serum
detergent samples diluted to working strength and filter sterilized (see below; for fixation and staining cells, see Chapter 15)

Protocol

(i)
Set up microtitration plate with suitable target cells 1,000/well, incubate to 20% confluence, and change to fresh medium, 100 μl/well
(ii)
Add 100 μl of a diluted detergent to the first well in each row and dilute serially across plate in that row, leaving the last two wells in each row as untouched controls
(iii)
Continue growth for 3–5 d but *not beyond* the time that the control wells show confluent monolayers
(iv)
Wash, fix, and stain plate
(v)
Determine titration point (point at which cell number per well is reduced by approximately 50%) and record. A high detergent concentration at this point means a low cytotoxicity

Miscellaneous Equipment

Cleaning

All new apparatus and materials (silicone tubing, filter holders, instruments, etc.) should be soaked in detergent overnight, thoroughly rinsed, and dried. Anything which will corrode in the detergent—mild steel, aluminum, copper, or brass, etc.—should be washed directly by hand, without soaking, or soaking for 30 min only, using detergent if necessary, then rinsed and dried.

Used items should be rinsed in tap water and immersed in detergent immediately after use. Allow to soak overnight, rinse, and dry. Again, do not expose materials which might corrode to detergent for longer than 30 min. Aluminum centrifuge buckets and rotors must never be allowed to soak in detergent.

Particular care must be taken with items treated with silicone grease or silicone fluids. They must be treated separately and the silicone removed, if necessary, with carbon-tetrachloride. Silicones are very difficult to remove if allowed to spread to other apparatus, particularly glassware.

Packaging

Ideally, all apparatus for sterilization should be wrapped in a covering which will allow steam penetration but be impermeable to dust, microorganisms, and mites. Proprietary bags are available bearing sterile indicating marks which show up after sterilization. Semi-permeable transparent nylon film (Portex Plastics, Cedanco) is sold in rolls of flat tube of different diameters and can be made up into bags with sterile-indicating tape. Although expensive, it can be reused several times before becoming brittle.

Tubes and orifices should be covered with tape and paper or nylon film before packaging, and needles or other sharp points should be shrouded with a glass test tube or other appropriate guard.

Sterilization

The type of sterilization used will depend on the material (see Table 10.1). Metallic items are best sterilized by dry heat. Silicone rubber (which should be used in preference to natural rubber), Teflon, polycarbonate, cellulose acetate, and cellulose nitrate filters (see below for filters in holders), etc., should be autoclaved for 20 min at 121°C, 100 kPa (15 lb/in^2) with preevacuation and postevacuation steps in the cycle. In small bench-top varieties and pressure-cookers, make sure that the autoclave boils vigorously for 10–15 min before placing on the weight, to displace all the air. (Take care that enough water is put in at the start to allow for this.) After sterilization, the steam is released and the weight replaced or valve closed so that when the autoclave cools, a negative pressure is generated, helping to dry off the contents.

Sterilizing Filters

Reusable filter holders should be made up and sterilized as follows: 1. After thorough washing in detergent (see above) rinse in water, then deionized water and dry

2.

Insert support grid in filter and place filter membrane on grid. If polycarbonate, apply wet to counteract static electricity

3.

Place prefilters (glass fiber and others as required, see below) on top of filter

4.

Reassemble filter holder, but do not tighten up completely (leave about one half turn on collars, one whole turn on bolts)

5.

Cover inlet and outlet of filter wih aluminum foil

6.

Pack filter in sterilizing paper or steam-permeable nylon film and close with sterile-indicating tape

7.

Autoclave at 121°C, 100 kPa (15 lb/in^2) with no pre- or postevacuation ("liquids cycle" in automatic autoclaves)

8.

Remove and allow to cool

9.

Do not tighten filter holder completely until the filter is wetted at the beginning of filtration (see below)

Alternative methods of sterilization. Many plastics cannot be exposed to the temperature required for autoclaving or dry heat sterilization. To sterilize such items, immerse in 70% alcohol for 30 min and dry off under uv light in a laminar flow cabinet. Care must be taken with plexiglass (Perspex, Lucite) as it may crack in alcohol or uv treatment due to release of stresses built in during manufacture.

Ethylene oxide may be used to sterilize plastics, but 2–3 wk are required for the Et_2O to clear from the plastic surface.

γ-irradiation, 20,000–100,000 rad, is the best method for plastics. Items should be packaged and sealed. Polythene may be used and sealed by heat welding.

REAGENTS AND MEDIA

The ultimate objective in preparing reagents and media is to produce them in a pure form (1) to avoid the accidental inclusion of inhibitors and subtances toxic to cell survival, growth, and expression of specialized functions and (2) to enable the reagent to be totally defined and the functions of its constituents fully understood. The requirement for various trace metals has been difficult to determine because of their inclusion as contaminants in other reagents.

Most reagents or media can be sterilized either by autoclaving if they are heat stable (water, salt solutions, amino acid hydrolysates) or by membrane filtration if heat labile. During autoclaving the container should be kept sealed if borosilicate glass or polycarbonate. Soda glass bottles are better left with the caps slack to minimize breakage. Evolution of vapor will help to prevent ingress of steam from the autoclave, but the liquid level will need to be restored with sterile distilled water later.

Water

High purity water can be produced in a number of ways.

Double glass distillation. The still should be electrically operated, preferably automatic, and be able to feed the first distillate into the second boiler. The heating elements should be glass or silica sheathed.

Deionization and glass distillation. In this case, tap water is fed into a deionizer and the effluent fed into the boiler of the still, again with a glass-sheathed element (see Fig. 4.4). The quality of the deionized water should be monitored by conductivity at regular intervals and the cartridge changed when an increase in conductivity is observed.

Reverse osmosis, high-efficiency deionization, and carbon filtration. Ionic and non ionic solutes are removed by reverse osmosis (Milli-Q Fig. 10.7) (Millipore, Fisons); further removal of ionic contaminants is achieved by high-efficiency ion exchange, and the

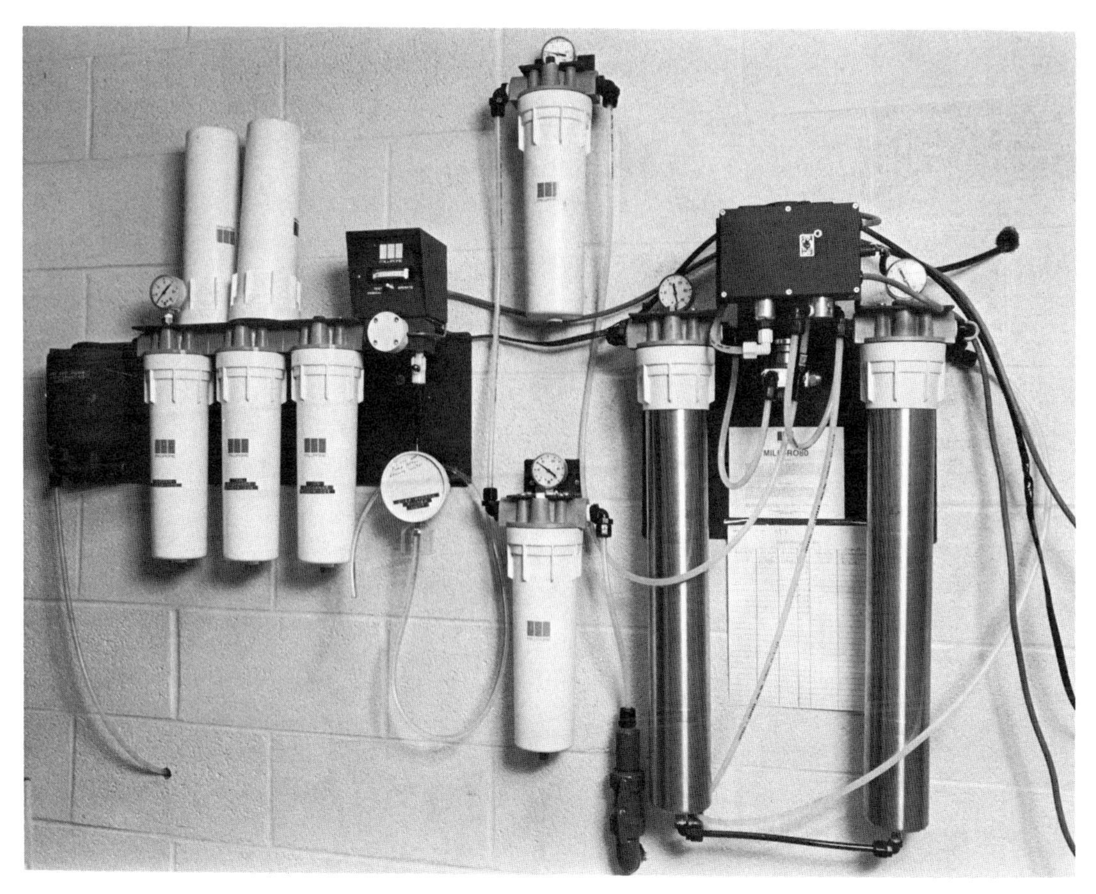

Fig. 10.7. *Water purification by reverse osmosis, ion exchange and carbon filtration (Millipore). Water enters (center, bottom) and passes through two preliminary ion exchangers (center and top) to reverse osmosis unit (right). Water from reverse osmosis may be stored (in this case on the other side of the wall) and returns (left of picture) to further ion exchange and carbon filtration. Final delivery is via a membrane filter (circular disc with delivery tube hooked over it).*

product is then further purified by filtration through charcoal and a 0.2-μm cellulose acetate filter. The water is not stored except after reverse osmosis, and is delivered directly on demand. This is designed to minimize pollution during storage. Regular distillation on a metal element still can be substituted for the first stage (instead of reverse osmosis) but will increase the cost per liter.

Water is sterilized by autoclaving at 121°C, 100 kPa (15 lb/in^2) for 10–15 min. It should be dispensed in premeasured amounts into borosilicate glass (e.g., Pyrex) bottles, plus 10% to allow for evaporation, to make it easier to use as a diluent for media concentrates.

Balanced Salt Solutions

A selection of formulations are given in Chapter 9. The formula for Hanks's BSS [after Paul, 1975] contains magnesium chloride in place of some of the sulphate originally recommended and should be autoclaved below pH 6.5 and neutralized before use. Similary, Dulbecco's PBS is made up without calcium and magnesium (PBSA), which are made up separately (PBSB) and added just before use. These precautions are designed to minimize precipitation of calcium and magnesium phosphate, during autoclaving and storage.

Most balanced salt solutions contain glucose. Since glucose may caramelize on autoclaving, it is best omitted and added later. If prepared as a × 10 concentrate (20%) caramelization during autoclaving is reduced.

Salt solutions may be sterilized by autoclaving for 10–15 min. at 121°C, 100 kPa (15 lb/in^2) (Fig. 10.8). Since evaporation may occur during autoclaving, mark the level at the start and make up with sterile distilled water later if necessary. Evaporation may be prevented by autoclaving with the bottles sealed. However, this must be done in Pyrex or equivalent bottles, or breakages will occur.

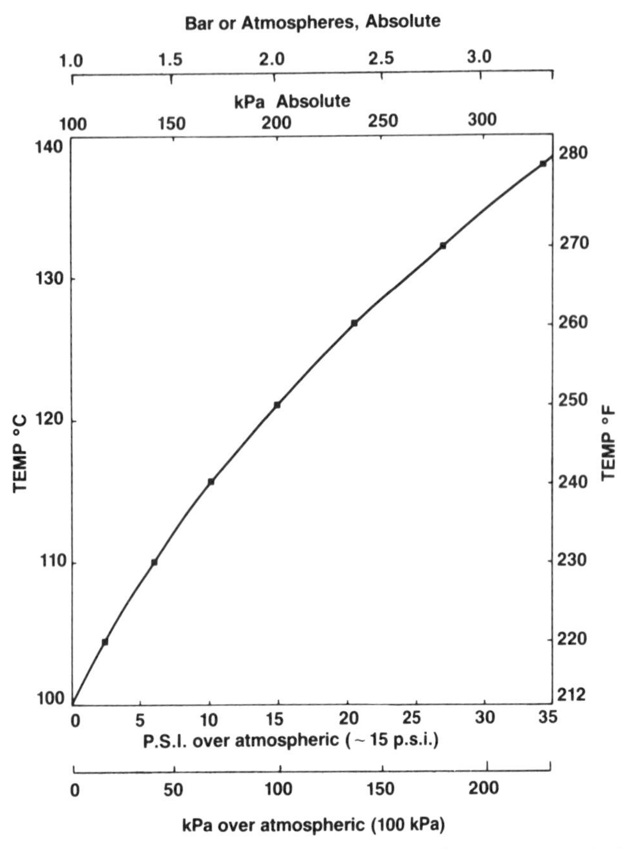

Fig. 10.8. *Relationship between pressure and temperature: 121° C and 100 kPa (250°F, 15 lb/in^2) for 15–20 min are the conditions usually recommended.*

Media

During the preparation of complex solutions care must be taken to ensure that all the constituents dissolve and do not get filtered out during sterilization, and that they remain in solution after autoclaving or storage. Concentrated media are often prepared at a low pH (between 3.5 and 5.0) to keep all the constituents in solution, but even then some precipitation may occur. If properly resuspended this will usually redissolve on dilution; but if the precipitate has been formed by degradation of some of the constituents of the medium, then the quality of the medium may be reduced. If a precipitate forms, the medium performance should be checked by cell growth, cloning, and assay of special functions (see below).

The preparation of media is rather complex. It is convenient to make up a number of concentrated stocks, essential amino acids at × 50 or × 100 concentrated, vitamins at × 1,000 concentrated, tyrosine and tryptophan at × 50 concentrated in 0.1 N HCl, glucose, 200 g/l, and single strength BSS. The requisite amount of each concentrate is then mixed and filtered through a 0.2-μm porosity cellulose acetate, cellulose nitrate, or polycarbonate (Nucleopore) filter and diluted with the BSS (see below and Fig. 10.9, Table 10.2), e.g.,

Amino acid concentrate, × 100 (in water)	100 ml
Tyrosine and Tryptophan, × 50 (in 0.1 N HCl)	200 ml
Vitamins, × 1,000 (in water)	10 ml
Glucose, × 100/200g/l (in BSS)	100 ml

Mix and sterilize by filtration. Store frozen.

For use dilute 41 ml concentrate mixture with 959 ml sterile × 1 BSS.

In practice this procedure is so laborious and time consuming that few laboratories make up their own

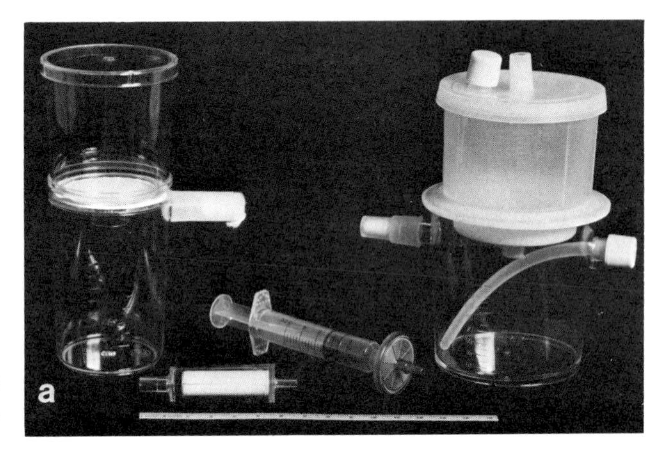

Fig. 10.9. *Selection of disposable filters. a. Syringe type (e.g., Millex, 25 mm), 47 mm with reservoirs (Nalgene, left, and Falcon, right), and a cylindrical syringe type (Millipore, bottom) with equivalent surface area to 47-mm-diameter disc filter. b. Millipore "Twin-90"; two 90-mm discs in parallel.*

media from basic constituents unless they wish to alter individual constituents regularly. Commercial media, which were quite unreliable in the 1950s and early 1960s, have now improved greatly, largely due to the introduction of appropriate quality control measures. There are now several reputable suppliers (see Trade Index) of standard formulations, and many of them will prepare media to your own formulation.

Commercial media are supplied as (1) working strength solutions, complete with sodium bicarbonate and glutamine, (2) × 10 concentrates without $NaHCO_3$ and glutamine, which are available as separate concentrates or (3) powdered media, complete. Powdered media is the cheapest, and not a great deal more expensive than making up your own, if you include time for preparation, sterilization, and quality control, cost of raw materials of high purity, and overheads such as power and wages. Powdered media are quality controlled by the manufacturer for their growth-promoting properties but not, of course, for sterility. Tenfold concentrates cost about twice as much per liter of working strength medium but are purchased sterile. Buying media at working strength is the most expensive (about five times the cost of a × 10 concentrate) but is the most convenient as no further preparation is required other than the addition of serum, if required.

Preparation of medium from × 10 concentrate.

Sterilize aliquots of deionized distilled water (see above) of such a size that one aliquot will last from 1 to 3 wk. Add concentrated medium and other constituents as follows:

1.

For sealed culture flask: gas phase air, low buffering capacity, and low CO_2/HCO_3 concentration

Water	884
× 10 concentrate	100
200 m*M* glutamine	10
5.6% $NaHCO_3$	6
	1,000 ml

1 N NaOH to pH 7.2 at 20° C, 7.4 at 36.5° C. If extra constituents are added, e.g., HEPES buffer, extra amino acids, this becomes:

2.

For sealed culture flask; high buffering capacity, gas phase air; may be vented to atmosphere by slackening cap for some cell lines at a high cell density if a lot of acid is produced; atmospheric CO_2/HCO_3^-

Water	854
× 10 concentrate	100
200 m*M* glutamine	10
× 100 nonessential amino acids	10
1 M HEPES	20
5.6% $NaHCO_3$	6
	1,000 ml

3.

For cultures in open vessels in a CO_2 incubator or under CO_2 in sealed flasks, two suggested formulations are as follows:

	(1) 2% CO_2	(2) 5% CO_2
Water	858	854
× 10 concentrate	100	100
200 m*M* glutamine	10	10
1 M HEPES	20	—
5.6% $NaHCO_3$	12	36
	1,000 ml	1,000 ml

1 N NaOH to pH 7.4 at 36.5° C (1) good buffering capacity, moderate CO_2/HCO_3^- concentration, (2) moderate buffering capacity, high CO_2/HCO_3 concentration. Always equilibrate at 36.5° C as the solubility of CO_2 decreases with increased temperature and the pKa of the HEPES will change.

The amount of alkali needed to neutralize a × 10 concentrated medium (which is made up in acid to maintain solubility of the constituents) may vary from batch to batch and from one medium to another, and, in practice, titrating medium to pH 7.4 at 36.5° C can sometimes be a little difficult. When making up a new medium for the first time, add the stipulated amount of $NaHCO_3$, and allow samples with varying amounts of alkali to equilibrate overnight at 36.5° C in the appropriate gas phase. Check the pH the following morning, select the correct amount of alkali, and prepare the rest of the medium accordingly.

Some media are designed for use with a high bicarbonate concentration and elevated CO_2 in the atmosphere, e.g., Ham's F12, while others have a low bicarbonate concentration for use with a gas phase of air, e.g., Eagle's MEM with Hanks' salts. If the use of a medium is changed, and the bicarbonate concentration altered, it is important to check that the osmolality is still within an acceptable range.

The bicarbonate concentration is important in establishing a stable equilibrium with atmospheric CO_2, but regardless of the amount of bicarbonate used, if the medium is at pH 7.4 and 36.5° C, the bicarbonate concentration at each concentration of CO_2 will be as in Table 8.3 (see Chapter 9).

The osmolality should be checked (see Chapter 9), where alterations are made to a medium that are not in the original formulation.

If your consumption of medium is great (> 20 l/yr) and you are buying medium ready-made, then it may be better to get extra constituents included in the formulation as this will work out to be cheaper. HEPES particularly is very expensive to buy separately.

Glutamine is supplied separately as it is unstable and should be kept frozen. The half-life in medium at 4° C is about 3 wk, and at 36.5° C, about 1 wk.

Care should be taken with × 10 concentrates to ensure that all of the constituents are in solution, or at least evenly suspended before dilution. Some constituents, e.g., folic acid or tyrosine, can precipitate and be missed at dilution. Incubation at 36.5° C for several hours may overcome this.

The final step is the addition of serum which, since it is close to isotonic, is added to the final volume, e.g.,

Complete medium	1,000 ml
Serum	111 ml (10% final)
	1,111 ml

Remember to allow space for all additions when choosing the container and the volume of H_2O to be sterilized.

Having once tested a batch of medium for growth promotion, etc., and found it to be satisfactory, then it need not be tested each time it is made up to working strength. The sterility should be checked, however, by incubating the medium at 36.5° C for 48 hr before adding glutamine, serum, or antibiotic.

Powdered media. Select a formulation lacking glutamine. If there are other unstable constituents, they also should be omitted and added later as a sterile concentrate, just before use. Dissolve the powder in the recommended amount of water (choose a pack size that you can make up all at once and use within 3 months), taking care that all the constituents are completely dissolved. (Follow the manufacturers instructions.) Do not store, but filter sterilize immediately. Precipitation may occur on storage, and microbiological contamination may also appear.

Sterilization. Filter through 0.2 μm polycarbonate, cellulose acetate, or cellulose nitrate filter (see below). Add glutamine before using.

Autoclavable Media

Some commercial suppliers offer autoclavable versions of Eagle's MEM and other media. Autoclaving is much less labor intensive, less expensive, and has a much lower failure rate than filtration. The procedure to follow is supplied in the manufacturer's instructions and is similar to that described above for BSS. The medium is buffered to pH 4.25 with succinate to stabilize the B vitamins during autoclaving and subsequently neutralized. Glutamine is replaced by glutamate or added sterile after autoclaving. As with BSS, care should be taken with evaporation and any deficit made up with sterile deionized distilled water.

Filter Sterilization

This method is suitable for filtering heat-labile solutions (Figs. 10.9, 10.10, 10.11). Reusable filters are made up and sterilized by autoclaving, or presterilized disposable filters may be used. The latter are more expensive but less time consuming to use and give fewer failures.

Preparation and sterilization of filter—see above.

Materials for filtration

Pressure vessel
pump
clamp to secure filter

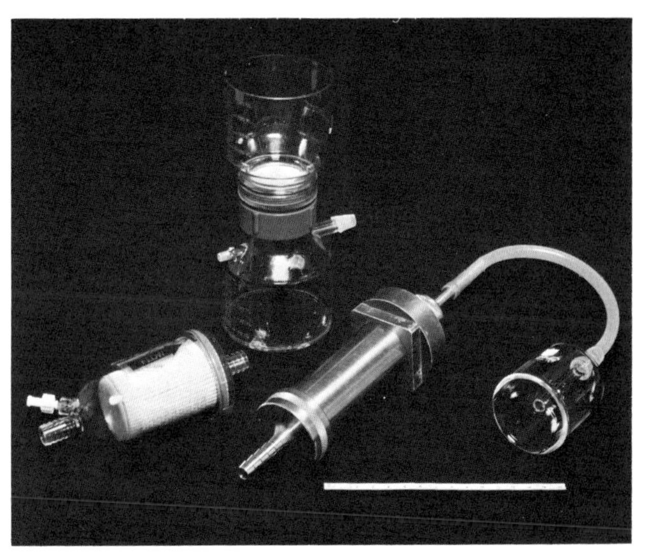

Fig. 10.10. *Reusable metallic and plastic filter holders (Gelman/Flow) (Millipore). The metallic holders must be disassembled, the filter replaced, and the whole resterilized each time it is used. The plastic cartridge filter (650 cm² surface area) is rinsed by back filtration with pure, particle-free water after use and resterilized by autoclaving. It will not last indefinitely, however, and eventually needs to be replaced.*

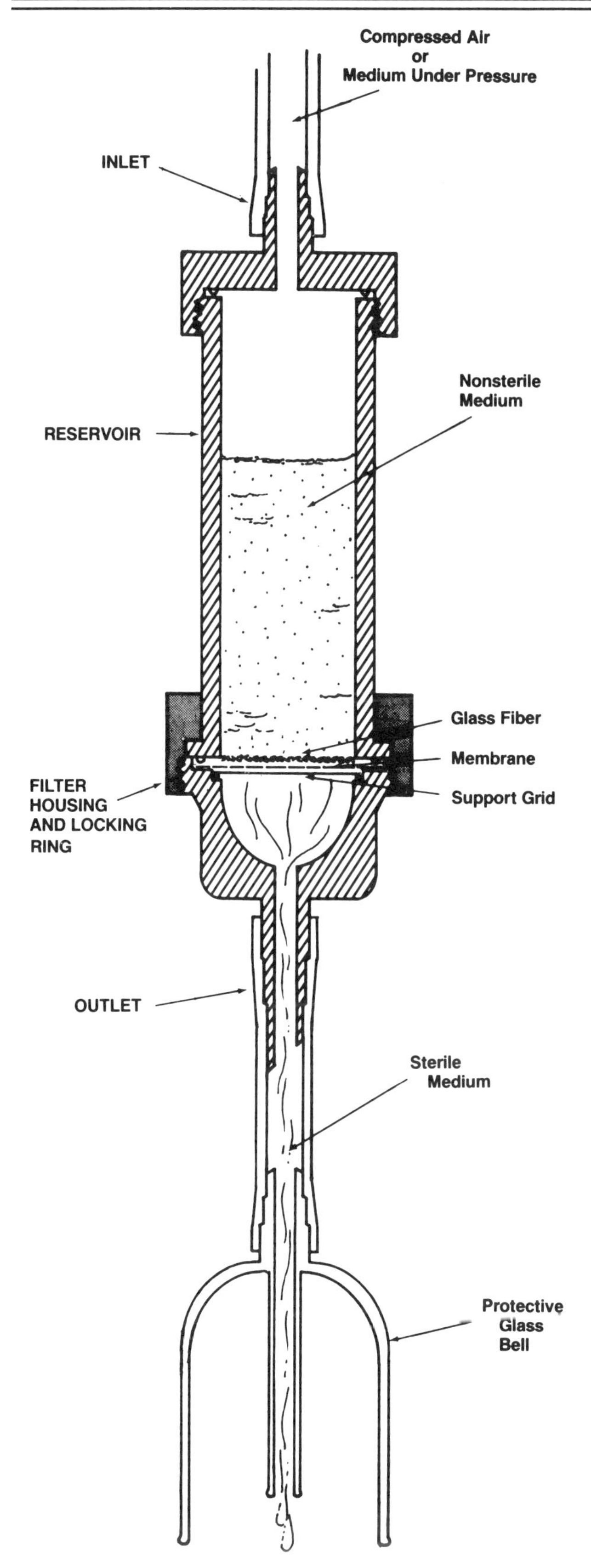

Fig. 10.11. *Longitudinal section through 47-mm reusable filter holder. The whole apparatus is sterilized. Fluid may be added directly to the reservoir or may be delivered from a pressurized tank (see Figs. 10.12 and 10.13).*

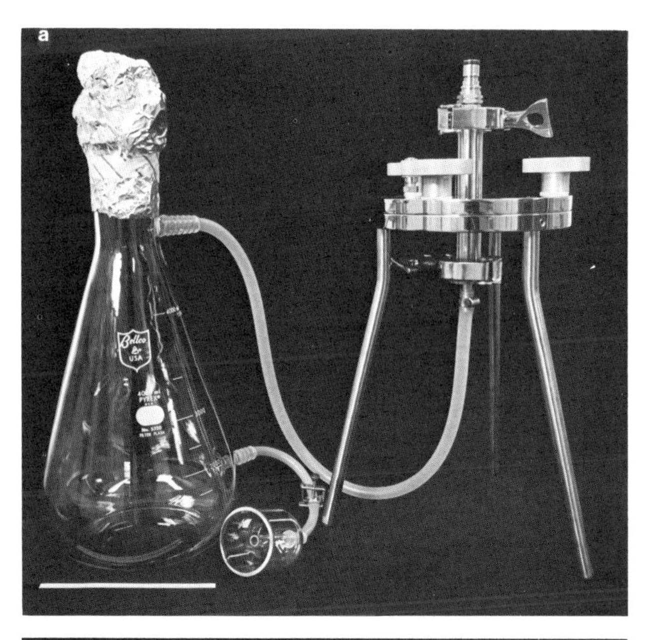

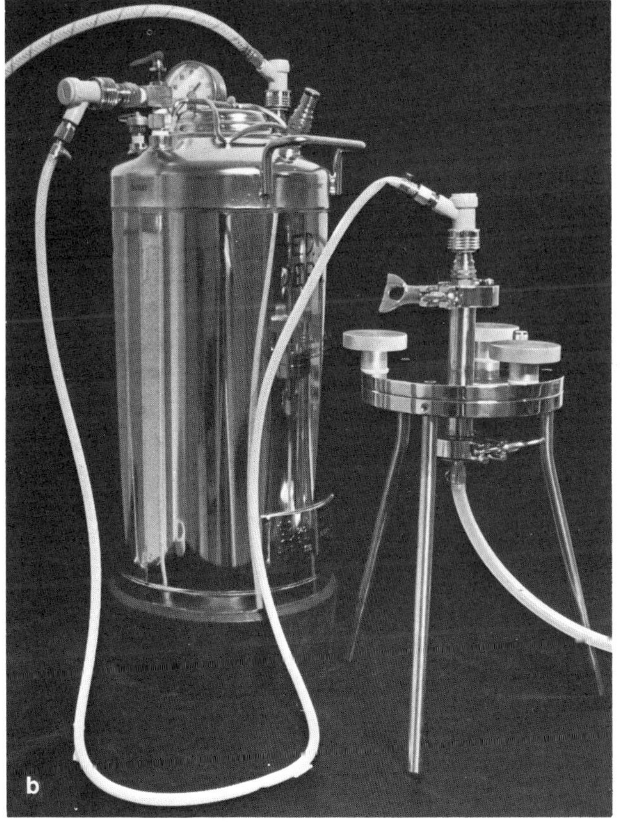

Fig. 10.12. *In-line filter assembly connected to receiver flask (a) and to pressurized reservoir (b). Only those items in (a) need be sterilized. Normally the glass bell would be covered in protective foil (left off here for purposes of illustration).*

sterile filter holder with filter
sterile container to receive filtrate with outlet at the base
tubing
spring clip and glass bell (see Figs. 10.12, 10.13)
sterile bottles with caps and foil

Protocol

1.
Choose an appropriate filter size from the range available (Table 10.3), secure the filter holder in position, assemble the sterile and nonsterile components, and make the necessary connexions (Figs. 10.12, 10.13)

2.
Decant the medium into the pressure vessel

TABLE 10.3. Filter Size and Fluid Volume

Filter size, diameter (mm)	Approx. area (cm^2)	Disposable reusable	Approximate volume which may be filtered	
			Crystalloid	Colloid
25	4.5	D	2–100 ml	2–25 ml
47*	14	D & R	100 ml–1 l	50–200 ml
90	50	R	1–10 l	200 ml–2 l
Twin 90	100	D	2–50 l	Not recommended
142	140	R	10–50 l	1–5 l
10 × 76	250	D	20–250 l	2–10 l
293*	650	R	50–500 l	5–20 l

*Cartridge filters available with equivalent surface area.

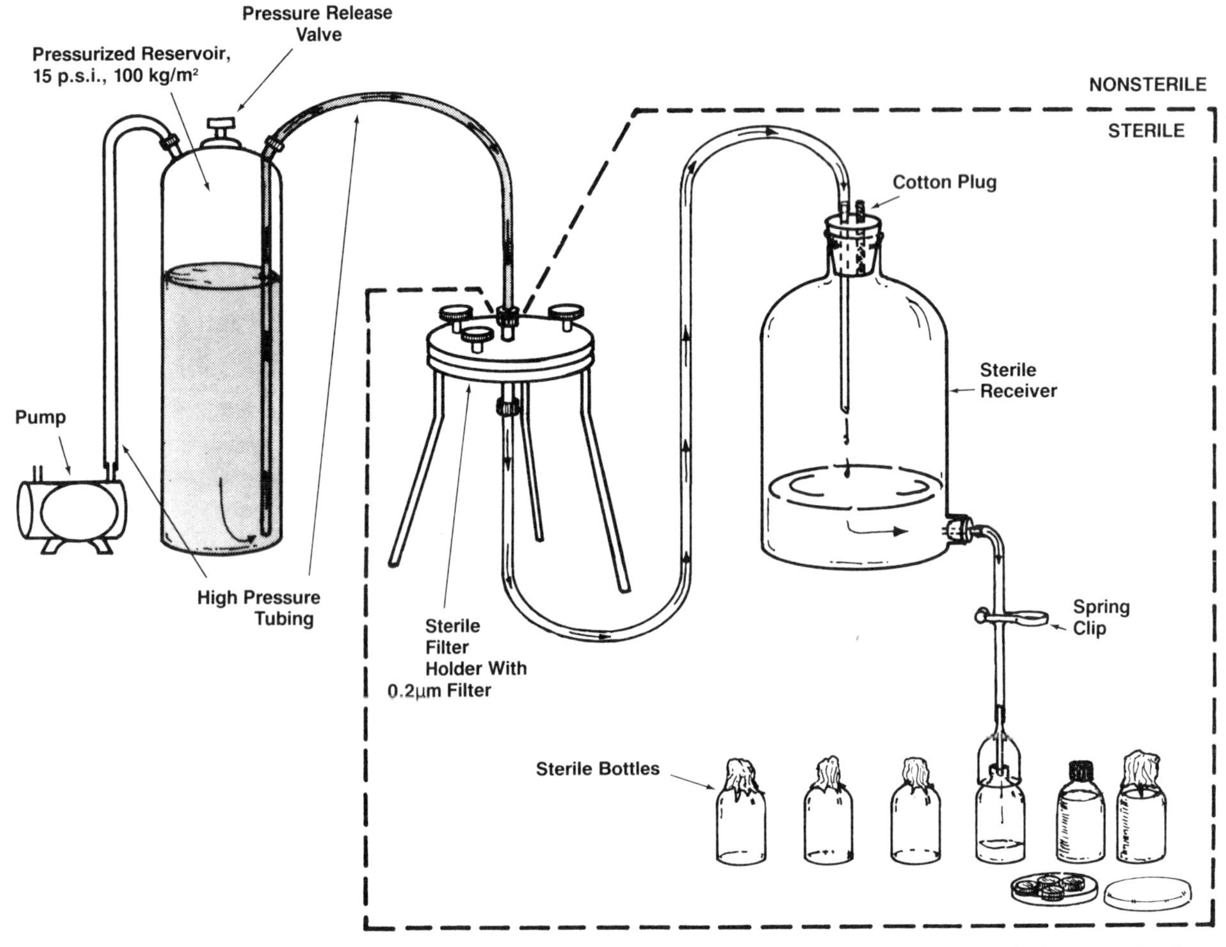

Fig. 10.13. *Diagram of complete system illustrated in Figure 10.12. The broken line encloses those items which are sterilized.*

3.
Position the sterile receiver under the outlet of the filter
4.
For reusable filters, turn on the pump just long enough to wet the filter. Stop the pump and tighten up the filter holder.
5.
Switch on the pump to deliver 100 kPa (15 lb/in^2). When the receiver starts to fill, draw off aliquots into medium stock bottles of the desired volume, cap, and store

When sterilizing small volumes (100–200 ml), the solution may fit within the filter holder and a pressure vessel and receiver will not be required. Collect directly into storage bottle.

Positive pressure is recommended for optimum filter performance and to avoid removal of CO_2 which results from negative pressure filtration. However, the latter method is often used for filtering small volumes as the equipment is simple. It is important to incorporate a filter and trap on the outlet from the filter flask (Fig. 10.14).

Sterility testing. (1) When filtration is complete and all the liquid has passed through the filter, disconnect the outlet, and raise the pump pressure until bubbles form in the effluent from the filter. This is the "bubble point" and should occur at more than twice the pressure used for filtration (see manufacturer's instructions). If the filter bubbles at the sterilizing pressure (100 kPa) or lower, then it is perforated and should be discarded. Any filtrate which has been collected should then be regarded as nonsterile and refiltered.

(2) If the filter passes the "bubble point" test, withdraw aliquots from the beginning, middle, and end of the run and test for sterility by incubating at 36.5° C for 72 hr. If the samples become cloudy, discard them

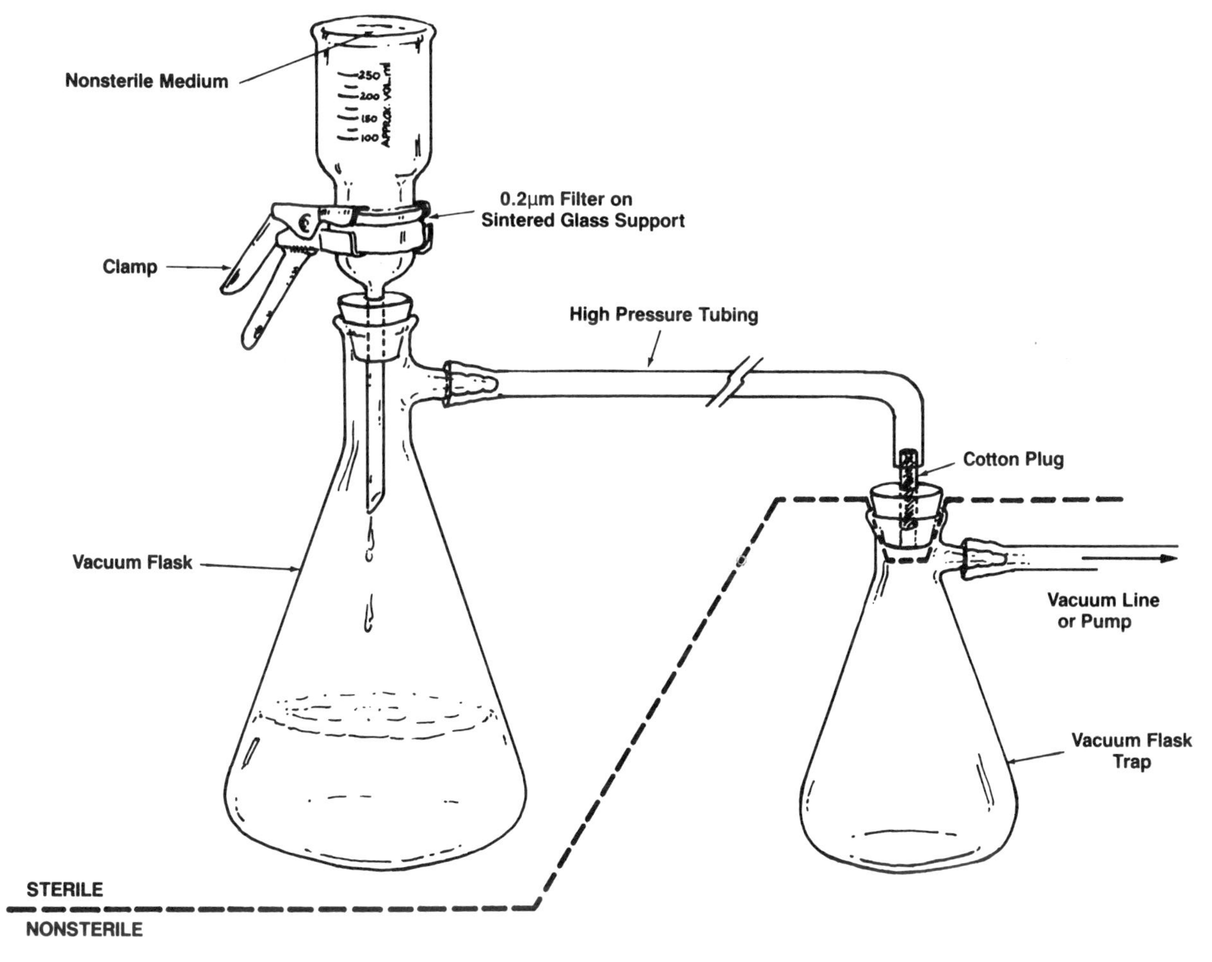

Fig. 10.14. *Apparatus for vacuum filtration. Only those items above the broken line need be sterilized. Vacuum filtration is also possible with some disposable filters (Nalgene and Falcon, see Fig. 10.8).*

and resterilize the batch. If there are signs of contamination in the other stored bottles, the whole batch should be discarded.

For a more thorough test, take samples, as in 2, dilute one-third of each into nutrient broths, e.g., beef heart hydrolysate and thioglycollate. Divide each in two and incubate one at 36.5° C and one at 20° C for 10 days, with uninoculated controls. If there is any doubt after this incubation, mix and plate out aliquots on nutrient agar.

Alternatively, place a demountable sterile filter in the effluent line from the main sterilizing filter. Any contamination that passes due to failure in the first filter will be trapped in the second. At the end of the run, remove the filter and place on nutrient agar. If colonies grow, discard or refilter. This method has the advantage that it monitors the entire filtrate and not just a small fraction, but it does not cover risks of contamination during bottling and capping.

Sterility testing of autoclaved stocks is much less essential, provided proper monitoring of the autoclave is carried out.

Culture testing. Media which have been produced commercially will have been tested for their ability to sustain growth of one or more cell lines. (If they have not, then you should change your supplier!) However, there are certain circumstances when you may wish to test your own media for quality, e.g., (1) if it has been made up in the laboratory from basic constituents, (2) if any additions are made to the medium, (3) if the medium is for a special purpose that the commercial supplier is not able to test, and (4) if the medium is made up from powder and there is a risk of losing constituents during filtration.

Contamination of medium with toxic substances can arise during filtration. Some filters are treated with traces of detergent to facilitate wetting, and this may leach out into the medium during filtration. Such filters should be washed by passing PBS or BSS through before use, or by discarding the first aliquot of filtrate. Polycarbonate filters, e.g., Nucleopore, are wettable without detergents and are preferred by some workers, particularly where the serum concentration in the medium is low.

There are three main types of culture test: (1) plating efficiency, (2) growth curve at regular passage densities and up to saturation density, and (3) expression of a special function, e.g., differentiation in the presence of an inducer, virus propagation, or expression of a specific antigen. All of these should be performed on your regular medium as a control.

Plating efficiency (see Chapter 19). This is the most sensitive test as it will detect minor deficiencies and low concentrations of toxins not apparent at higher cell densities. Ideally it should be performed at a range of serum concentrations from 0 to 20% as serum may mask deficiencies in the medium.

Growth curve. Clonal assay will not always detect insufficiencies in the amount of particular constituents (see Chapter 19). For example, if the concentration of one or more amino acids is low, it may not affect clonal growth but could influence the maximum cell concentration attainable.

A growth curve gives three parameters of measurement: (1) the lag phase before cell proliferation is initiated after transfer, indicating whether the cells are having to adapt to different conditions, (2) the doubling time in the middle of the exponential growth phase, indicating the growth-promoting capacity, and (3) the terminal cell density. In cell lines which are not sensitive to density limitation of growth, e.g., continuous cell lines (see Chapter 16), this indicates the total yield possible and usually reflects the total amino acid or glucose concentration. Remember that a medium which gives half the terminal cell density is costing twice as much per cell produced.

Special functions. In this case a standard test from the experimental system you are using, e.g., virus titer in medium after a set number of days, should be performed on the new medium alongside the old.

A major implication of these tests is that they should be initiated well in advance of the exhaustion of the current stock of medium (1) so that proper comparisons may be made and (2) so that there is time to have fresh medium prepared if the medium fails any of these tests.

Storage

Opinions differ as to the shelf-life of different media. As a rough guide, media made up without glutamine should last 2–3 months at 4° C. Once glutamine is added, storage time is reduced to 2–3 wk. Media which contain labile constituents should either be used within 2–3 wk of preparation or stored at −20° C.

Some forms of room fluorescent lighting will cause deterioration of riboflavin and tryptophan into toxic by-products [Wang, 1976]. Incandescent lighting should be used in cold-rooms where media is stored and the light extinguished when the room is not occupied. Bottles should not be exposed to fluorescent lighting for longer than a few hours. A dark freezer is recommended for long-term storage but is not always practicable.

Serum

This is one of the more difficult preparative procedures in tissue culture because of variations in the quality and consistency of the raw material and because of the difficulties encountered in sterile filtration. However, it is also the highest constituent of the cost of doing tissue culture, accounting for 30–40% of the total budget if bought from a commercial supplier. Buying sterile serum is certainly the best approach from the point of view of consistency and simplicity (see Chapter 9) but it may be prepared as follows.

Outline

Collect blood, allow to clot, and separate the serum. Filter serum through gradually reducing porosity of filters, bottle and freeze.

Protocol

Large scale, 20–100 l per batch.

Collection

Arrangements may be made to collect whole blood from a slaughterhouse. It should be collected directly from the bleeding carcass and not allowed to lie around after collection. Alternatively, blood may be withdrawn from live animals under proper supervision. The second routine, if performed consistently on the same group of animals, gives a more reproducible serum but a lower volume for greater expenditure of effort. If done carefully, it may be collected aseptically.

Clotting

Allow the blood to clot by standing overnight in a covered container at 4° C. This so called "natural clot" serum is superior to serum physically separated from the blood cells by centrifugation and defibrination, as platelets release growth factor into the serum during clotting. Separate the serum from the clot and centrifuge at 2,000 *g* for 1 hr to remove sediment.

Sterilization. Serum is usually sterilized by filtration through a 0.2-μm-porosity sterilizing filter, but because of its viscosity and high particulate content, it should be passed through a graded series of cartridge-type glass-fiber prefilters before passing through the final sterilizing filter. Only the last filter, a 350-mm in-line disc filter or equivalent, need be sterilized.

The prefilter assemblies may be of stainless steel with replaceable cartridges (Pall, Europe, Ltd.) or may be a single bonded unit (Gelman). The latter are easier to use but more difficult to clean and reuse. Reuse is possible, however. For smaller volumes (<1l) graded filters may be stacked in a single unit (Fig. 10.15).

Materials

Pump
pressure vessel
clamp
filter holder with series of nonsterile stacked filters, e.g., glass fiber, 5, 1.2, 0.45 μm
Twin-90 disposable filter or equivalent
sterile receiving vessel with outlet at base
sterile bottles with caps and foil

For volumes of 5–20 l

1. Connect a nonsterile reusable filter holder (142 mm) or cartridge-type filter in line with a sterile Twin-90 (disposable) or sterile 142 mm (reusable) filter holder (Fig. 10.16) with a 0.2-μm-porosity filter, and connected to a sterile receiver
2. Place a 0.45-μm-porosity filter on the support screen of the nonsterile holder, a 1.2-μm filter on top of it, and a 5-μm filter on top of that. Finally place a glass-fiber filter on the top of the 5-μm filter, wet the filters with sterile water, and close up the holder
3. Connect a pressure reservoir (see Figs. 10.12, 10.13) to the top of the nonsterile filter
4. Add serum to the pressure vessel, close, and apply pressure (15 kPa, 2 lb/in^2) until the first sign of liquid appears leaving sterile filter. Stop and tighten sterile filter
5. Reapply pressure and continue filtration, checking for leaks or blockages. Increasing the filtration pressure will increase the rate of filtration but may cause packing or clogging of the filters
6. When all serum is through, bubble-point test the filter (see above), then bottle and freeze the serum, taking samples out previously to test (see above)

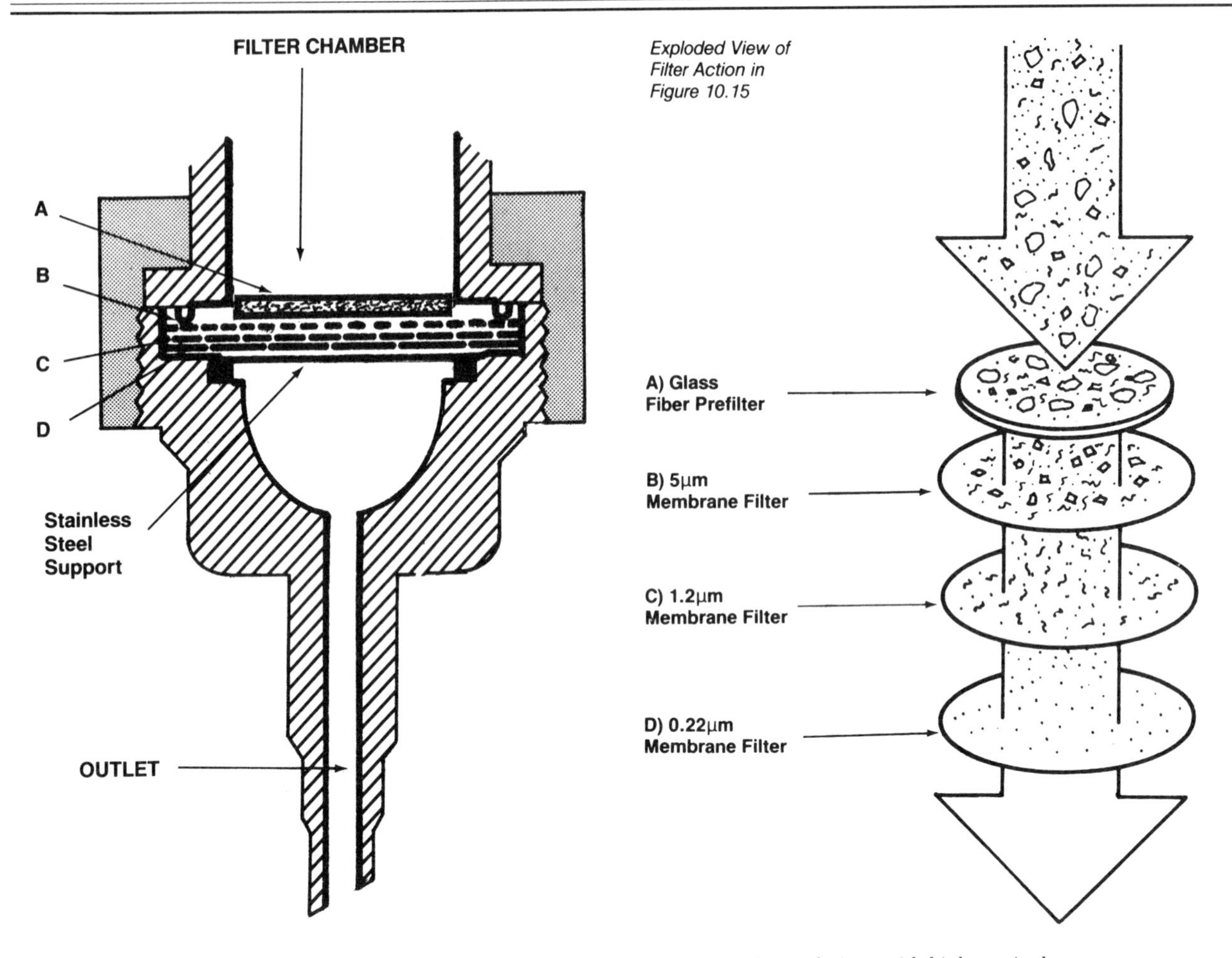

Fig. 10.15. *Stacking filters for filtering colloidal solutions (e.g., serum) or solutions with high particulate content.*

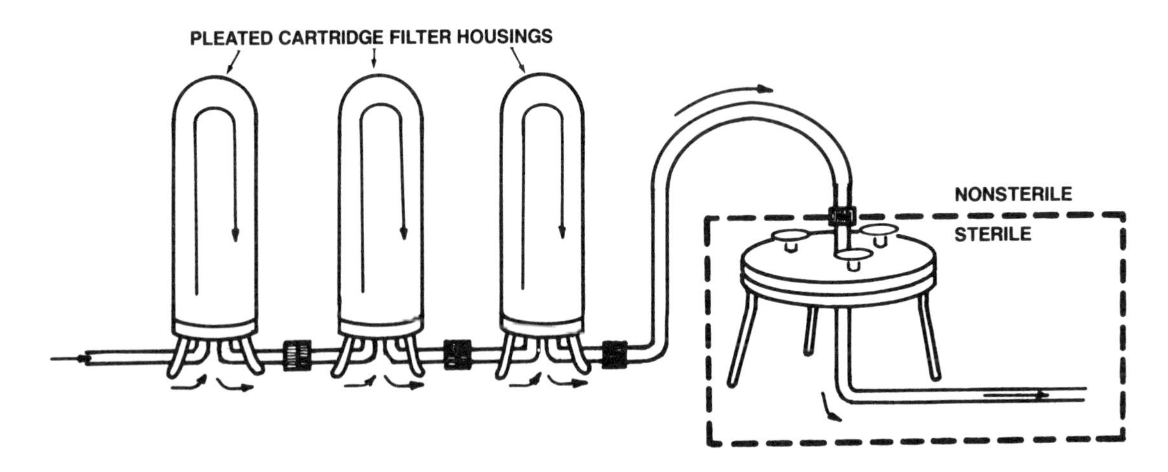

Fig. 10.16. *Series of pleated cartridge filters for large-scale filtration of colloidal solutions. Sterile filtration is completed by a conventional membrane disc, in-line filter.*

Small-scale serum processing. If small amounts (< 1 l) of serum are required, then the process is similar but can be scaled down. After clot retraction (see above) small volumes of serum may be centrifuged (5–10,000 *g*), and then filtered through a reusable in-line filter assembly (47 or 90 mm) containing glass fiber, 5-μm, 1.2-μm, and 0.45-μm filters as described above and finally through a 47-mm, 0.2-μm-porosity sterile disposable filter.

With very small volumes (10–20 ml), centrifuge at 10,000 *g* and filter directly through one or more disposable 25 mm, 0.2-μm filters. A graded series of the syringe-type filters (e.g., Millex, Millipore) is now available.

Storage. Bottle the serum in sizes that will be used up within 2–3 wk after thawing. Freeze the serum as rapidly as possible and if thawed, do not refreeze unless further prolonged storage is required.

Serum is best used within 6–12 months of preparation if stored at −20° C, but more prolonged storage may be possible at −70° C. The bulk of serum stocks usually make this impractical. Polycarbonate or high-density polypropylene bottles will eliminate the risk of breakage if storage at −70° C is desired. Regardless of the temperature of the freezer, or the nature of the bottles, do not fill them completely. Allow for the anomalous expansion of water during freezing.

Quality Control. Use same procedures as for medium.

Dialysis. For certain studies, the presence of small molecular weight constituents (amino acids, glucose, nucleosides, etc.) may be undesirable. These may be removed by dialysis through conventional dialysis tubing.

Materials

Dialysis tubing
beaker with distilled water
bunsen
tripod with wire gauze
serum to be dialyzed
HBSS at 4° C
magnetic bar and stirrer
measuring cylinder
sterile stacked filters: 0.22, 0.45, 1.2, 5 μm, glass fiber
sterile bottles with caps

Protocol

1. Boil five pieces, 30-mm × 500-mm dialysis tubing, in three changes of distilled water
2. Transfer to Hanks' Balanced Salt Solution (HBSS) and allow to cool
3. Tie double knots at one end of each tube
4. Half-fill each dialysis tube with serum (20 ml)
5. Express air and knot other end leaving a space between the serum and the knot of about half the tube
6. Place in 5 l of HBSS and stir on a magnetic stirrer overnight at 4° C
7. Change HBSS and repeat twice
8. Collect serum into measuring cylinder and note volume. (If volume is reduced, add HBSS to make up to starting volume. If increased, make due allowance when adding to medium later)
9. Sterilize through graded series of filters (see above)
10. Bottle and freeze

Preparation and Sterilization of Other Reagents

Individual recipes and procedures are given in the Appendix. On the whole most reagents are sterilized by filtration if heat labile and by autoclaving if stable (see summary Table 10.2).

Chapter 11
Disaggregation of the Tissue and Primary Culture

A primary cell culture may be obtained either by allowing cells to migrate out of fragments of tissue adhering to a suitable substrate or by disaggregating the tissue mechanically or enzymatically to produce a suspension of cells, some of which will ultimately attach to the substrate. It appears to be essential for most normal untransformed cells (with the exception of hemopoietic cells) to attach to a flat surface in order to survive and proliferate with maximum efficiency. Tumor cells, on the other hand, particularly cells from transplantable animal tumors, are often able to proliferate in suspension.

The enzymes used most frequently are crude preparations of trypsin, collagenase, elastase, hyaluronidase, DNase, pronase, dispase, or various combinations. Trypsin and pronase give the most complete disaggregation but may damage the cells. Collagenase and dispase give incomplete disaggregation but are less harmful. Hyaluronidase can be used in conjunction with collagenase, and DNase is employed to disperse DNA released from lysed cells as it tends to impair proteolysis and promote reaggregation.

Although each tissue may require a different set of conditions, certain common requirements are shared by most primary cultures.

(1) Fat and necrotic tissue are best removed during dissection.

(2) The tissue should be chopped finely with minimum damage.

(3) Enzymes used for disaggregation should be removed by gentle centrifugation afterward.

(4) The concentration of cells in the primary culture should be much higher than that normally used for subculture, since the proportion of cells from the tissue which survive primary culture may be quite low.

(5) A rich medium, such as Ham's F10 or F12, should be used in preference to a simple, basal medium, such as Eagle's BME; and if serum is required, fetal bovine often gives better survival than calf or horse.

(6) Embryonic tissue disaggregates more readily, yields more viable cells, and proliferates more rapidly in primary culture than adult.

ISOLATION OF THE TISSUE

An attempt should be made to sterilize the site of the dissection if necessary. Remove tissue aseptically and transfer to tissue culture laboratory in BSS or medium as soon as possible. If a delay is unavoidable, refrigerate the tissue. (Viable cells can be recovered from chilled tissue several days after explantation.)

Mouse Embryos

Outline

Remove uterus aseptically from timed pregnant mouse and dissect out embryos.

Materials

70% alcohol in wash bottle
bunsen burner
70% alcohol to sterilize instruments
sterile BSS in 50-ml sterile beaker to cool instruments after flaming
HBSS (with antibiotics if required) in 25–50 ml screw-capped vial or tube
timed pregnant mice

Protocol

1.
If males and females are housed separately, when they are put together for mating, estrous will be induced in the female 3 d later, when the maximum number of successful matings will occur. This enables the planned production of embryos at the appropriate time. The timing of successful matings

may be determined by examining the vaginas each morning for a hard mucous plug. The day of detection of the plug (the "plug date") is noted as day zero, and the development of the embryos timed from this date. Full term is about 19–21 d. The optimal age for preparing cultures from whole disaggregated embryo is around 13 d, when the embryo is relatively large (Figs. 11.1, 11.2) but still contains a high proportion of undifferentiated mesenchyme. It is from this mesenchyme that most of the culture will be derived.

Most individual organs, with the exception of brain and heart, begin to form about the 9th d of gestation but are difficult to isolate until about the 11th d. Dissection is easier by 13–14 d, and most of the organs are completely formed by the 18th d

2. Kill the mouse by cervical dislocation and swab the ventral surface liberally with 70% alcohol. Tear

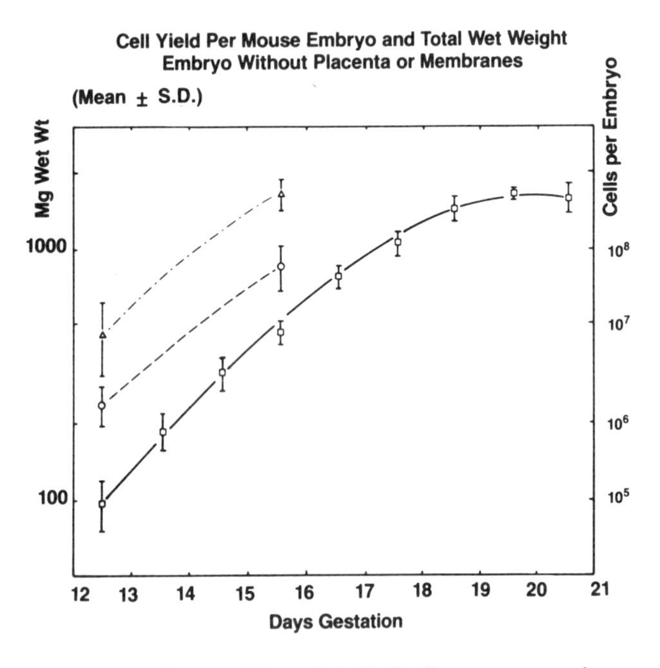

Fig. 11.1. *Total wet weight and yield of cells per mouse embryo. Total wet weight of embryo without placenta or membranes, mean ± standard deviation (squares) [From Paul et al., 1969]. Cell yield per embryo after incubation in 0.25% trypsin at 36.5°C for 4 hr (circles). Cell yield per embryo after soaking in 0.25% trypsin at 4°C for 5 hr and incubation at 36.5°C for 30 min (triangles; see text).*

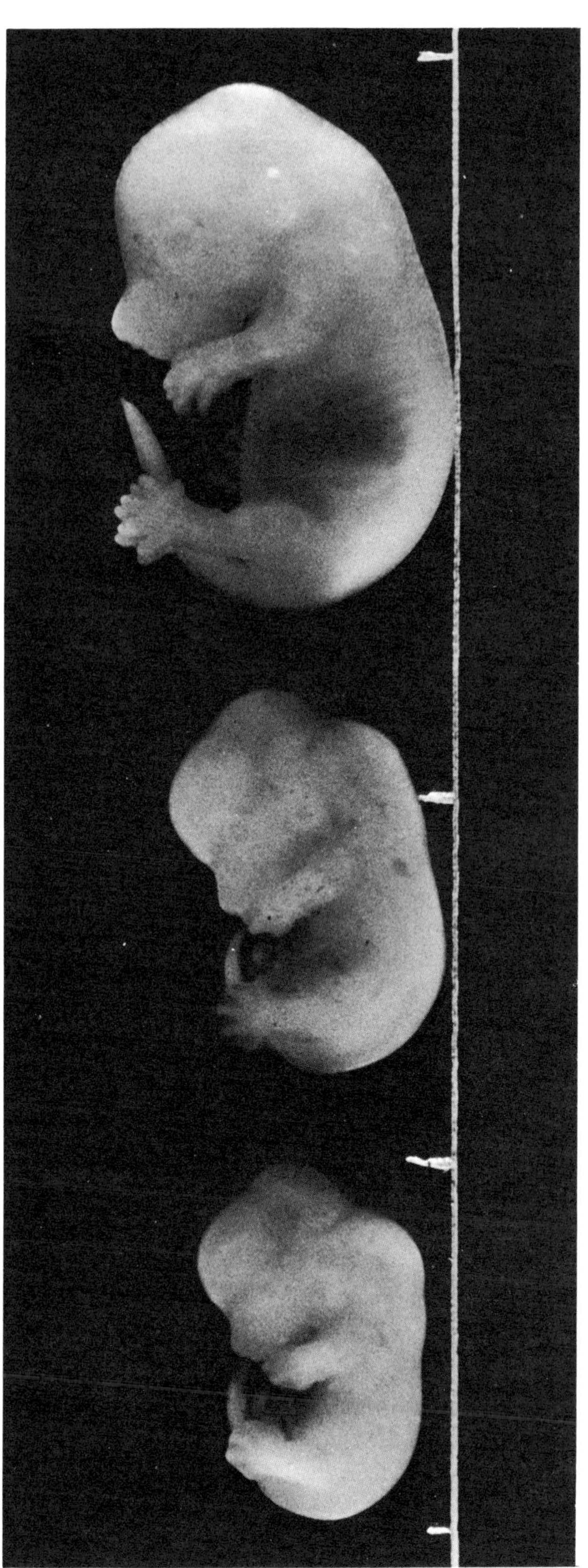

Fig. 11.2. *Mouse embryos from the 12th, 13th and 14th d of gestation. The 12-d embryo (bottom) came from a small litter (three) and is larger than would normally be found at this stage. Scale, 10 mm between marks.*

the ventral skin transversely at the median line just over the diaphragm and, grasping the skin on both sides of the tear, pull in opposite directions to expose the untouched ventral surface of the abdominal wall (Fig. 11.3a–c)

3.
Cut longitudinally along the median line with sterile scissors, revealing the viscera. At this stage, the uteri filled with embryos will be obvious posteriorly and may be dissected out into a 25-ml or 50-ml screw-capped tube containing 10 or 20 ml BSS (Fig. 11.3d–f). Antibiotics may be added to the BSS where there is a high risk of infection (see Appendix, Dissection BSS (DBSS).

Note. All of the preceding steps should be done outside the tissue culture laboratory; a small laminar flow hood and rapid technique will help to maintain sterility. Do not take live animals into the tissue culture laboratory; they may carry contamination. If the carcass must be handled in the tissue culture area, make sure it is immersed in alcohol briefly, or thoroughly swabbed, and disposed of quickly after use

4.
Take the intact uteri to the tissue culture laboratory and transfer to a fresh dish of sterile BSS

5.
Dissect out the embryos (Fig. 11.3g–i). Tear the uterus with two pairs of sterile forceps, keeping their points close together to avoid distorting the uterus and bringing too much pressure to bear on the embryos. As the uterus is torn apart, the embryos may be freed from the membranes and placenta and placed to one side of the dish to bleed, then transferred to a fresh dish. If a large number of embryos is required (more than four or five litters), it may be helpful to place the last dish on ice (for subsequent dissection and culture, see below)

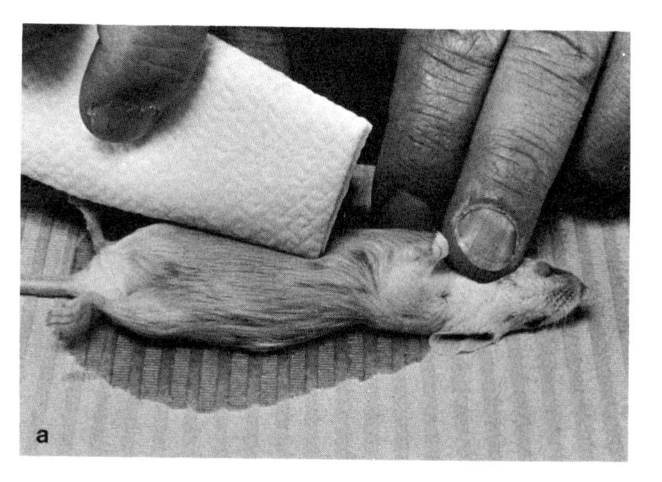

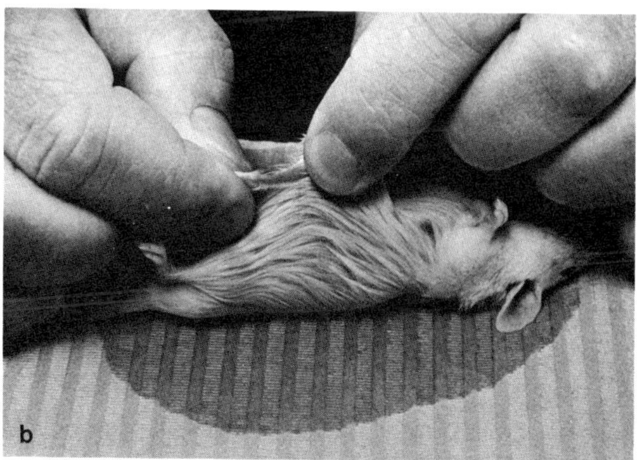

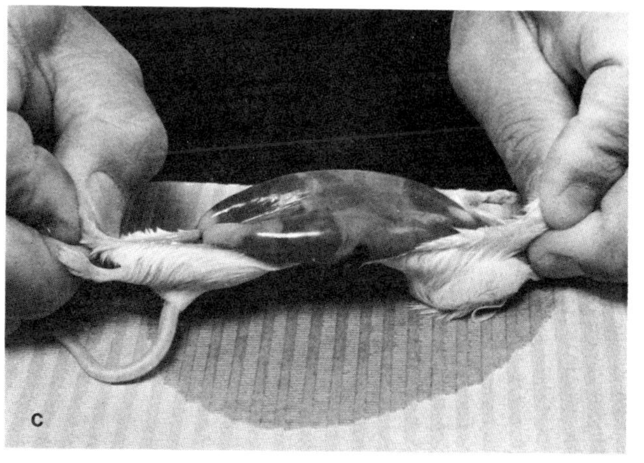

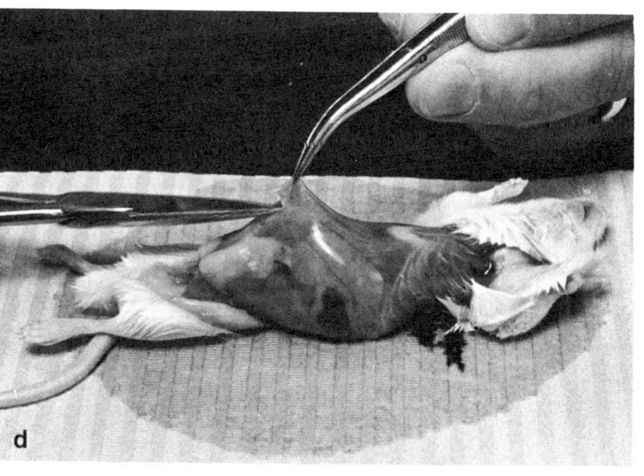

Hen's Egg

Outline

Remove embryo from egg and transfer to dish.

Fig. 11.3. *Stages in the aseptic removal of mouse embryos for primary culture (see text). a. Swabbing the abdomen. b,c. Tearing the skin to expose the abdominal wall. d. Opening the abdomen. e. Uterus* in situ. *f. Removing the uterus. g,h. Dissecting embryos from uterus. i. Removing membranes. j. Chopping embryos.*

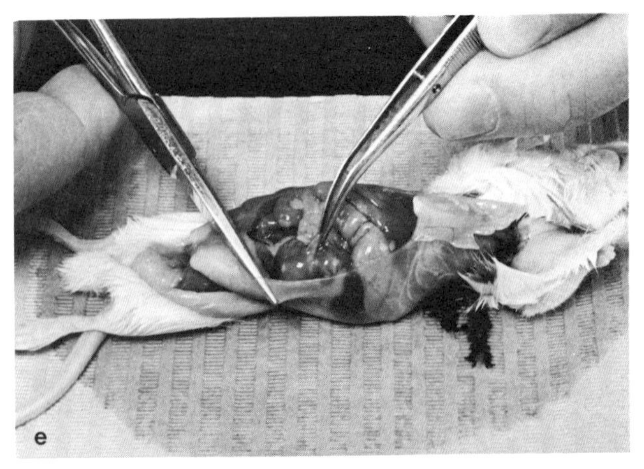

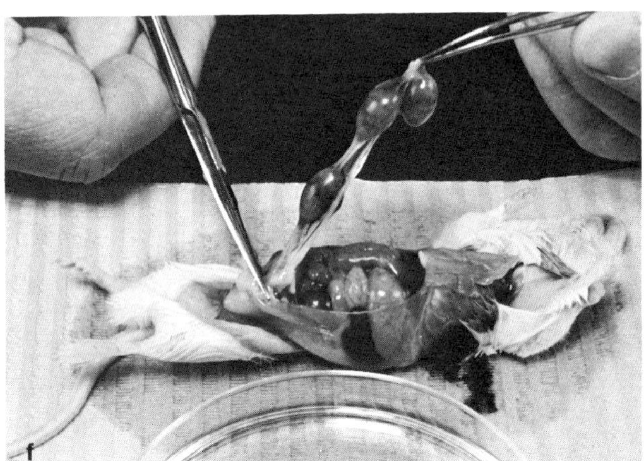

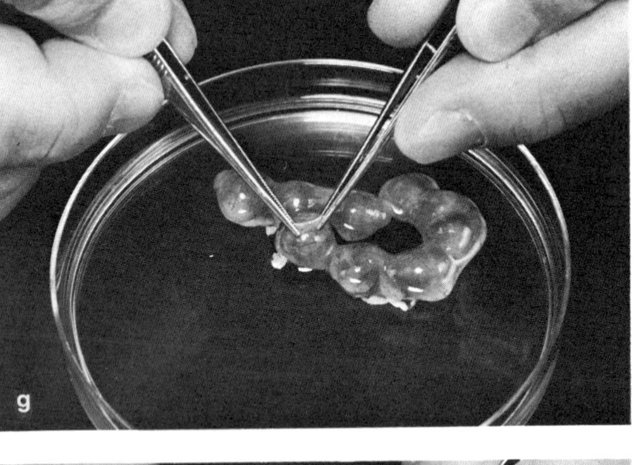

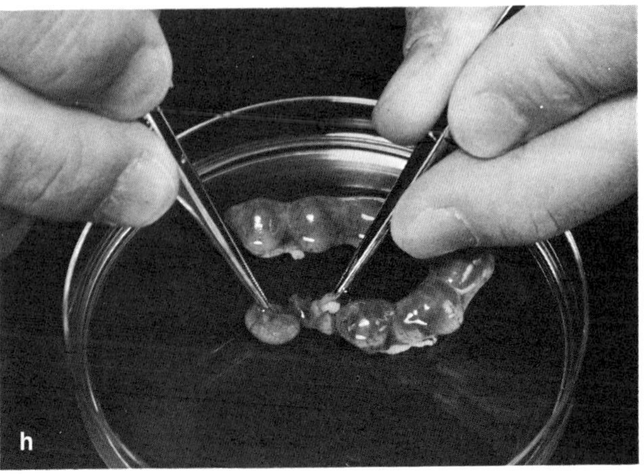

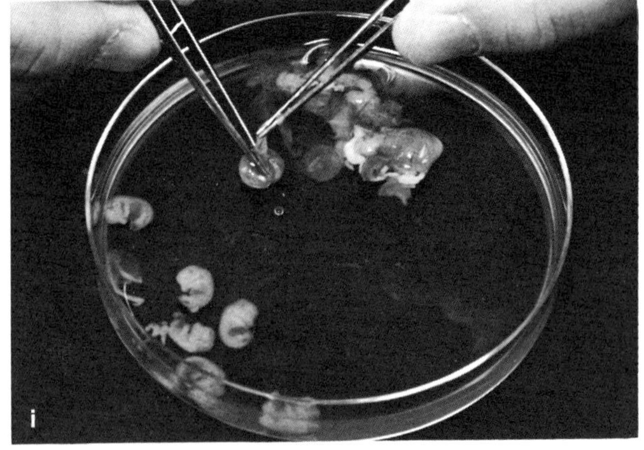

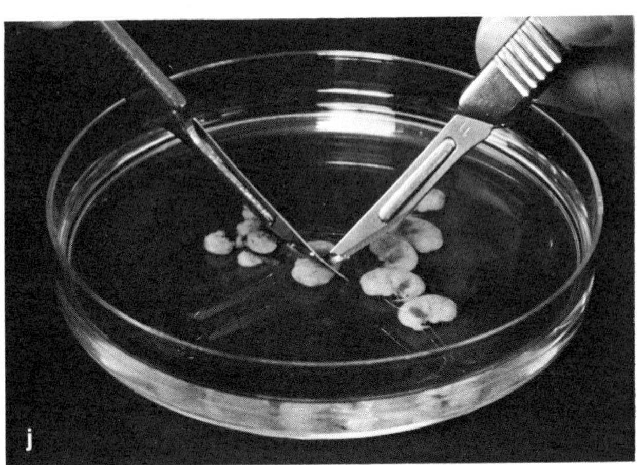

Figure 11.3. *(continued; see legend and text on page 101)*

Materials

70% alcohol
swabs
small beaker 20–50 ml
forceps—straight and curved
9-cm petri dishes
BSS
11-day embryonated eggs
humid incubator (no CO_2 above atmospheric)

Protocol

1\.
Incubate the eggs at 38.5° C in a humid atmosphere and turn through 180° daily. Although hen's eggs hatch at around 20 to 21 d, the lengths of the developmental stages are different from the mouse. For culture of dispersed cells from the whole embryo, the egg should be taken at about 8 d, from isolated organ rudiments, at about 10 to 13 d

2\.
Swab the egg with 70% alcohol and place with blunt end uppermost in a small beaker (Fig. 11.4a)

Fig. 11.4. *Stages in the aseptic removal of chick embryo from the egg (see text).*

3.
Crack the top of the shell and peel off to the edge of the air sac with sterile forceps (Fig. 11.4b)
4.
Resterilize the forceps (dip in a beaker of alcohol, burn off alcohol and cool in sterile BSS, and peel off the white shell membrane to reveal the chorioallantoic membrane (CAM) below, with its blood vessels (Fig. 11.4c,d)
5.
Pierce the CAM with sterile curved forceps and lift out the embryo by grasping gently under the head. Do not close the forceps completely or the neck will sever (Fig. 11.4e–g)
6.
Transfer embryo to a 9-cm petri dish containing 20 ml DBSS. (For subsequent dissection and culture, see below.)

Human Biopsy Material

There are four main problems in collecting human biopsy material. The first two relate to the common disadvantages in handling most human material; i.e., that the tissue sample is usually taken for another reason, e.g., to alleviate a particular problem in the patient, and to provide material for the pathologist to aid diagnosis. Your collection of samples must not interfere in any way with patient care; whatever your needs, the needs of the pathologist must be met first. So consult the surgeon in charge, the head operating room nurse, and the pathologist as to the best method of collecting samples without inconveniencing them or rendering the sample useless.

Outline

Provide labeled container(s) of medium, consult with hospital staff, and collect sample from operating room or pathologist.

Materials

specimen tubes (15–30 ml) with leakproof caps about one-half full with culture medium containing antibiotics (see Appendix, Collection Medium), and labeled with your name, address, and telephone number.

Protocol

1.
Provide a container of collection medium clearly labeled, and either arrange to be there to collect it after surgery, or have your name and address on it and, preferably, your telephone number, so that it can be sent to you immediately and you can be informed easily when it has been dispatched
2.
Transfer sample to tissue culture laboratory. Usually, if kept at 4°C, biopsy samples will survive for at least 24 hr and even up to 3 or 4 d, although the longer the time from surgery to culture, the more the deterioration that may be expected
3.
Decontamination. A disinfectant wash is given before skin biopsy, and an oral antibiotic before gut surgery. Most surgical specimens, however, from the needs of surgery, are sterile when removed though problems may arise with subsequent handling. Superficial (skin biopsies, melanomas, etc.) and gastrointestinal tract (colon and rectal samples) are particularly prone to infection. It may be advantageous to consult a medical microbiologist to determine what flora to expect in a given tissue and choose your antibiotics accordingly. If the surgical sample is large enough, (200 mg or more), a brief dip (30 s–1 min) in 70% alcohol will help to reduce superficial contamination without causing much harm to the center of the tissue sample

PRIMARY CULTURE

Primary Explant Technique

This was the original method developed by Harrison [1907], Carrel [1912], and others for initiating a tissue culture. A fragment of tissue was imbedded in blood plasma or lymph, mixed with embryo extract and serum, and placed on a slide or coverslip. The plasma clotted and held the tissue in place. The embryo extract plus serum both supplied nutrients and stimulated migration out of the explant across the solid substrate. Heterologous serum was often used to promote clotting of the plasma. This technique is still used but has largely been replaced by the simplified method below.

Outline

The tissue is chopped finely, rinsed, and the pieces seeded onto the culture surface in a small volume of medium with a high concentration (40–50%) of serum, such that surface tension holds the pieces in place until they adhere spontaneously to the surface (Fig. 11.5). Once this is achieved, outgrowth of cells usually follows (Fig. 11.6).

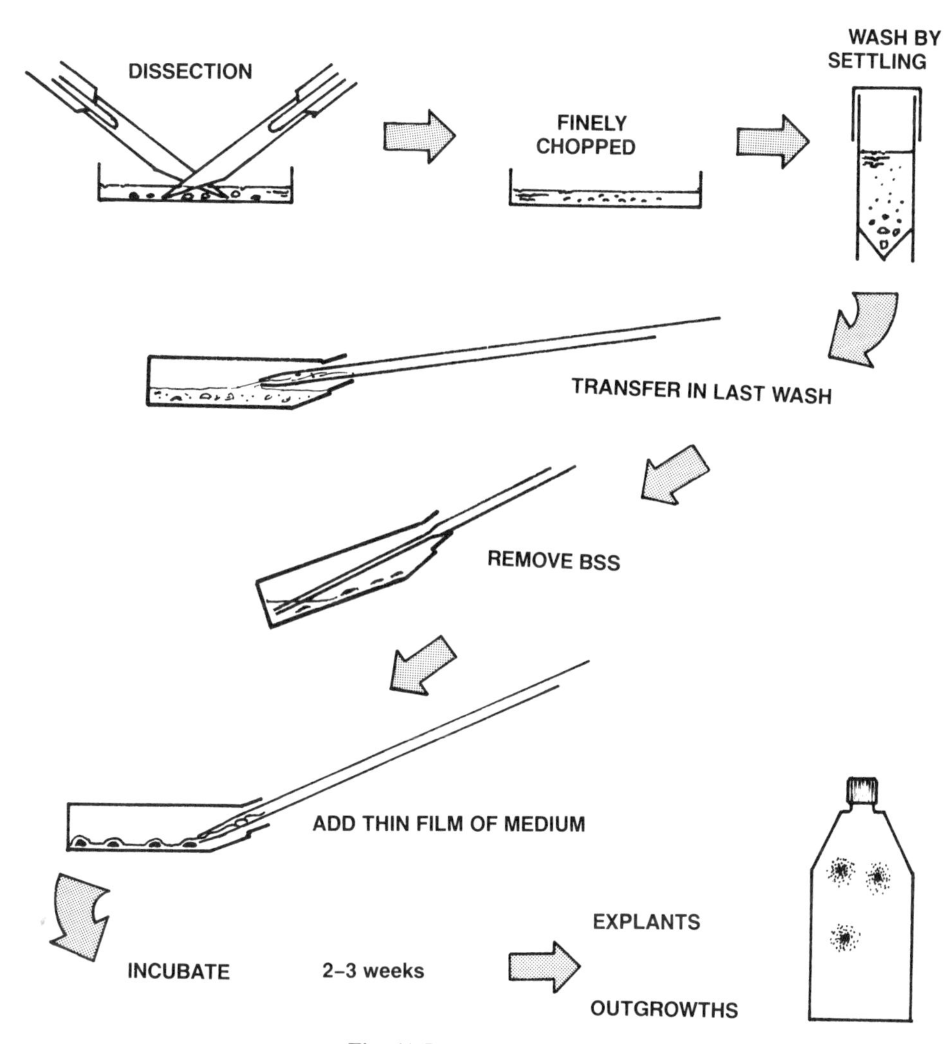

Fig. 11.5. *Primary explant technique.*

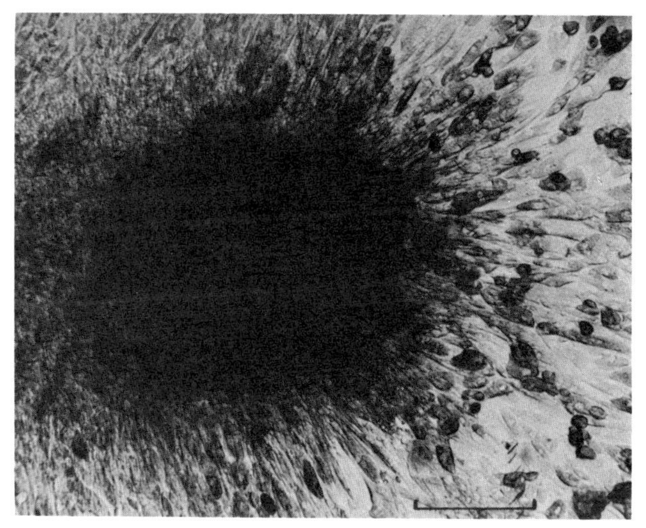

Fig. 11.6. *Primary explant from human mammary carcinoma. Dense area in center is an undisaggregated tissue fragment and cells are seen migrating radially from the explant. This is a good size for an explant, promoting maximal radial outgrowth, but routinely it is difficult to dissect below 0.5 mm. Scale bar 100 μm.*

Materials

- petri dishes
- 100 ml BSS
- forceps
- scalpels
- 10-ml pipettes
- 15- or 20-ml centrifuge tubes or universal containers
- culture flasks
- growth medium

The size of flasks and volume of growth medium will depend on the amount of tissue—roughly five 25-cm^2 flasks per 100 mg tissue, and initially 1 ml medium per flask, building up to 5 ml per flask over the first 3–5 d.

Protocol

1.

Transfer tissue to fresh sterile BSS and rinse

2. Transfer to a second dish, dissect off unwanted tissue such as fat or necrotic material, and chop finely with crossed scalpels (Fig 11.5) to about 1-mm cubes
3. Transfer by pipette to a 15- or 50-ml sterile centrifuge tube or universal container (wet the inside of the pipette first or the pieces will stick). Allow the pieces to settle
4. Wash by resuspending the pieces in BSS, allowing the pieces to settle, and removing the supernatant fluid, two or three times
5. Transfer the pieces (remember to wet the pipette) to a culture flask, about 20–30 pieces per 25-cm^2 flask
6. Remove most of the fluid and add about 1 ml growth medium per 25-cm^2 growth surface
7. Tilt the flask gently to spread the pieces evenly over the growth surface
8. Cap the flask and place in incubator or hot room at 36.5°C for 18–24 hr
9. If the pieces have adhered, the medium volume may be made up gradually over the next 3–5 d to 5 ml per 25 cm^2 and then changed weekly until a substantial outgrowth of cells is observed (see Fig. 11.6)
10. The explants may then be picked off from the center of the outgrowth with a scalpel and transferred by prewetted pipette to a fresh culture vessel. (return to step 7 above)
11. Replace medium in the first flask until the outgrowth has spread to cover at least 50% of the growth surface, at which point the cells may be passaged (see Chapter 12)

This technique is particularly useful for small amounts of tissue such as skin biopsies where there is a risk of losing cells during mechanical or enzymatic disaggregation. Its disadvantages lie in the poor adhesiveness of some tissues and the selection of cells in the outgrowth. In practice, however, most cells, fibroblasts, myoblasts, glia, epithelium, particularly from the embryo, will migrate out successfully.

Both adherence and migration may be stimulated by a plasma clot:

1. Place the tissue pieces in position in the culture flask as in step 7 above
2. Mix 2 parts of chicken plasma with 1 part chicken embryo extract and 1 part fetal bovine serum. Immediately pipette gently over the tissue pieces, spacing the pieces evenly on the surface of the dish as you do so
3. Allow to clot and place at 37°C [see Paul, 1975, for further description]

Enzymatic Disaggregation

Mechanical and enzymatic disaggregation of the tissue avoids problems of selection by migration, but more important perhaps, yields a higher number of cells, more representative of the whole tissue, in a shorter time. Just as the primary explant technique selects on the basis of cell migration, dissociation techniques select cells resistant to the method of disaggregation and still capable of attachment.

Embryonic tissue disperses more readily and gives a high yield of proliferating cells than newborn or adult. The increasing difficulty in obtaining viable proliferating cells with increasing age is due to several factors including the onset of differentiation, an increase in fibrous connective tissue, and a reduction of the undifferentiated proliferating cell pool. Where procedures of greater severity are required to disaggregate the tissue, e.g., longer trypsinization or increased agitation, the more fragile components of the tissue may be destroyed. In fibrous tumors, for example, it is very difficult to obtain complete dissociation with trypsin while still retaining viable carcinoma cells.

Crude trypsin is by far the most common enzyme used in tissue disaggregation [Waymouth, 1974] as it is tolerated quite well by many cells, is effective for many tissues, and any residual activity left after washing is neutralized by the serum of the culture medium. A trypsin inhibitor (e.g., soya bean trypsin inhibitor, Sigma) must be included when serum-free medium is used.

It is important to minimize the exposure of cells to active trypsin to preserve maximum viability. Hence when trypsinizing whole tissue at 36.5°C dissociated cells should be collected every half hour, the trypsin removed by centrifugation and neutralized with serum in medium. Soaking the tissue for 6–18 hr in trypsin at 4°C (see below) allows penetration with minimal tryptic activity, and digestion may then proceed for a much shorter time (20–30 min) at 37°C [Cole and Paul, 1966]. Although the cold trypsin method gives a higher yield of viable cells and is less effort, the warm trypsin method is still used extensively and is presented here for comparison.

Disaggregation in Warm Trypsin

Outline

The tissue is chopped and stirred in trypsin for a few hours, collecting dissociated cells every half hour, and the dissociated cells are then centrifuged and pooled in medium containing serum (Fig. 11.7).

Materials

250-ml Erlenmeyer flask
magnetic bar
magnetic stirrer
100 ml 0.25% trypsin (crude: Difco 1:250, Flow or Gibco) in CMF or PBSA (see Appendix)
two 50-ml centrifuge tubes
hemocytometer or cell counter
growth medium with serum (e.g., Ham's F12 with 10% fetal bovine serum)
culture flasks, 5–10 g tissue (will vary depending on cellularity of tissue)

Protocol

1. Proceed as for primary explant, although the pieces need only to be chopped to about 3-mm diameter. As described here, this method requires about 20 times as much tissue as the primary explant technique, although it can be scaled down if desired
2. After washing as in step 4 for primary explantation, transfer the chopped pieces to the trypsinization flask (Bellco makes one specially designed for the purpose, Fig 11.8, but a 250-ml conical Erlenmeyer flask will do), and add 100 ml trypsin
3. Stir at about 200 rpm for 30 min at 36.5°C
4. Allow pieces to settle, collect supernatant, centrifuge at approximately 500 *g* for 5 min, resuspend pellet in 10 ml medium with serum, and store cells on ice
5. Add fresh trypsin to pieces and continue to stir and incubate for a further 30 min. Repeat steps 3 to 5 until complete disaggregation occurs or until no further disaggregation is apparent
6. Collect and pool chilled cell suspensions, count by hemocytometer or electronic cell counter (see Chapter 19)
Remember you are dealing with a very heterogeneous population of cells; electronic cell counting will initially require confirmation with a hemocytometer, as a "plateau" (see Chapter 19) is rather difficult to obtain.
7. Dilute to 10^6 per ml in growth medium and seed as many flasks as are required with approximately 2×10^5 cells per cm^2. Where the survival rate is unknown or unpredictable, a cell count is of little value (e.g., tumor biopsies where the proportion of necrotic cells may be quite high). In this case, set up a range of concentrations from about 5 to 25 mg tissue per ml. Change the medium at regular intervals (2 to 4 d as dictated by depression of pH)

This technique is useful for the dissociation of large amounts of tissue in a relatively short time, particularly whole mouse embryo or chick embryo. It does not work as well with adult tissue where there is a lot of fibrous connective tissue, and mechanical agitation can be damaging to some of the more sensitive cell types such as epithelium.

Trypsin at 4°C. One of the disadvantages of using trypsin to disaggregate tissue is the damage that may result from prolonged exposure to trypsin at 36.5°C—hence, the need to harvest cells after 30-min incubations in the warm trypsin method, rather than have them exposed for the full time (3 to 4 hr) required to disaggregate the whole tissue. A simple method of minimizing damage to the cells during disaggregation is to soak the tissue in trypsin at 4°C to allow penetration of the enzyme with little tryptic activity (Table 11.1). Following this, the tissue will require much shorter incubation at 36.5°C for disaggregation [Cole and Paul, 1966].

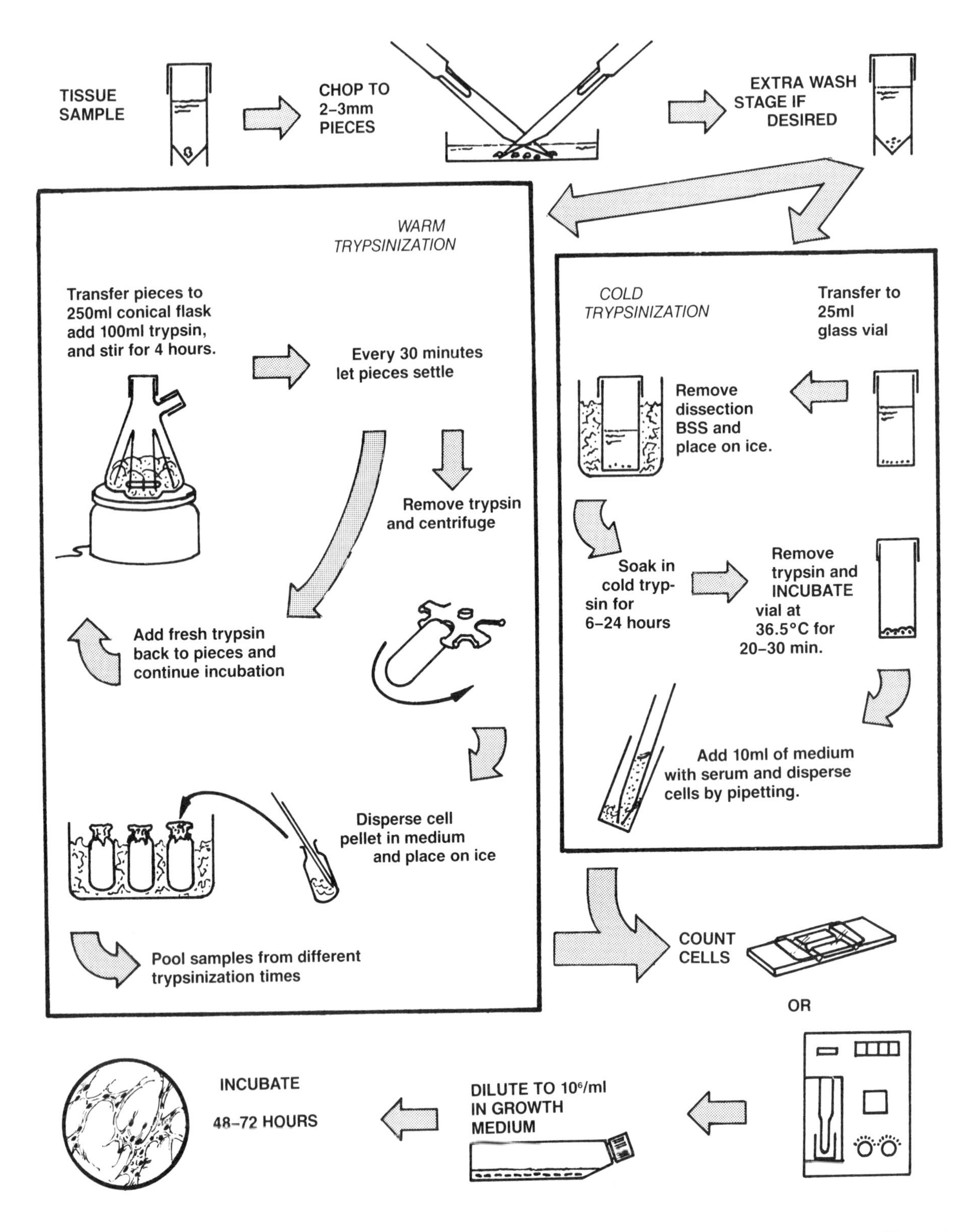

Fig. 11.7. *Preparation of primary culture by disaggregation in trypsin. Warm trypsin method on the left, cold trypsin method on the right.*

TABLE 11.1. Relative Cell Yield From 12-d Mouse Embryos by Warm or Cold Trypsin Methods

Trypsin temperature and duration of exposure (hr)		Total cell number recovered per embryo ($\times 10^{-7}$)	Viable cells per embryo		Percentage of total recovered after 24 hr in culture	Percentage of viable cells recovered
4°C	36.5°C		No. ($\times 10^{-7}$)	% Viability by dye exclusion (Trypan blue)		
—	4*	1.69	1.45	86	47.2	54.9
5.5	0.5†	3.32	1.99	60	74.5	124
24	0.5†	3.40	2.55	75	60.3	80.2

*Stirred continuously at 36.5°C; fractions were *not* collected at 30-min intervals.
†Incubated without agitation.

Outline

Chop tissue and place in trypsin at 4°C for 6 to 18 hr. Incubate after removing the trypsin, and disperse the cells in warm medium (Fig. 11.7).

Materials

- petri dish
- BSS
- forceps—straight and curved
- 0.25% crude trypsin
- scalpels
- 25-ml screw-capped Erlenmeyer flask(s), 25-cm^2 or 75-cm^2 culture flasks
- pipettes
- culture medium

Protocol

1. Follow steps 1–4 as for primary explants, but collect tissue in glass tube or vial to facilitate chilling (see below). Tissue need only be chopped to 3–4-mm pieces. Embryonic organs, if they do not exceed this size, are better left whole
2. After washing, place the container on ice, remove the last BSS wash, and replace with 0.25% trypsin in CMF at 4°C (approximately 1 ml for every 100 mg of tissue)
3. Place at 4°C for 6–18 hr
4. Remove and discard the trypsin carefully, leaving the tissue with only the residual trypsin
5. Place tube at 36.5°C for 20–30 min
6. Add warm medium, approximately 1 ml for every 100 mg, and gently pipette up and down until the tissue is completely dispersed

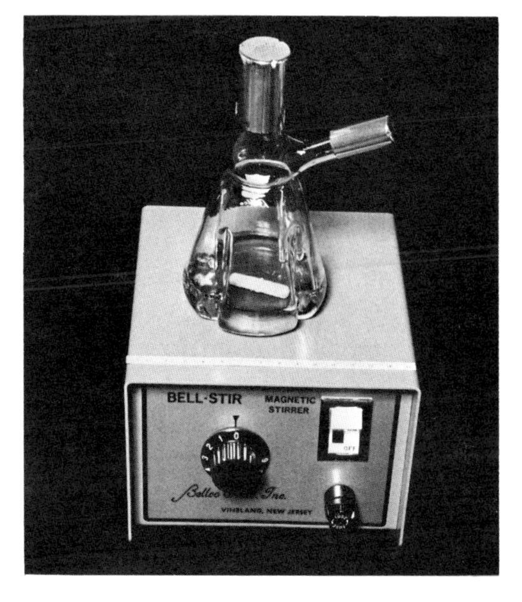

Fig. 11.8. *Trypsinization flask (Bellco). The indentations in the side of the flask improve mixing, and the rim around the neck, below the side arm, allows the cell suspension to be poured off while leaving the stirrer bar and any larger fragments behind.*

7. If some tissue is left undispersed, the cell suspension may be filtered through sterile muslin or stainless steel mesh (100–200 μm), or the larger pieces may simply be allowed to settle. Where there is a lot of tissue, increasing the volume of suspending medium to 20 ml for each gram of tissue will facilitate settling and subsequent collection of supernatant fluid. Two to 3 min should be sufficient to get rid of most of the larger pieces
8. Determine the concentration of the cell suspension and seed the vessels at 10^6 cells per ml (2×10^5 cells per cm^2

The cold trypsin method usually gives a higher yield of cells (see Fig. 11.1) and preserves more different cell types than the warm method. Cultures from mouse embryos contain more epithelial cells when prepared by the cold method, and erythroid cultures from 13-d fetal mouse liver respond to erythropoietin after this treatment but not after warm trypsin or mechanical disaggregation [Cole and Paul, 1966; Conkie, personal communication]. The method is also convenient as no stirring or centrifugation is required, and the incubation at 4°C may be done overnight. This method does take longer, however, and is not as convenient where large amounts of tissue (greater than 10 g) are being handled. A particular advantage in the cold trypsin method is in the handling of small amounts of tissue, such as embryonic organs. Taking 10–13-day chick embryo as a starting point, the following procedure gives good reproducible cultures with evidence of several different cell types characteristic of the tissue of origin.

Chick Embryo Organ Rudiments

Outline

Dissect out individual organs or tissues, and place, preferably whole, in cold trypsin overnight. Remove the trypsin, incubate briefly, and disperse in culture medium. Dilute and seed cultures.

Materials

petri dish
BSS
scalpels (No. 11 blade for most steps)
iridectomy knives (Beaver, blade 21)
curved and straight fine forceps
pipettes (Pasteur, 2 ml, 10 ml)
0.25% crude trypsin in CMF or PBS ice
10–15-ml test tubes with screw caps
25-cm² culture flasks
culture medium (e.g., Ham's F12 + 10% fetal bovine serum)

Protocol

1. Remove the embryo from the egg as described above and place in sterile BSS
2. Remove the head (Fig. 11.9a,b)
3. Remove an eye and open carefully, releasing the lens and aqueous and vitreous humors (Fig. 11.9c,d)
4. Grasp the retina in two pairs of fine forceps and gently peel the pigmented retina off the neural retina and connective tissue (Fig. 11.9e). (A brief exposure to 0.25% trypsin in 1 m*M* EDTA will separate the two tissues more easily.) Put tissue to one side
5. Pierce the top of the head with curved forceps and scoop out the brain (Fig. 11.9f)
6. Halve the trunk transversely where the pink color of the liver shows through the ventral skin (Fig. 11.9g). If the incision is made on the line of the diaphragm, it will pass between the heart and the liver; but sometimes the liver will go to the anterior instead of the posterior half
7. Gently probe into the cut surface of the anterior half and draw out the heart and lungs (Fig. 11.9h; tease the organs out and do not cut until you have identified them). Separate
8. Probe the posterior half, and draw out the liver with the folds of the gut enclosed in between the lobes (Fig. 11.9i). Separate
9. Fold back the body wall to expose the inside of the dorsal surface of the body cavity in the posterior half. The elongated lobulated kidneys should be visibly parallel to and on either side of the midline
10. Gently slide the tip of the scalpel under each kidney and tease away from the dorsal body wall (Fig. 11.9j). Carefully cut free and place on one side
11. Place the tips of the scalpels together on the midline at the posterior end and, advancing the tips forward, one over the other, express the spinal cord like toothpaste from a tube (Fig. 11.9k)
12. Turn the posterior trunk of the embryo over and strip the skin off the back and upper part of the legs (Fig. 11.9l). Collect and place on one side
13. Dissect off muscle from each thigh and collect together (Fig. 11.9m)

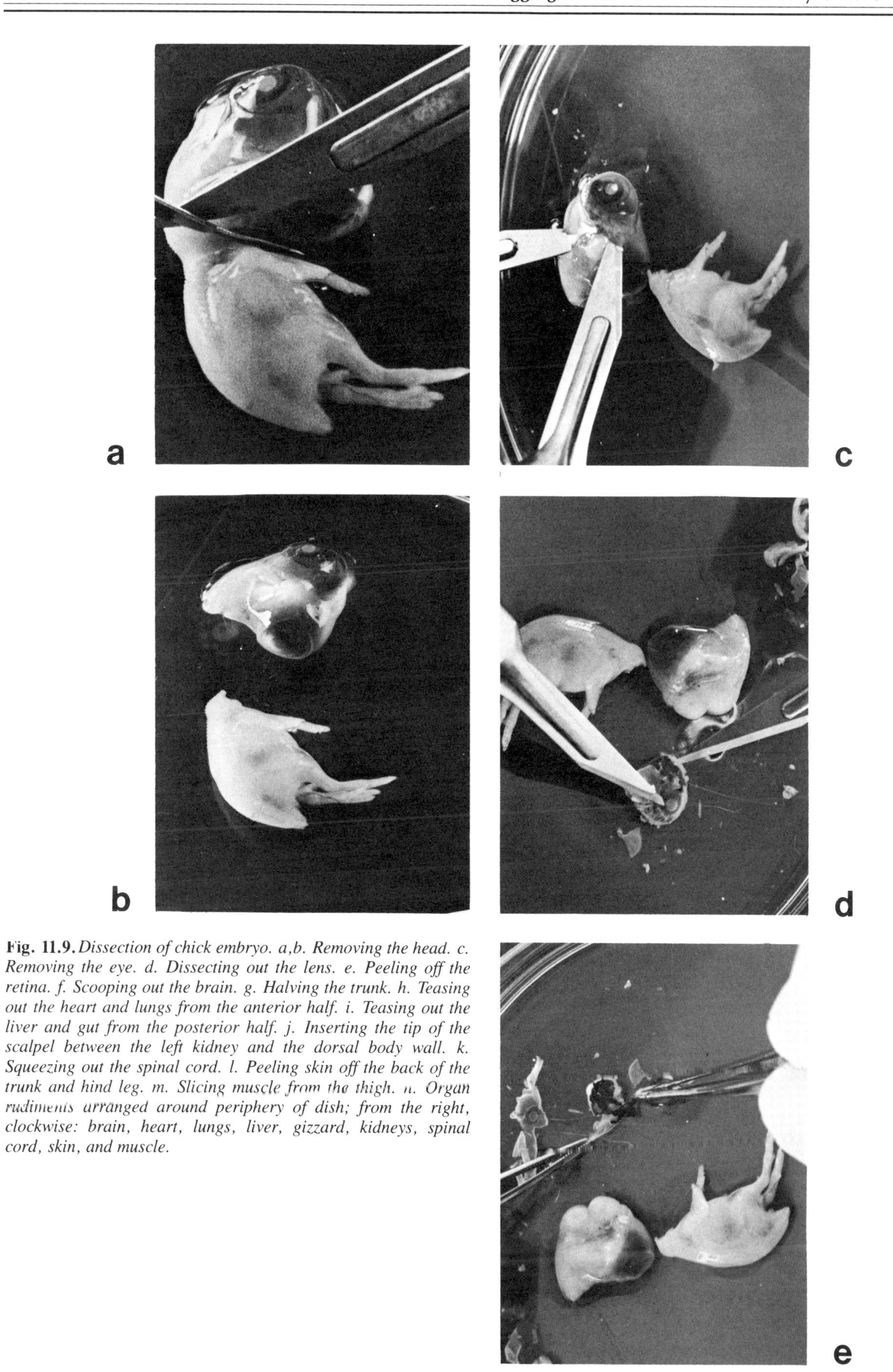

Fig. 11.9. *Dissection of chick embryo. a,b. Removing the head. c. Removing the eye. d. Dissecting out the lens. e. Peeling off the retina. f. Scooping out the brain. g. Halving the trunk. h. Teasing out the heart and lungs from the anterior half. i. Teasing out the liver and gut from the posterior half. j. Inserting the tip of the scalpel between the left kidney and the dorsal body wall. k. Squeezing out the spinal cord. l. Peeling skin off the back of the trunk and hind leg. m. Slicing muscle from the thigh. n. Organ rudiments arranged around periphery of dish; from the right, clockwise: brain, heart, lungs, liver, gizzard, kidneys, spinal cord, skin, and muscle.*

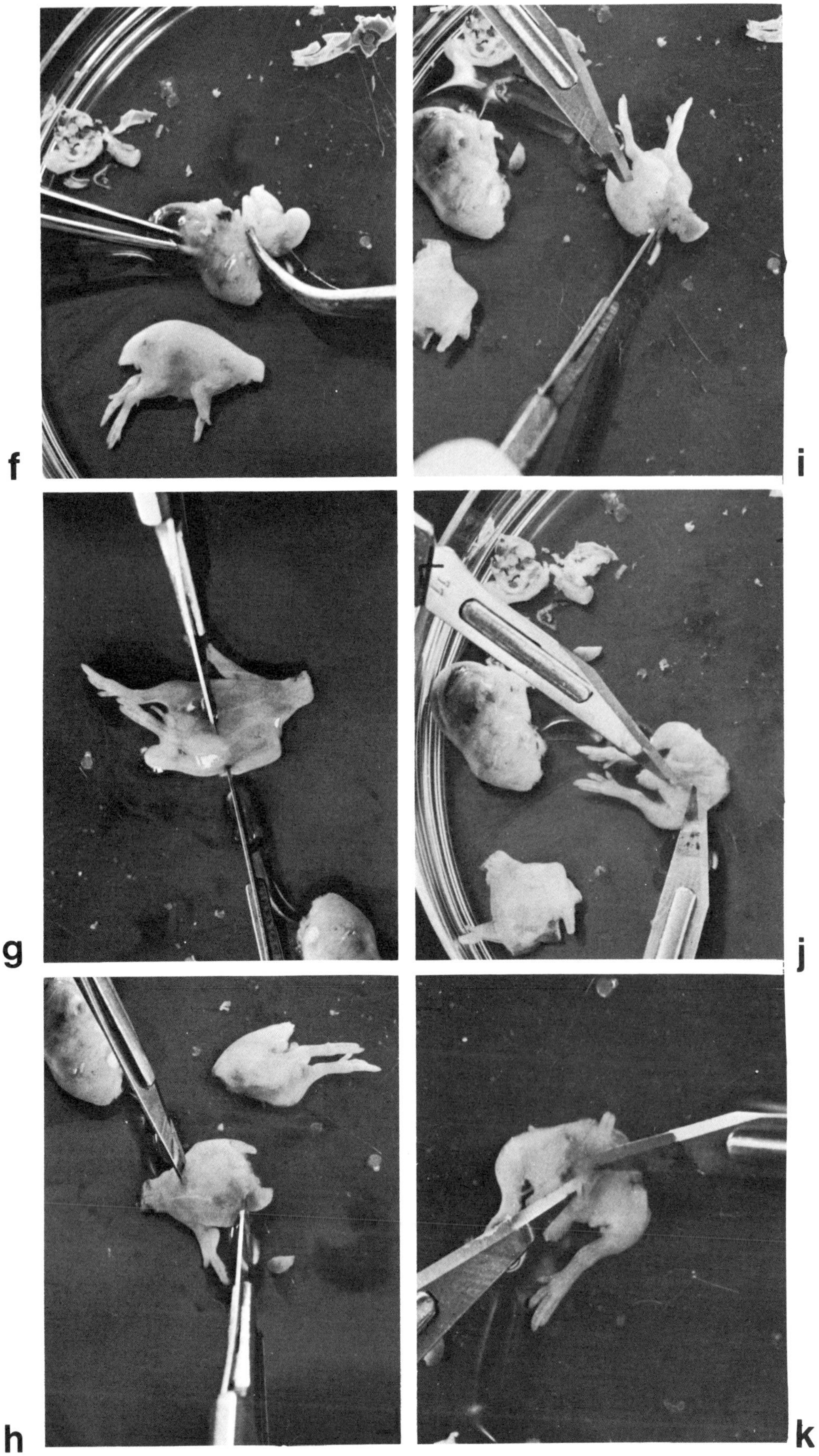

f

i

g

j

h

k

Figure 11.9 *(continued)*

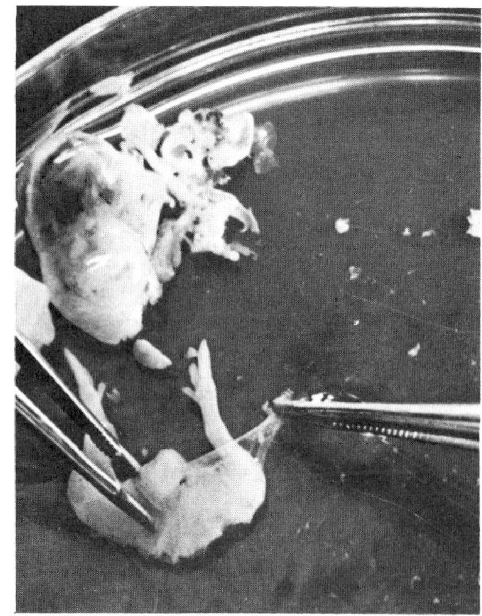
l

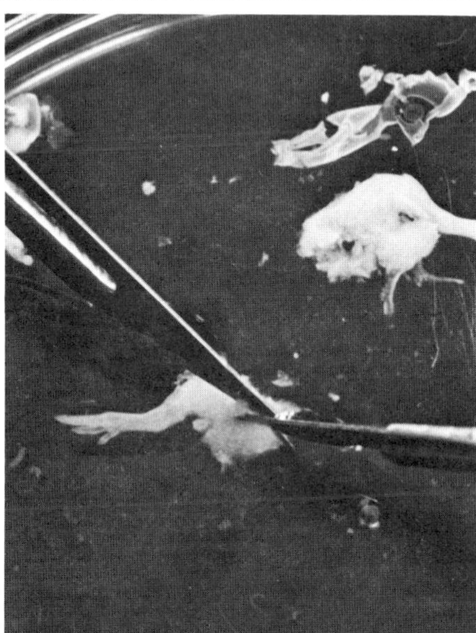
m

n

Figure 11.9 *(continued)*

14.
Transfer all of these tissues, and any others you may want, to separate test tubes containing 1 ml of 0.25% trypsin in CMF and place on ice. Make sure the tissue slides right down the tube into the trypsin
15.
Leave 6–18 hr at 4°C
16.
Remove trypsin carefully; tilting and rolling the tube slowly will help
17.
Incubate in the residual trypsin for 15–20 min at 36.5°C
18.
Add 4 ml medium to each of two 25-cm^2 flasks for each tissue to be cultured
19.
Add 2 ml medium to tubes after step 18, and pipette up and down gently to disperse the tissue
20.
Allow any large pieces to settle, pipette off supernatant fluid into the first flask, mix, and transfer 1 ml of diluted suspension to second flask. This gives two flasks at different cell concentrations and avoids the need to count the cells. Experience will determine the appropriate cell concentration to use in subsequent attempts
21.
Change the medium as required (e.g., with brain it may need to be changed after 24 hr, but pigmented retina will probably last 5–7 d), and check for characteristic morphology and function. After 3 to 5 d, contracting cells may be seen in the heart cultures, colonies of pigmented cells in the pigmented retina culture, and the beginning of myotubes in skeletal muscle cultures

Other Enzymatic Procedures

Disaggregation in trypsin can be damaging (e.g., to some epithelial cells) or ineffective (e.g., for very fibrous tissue such as fibrous connective tissue), so attempts have been made to utilize other enzymes. Since the intracellular matrix often contains collagen, particularly in connective tissue and muscle, collagenase has been the obvious choice [Coon and Cahn, 1966; Lasfargues, 1973; Freshney, 1972]. Other bacterial proteases such as pronase [Wiepjes and Prop, 1970; Prop and Wiepjes, 1973; Gwatkin, 1973] and dispase [Matumura, et al., 1975]; (Boehringer-Mannheim Biochemicals) have also been used with varying degrees of success. The participation of carbohydrate in intracellular adhesion has led to the use of hyaluronidase [Berry and Friend, 1969] and neuraminidase in conjunction with collagenase. It is not possible here to describe all the primary disaggregation techniques that have been used, but the following are useful examples.

Collagenase Alone

This technique is very simple and effective for embryonic and normal and malignant adult tissue. It is of greatest benefit where the tissue is either too fibrous or too sensitive to allow the successful use of trypsin [Freshney, 1972]. Crude collagenase is often used and may depend for some of its action on contamination with other nonspecific proteases. More highly purified grades are available if nonspecific proteolytic activity is undesirable.

Outline

Place finely chopped tissue in complete medium containing collagenase and incubate. When tissue is disaggregated, remove collagenase by centrifugation, seed cells at a high concentration, and culture (Fig. 11.10)

Materials

- pipettes
- 25-cm^2 culture flasks
- growth medium
- collagenase (2,000 units/ml)
- centrifuge tubes
- centrifuge

Protocol

1. Proceed as for primary explant up to step five, but transfer 20–30 pieces to one 25-cm^2 flask and 100 to 200 pieces to a second
2. Drain off BSS and add 4.5 ml growth medium with serum to each flask
3. Add 0.5 ml crude collagenase, 2,000 units/ml (Worthington CLS or Sigma 1A grade), to give a final concentration of 200 units/ml collagenase
4. Incubate at 36.5 °C for 4 to 48 hr without agitation. Tumor tissue may be left up to 5 d or more if disaggregation is slow, although it may be necessary to centrifuge the tissue and resuspend in fresh medium and collagenase before then if an excessive drop in pH is observed (to less than pH 6.5)
5. Check for effective disaggregation by gently moving the flask; the pieces of tissue will "smear" on the bottom of the flask and, with moderate agitation, will break-up into single cells and small clusters (Fig. 11.11). With others, small clusters of epithelial cells can be seen to resist the collagenase and may be separated from the rest by allowing them to settle for about 2 min. If these clusters are further washed with BSS by resuspension and settling and the sediment seeded in medium, they will form healthy islands of epithelial cells. Epithelial cells generally survive better if not completely dissociated
6. Where complete disaggregation has occurred, or when the supernatant cells are collected after allowing clusters to settle, centrifuge at 50–100 *g* for 3 min. Discard supernatant DBSS, resuspend, and combine pellets in 5 ml medium, and seed in a 25-cm^2 flask. If the pH fell during collagenase treatment (to pH 6.5 or less by 48 hr), dilute two- to threefold in medium after removing the collagenase.
7. Replace medium after 48 hr

Some cells, particularly macrophages, may adhere to the first flask during the collagenase incubation. Transferring the cells to a fresh flask after collagenase treatment (and removal) removes the macrophages from the culture. The first flask may be cultured as well if required. Trypsinization will remove any adherent cells other than macrophages.

Disaggregation in collagenase has proved particularly suitable for the culture of human tumors, mouse kidney, human adult and fetal brain, and many other tissues, particularly epithelium. It is gentle, requires no mechanical agitation or special equipment. With more than 1 g of tissue, however, it becomes tedious at the dissection stage and can be expensive due to the amount of collagenase required. It will also release most of the connective tissue cells, accentuating the problem of fibroblastic overgrowth; so it may require to be followed by selective culture or cell separation (see Chapters 13,14).

Many epithelial tissues (e.g., kidney tubules, clusters of carcinoma cells of breast and gastrointestinal

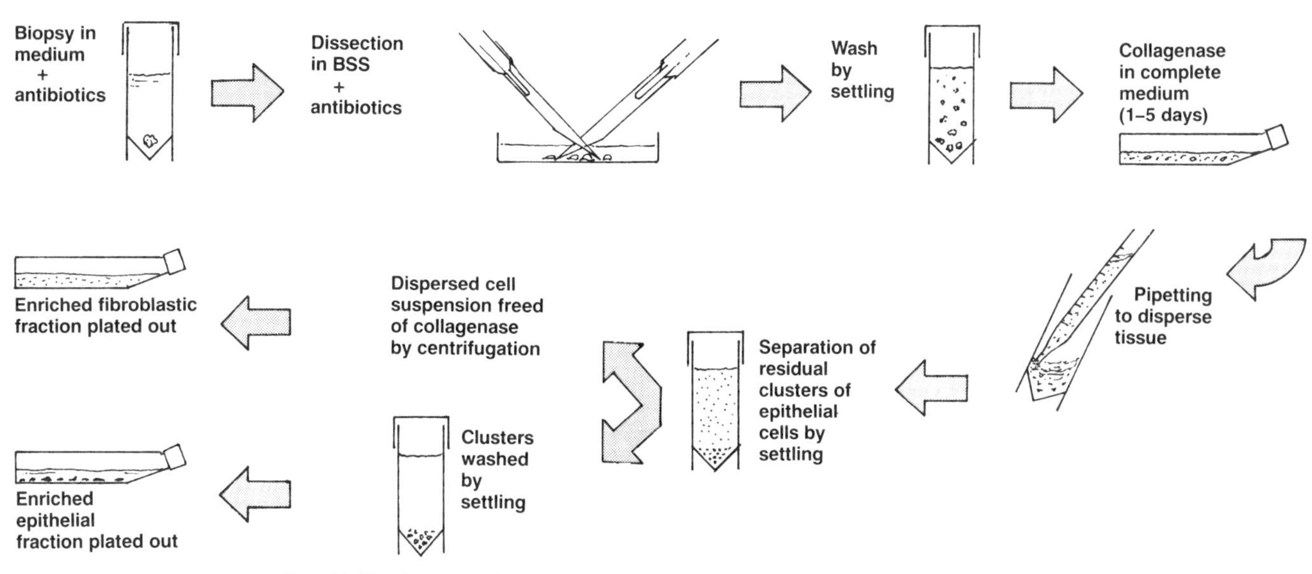

Fig. 11.10. *Stages in disaggregation of tissue for primary culture by collagenase.*

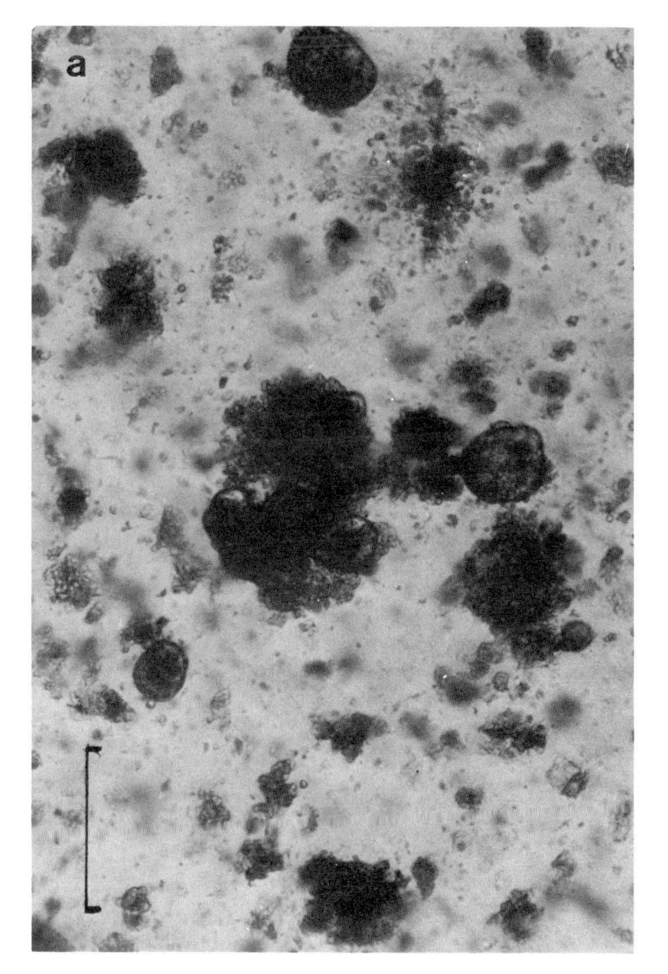

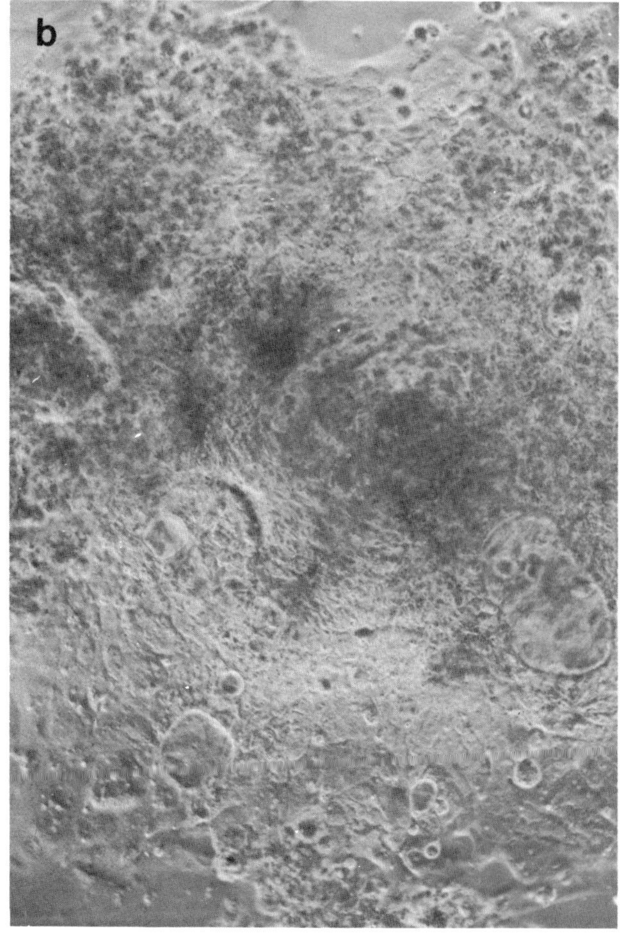

Fig. 11.11. *Cells and cell clusters from human colonic carcinoma after 48-hr dissociation in crude collagenase (Worthington CLS grade). a. Before removal of collagenase. b. After removal of collagenase, further disaggregation by pipetting and culture for 48 hr. The clearly defined clusters in a form epitheliallike sheets in b and the more irregularly shaped clusters form fibroblastic areas.*

tract) are not disaggregated by collagenase and may be separated from connective tissue cells by allowing the epithelial clusters or tubules to sediment for 5–10 min. Connective tissue cells are completely dissociated by the collagenase and remain in suspension. If the sediment is resuspended in BSS and allowed to settle twice more, the final sediment is enriched for epithelial cells. The survival of this epithelium is probably enhanced by culturing the cells as undissociated clusters and tubule fragments.

The discrete clusters of epithelial cells produced by disaggregation in collagenase can be selected under a dissection microscope and transferred to individual wells in a microtitration plate, alone or with irradiated or mitomycin-C-treated feeder cells (see Chapter 13).

The addition of hyaluronidase aids disaggregation by attacking terminal carbohydrate residues on the surface of the cells. This combination has been found to be particularly effective for dissociating rat or rabbit liver, by perfusing the whole organ *in situ* [Berry and Friend, 1969] and completing the disaggregation by stirring the partially digested tissue in the same enzyme solution for a further 10–15 min, if necessary.

This technique gives a good yield of viable hepatocytes and is a good starting point for further culture (see Chapter 20).

Collagenase may also be used in conjunction with trypsin. Cahn and others [Cahn et al., 1967] developed a formulation including chick serum, collagenase, and trypsin. Chick serum has a moderating effect on the activity of the trypsin but does not inhibit it to the extent expected of other sera.

Mechanical Disaggregation

The outgrowth of cells from primary explants is a relatively slow process and can be highly selective. Enzymatic digestion (see above) is rather more labor intensive, though potentially it gives a culture which is more representative of the tissue. As there is a risk of damaging cells during enzymatic digestion, many people have chosen the alternative of mechanical disaggregation, e.g., collecting the cells which spill out when the tissue is carefully sliced [Lasfargues, 1973], pressing the cells through sieves of gradually reduced mesh, or forcing cells through a syringe and needle [Zaroff et al., 1961], or simply repeatedly pipetting. This gives a cell suspension more quickly than with enzymatic digestion but may cause mechanical damage. The following is one method of mechanical disaggregation found to be moderately successful with soft tissues such as brain.

Outline

The tissue in culture medium is forced through sieves of gradually reduced mesh until a reasonable suspension of single cells and small aggregates is obtained. The suspension is then diluted and cultured directly.

Materials

forceps
sieves (1 mm, 100 μm, 20 μm; Fig. 11.12a)
9-cm petri dishes
scalpels
medium
disposable plastic syringes (2 ml or 5 ml)
culture flasks

Protocol

1. After washing and preliminary dissection (see above) chop tissue into pieces about 5–10 mm across, and place a few at a time into a stainless steel sieve of 1-mm mesh in a 9-cm petri dish (Fig. 11.12,b)
2. Force the tissue through the mesh into medium by applying gentle pressure with the piston of a disposable plastic syringe. Pipette more medium into the sieve to wash the cells through
3. Pipette the partially disaggregated tissue from the petri dish into a sieve of finer porosity, perhaps 100-μm mesh and repeat step 2
4. The suspension may be diluted and cultured at this stage or sieved further through 20-μm mesh if it is important to produce a single cell suspension. In general, the more highly dispersed, the higher the sheer stress required, and the lower the resulting viability
5. Seed culture flasks at 10^6 cells/ml and 2×10^6 cells/ml by dilution of cell suspension in medium

Only such soft tissues as embryonic liver, embryonic and adult brain, and some human and animal soft tumors respond at all well to this technique. Even with brain, where fairly complete disaggregation can be obtained easily, the viability of the resulting suspension is lower than that achieved with enzymatic diges-

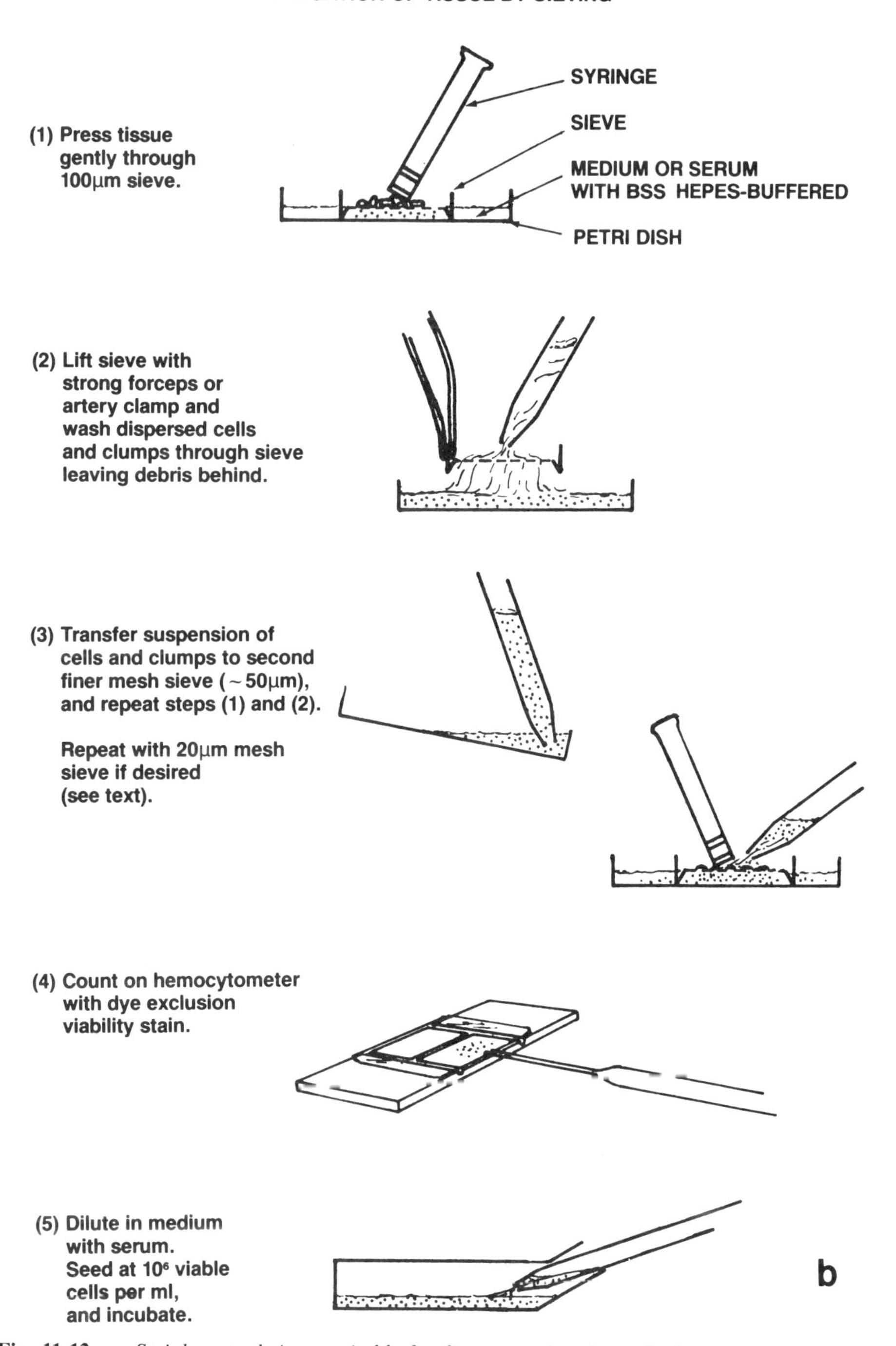

Fig. 11.12. *a. Stainless steel sieves suitable for disaggregating tissue. b. Disaggregation of tissue by sieving.*

tion, although the time taken may be very much less. Where the availability of tissue is no limitation and the efficiency of the yield unimportant, it may be possible to produce as many viable cells as with enzymatic digestion in a shorter time but at the expense of very much more tissue.

Separation of Viable and Nonviable Cells

When an adherent primary culture is prepared from dissociated cells, nonviable cells will be removed at the first medium change. With primary cultures maintained in suspension, nonviable cells are gradually diluted out when cell proliferation starts. If necessary, however, nonviable cells may be removed from the primary disaggregate by centrifuging the cells on a mixture of Ficoll and sodium metrizoate (e.g., Hypaque or Triosil) [Vreis et al., 1973]. This technique is similar to the preparation of lymphocytes from peripheral blood described in Chapter 21. Up to 2×10^7 cells in 9 ml medium may be layered on top of 6 ml Ficoll/Hypaque ("Lymphoprep," Flow Laboratories) in a 25-ml screw-capped centrifuge bottle, centrifuged, and viable cells collected from the interface.

The disaggregation of tissue and preparation of the primary culture is the first, and perhaps most vital, stage in the culture of cells with specific functions. If the required cells are lost at this stage, then the loss is irrevocable. Many different cell types may be cultured by choosing the correct techniques (see Chapters 13, 14, 20). In general, trypsin is more severe than collagenase but sometimes more effective in creating a single cell suspension. Collagenase does not dissociate epithelial cells readily, but this can be an advantage in separating them from stromal cells. Mechanical disaggregation is much quicker but will damage more cells. The best approach is to try out the techniques described above and select the method which works best in your system. If none of these is successful, try additional enzymes such as pronase, dispase, and DNase and consult the literature.

Chapter 12
Maintenance of the Culture—Cell Lines

The first subculture represents an important transition for the culture. The need to subculture implies that the primary culture has increased to occupy all of the available substrate. Hence cell proliferation has become an important feature. While the primary culture may have variable growth fraction (see Chapter 19) depending on the type of cells present in culture, after the first subculture, the growth fraction is usually high (80% or more).

From a very heterogeneous primary culture, containing many of the cell types present in the original tissue, a more homogeneous cell line emerges. In addition to its biological significance, this has also considerable practical importance as the culture can now be propagated, characterized, and stored, and the potential increase in cell number and uniformity of the cells opens up a much wider range of experimental possibilities.

Once a primary culture is subcultured (or "passaged," or "transferred"), it becomes known as a "cell line." This term implies the presence of several cell lineages either of similar or distinct phenotypes. If one cell lineage is selected, by cloning (see Chapter 13), by physical cell separation (see Chapter 14), or by any other selection technique, with certain specific properties which have been identified in the bulk of the cells in the culture, this becomes known as "cell strain."

NOMENCLATURE

The first subculture gives rise to a "secondary" culture, the secondary to a "tertiary" and so on, although in practice this nomenclature is seldom used beyond tertiary. Since the importance of culture lifetime was highlighted by Hayflick and others with diploid fibroblasts [Hayflick and Moorhead, 1961], where each subculture divided the culture in half ("split ratio"—1:2), passage number has often been confused with "generation number." Cell lines with limited culture life-spans ("finite" cell lines) behave in a fairly reproducible fashion (see Chapter 2). They will grow through a limited number of cell generations, usually between 20 and 80 population doublings, before extinction. The actual number depends on strain differences and culture conditions but is consistent for one cell line grown under the same conditions. It is, therefore, important that reference to a cell line should express the approximate generation number or number of doublings since explantation, "approximate" because the number of generations which have elapsed in the primary culture is difficult to assess.

The cell line should also be given a code or designation (e.g., NHB, normal human brain), a cell strain or cell line number (if several cell lines were derived from the same source) NHB1, NHB2, etc., and if cloned, a clone number, NHB2-1, NHB2-2, etc. Together with the number of population doublings, this becomes, for example, NHB2/2 and will increase by one for a split ratio of 1:2 (NHB2/2, NHB2/3, etc.), by a two for a split ratio of 1:4 (NHB2/2, NHB2/4, etc.), and so on. For publication, each cell line should be prefixed with a code designating the laboratory in which it was derived, e.g., WI, Wistar Institute [Federoff, 1975].

ROUTINE MAINTENANCE

Once a culture is initiated, whether it be a primary culture or a subculture of a cell line, it will need a periodic medium change ("feeding") followed eventually by subculture if the cells are proliferating. In nonproliferating cultures, the medium will still need to be changed periodically as the cells will still metabolize, and some constituents of the medium will become exhausted or will degrade spontaneously. Intervals between medium changes and between subcultures vary from one cell line to another depending on the rate of growth and metabolism; rapidly growing cell lines such as HeLa are usually subcultured once per week and the medium changed 4 d later. More slowly growing cell lines may only require to be subcultured every 2, 3, or even 4 wk, and the medium changed weekly between subcultures. (For a more detailed discussion of the growth cycle, see below.)

Replacement of Medium

Four factors indicate the need for the replacement of culture medium (see also Chapter 9).

A drop in pH. The rate of fall and absolute level should be considered. Most cells will stop growing as the pH falls from pH 7.0 to pH 6.5 and will start to lose viability between pH 6.5 and pH 6.0, so if the medium goes from red through orange to yellow, the medium should be changed. Try to estimate the rate of fall; a culture which falls at 0.1 pH units in 1 d will not come to harm if left a day or two longer before feeding, but a culture that falls 0.4 pH units in 1 d will need to be fed 24–48 hr later and cannot be left over a weekend.

Cell concentration. Cultures at a high cell concentration will use up the medium faster than at a low concentration. This is usually evident in the pH change but not always.

Cell type. Normal cells (e.g., diploid fibroblasts) will usually stop dividing at high cell concentrations (density limitation of growth; see Chapter 16) due to growth factor depletion and other factors. The cells block in the G1 phase of the cell cycle and deteriorate very little even if left for 2–3 wk. Transformed cells, continuous cell lines, and some embryonic cells, however, will deteriorate rapidly at high cell densities unless the medium is changed daily or they are subcultured.

Cell morphology. When checking a culture for routine maintenance, be alert to signs of morphological deterioration: granularity around the nucleus, cytoplasmic vacuolation, and rounding up of the cells with detachment from the substrate (Fig. 12.1). This may

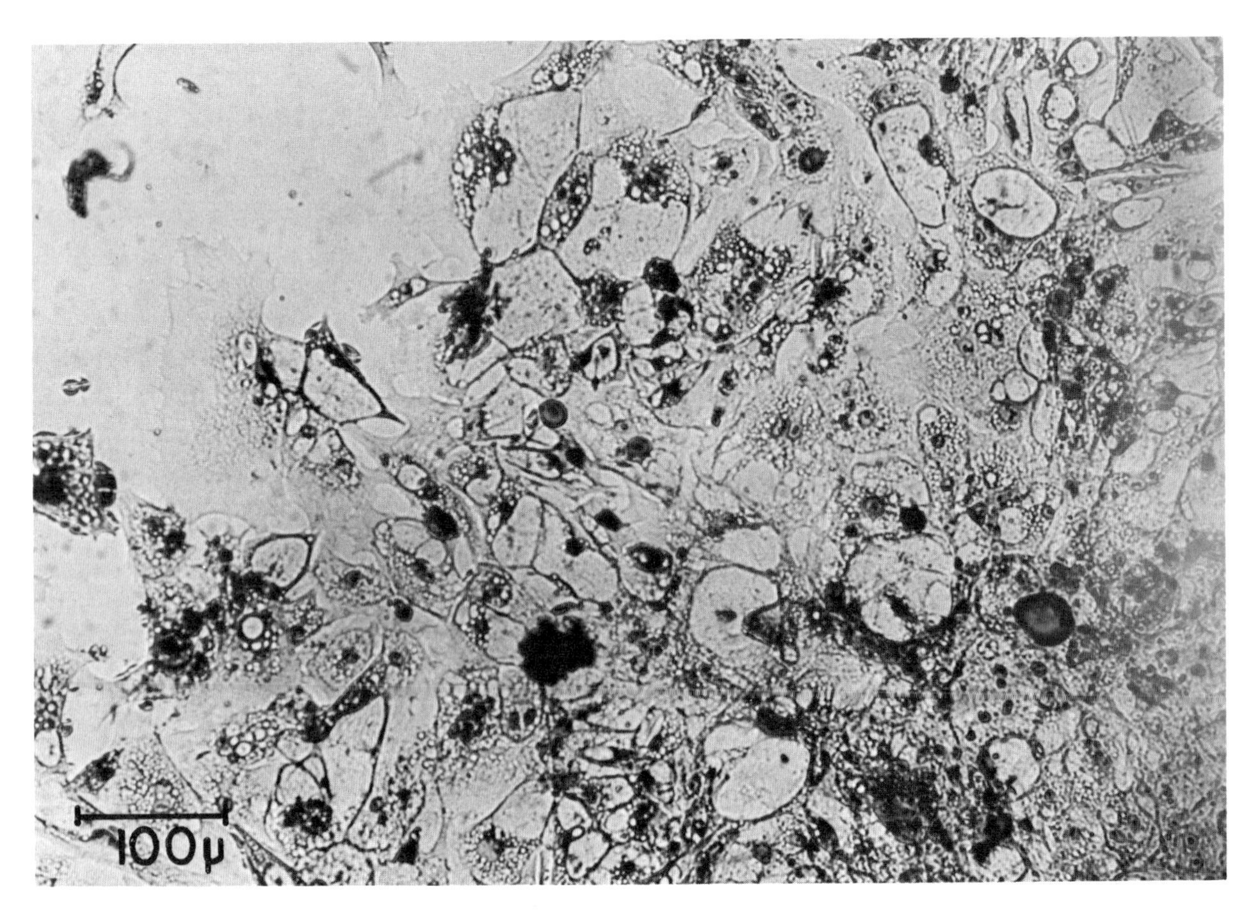

Fig. 12.1. *Signs of deterioration of the culture. Cytoplasm of cells becomes granular, particularly around the nucleus, and vacuolation occurs. Cells may become more refractile at the edge if cell spreading is impaired.*

imply that the culture requires a medium change, or may indicate a more serious problem, e.g., inadequate or toxic medium or serum, microbiological contamination, or senescence of the cell line. During routine maintenance, the medium change or subculture frequency should prevent such deterioration.

Volume, Depth, and Surface Area

The usual ratio of medium volume to surface area is 0.2–0.5 ml cm^2. The upper limit is set by gaseous diffusion through the liquid layer and the optimum will depend on the oxygen requirement of the cells. Cells with a high O_2 requirement will be better in shallow medium (2 mm) and those with a low requirement may do better in deep medium (5 mm). If the depth is greater than 5 mm, then gaseous diffusion may become limiting. With monolayer cultures this can be overcome by rolling the bottle or perfusing the culture with medium and arranging for gas exchange in an intermediate reservoir (see Chapter 21). When the depth of suspension culture is increased, it should be stirred with a bar magnet (see Chapter 21). To prevent frothing, the depth of stirrer cultures must be a minimum of 5 cm. For intermediate depths of medium between 5 mm and 5 cm, use a roller bottle (see Table 8.1).

"Holding Medium"

A holding medium may be used where stimulation of mitosis, which usually accompanies a medium change, even at high cell densities, is undesirable. Holding media are usually regular media with the serum concentration reduced to 1 or 2%. This will not stimulate mitosis in most untransformed cells. Transformed cell lines are unsuitable for this procedure as they may either continue to divide successfully or the culture may deteriorate as transformed cells do not block in a regulated fashion in G_1.

Holding medium is also used to maintain cell lines with a finite life-span without using up the limited number of cell generations available to them (see Chapter 2). Reduction of serum and cessation of cell proliferation also promotes expression of the differentiated phenotype in some cells [Schousboe et al., 1979; Maltese and Volpe, 1979].

Changing the Medium or "Feeding" a Culture

Outline

Examine the culture on an inverted microscope. Then, if indicated, remove the old medium and add fresh medium. Return the culture to the incubator.

Materials

Pipettes
medium
(both sterile)

Protocol

1. Examine culture carefully for signs of contamination or deterioration (see Fig. 17.1 and 12.1)
2. Check the criteria described above—pH, cell density, or concentration, and, based on your knowledge of the behavior of the culture, decide whether or not to replace the medium. If feeding is required, proceed as follows
3. Take to sterile work area, remove and discard medium (see Chapter 5)
4. Add same volume of fresh medium, prewarmed to 36.5°C if it is important that there is no check in cell growth
5. Return culture to incubator

Note.
Where a culture is at a low density and growing slowly, it may be preferable to "half-feed." In this case, remove only half the medium at step 3 and replace it in step 4 with the same volume as was removed.

Subculture

When all the available substrate is occupied, or when the cell concentration exceeds the capacity of the medium, either the frequency of medium changing must increase or the culture must be divided. The usual practice in subculturing an adherent cell line involves removal of the medium and dissociation of the cells in the monolayer with trypsin, although some loosely adherent cells (e.g., HeLa or Chang liver) may be subcultured by shaking the bottle and collecting the cells in the medium, and diluting as appropriate in fresh medium in new bottles. Exceptionally, some cell monolayers cannot be dissociated in trypsin and require the action of alternative proteases such as pronase, dispase, or collagenase (Table 12.1) [Foley and Aftonomos, 1973].

TABLE 12.1. Cell Dissociation for Transfer or Counting—Procedures of Gradually Increasing Severity

1.	Shake-off	Mitotic or other loosely adherent cells
2.	Trypsin* in PBS (0.01–0.5% as required, usually 0.25%, 5–15 min)	Most continuous cell lines
3.	Prewash with PBS or CMF, then 0.25% trypsin* in PBS or saline-citrate	Some strongly adherent continuous cell lines and many cell lines at early passage stages
4.	Prewash with 1 m*M* EDTA in PBS or CMF then 0.25% trypsin* in citrate	Some strongly adherent early passage cell lines
5.	Prewash with 1 m*M* EDTA, then EDTA 2nd rinse, and leave on, 1 ml/5 cm	Epithelial cells, although some may be sensitive to EDTA
6.	EDTA prewash, then 0.25% trypsin* with 1 m*M* EDTA	Strongly adherent cells, particulary epithelial and some tumor cells (note: EDTA can be toxic to some cells)
7.	1 m*M* EDTA prewash, then 0.25% trypsin* and collagenase*, 200 units/ml PBS or saline-citrate or EDTA/PBS	Thick cultures, multilayers, particularly collagen-producing dense cultures
8.	Scraping	All cultures, but may cause mechanical damage and usually will not give a single cell suspension
9.	Add dispase (0.1–1.0 mg/ml) or pronase (0.1–1.0 mg/ml) to medium and incubate till cells detach	Will dislodge most cells, but requires centrifugation step to remove enzyme not inactivated by serum. May be harmful to some cells

*Digestive enzymes are available (Difco, Worthington, Boehringer Mannheim, Sigma) in varying degrees of purity. Crude preparations, e.g., Difco trypsin 1:250 or Worthington CLS grade collagenase, contain other proteases which may be helpful in dissociating some cells but may be toxic to others. Start with a crude preparation and progress to purer grades if necessary. Purer grades are often used at a lower concentration (mg/ml) as their specific activities (enzyme units/g) are higher. Purified trypsin at 4° C has been recommended for cells grown in low serum concentrations or in the absence of serum [McKeehan, 1977].

The attachment of cells to each other and to the culture substrate is mediated by cell surface glycoproteins and Ca^{2+} and Mg^{2+} ions. Other proteins, derived from the cells and from the serum, become associated with the cell surface and the surface of the substrate and facilitate cell adhesion.

Outline

Remove medium, expose cells briefly to trypsin, incubate and disperse cells in medium.

Materials

Pipettes (sterile)
medium (sterile)
PBSA (sterile)
0.25% trypsin in PBSA, saline citrate, or EDTA (sterile) (see Table 12.1)
culture flasks
hemocytometer or cell counter

Protocol

1. Withdraw medium and discard
2. Add PBSA prewash (5 ml/25 cm^2) to the side of the flask opposite the cells, so as to avoid dislodging cells, rinse the cells, and discard rinse. This step is designed to remove traces of serum which would inhibit the action of the trypsin
3. Add trypsin (3 ml/25 cm^2) to the side of the flask opposite the cells. Turn the flask over to cover the monolayer completely. Leave 15–30 sec and withdraw the trypsin, making sure beforehand that the monolayer has not detached. Using trypsin at 4° C helps to prevent this
4. Incubate until cells round up; when the bottle is tilted, the monolayer should slide down the surface (this usually occurs after 5–15 min). Do not leave longer than necessary, but do not force the cells to detach before they are ready to do so, or clumping may result

Note. If difficulty is encountered in getting cells to detach, and, subsequently, in preparing a single cell suspension, alternative procedures, as described in Table 12.1, may be employed. In each case the main dissociating agent, be it trypsin or EDTA, is present only briefly and the incubation is performed in the residue after most of the dissociating agent has been removed.

5. Add medium (0.1–0.2 ml/cm^2) and disperse cells by repeated pipetting over the surface bearing the monolayer. Finally, pipette the cell suspension up and down a few times, with the tip of the pipette resting on the bottom corner of bottle. The degree of pipetting required will vary from one cell line to

another; some disperse easily, others require vigorous pipetting. Almost all will incur mechanical damage from shearing forces if pipetted too vigorously; primary suspensions and early passage cultures are particularly prone to damage due partly to their greater fragility and partly to their larger size. Pipette up and down sufficiently to disperse the cells into a single cell suspension. If this is difficult, apply a more aggressive dissociating agent (see Table 12.1) [Toshiharu et al., 1975]

A single cell suspension is desirable at subculture to ensure an accurate cell count and uniform growth on reseeding. It is essential where quantitative estimates of cell proliferation or of plating efficiency are being made and where cells are to be isolated as clones

6.
Count cells by hemocytometer or electronic particle counter (see Chapter 19)

7.
Dilute to the appropriate seeding concentration (a) by adding the appropriate volume of cells to a premeasured volume of medium in a culture flask, or (b) by diluting the cells to the total volume required and distributing that among several flasks. Procedure (a) is useful for routine subculture when only a few flasks are used and precise cell counts and reproducibility are not critical, but (b) is preferable when setting up several experimental replicate samples as the total number of manipulations is reduced and the concentration of cells in each flask will be identical

8.
Cap the flask(s) and return to the incubator. Check after about 1 hr for pH change. If the pH rises, return to aseptic area and gas culture lightly with 5% CO_2. Since each culture will behave predictably in the same medium, you will know eventually which cells to gas when they are passaged, without having to incubate them first.

Expansion of air inside plastic flasks causes the flasks to swell and prevents them from lying flat. Release the pressure by slackening the cap briefly. This may be prevented by compressing the top and bottom of large flasks before sealing them. Incubation then restores the correct shape. Care must be taken not to crack the flasks

For finite cell lines, it is convenient to reduce the cell concentration at subculture by two-, four-, eight-, or 16-fold, making the calculation of the number of population doublings easier ($2 \equiv 1$, $4 \equiv 2$, $8 \equiv 3$, $16 \equiv 4$), e.g., a culture divided eightfold will require three doublings to achieve the same cell density. With continuous cell lines, where generation number is not usually recorded, the cell concentration is more conveniently reduced to a round figure, e.g., 5×10^4 cells/ml. In both cases, the cell count should be recorded so that growth rate can be estimated at each subculture and consistency monitored (see below under Growth cycle).

Propagation in Suspension

The preceding instructions refer to subculture of monolayers, as most primary cultures or continuous lines grow in this way. Cells which grow continuously in suspension, either because they are nonadhesive (e.g., many mouse leukemias and ascites tumors) or because they have been kept in suspension mechanically, or selected (see also Chapter 21), may be subcultured like micro-organisms. Trypsin treatment is not required and the whole process is quicker and less traumatic for the cells. Medium replacement is not usually carried out with suspension cultures as this would require centrifugation of the cells. Routine maintenance is, therefore, reduced to one of two alternative procedures, i.e., subculture by dilution, or increase of the volume without subculture.

Outline

Count cells, withdraw cell suspension, and add fresh medium to restore cell concentration to starting level.

Materials

Culture flasks (sterile)
medium (sterile)
pipettes (sterile)
bar magnet (sterile)
magnetic stirrer
hemocytometer or cell counter

Protocol

1.
Mix cell suspension and disperse any clumps by pipetting

2.
Remove sample and count

3.
Add medium to fresh flask

Note.
Any culture flask with a reasonably flat surface

may be used for cells which grow spontaneously in suspension. Where stirring is required, e.g., for larger cultures, or cells which would normally attach, use standard round reagent bottles, or aspirators, siliconized, if necessary (see Appendix), and insert a magnetic stirrer bar, Teflon coated, and with a ridge around the middle (Fig. 12.2) or suspended from the top of the bottle. Select the appropriate size of bottle to give between 5 and 8 cm depth with the volume of medium that you require

4.

Add sufficient cells to give a final concentration of 10^5 cells/ml for slow-growing cells (24–48-hr doubling time) or 2×10^4/ml for rapidly growing cells (12–18-hr doubling time)

5.

Cap and return culture to incubator

6.

Culture flasks should be laid flat as for monolayer culture. Stirrer bottles should be placed on a magnetic stirrer and stirred at 60–100 rpm. Take care that the stirrer motor does not overheat the culture. Insert a polystyrene foam mat under the bottle if necessary. Induction-drive stirrers generate less heat and have no moving parts

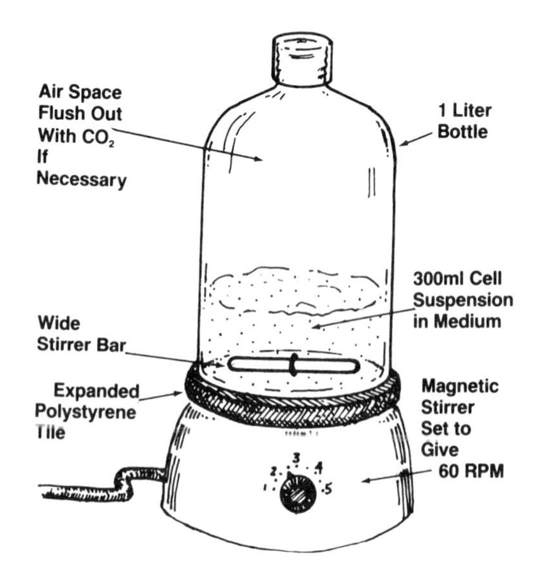

Fig. 12.2. *Simple stirrer culture for cells growing in suspension. An expanded polystyrene mat (dark shaded area below bottle) should be interposed between the bottle and the magnetic stirrer to avoid heat transfer from the stirrer motor.*

Suspension cultures have a number of advantages (see Table 12.2). The production and harvesting of large quantities of cells may be achieved without increasing the surface area of the substrate (see Chapter 21). Furthermore, if dilution of the culture is continuous and the cell concentration kept constant, a steady state can be achieved; this is not readily achieved in monolayer culture. Maintenance of monolayer cultures is essentially cyclic with the result that growth rate varies depending on the phase of the growth cycle.

Monolayers are convenient for cytological and immunological observations, cloning, mitotic "shake off" (for cell synchronization of chromosome preparation) and *in situ* extractions without centrifugation.

TABLE 12.2. Properties of Monolayer and Suspension Cultures

	Monolayer	Suspension
Maintenance	Cyclic pattern of propagation (see text) Require dissociation Dependent on availability of substrate	Can be maintained at "steady state" Simple dilution at passage Dependent on medium volume only (with adequate gas exchange)
Results of differences in Geometry	Cell Interaction: metabolic cooperation, contact inhibition of movement and membrane activity, density limitation of growth Diffusion boundary effects (see text) Cell shape and cytoskeleton—spreading, motility, overlapping, underlapping	Homogeneous suspension Cell density limited by nutrient and hormonal concentration of the medium only Shearing effects in stirred cultures may damage some cells
Sampling and analysis	Good cytological preparations, chromosomes, immunofluorescence, histochemistry Enrichment of mitoses by "shake-off" (see Chapter 22) Serial extractions *in situ* possible without centrifugation	Bulk production of cells Ease of harvesting (no trypsinization required)
Which cells?	Most cell types except some hemopoietic cells and ascites tumors	Transformed cells and lymphoblastoid cell lines

At subculture a fragile or slowly growing line should be split 1:2; and a vigorous, rapidly growing line, 1:8 or 1:16. Once a cell line becomes continuous (usually taken as beyond 150 or 200 generations) the generation number is disregarded and the culture should simply be cut back to between 10^4 and 10^5 cells/ml. The split ratio or dilution is also chosen to establish a convenient subculture interval (perhaps 1 or 2 wk), and to ensure that the cells (1) are not diluted below that concentration which permits them to reenter the growth cycle with a lag period of 24 hr or less, and (2) do not enter plateau before the next subculture.

GROWTH CYCLE

Routine passage leads to the repetition of a standard growth cycle. It is essential to become familiar with this cycle for each cell line that is handled as this controls the seeding concentration, the duration of growth before subculture, the duration of experiments, and the appropriate times for sampling to give greatest consistency. Cells at different phases of the growth cycle behave differently with respect to proliferation, enzyme activity, glycolysis and respiration, synthesis of specialized products, and many other properties.

Outline

Set up a series of cultures at three different cell concentrations and count the cells at daily intervals until they "plateau."

Materials

Cell culture
24-well plates (sterile)
100 ml growth medium (sterile)
0.25% crude trypsin (for monolayer cultures only) (sterile)
plastic box to hold plates
CO_2 incubator or CO_2 supply to gas box

Protocol

1.
Trypsinize cells as for regular subculture (see above)
2.
Dilute cell suspension to 10^5 cells/ml, 3×10^4 cells/ml, and 10^4cells/ml, in 25 ml medium for each concentration
3.
Seed three 24-well plates, one at each cell concentration, with 1 ml per well. Add cell suspension slowly from the center of the well so that it does not swirl around the well. Similarly, do not shake the plate to mix the cells, as the circular movement of medium will concentrate the cells in the middle of the well
4.
Place in a humid CO_2 incubator or sealed box gassed with CO_2
5.
After 24 hr, remove plates from incubator and count the cells in three wells of each plate: (a) Remove medium completely from wells to be counted; (b) add 1 ml trypsin to each well; (3) incubate with trypsin; (4) after 15 min, disperse cells in trypsin and transfer 0.4 ml to counting fluid and count on cell counter
Note.
Hemocytometer counting may be used but may be difficult at lower cell concentrations. Reduce trypsin volume to 0.1 ml and disperse cells carefully without frothing using a micropipette and transfer to hemocytometer.
6.
Return plate to incubator as soon as cell samples in trypsin are removed. The plate must be out of the incubator for the minimum length of time, to avoid disruption of normal growth
7.
Repeat sampling at 48 and 72 hr as in steps 5 and 6
8.
Change medium at 72 hr or sooner if indicated by pH drop (see above)
9.
Continue sampling at daily intervals for rapidly growing cells (doubling time 12–14 hr) but reduce frequency of sampling to every 2 days for slowly growing cells (doubling time >24 hr), until plateau is reached
10.
Keep changing medium every 1, 2, or 3 days as indicated by pH

Analysis. 1. Calculate cell number per well, per ml of culture medium (same figure), and per cm^2 of available growth surface in well. (Stain one or two wells (see Chapter 15) at each density to determine whether distribution of cells in wells is uniform and whether they grow up the sides of the well)

2. Plot cell density (per cm^2) and cell concentration (per ml), on a log scale, against time on a linear scale (Fig. 12.3)

3. Determine the lag time, population doubling time, and plateau density (see below and Fig 12.3)

4. Establish which is the appropriate starting density for routine passage. Repeat growth curve at intermediate cell concentrations if necessary

Variations. 1. Different culture vessels may be used, e.g., 25 cm^2 flasks, although more cells and medium will be required

2. Frequency of medium changing may be altered

3. Different media or supplements may be tested

Suspension cultures

1.
Add cell suspension in growth medium to wells at a range of concentrations as for monolayer

2.
Sample 0.4 ml at intervals as per trypsin samples. Alternatively, seed two 75-cm^2 flasks with 20 ml for each cell concentration and sample 0.4 ml from each flask daily or as required. Mix well before sampling and keep flasks out of incubator for the minimum length of time. Do not feed cultures during growth curve

The growth cycle (Fig. 12.3) is conventionally divided into three phases.

The Lag Phase

This is the time following subculture and reseeding during which there is little evidence of an increase in cell number. It is a period of adaptation during which the cell replaces elements of the glycocalyx lost during trypsinization, attaches to the substrate, and spreads out. During spreading the cytoskeleton reappears and its reappearance is probably an integral part of the spreading process. Enzymes, such as DNA polymerase, increase, followed by the synthesis of new DNA and structural proteins. Some specialized cell products may disappear and not reappear until cessation of cell proliferation at high cell density.

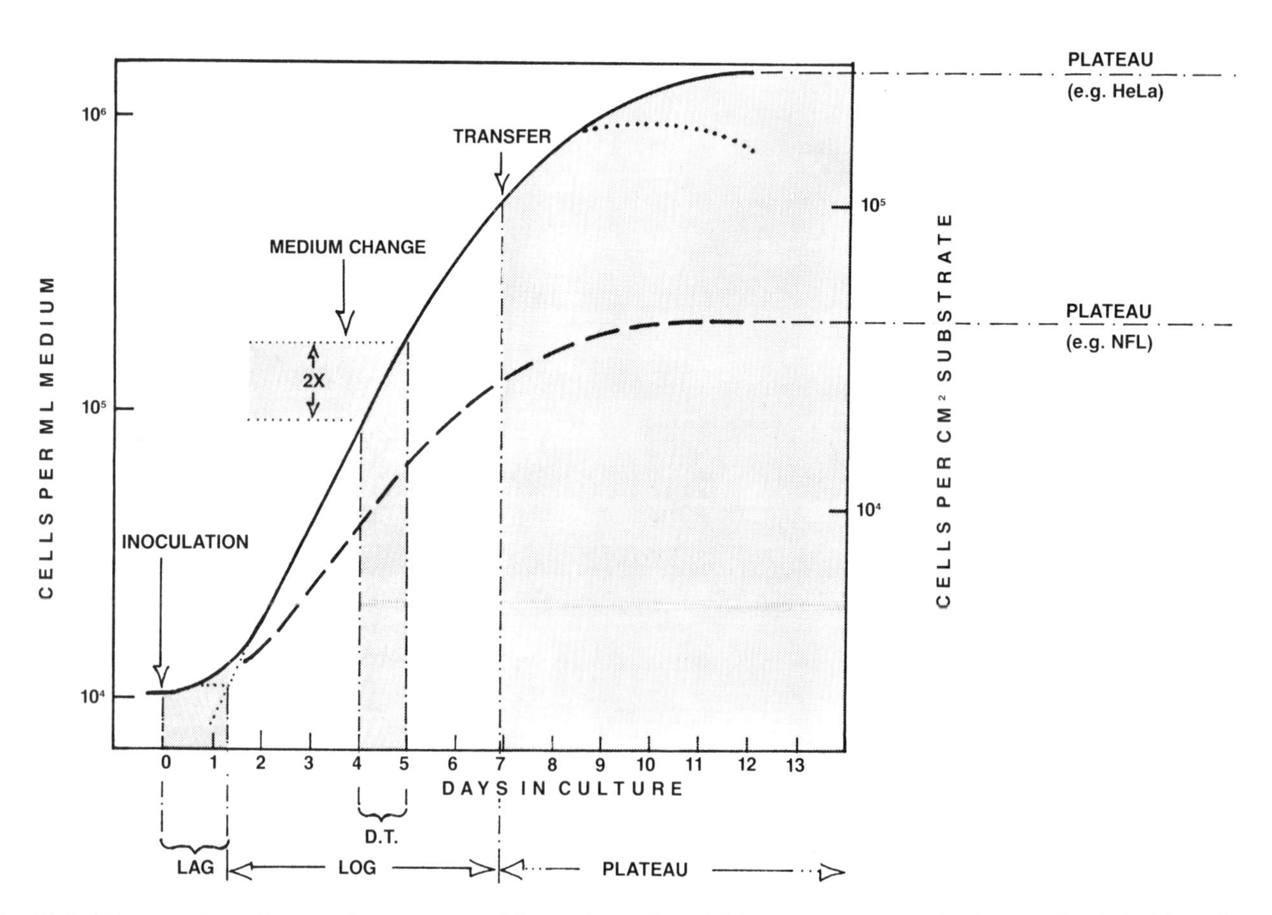

Fig. 12.3. *Diagram of growth curve of a continuous cell line such as HeLa and a finite cell line, NFL (normal fetal lung fibroblasts). The solid line represents growth of HeLa; the dashed line illustrates the lower plateau obtained with NFL (see text).*

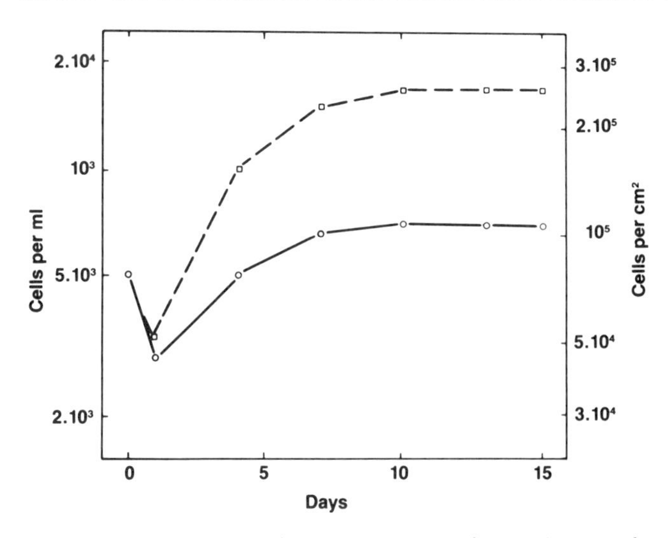

Fig. 12.4. *Difference in plateaus (saturation densities) attained by cultures from normal brain (circles, solid line) and a glioma (squares, broken line). Cells were seeded onto 13-mm coverslips and 48 hr later the coverslips were transferred to 9-cm petri dishes with 20 ml growth medium, to minimize exhaustion of the medium.*

The Log Phase

This is the period of exponential increase in cell number following the lag period and terminating one or two doublings after confluence is reached. The length of the log phase depends on the seeding density, the growth rate of the cells and the density at which cell proliferation is inhibited by density. In the log phase the growth fraction is high (usually 90–100%) and the culture is in its most reproducible form. It is the optimal time for sampling since the population is at its most uniform and viability is high. The cells are, however, randomly distributed in the cell cycle and, for some purposes, may need to be synchronized (see Chapter 22).

The Plateau Phase

Toward the end of the log phase, the culture becomes confluent—i.e., all the available growth surface is occupied and all the cells are in contact with surrounding cells. Following confluence the growth rate of the culture is reduced, and in some cases, cell proliferation ceases almost completely after one or two further population doublings. At this stage, the culture enters the plateau (or stationary) phase, and the growth fraction falls to between 0 and 10%. The cells may become less motile; some fibroblasts become oriented with respect to one another, forming a typical parallel array of cells. "Ruffling" of the plasma membrane is reduced, and the cell both occupies less surface area of substrate and presents less of its own surface to the medium. There may be a relative increase in the synthesis of specialized versus structural proteins and the constitution and charge of the cell surface may be changed.

The phenomenon of cessation of motility, membrane ruffling, and growth was described originally by Abercrombie and Heaysman [1964] and designated *"contact inhibition."* It has since been realized that the reduction of the growth of normal cells after confluence is reached is not due solely to contact but may also involve reduced cell spreading [Stoker et al., 1968; Folkman and Moscona, 1978] and depletion of nutrients, and, particularly, growth factors [Stoker, 1973; Dulbecco and Elkington, 1973; Westermark and Wasteson, 1975] in the medium [Holley et al., 1978]. The term *"density limitation"* of growth has, therefore, been used to remove the implication that cell-cell contact is the major limiting factor [Stoker and Rubin, 1967], and "contact inhibition" is best reserved for those events directly contingent on cell contact, i.e., reduced cell motility and membrane activity, resulting in the formation of a strict monolayer and orientation of the cells with respect to each other.

Cultures of normal epithelial and endothelial cells will stop growing after confluence and remain as a monolayer, implying that they are dependent on anchorage to the substrate for continued growth. (This may be analogous to anchorage to basement membrane *in vivo*.) Most cultures, however, with regular replenishment of medium, will continue to proliferate, although at a reduced rate, well beyond confluence, resulting in multilayers of cells. Human embryonic lung, or adult skin, fibroblasts, which express contact inhibition of movement, will continue to proliferate, laying down layers of collagen between the cell layers, until multilayers of six or more cells can be reached under optimal conditions [Kruse et al., 1970]. They still retain an ordered parallel array, however. The terms "plateau" and "stationary" are not strictly accurate, therefore, and should be used with caution.

Cultures which have transformed spontaneously or have been transformed by virus or chemical carcinogens will usually reach a higher cell density in the plateau phase than their normal counterparts [Westermark, 1974] (Fig. 12.4). This is accompanied by a higher growth fraction and loss of density limitation. These cultures are often *anchorage independent* for growth—i.e., they can easily be made to grow in suspension (see Density Limitation of Growth in Chapter 16; also see Chapter 2).

The construction of a growth curve from cell counts performed at intervals after subculture enables the measurement of a number of parameters which should be found to be characteristic of the cell line under a given set of culture conditions. The first of these is the duration of the *lag period* or "lag time" obtained by extrapolating a line drawn through the points on the exponential phase until it intersects the seeding or inoculum concentration (see Fig. 12.3), and reading off the elapsed time since seeding equivalent to that intercept. The second is the *doubling time,* i.e., the time taken for the culture to increase two-fold in the middle of the exponential, or "log", phase of growth. This should not be confused with the *generation time* or *cell cycle time (see Chapter 19),* which are determined by measuring the transit of a population of cells through the cell cycle until they return to the same point in the cell cycle.

The last of the commonly derived measurements from the growth cycle is the *"plateau level"* or *"saturation density."* This is the cell concentration in the plateau phase and is dependent on cell type and frequency of medium replenishment. It is difficult to measure accurately as a steady state is not achieved as easily as in the log phase. Ideally the culture should be perfused; but a reasonable compromise may be achieved by growing the cells on a restricted area, say a small-diameter coverslip (15 mm) in a large-diameter petri dish (90 mm) with 20 ml of medium replaced daily (see Chapter 16). Under these conditions, medium limitation of growth is minimal, and cell density exerts the major effect. Counting the cells under these conditions gives a more accurate and reproducible measurement. "Plateau" does not imply complete cessation of cell proliferation but represents a steady state where cell division is balanced by cell loss.

The maximum cell concentration in suspension cultures, which are not limited by available substrate, is usually limited by available nutrients. By fortifying the medium with a higher concentration of amino acids, Pirt and others [Birch and Pirt, 1971; Blaker et al., 1971] were able to obtain a maximum cell concentration of 5×10^6 cells/ml for L"S" cells, far in excess of what can be achieved with attached cells.

SLOW CELL GROWTH

Even in the best-run laboratories, problems may arise in routine cell maintenance. Some may be attributed to microbiological contamination (see Chapter 17), but often the cause lies in one or more alterations in culture conditions. The following check list may help to track these down:

1. Any change in procedure or equipment?
2. Medium:
   a. Medium adequate?—check against other media (see Chapter 9).
   b. Frequency of changing adequate?
   c. pH: Check that it is within 7.0–7.4 during culture.
   d. Osmolality: check on osmometer.
   e. Component missed out: make up fresh batch.
   f. New batch of stock medium which is faulty?
   g. If BSS-based, is BSS satisfactory? (Check with other users.)
   h. If water-based, is water satisfactory (check with other users, or against fresh IX medium, bought in).
   i. Check still—deionizer—conductivity, contamination—glass boiler—residue.
   j. Storage vessel, for algal or fungal contamination: chemical traces in plastic.
   k. HCO_3^-.
   l. Antibiotics.
3. Serum:
   a. New batch? Check supplier's quality control.
   b. Check concentration. Too low or too high?
   c. Reconfirm lack of toxicity, growth promotion and plating efficiency.
4. Glassware or plastics:
   a. If new, check against previous stock.
   b. Wash-up—other cells showing symptoms? Other users having trouble?
   c. Trace contamination of glass? Check growth on plastic.
5. Cells: If other people's cells are all right.
   a. Contamination (see also Chapter 17).
      i) Bacterial, fungal—grow up without antibiotics.
      ii) Mycoplasma:
         (A) Stain Culture with Hoechst 33258 (Chapter 17).
         (B) Check for cytoplasmic DNA (incorporation of radioactive thymidine) by autoradiography.
         (C) Get commercial test done (e.g., Flow Laboratories or Microbiological Associates).
      iii) Viral—difficult to detect—try E M or fluorescent antibody.
   b. Seeding density too low at transfer.
   c. Transferred too frequently.
   d. Allowed to remain for too long in plateau before transfer.
6. Hot room and incubators: check temperature and stability.

Chapter 13
Cloning and Selection of Specific Cell Types

It can be seen from the preceding two chapters that a major recurrent problem in tissue culture is the preservation of a specific cell type and its specialized properties. While environmental conditions undoubtedly play a significant role in maintaining the differentiated properties of a culture, the selective overgrowth of unspecialized cells is still a major problem.

CLONING

The traditional microbiological approach to the problem of culture heterogeneity is to isolate pure cell strains by cloning, but the success of this technique in animal cell culture is limited by the poor cloning efficiencies of most primary cultures.

A further problem of cultures derived from normal tissue is that they may only survive for a limited number of generations (see Chapter 2), and by the time a clone has produced a usable number of cells, it may already be near to senescence (Fig. 13.1). Even cultures which do not die out may have undergone alteration by the time the cloned cell line is established. Cloning is most successful in isolating variants from continuous cell lines, but even then considerable heterogeneity may arise within the clone as it is grown up for use (See Chapter 18).

Coon and Cahn [1966] were able to clone cartilage- and pigment-producing cell strains. Under the correct conditions, these cultures were able to retain their specialized functions over many generations. Similarly, Clark and Pateman [1978] isolated a Kupffer cell line from Chinese hamster liver by cloning the primary culture.

Cloning has also been used to isolate specific biochemical mutants and cell strains with marker chromosomes and may help to reduce the heterogeneity of a culture.

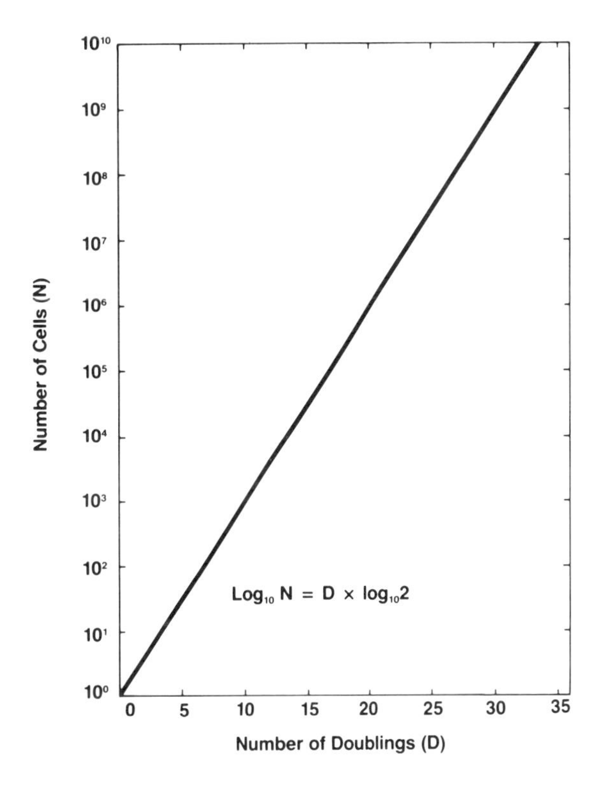

Fig. 13.1. *Relationship of cell yield in a clone to the number of population doublings—e.g., 20 doublings are required to produce 10^6 cells.*

Dilution Cloning [Puck and Marcus, 1955]

Outline

Seed cells at low density, incubate until colonies form, isolate and propagate into cell strain (Fig. 13.2).

Materials

pipettes
medium
trypsin

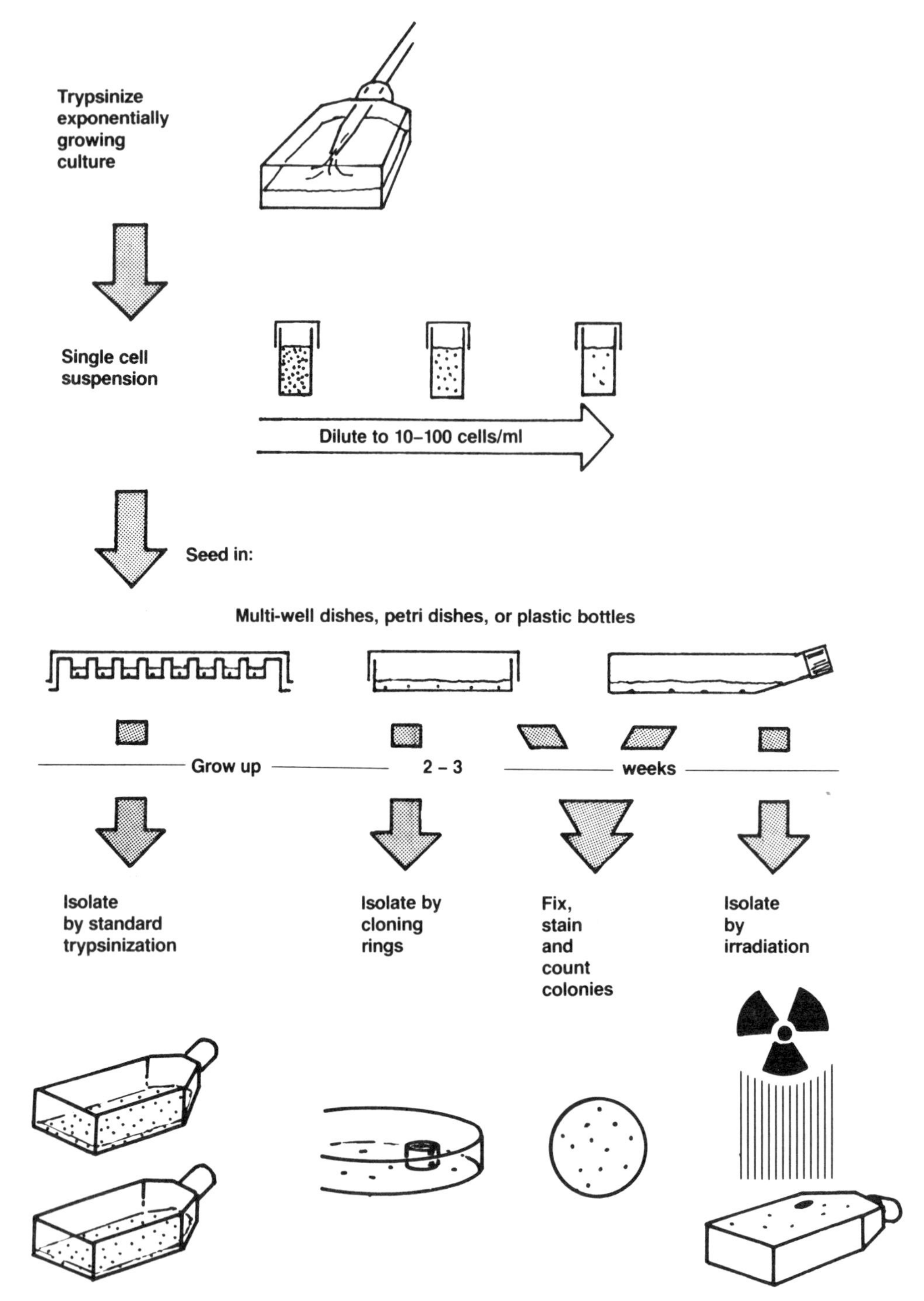

Fig. 13.2. *Cloning cells in monolayer culture. When clones form, they may be (1) isolated (a) directly, from multi-well dishes, (b) by the cloning ring technique (center left of figure), (c) by irradiating the flask while shielding one colony (bottom right-hand side of figure), or (2) fixed, stained, and counted (center right of figure) for analysis.*

culture flasks or dishes
tubes for dilution
hemocytometer or cell counter

Protocol

1. Trypsinize cells (see Chapter 12) to produce a single cell suspension. Undertrypsinizing will produce clumps; overtrypsinizing will reduce viability

2.
While cells are trypsinizing, number flasks or dishes and measure out medium for dilution steps (four dilution steps may be necessary to reduce a regular monolayer to a concentration suitable for cloning)
3.
When cells round up and start to detach, disperse the monolayer in medium containing serum or trypsin inhibitor, count and dilute to desired seeding concentration. If cloning the cells for the first time, choose a range of 10, 50, 100 and 200 cells/ml (Table 13.1)
4.
Seed petri dishes or flasks with the requisite amount of medium (see Chapter 12), place petri dishes in a humid CO_2 incubator or gassed sealed container (2–10% CO_2, see Chapter 9), or gas flasks with CO_2, seal with cap, and place in dry incubator
5.
Leave untouched for 1 wk. If colonies have formed, isolate (see below); if not, replace medium and continue to culture for a further week, or feed again and culture for a further week, or feed again and culture for 3 wk if necessary. If no colonies have appeared by 3 wk, it is unlikely that they will do so

Stimulation of Plating Efficiency

When cells are plated at low densities, the survival falls in all but a few cell lines. This does not usually present a severe problem with continuous cell lines where the plating efficiency seldom drops below 10%, but with primary cultures and finite cell lines, the plating efficiency may be quite low—0.5%–5% or even zero. Numerous attempts have been made to improve plating efficiencies, based on the assumption either that cells require a greater range of nutrients at low densities or that cell-derived diffusible signals or conditioning factors are present in high-density cultures and absent or too dilute at low densities. The intracellular metabolic pool of a leaky cell in a dense population will soon reach equilibrium with the surrounding medium, while that of an isolated cell never will. This was the basis of the capillary technique of Sandford et al. [1948], when the L929 clone of L-cells was first produced. The confines of the capillary tube allowed the cell to create a locally enriched environment mimicking the higher cell density state. In microdrop techniques developed later, the cells were seeded as a microdrop under liquid paraffin. Keeping one colony separate from another, as in the capillary techniques, colonies could be isolated subsequently. As media improved, however, plating efficiencies increased, and Puck and Marcus [1955] were able to show that cloning cells by simple dilution (as described above) in association with a feeder layer of irradiated mouse embryo fibroblasts (see below) gave acceptable cloning efficiencies, although subsequent isolation required trypsinization from within a collar placed over each colony.

TABLE 13.1. Relationship of Seeding Density to Plating Efficiency

Expected plating efficiency	Optimal cell number to be seeded	
	Per ml	Per cm^2
0.1%	10^4	2×10^3
1.0%	10^3	200
10%	100	20
50%	20	4
100%	10	2

Some modifications which may improve clonal growth are listed below.

Medium. Choose a rich medium such as Ham's F12 or one which has been optimized for the cell type in use, e.g., MCDB 105 [Ham and McKeehan, 1978] for human fibroblasts, Ham's F12 or MCDB 301 for CHO [Ham 1963; Hamilton and Ham, 1977].

Serum. Where serum is required, fetal bovine is generally better than calf or horse. Select a batch for cloning experiments which gives a high plating efficiency during tests.

Conditioning. (1) Grow homologous cells, embryo fibroblasts, or another cell line to 50% of confluence, change to fresh medium, incubate for a further 48 hr, and collect the medium. (2) Filter through a 0.2-μm sterilizing filter (the medium may need to be clarified first by centrifugation 10,000 g, 20 min, or filtration through 5-μm and 1.2-μm filters (see Chapter 10, section on sterilization of serum). (3) Add to cloning medium 1 part conditioned medium to 2 parts cloning medium.

Feeder layers (Fig. 13.3, regular feeder layer). (1) Trypsinize embryo fibroblasts from primary culture (see Chapter 11) and reseed at 10^5 cells/ml. (2) At 50% confluence, add mitomycin-C, 2 μg/10^6 cells, 0.25 μg/ml, overnight [MacPherson and Bryden, 1971], or ir-

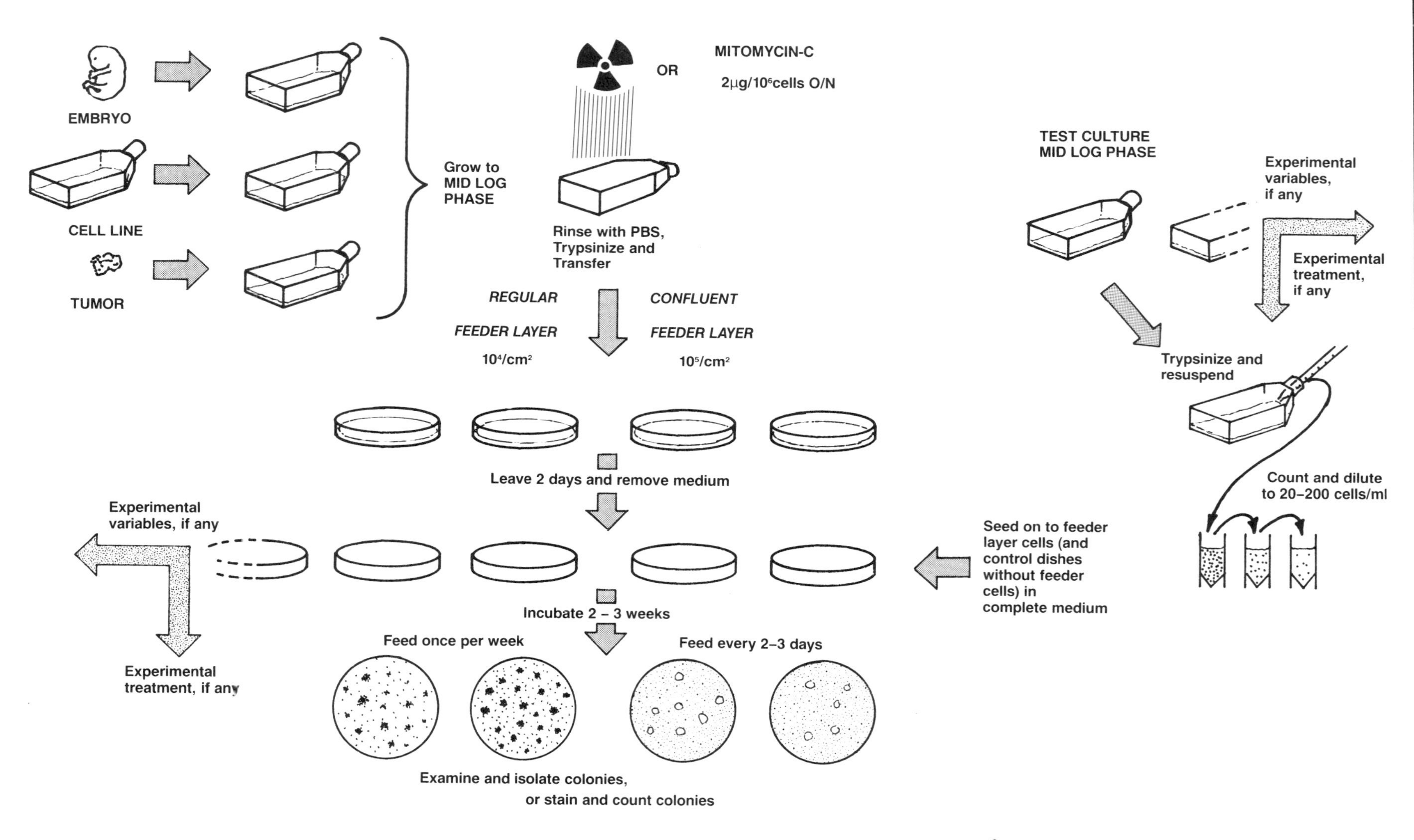

Fig. 13.3. *Cloning cells in the presence of a feeder layer. A low-density feeder layer (~ 10^4 cells/cm^2) (left-hand side of figure) is used to enhance cloning efficiency and clonal growth, while a high-density, confluent, feeder layer (e.g., 2×10^5 normal human diploid fibroblasts or glial cells/cm^2) can be used as a selective substrate to minimize fibroblastic over growth (see also Fig. 13.8 and text).*

radiate culture with 3,000 rad. (3) Change the medium after treatment, and after a further 24 hr, trypsinize the cells and reseed in fresh medium at 5×10^4 cells/ml (10^4 cells/cm^2). (4)Incubate for a further 24–48 hr and then seed cells for cloning. The feeder cells will remain viable for up to 3 wk but will eventually die out and are not carried over if the colonies are isolated.

Other cell lines or homologous cells may be used to improve the plating efficiency but heterologous cells have the advantage that if clones are to be isolated later, chromosome analysis will rule out accidental contamination from the feeder layer.

Hormones. Insulin, 1–10 IU//ml has been found to increase the plating efficiency of several cell types [Hamilton and Ham, 1977]. Dexamethasone, 2.5×10^{-5} M, ~10 μg/ml (a soluble synthetic hydrocortisone analogue) improves the plating efficiency of human normal glia, glioma, fibroblasts, and melanoma, and chick myoblasts, and will give increased clonal growth (colony size) if removed 5 d after plating [Freshney et al., 1980a,b].

Intermediary metabolites. Keto acids, e.g., pyruvate or α-ketoglutarate, [Griffiths and Pirt, 1967; McKeehan and McKeehan, 1979] and nucleosides [α-medium, Stanners et al., 1971], have been used to supplement media and are already included in the formulation of a rich medium like Ham's F12. Pyruvate is also added to Dulbecco's modification of Eagle's MEM [Dulbecco and Freeman, 1959; Morton, 1970].

Carbon dioxide. CO_2 is essential to obtain maximum cloning efficiency for most cells. While 5% is most usual, 2% is sufficient for many cells, and may even be slightly better for human glia and fibroblasts. HEPES (20 mM) may be used with 2% CO_2, protecting the cells against pH fluctuations during feeding and in the event of failure of the CO_2 supply. (Using 2% CO_2 also cuts down in the consumption of CO_2.) At the other extreme, Dulbecco's modification of Eagle's MEM is normally equilibrated with 10% CO_2 and is frequently used for cloning myeloma hybrids for monoclonal antibody production. The concentration of bicarbonate must be adjusted if the CO_2 tension is altered so that equilibrium is reached at pH 7.4 (see Table 8.2).

Treatment of substrate. Polylysine improves the plating efficiency of human fibroblasts in low serum concentrations [McKeehan and Ham, 1976] (see Chapter 8). (1) Add 1 mg/ml poly-D-lysine in water to plates (~5 ml/25 cm^2). (2) Remove and wash plates with 5 ml PBSA per 25 cm^2. The plates may be used immediately or stored for several weeks before used.

Fibronectin also improves the plating of many cells [Barnes and Sato, 1980]. The plates should be pretreated with 5 μg/ml fibronectin incorporated in the medium.

Trypsin. Pure, twice recrystallized, trypsin used at 0.05 μg/ml may be preferable to crude trypsin, but there are conflicting reports on this. McKeehan [1977] noted a marked improvement in plating efficiency when trypsinization (pure trypsin) was carried out at 4° C.

Multiwell Dishes

If clones are to be isolated, cloning by dilution directly into microwells (microtitration dishes or 24-well plates, see Fig. 8.4) makes subsequent harvesting easier. The plates must be checked regularly after seeding, however, to confirm that either only one cell is present per well at the start or, if there is more than one cell per well, they are not clumped and that only one cell gives rise to a colony, i.e., that the colonies which form are truly clonal in origin, and only one colony forms in the well.

Semisolid Media

Some cells, particularly virally transformed fibroblasts, will clone readily in suspension. To hold the colony together and prevent mixing, the cells are suspended in agar or methocel and plated out over an agar underlay or into nontissue culture grade dishes.

Cloning in agar. See Figure 13.4 and see also Metcalf [1970], Pike and Robinson [1970], and MacPherson [1973].

Outline

Agar is liquid at high temperatures but is a gel at 36.5° C. Cells are suspended in warm agar, and, when incubated after the agar gels, will form discrete colonies which may be isolated easily.

Materials

- 2% agar
- medium, e.g., Ham's F12, RPMI 1640, or CMRL 1066
- fetal bovine serum
- pipettes
- 35-mm petri dishes
- 50-mm × 8-mm test tubes
- boiling water bath
- water bath at 45° C
- ice tray

Note. preparing medium and cells, work out cell dilutions and label petri dishes or multiwell plates.

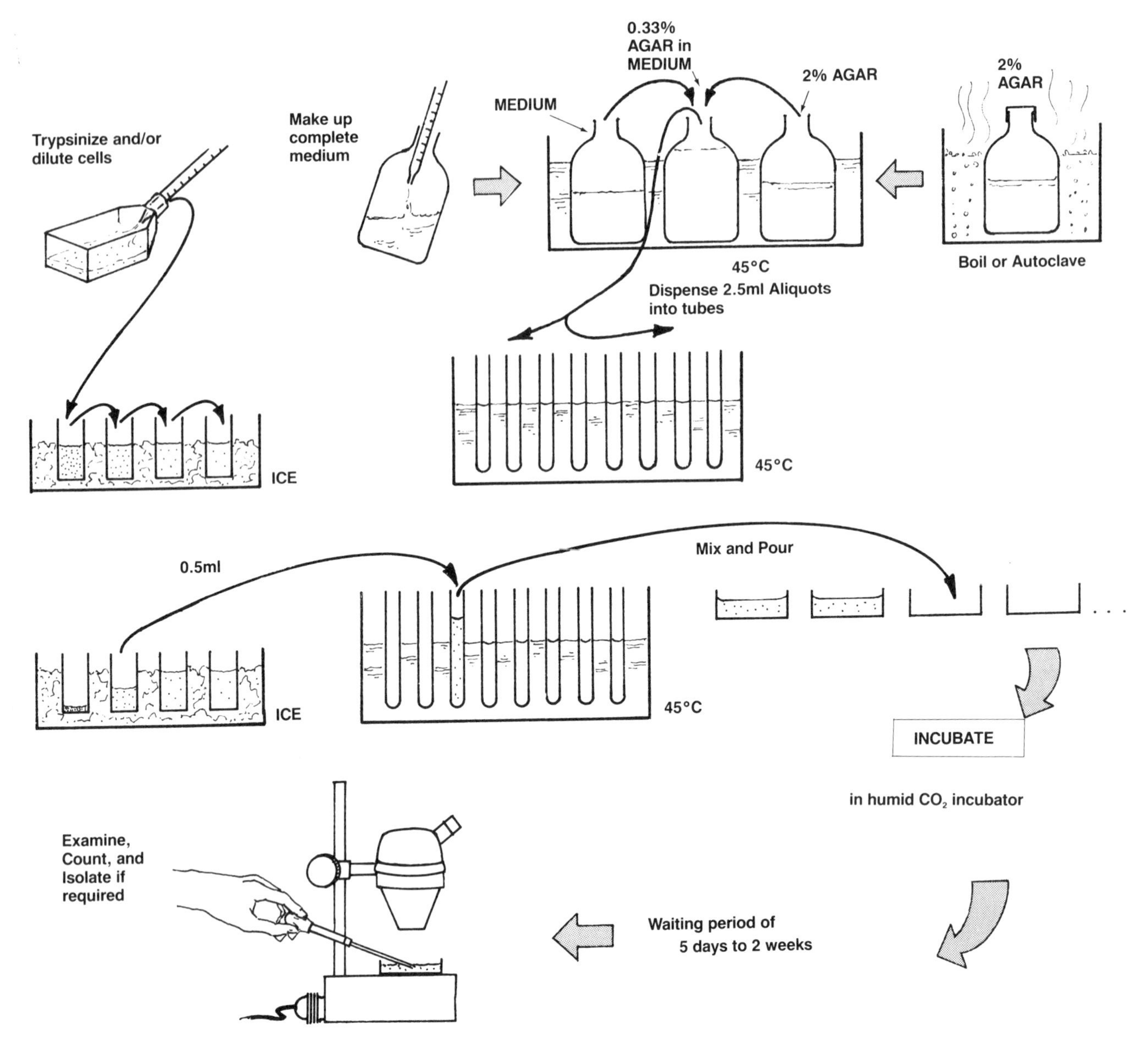

Fig. 13.4. *Cloning cells in suspension in agar.*

Convenient cell numbers per 35-mm dish are 1,000, 333, 111, 37.

Protocol

1.
Dissolve agar by placing bottle in a boiling water bath or autoclave. Cool to 80° C and transfer to 45° C water bath. (If you are delayed, keep agar at 60° C)
2.
Make up medium A: 70 ml medium and 30 ml FBS, keep at 45° C
3.
Make up medium B: 60 ml medium A and 12 ml 2% agar, keep at 45° C
4.
Count the cell suspension and dilute to give 2,000/ml, 667/ml (1:3), 222/ml (1:3), 74/ml (1:3). Place on ice
5.
Place 5-ml test tubes in rack and keep at 45° C
6.
Add 2.5 ml medium B to test tubes
7.
Add 0.5-ml cell suspension to one tube, mix, and pour into dish immediately
8.
After all dishes have been poured, place them at

4° C for 10 min

9.

Before incubating the dishes in a humid CO_2 incubator, it is advisable to put them into another container to try to avoid contamination of the cultures from the moist atmosphere of the incubator. (a) Place petri dishes or multiwell plates in a plastic box with a lid and containing a dish of water. (The box should be washed first with 70% alcohol and allowed to dry.) (b) When using 35-mm petri dishes, two can be put into a 10-cm petri dish (nontissue culture grade) with a third 35-mm petri

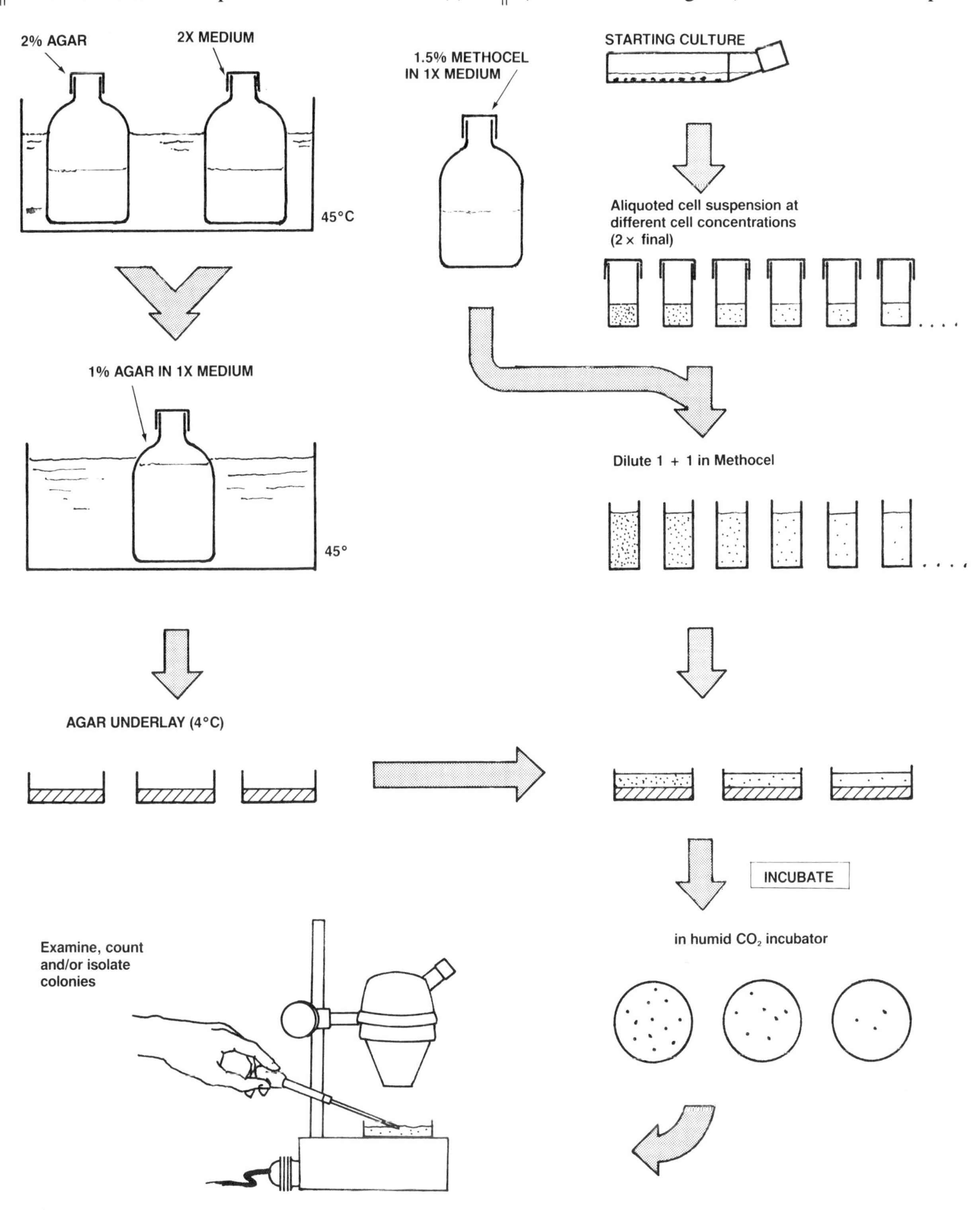

Fig. 13.5. *Cloning cells in suspension in Methocel over an agar underlay.*

dish containing 3 ml sterile water
Note. Rinse all pipettes used for agar with hot water before discarding.

Cloning in methocel over agar base [Buick et al., 1979].

Outline

Suspend cells in medium containing Methocel and seed into dishes containing gelled agar medium (Fig. 13.5).

Materials

As for agar cloning; 1.36% Methocel (4,000 counts/sec) in deonized distilled water.

Protocol

1. Prepare agar underlay by heating sterile 1.0–2.0% agar to 100° C, bring to 45° C, and dilute with an equal volume of double-strength medium at 45° C (prepare from × 10 concentrate to half the recommended final volume and add twice the normal concentration of serum). Plate out 1 ml immediately into 35-mm dishes or 6 × 35 mm multiwell plates, and allow to gel at 4° C for 10 min
2. Trypsinize or collect cells from suspension and dilute to double the required final concentration
3. Dilute the cells with an equal volume of methocel and plate out 1 ml over the agar underlay (10–1,000 cells per dish for continuous cell strains but up to 5 × 10^5 per dish may be needed for primary cultures)
4. Incubate until colonies form. Since the colonies form at the interface between the agar and the Methocel, fresh medium may be added, 1 ml per dish or well, after 1 wk and removed and replaced with more fresh medium after 2 wk without disturbing the colonies

Many of the recommendations applying to medium supplementation for monolayer cloning also apply to suspension cloning. In addition, sulphydryl compounds such as mercaptoethanol (5 × 10^{-5} M), glutathione (1 mM), or α-thioglycerol (7.5 × 10^{-5} M) [Iscove et al., 1980] are sometimes used. Macpherson [1973] found the inclusion of DEAE dextran was beneficial for cloning.

Most cell types clone in suspension with a lower efficiency than in monolayer, some cells by two or three orders of magnitude. Isolation of colonies is, however, much easier.

Isolation of Clones

Monolayer clones—multiwell plates. If cells are cloned directly into multiwell plates (see above), colonies may be isolated by trypsinizing individual wells. It is necessary to confirm the clonal origin of the colony during its formation by regular microscopic observation.

Cloning rings. If cloning is performed in petri dishes, there is no physical separation between colonies. This must be created by removing the medium and placing a ring around the colony to be isolated (Fig. 13.6).

Outline

The colony is trypsinized from within a porcelain, Teflon, or stainless steel ring and transferred to one of the wells of a 24- or 12-well plate, or directly to a 25-cm^2 flask (see step 3 above) (Fig. 13.7).

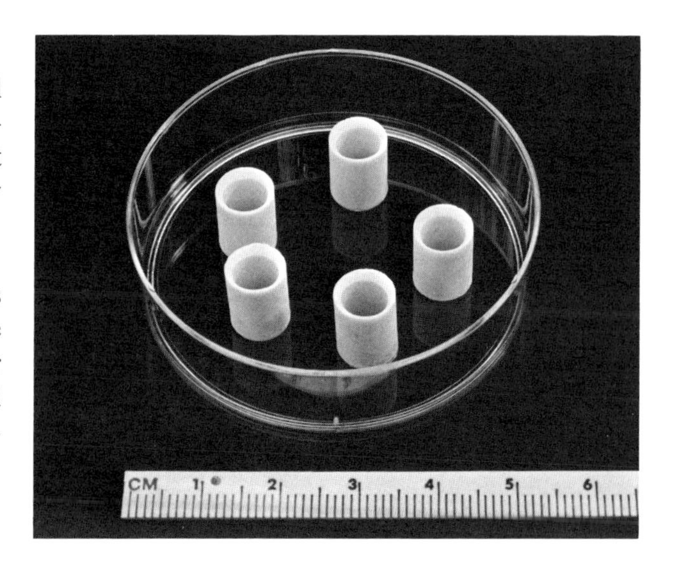

Fig. 13.6. *Cloning rings. Porcelain rings (Fisher) are illustrated, but thick-walled stainless steel rings (e.g., roller bearings) or plastic (e.g., cut from nylon or Teflon thick-walled tubing) can be used. Whatever the material, the base must be smooth, to seal with silicone grease onto the base of the petri dish, and the internal diameter just wide enough to enclose one whole clone, without overlapping adjacent clones.*

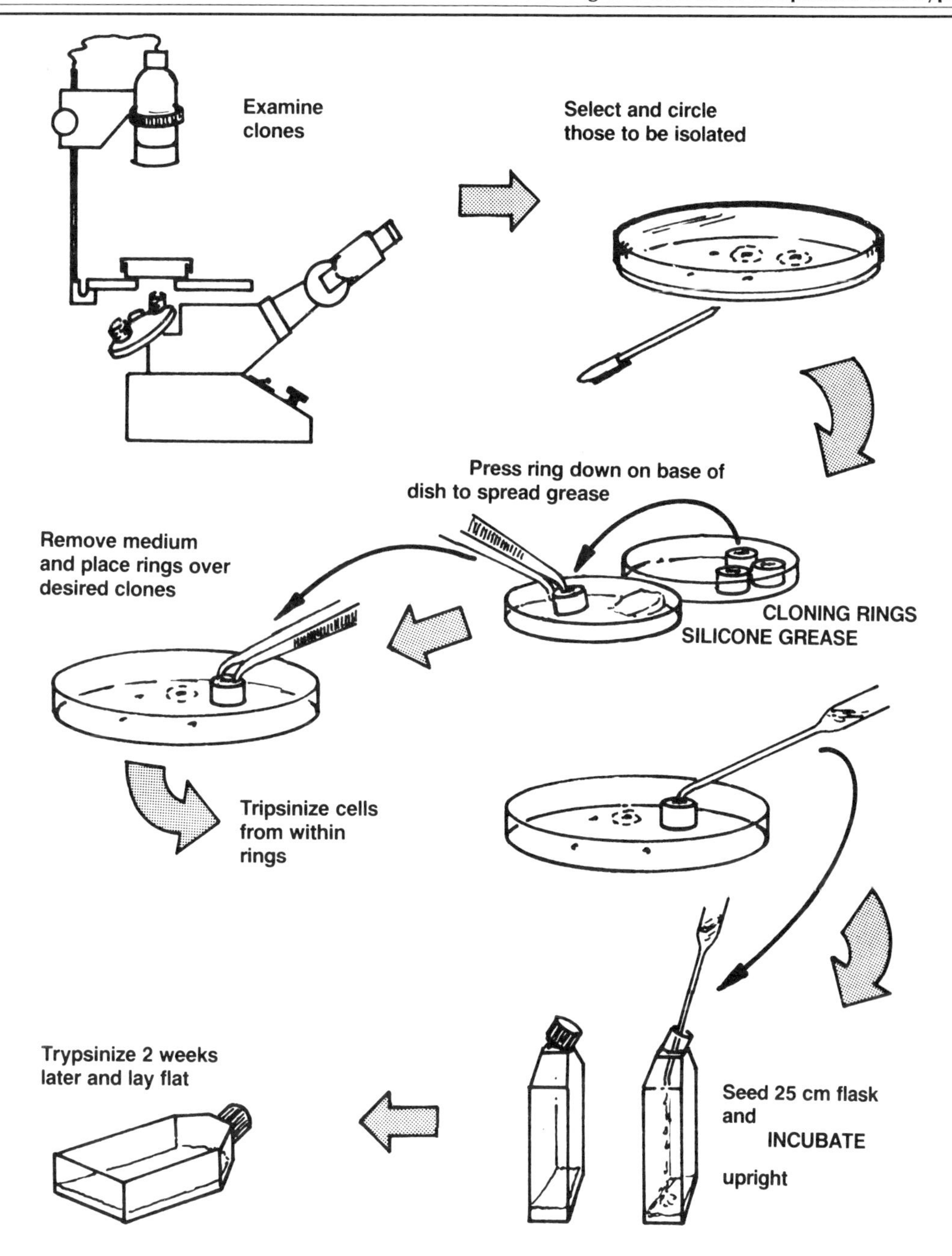

Fig. 13.7. *Isolation of clones with cloning rings.*

Materials

cloning rings
silicone grease
Pasteur pipettes with bent end
0.25% trypsin
medium
24-well plate and/or 25 cm^2 flasks
sterile forceps

Protocol

1.
Sterilize cloning rings and silicone grease separately in glass petri dishes, by dry heat

2.
Prepare about 20 bent Pasteur pipettes by heating briefly in a bunsen flame and allowing about 12 mm of the tip to drop under gravity. If the pipette

is held at 30° above horizontal, the bend will be 120°. Place pipettes in sterile test tubes and allow to cool before use

3. Examine clones and mark those that you wish to isolate with a felt tip marker on the underside of the dish

4. Remove medium from dish

5. Using sterile forceps, take one cloning ring, dip in silicone grease, and press down on dish alongside silicone grease to spread the grease round the base of the ring

6. Place ring around desired colony

7. Repeat steps 5 and 6 for two or three other colonies in same dish

8. Add sufficient 0.25% trypsin to fill the hole in ring (~0.4 ml), leave 20 sec, and remove

9. Close dish and incubate for 15 min

10. Add 0.4 ml medium to each ring

11. Taking each clone in turn, pipette medium up and down to disperse cells, and transfer to a well of a 24-well plate, or to a 25-cm^2 flask standing on end. Use a separate pipette for each clone

12. Wash out ring with a second 0.4 ml medium, and transfer to same well

13. Close plate and incubate, or if using flasks, add 1 ml medium and incubate standing on end

14. When clone grows to fill well, transfer up to 25-cm^2 flask, incubated conventionally with 5 ml medium. If using up-ended flask technique, remove medium when end of flask confluent, trypsinize cells, resuspend in 5 ml medium, and lay flask down flat. Continue incubation

The cloning ring technique may be applied when cells are cloned in a plastic flask by swabbing the flask with alcohol and slicing the top off with a heated sterile scalpel or hot wire. Thereafter proceed as for petri dishes.

Irradiation. Alternatively, where an irradiation source is available, clones may be isolated by shielding one and irradiating the rest of the monolayer.

Outline

Invert the flask under an x-ray machine or ^{60}Co source, screening the desired colony with lead.

Materials

x-ray or cobalt source
piece of lead 2 mm thick
PBSA
0.25% trypsin
medium

Protocol

1. Select desired colony

2. Invert flask under x-ray or cobalt source

3. Cover colony with a piece of lead 2 mm thick

4. Irradiate with 3,000 rad

5. Return to sterile area and remove medium, trypsinize, and allow cells to reestablish in the same bottle, using the irradiated cells as a feeder layer

Fig. 13.8. *Selective cloning of breast epithelium on a confluent feeder layer. a. Colonies forming on plastic alone after seeding 4,000 cells from a breast carcinoma culture/cm^2 (2 × 10^4 cells/ml). Small dense colonies are epithelial cells, larger stellate colonies are fibroblasts. b. Colonies of cells from the same culture, seeded at 400 cells/cm^2 (2,000 cells/ml) on a confluent feeder layer of FHS 74 Int cells [Owens et al., 1974]. The epithelial colonies are much larger than in a, the plating efficiency is higher, and there are no fibroblastic colonies. c. Colonies from a different breast carcinoma culture plated onto the same feeder layer. Note different colony morphology with lighter stained center and ring at point of interaction with feeder layer. d. Colonies from normal breast culture seeded onto FHI cells (fetal human intestine similar to FHS 74 Int). There are a few small fibroblastic colonies present in c and d (after a technique described by Dr. A.J. Hackett, personal communication).*

Fig. 13.9. *Selective growth of glioma on confluent feeder layer. Cells were seeded at 2 × 10^4/ml (4 × 10^3/cm^2) onto confluent, mitomycin-C treated feeder layers (see text) of FHS 74 Int cells [Owens et al., 1974] and labeled at intervals thereafter with ^{3}H-thymidine (see text), extracted, and counted.*

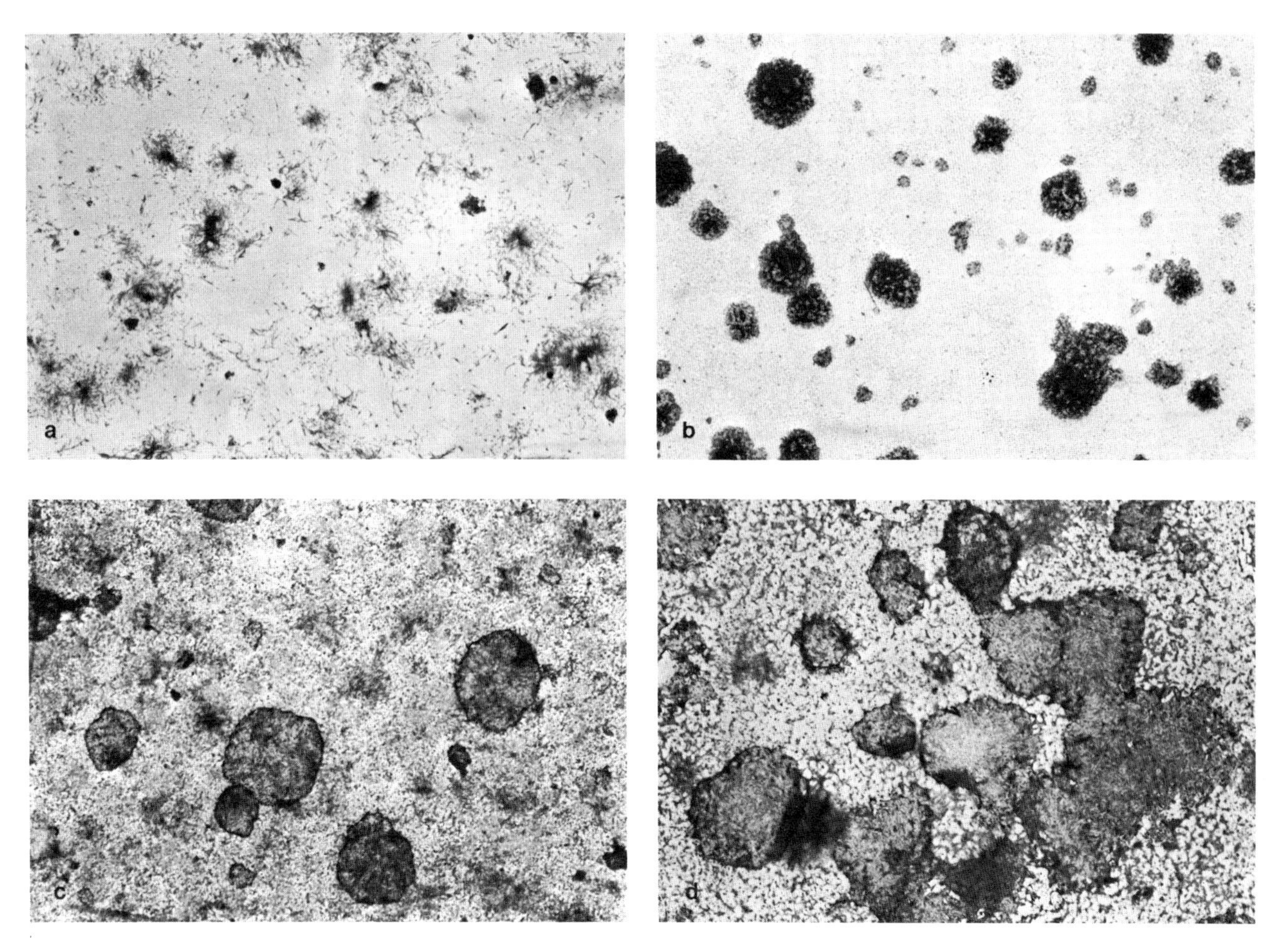

Figure 13.8. *Legend on facing page.*

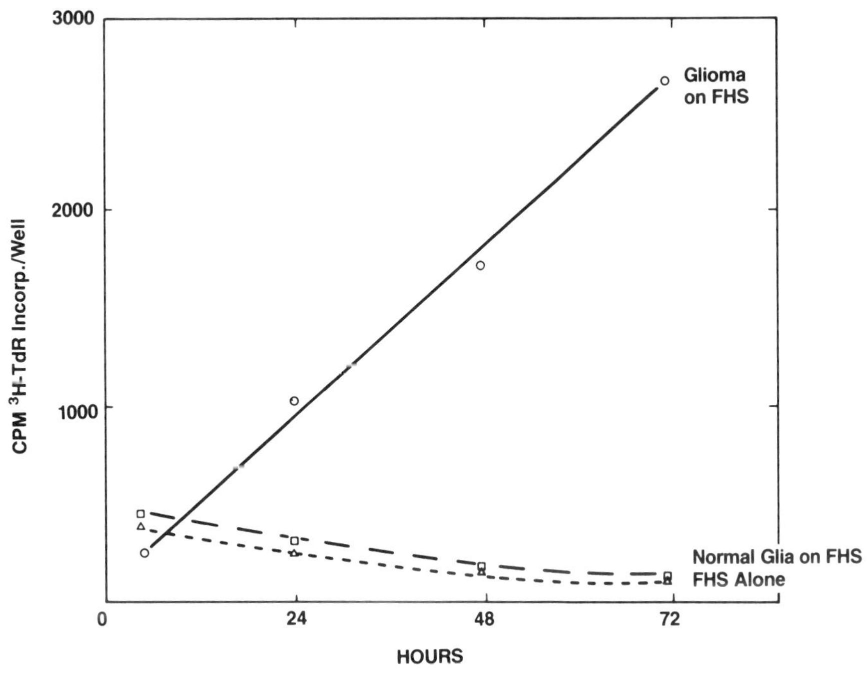

Figure 13.9. *Legend on facing page.*

If irradiation and trypsinization is carried out when the colony is about 100 cells in size, then the trypsinized cells will reclone. Three serial clonings may be performed within 6 wk by this method.

Other isolation techniques include: (1) distributing small coverslips or broken fragments of coverslips on the bottom of a petri dish. When plated out at the correct density, some colonies are found singly distributed on a piece of glass and may be transferred to a fresh dish or multiwell plate. (2) Capillary technique of Sanford et al. [1948]. A dilute cell suspension is drawn into a glass capillary tube (e.g., 50 μl Drummond Microcap) allowing colonies to form inside the tube. The tube is then carefully broken on either side of a colony and transferred to a fresh plate. (3) Petriperm dish. This is a petri dish with a thin gas-permeable base (see Chapter 8), which may be cut with scissors or a scalpel to isolate colonies. Since this means keeping the outside of the dish sterile, it needs to be handled aseptically and kept inside a larger sterile petri dish.

Suspension Clones

Outline

Draw colony into micropipette and transfer to a flask or the well of a multiwell plate.

Materials

24-well plates
medium
microcapillary pipettes
dissecting microscope
25-cm^2 culture flask

Protocol

Picking colonies is best done on a dissecting microscope.

1. Pipette 1 ml of medium into each well of a 24-well plate
2. Using a separate 50-μl microcapillary pipette for each clone, place the tip of the pipette against the colony to be isolated and gently draw in the colony
3. Transfer to a 24-well dish and flush out colony with medium. If from Methocel, the colony will settle, adhere, and grow out. If from agar, you

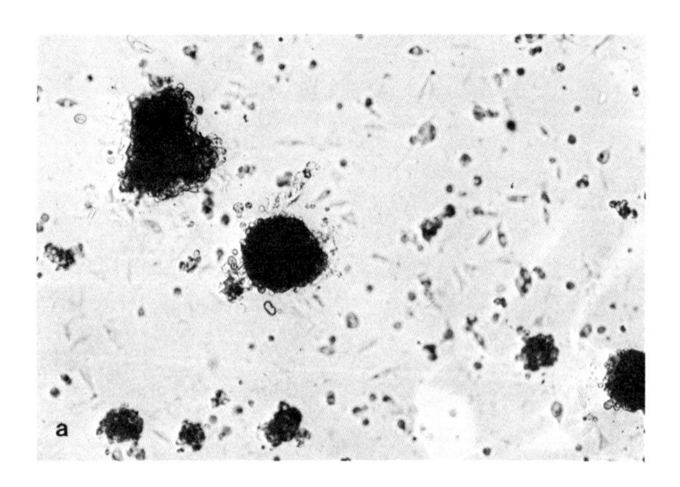

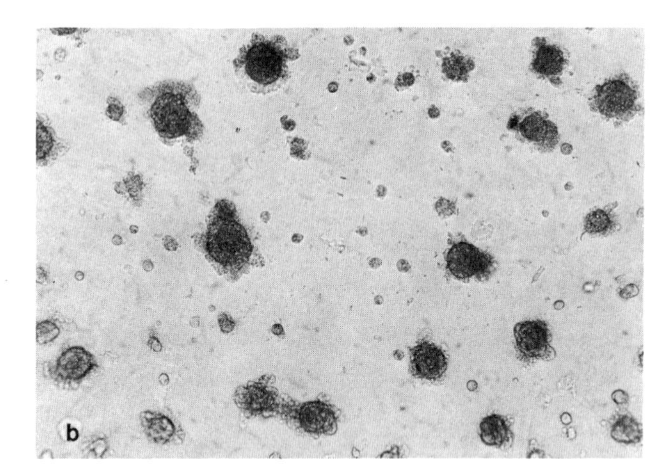

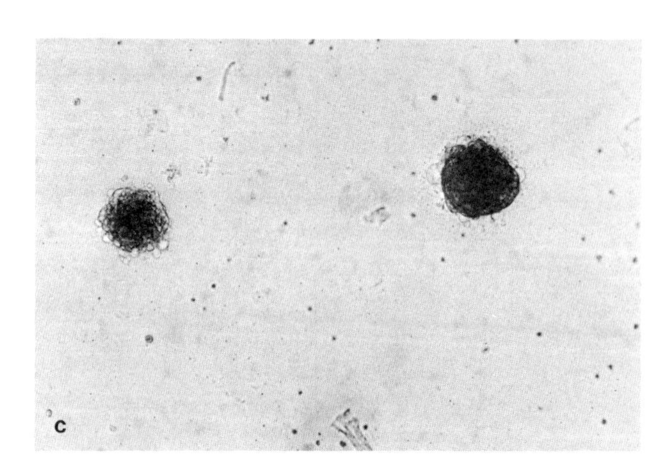

Fig 13.10. *Growth of melanoma, fibroblasts, and glia in suspension. Cells were plated out at 5 × 10^5 per 35 mm dish (2.5 × 10^5 cells/ml) in 1.5% methocel over a 1.25% agar underlay. Colonies were photographed after 3 wk. a. Melanoma. b. Human normal embryonic skin fibroblasts. c. Human normal adult glia. d. Colony-forming efficiency of normal and malignant glial cells in suspension. Unshaded bars, colonies of over eight cells (approximately), and stippled bars, colonies of over 32 cells (approximately). Colony counts were done on an Artek Colony Counter at different threshold settings.*

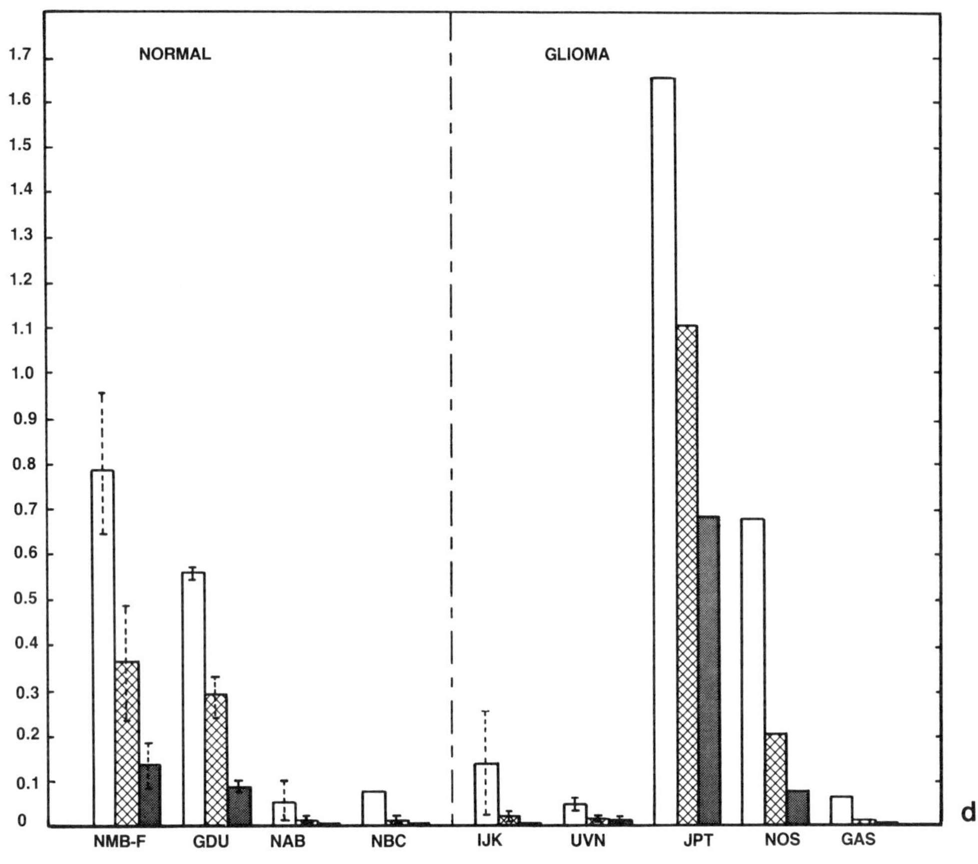

Figure 13.10. *(continued)*

may need to pipette the colony up and down a few times in the well to remove the agar

4. Clones may also be seeded directly into a 25-cm^2 plastic flask standing on end (see above)

SELECTIVE MEDIA

Manipulating the culture conditions by using a selective medium is a standard method for selecting microorganisms. Its application to animal cells in culture is limited, however, by the basic metabolic similarities of most cells isolated from one animal in terms of their nutritional requirements. The problem is accentuated by the effect of serum, which tends to mask the selective properties of different media. Peehl and Ham [1980] were able to demonstrate that by using two different media, MCDB 105 and MCDB 151, with minimal amounts of dialyzed serum, either fibroblasts or epithelial cells could be grown preferentially from human foreskin.

Gilbert and Migeon [1975, 1977] replaced the L-valine in the culture medium with D-valine and demonstrated that cells possessing D-amino acid oxidase would grow preferentially. Kidney tubular epithelium and epithelial cells from fetal lung and umbilical cord may be selected this way.

Much of the effort in developing selective conditions has been aimed at suppressing fibroblastic overgrowth. Whei-Yang Kao [1977] used cis-OH-proline for this purpose, although this substance can prove toxic to other cells. Fry and Bridges [1979] found phenobarbitone inhibited fibroblastic overgrowth in cultures of hepatocytes and Braaten et al. [1974] were able to reduce the fibroblastic contamination of neonatal pancreas by treating the culture with sodium ethylmercurithiosalicylate. One of the more successful approaches was the development of a monoclonal antibody to the stromal cells of a human breast carcinoma [Edwards et al., 1980]. Used with complement, this antibody proved cytotoxic to fibroblasts from several tumors and helped to purify a number of malignant cell lines.

Selective media are also commonly used to isolate hybrid clones from somatic hybridization experiments. HAT medium, a combination of hypoxanthine, aminopterin, and thymidine, selects hybrids with both hypoxanthine guanine phophoribosyltransferase and thymidine kinase from parental cells deficient in one or the other enzyme (see Chapter 21) [Littlefield, 1964].

INTERACTION WITH SUBSTRATE

Selective Adhesion

Different cell types have different affinities for the culture substrate and will attach at different rates. If a primary cell suspension is seeded into one flask and transferred to a second after 30 min, a third after 1 hr, and so on, the most adhesive cells will be found in the first flask and the least adhesive in the last. Macrophages will tend to remain in the first flask, fibroblasts in the next few flasks, then epithelial, and finally hemopoietic cells in the last flask. Polinger [1970] used a similar procedure for the separation of embryonic heart muscle cells from fibroblasts.

If collagenase in complete medium is used for primary disaggregation of the tissue (see Chapter 11), most of the cells released will not attach within 48 hr unless the collagenase is removed. However, macrophages migrate out of the fragments of tissue and attach during this period and can be removed from other cells by transferring the disaggregate to a fresh flask after 48–72 hr-treatment with collagenase. This technique works well during disaggregation of biopsy specimens from human tumors.

Selective Detachment

Treatment of a heterogeneous monolayer with trypsin or collagenase will remove some cells more rapidly than others. Periodic brief exposure to trypsin removed fibroblasts from cultures of fetal human intestine [Owens et al., 1974] and skin [Milo et al., 1980], and Lasfargues [1973] found exposure of cultures of breast tissue to collagenase for a few days at a time removed fibroblasts and left the epithelial cells. EDTA, on the other hand, may release epithelial cells more readily than fibroblasts [Paul, 1975].

Dispase II (Boehringer, Mannheim) selectively dislodges sheets of epithelium from human cervical cultures grown on feeder layers of 3T3 cells (see below) without dislodging the 3T3 cells [Stanley, personal communication]. This technique may be effective in subculturing epithelial cells from other sources, excluding stromal fibroblasts.

Nature of Substrate

The hydrophilic nature of most culture substrates (see also Chapter 8) appears to be necessary for cell attachment, but little is known about variations in charge distribution on the cell surface and how different mosaic patterns may interact with different substrates. Since cell sorting in the embryo is a highly selective process and probably relates to differences in the distribution of charged molecules and specific receptor sites on the cell surface, qualitative and quantitative variations in substrate affinity should be anticipated in cultured cells. The relative infrequency

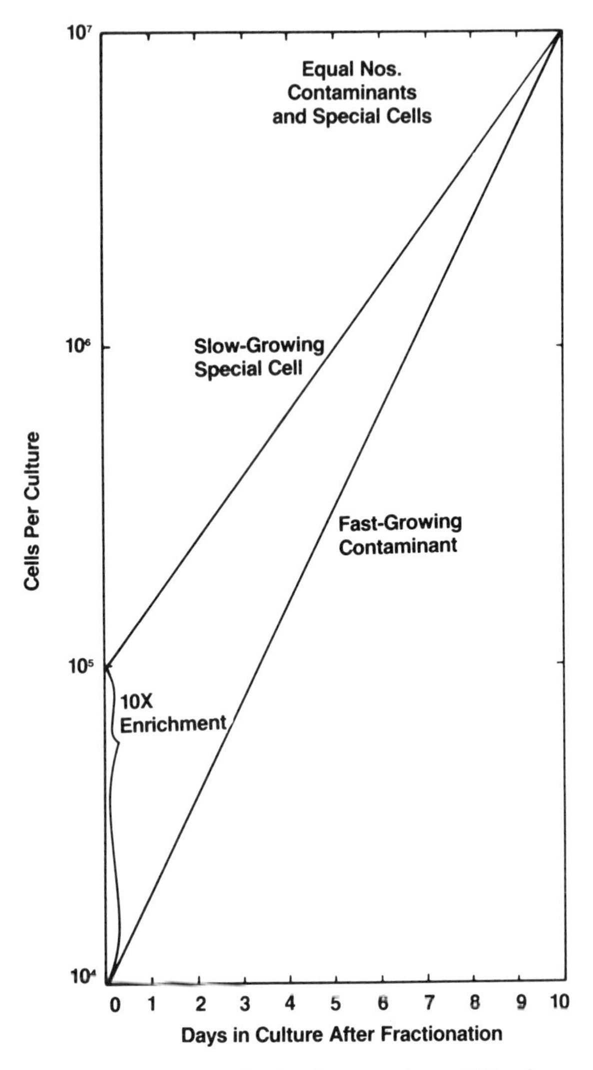

Fig. 13.11. *Overgrowth of a slow-growing cell line by a rapidly growing contaminant. This is a hypothetical example, but it demonstrates that a 10% contamination with a cell population which doubles every 24 hr will reach equal proportions with a cell population which doubles every 36 hr after only 10 d growth.*

with which this is actually found probably illustrates our ignorance of the subtlety of cell-cell and cell-substrate interactions.

The selective effect of substrates on growth may depend on both differential rates of attachment and growth, although in practice the two are indistinguishable. Polyacrylamide layers allow the cloning of tumor cells but not normal fibroblasts [Jones and Haskill, 1973, 1976]. Transformed cells proliferate on Teflon while most other cells will not [Parenjpe et al., 1975]. Macrophages will also attach to Teflon but do not proliferate. The dermal surface of freeze-dried pig skin was shown to allow growth of epidermal cells but not fibroblasts [Freeman et al., 1976]. Collagen, presumably the basis of the selection, has also been used in gel form to favor epithelial cell growth [Lillie et al., 1980] and in its denatured form to support endothelial outgrowth from aorta into a fibrin clot [Nicosia and Leighton, 1981].

Feeder layers. The conditioning of the substrate by feeder layers has been discussed already, in Chapter 8. Feeder layers can also be used for the selective growth of epidermal cells [Rheinwald and Green, 1975] and for repressing stromal overgrowth in cultures of breast (Fig. 13.8) and colon (see Fig. 20.1) carcinoma [Freshney et al., 1981]. The author has also been able to demonstrate that human glioma will grow on confluent feeder layers of normal glia while cells derived from normal brain will not [Freshney, 1980] (Fig. 13.9; see Chapter 16).

Semisolid supports. Transformation of many fibroblast cultures reduces anchorage dependence of cell proliferation (see Chapter 16) [Macpherson and Montagnier, 1964]. By culturing the cells in agar (see above) after viral transformation, it is possible to isolate colonies of transformed cells and exclude most of the normal cells. Most normal cells will not form colonies in suspension with the same high efficiency as virally transformed cells, although they will often do so with low plating efficiencies. Colonies of hemopoietic cells will form in semisolid media [Metcalf, 1970], but these usually mature to nondividing differentiated cells and cannot be subcultured (see also Chapter 16). The difference between virally transformed fibroblasts and untransformed cells is not seen as clearly in attempts at selective culture of spontaneously arising tumors. Experiments in the author's laboratory have shown that normal glia and fetal skin fibroblasts will form colonies in suspension just as readily as glioma and melanoma (Fig. 13.10).

Cell cloning and the use of selective conditions have a significant advantage over physical cell separation techniques (next chapter) in that contaminating cells are either eliminated entirely by clonal selection or repressed by constant or repeated application of selective conditions. Even the best physical cell separation techniques will still allow some overlap between cell populations such that overgrowth will recur. As long as this situation exists, a steady state cannot be achieved and the constitution of the culture is altering continuously. From Figure 13.11 it can be seen that a 90% pure culture of line A will be 50% overgrown by a 10% contamination with line B in 10 days, given that B grows 50% faster than A. For continued culture, therefore, selective conditions are required in addition to, or in place, of physical separation techniques.

Chapter 14
Physical Methods of Cell Separation

While cloning or selective culture conditions are the preferred methods for purifying a culture, there are occasions when cells do not grow with a high enough plating efficiency to make cloning possible or when appropriate selection conditions are not available. It may then be necessary to resort to a physical separation technique such as rate or density sedimentation. Physical separation techniques have the advantage that they give a high yield more quickly than cloning although not with the same purity.

The more successful separation techniques (Fig. 14.1) depend on differences in (1) cell size, (2) cell density (specific gravity), (3) cell surface charge, (4) cell surface chemistry (affinity for lectins, antibodies, or chromatographic media) (5) total light scatter per cell, and (6) fluorescence emission of one or more cellular constituents or adsorbed antibody. The apparatus required ranges from about $10-worth of glassware to $200,000-worth of complex laser and computer technology; the choice depends on the parameter that you are obliged to use (see Table 14.1) and on your budget.

CELL SIZE AND SEDIMENTATION VELOCITY

The relationship between particle size and sedimentation rate at 1*g*, though complex for submicron particles is fairly simple for cells and can be expressed approximately as

$$v = \frac{r^2}{4} \quad \text{(cq.14.1)}$$

[Miller and Phillips, 1969], where v = sedimentation rate in mm/hr and r = radius of the cell in μm.

Unit Gravity Sedimentation

The apparatus required is illustrated in Figure 14.2 and can be assembled from routine laboratory glassware. To ensure stability of the column of liquid supporting the cells in the sedimentation chamber, it is formed from a serum, Ficoll, or albumen gradient, and run into the chamber through a baffle to prevent turbulence.

The height of the separation chamber determines how long the sedimentation may run. Since this is usually 2–4 hr, in a low-viscosity medium, 10 cm is approximately correct. A longer sedimentation time may give better resolution but may cause deterioration of the cells.

The width of the chamber controls the number of cells that may be loaded onto the gradient as the cell layer should be kept thin (~5 mm) and the cell concentration low (~10^6/ml).

The height and width (i.e., volume) also affect the filling rate. The chamber must not be filled too rapidly or turbulence will result, and it cannot be filled too slowly or the cells will sediment faster than the liquid level rises. The dimensions given in Figure 14.3 are optimal for separating about 2×10^7 cells of 15–18-μm diameter or up to 10^8 cells of 10–12-μm diameter.

The procedure for separating a typical cell suspension of average cell diameter 15 μm is as follows.

Outline

Float cells on top of a gradient of serum in medium, allow cells to sediment through the gradient for about 3 hr, and run off gradient into culture vessels (see Fig. 14.2).

Materials

300 ml Eagle's MEMS (suspension salts, see Table 9.1) + 30% serum
300 ml MEMS + 15% serum
20 ml MEMS + 5% serum

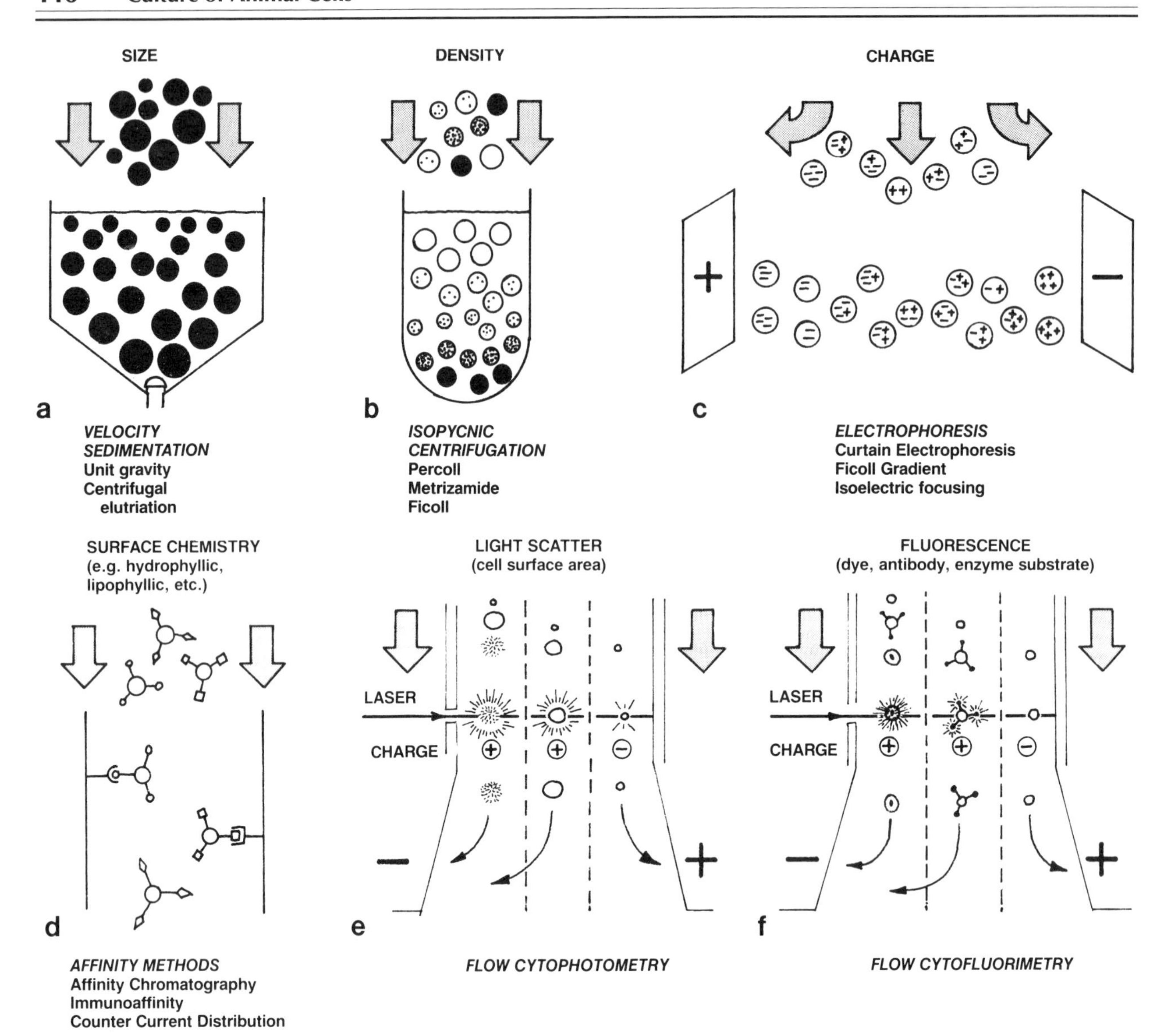

Fig. 14.1. *Cell separation techniques. a. Velocity sedimentation, influenced by cell size mainly, but also by cell density and surface area at elevated g. b. Isopycnic sedimentation. Cells sediment to a point in a density gradient equivalent to their own density. c. Electrophoresis. Cells migrate to either polarized plate according to net surface charge. d. Affinity methods. Cells are separated by their differential affinities for (1) chromotographic media, (2) antibodies or lectins bound to chromatographic media, or (3) two-phase aqueous polymer systems. e. Flow cytophotometry. Cells are diverted to either of two charged plates according to their light-scattering potential (proportional to surface area). f. Flow cytofluorimetry. Specific fluorochromes are used to label cells and electrophoretic separation is based on fluorescence emission. Both e and f are carried out on a single cell stream passing through the flow chamber of an instrument such as the Fluorescence-Activated Cell Sorter (FACS) (Becton Dickinson), the Cytofluorograph (Ortho), or the Coulter Cell Sorter.*

10 ml 0.25% trypsin-citrate (see Appendix)
30 ml PBS
20 ml MEMS + 3% serum
hemocytometer or cell counter
flotation medium (M sucrose or 20% Ficoll)
25-cm^2 flasks
growth medium

Protocol

1.

Prepare apparatus as in Figure 14.2. Incorporate Luer connections to allow for disassembly for sterilization. Package and autoclave

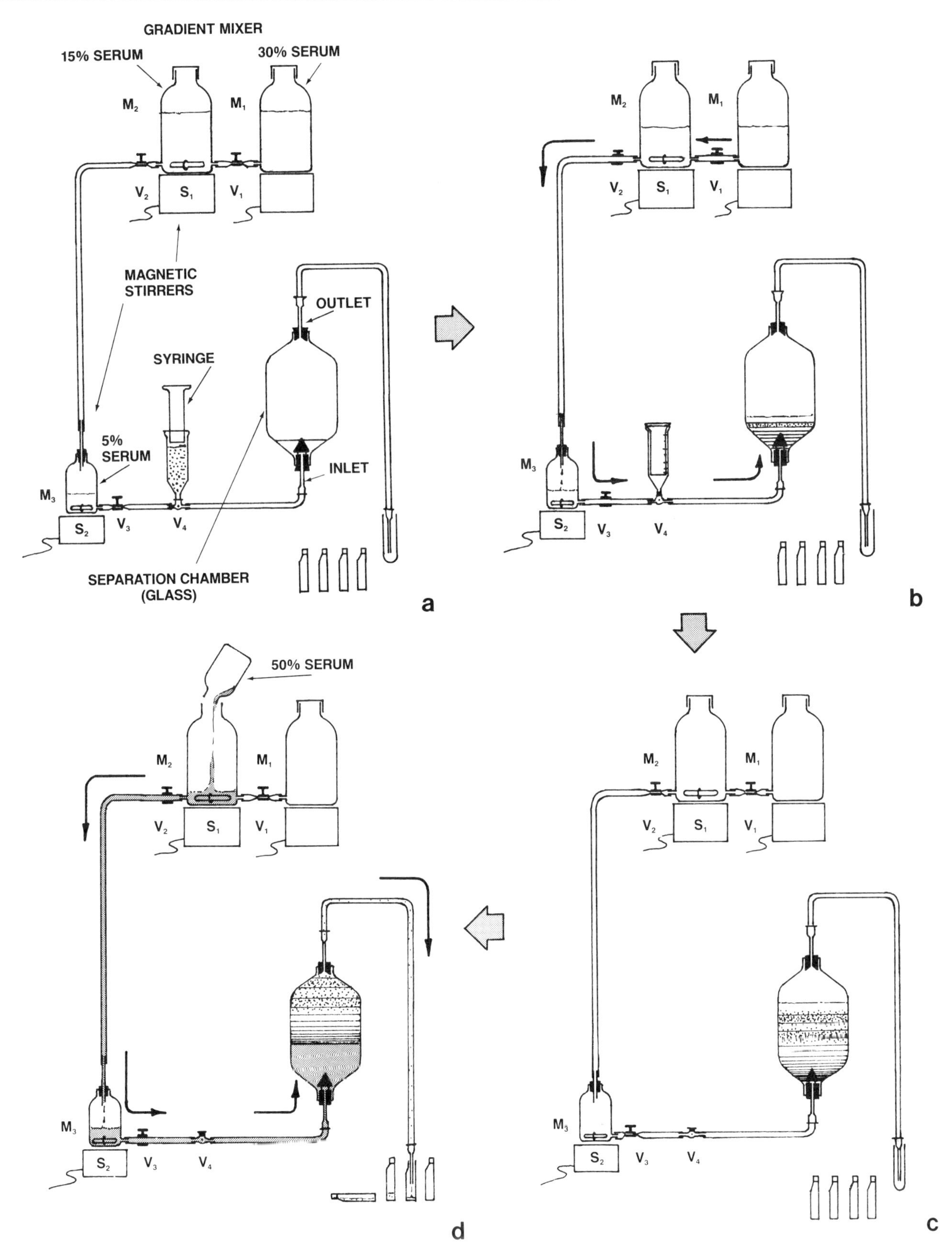

Fig. 14.2. *Cell separation by velocity sedimentation at unit gravity. Apparatus and position of valves at different stages of the procedure. a. At start, loading cells. b. Running in gradient. c. After cell sedimentation. d. Harvesting [Miller and Phillips, 1969]. V, valve; M, mixer vessel; S, stirrer.*

2.
Assemble and check that valves V_1, V_2, V_3, and V_4, are closed
3.
Add 300 ml 30% serum in medium to mixer vessel M_1 and 300 ml 15% serum in medium to mixer vessel M_2
4.
Check that stirrer S_1 is functioning
5.
Add 20 ml 5% serum in medium to mixer M_3 and check stirrer S_2 is functioning
6.
Open V_4 to connect syringe to separation chamber and insert 20 ml PBS into separation chamber
7.
Open V_4 to M_3 line, open V_3, and draw a little 5% serum into the syringe (just enough to fill line). Close V_3 and V_4
8.
Prepare cell suspension, e.g., by trypsinizing primary culture for 15 min in 0.25% trypsin-citrate. Disperse cells carefully in 3% serum in medium and check that a single cell suspension is formed
9.
Take up 20 ml at 10^6/ml (maximum) into syringe and connect to V_4 inlet
10.
With syringe held vertically, open V_4 to M_3 line, open V_3, and draw a little 5% serum into syringe to clear any bubbles from M_3 line and V_4
11.
Turn valve V_4 to separation chamber line and draw a little PBS into syringe to clear any bubbles from this line
12.
Insert cell suspension slowly into chamber; avoid mixing cell suspension with the overlaying PBS layer. Take care to stop while a little fluid is left in the syringe to avoid injecting any air bubbles back into the line. If difficulty is encountered injecting cells smoothly, without turbulence, remove piston from syringe and allow cells to run in under gravity alone by raising V_4
13.
Start stirrers S_1 and S_2
14.
Open V_4 to connect M_3 line to separation chamber
15.
Open V_1 and adjust flow rate by opening V_2 to give 15 ml/min (~5 drops/s) at M_3. Cell suspension will now float up into separation chamber on gradient of serum. Check for turbulence at baffle as suspension and gradient run in. If there is any, reduce flow rate at M_3 by closing V_2
16.
When gradient mixers M_1 and M_2 are empty, but before M_3 empties, close V_3 and V_2
17.
It should be possible to see the cell layer in the sedimentation chamber and to follow the cells as they sediment. As they do, the cell band will become wider and more diffuse. Check for signs of "streaming" in the early stages of sedimentation (tails of cells which sediment ahead of the main band). This occurs when the cell concentration is too high or the step between the cells and the gradient is too steep

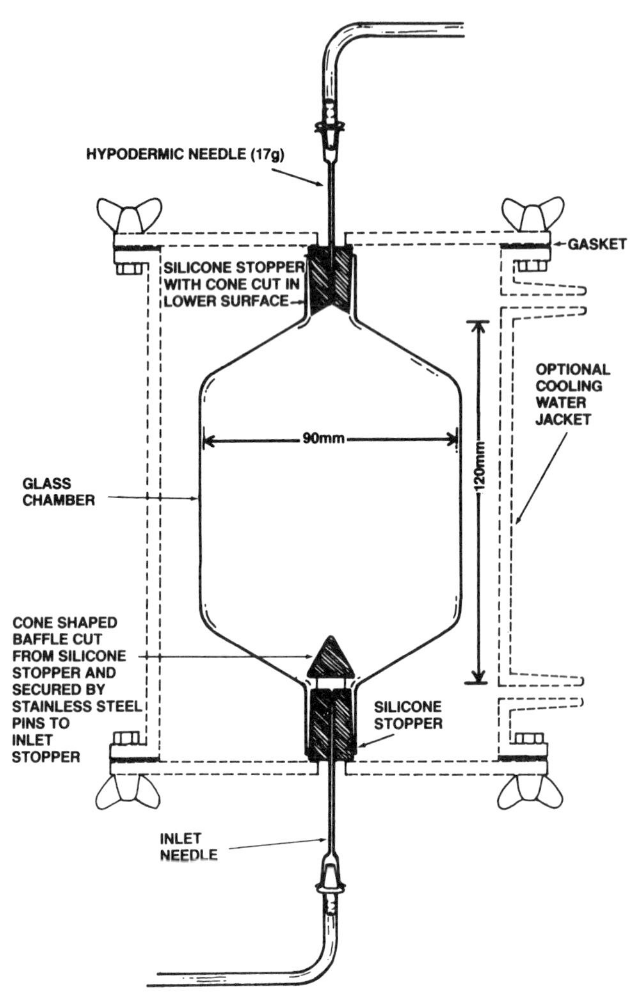

Fig. 14.3. *Separation chamber for unit gravity sedimentation. The dotted outline is a cooling jacket for carrying out sedimentations at 4° C [modified from Miller and Phillips, 1969].*

18.
After about 20 min, close V_1 and add 90 ml 50% serum to M_2. Open V_3 and adjust flow rate at M_3 to 15 ml/min. Stop when M_2 is empty but before M_3 empties, by closing V_3 and V_2

19.
Add 500 ml flotation medium (1 M sucrose or 20% Ficoll) to M_1 and M_2, open V_1 and V_2 and let some of the flotation medium run into M_3. Close V_2

20.
When sedimentation is complete, i.e., cell band midway down separation chamber, open V_3, and adjust V_2 to give a flow rate of 15 ml/min in M_3

21.
Collect eluate from top of chamber via elution line and run into graduated culture vessels, e.g., 25-cm^2 flasks, 10 ml per flask. Mix the contents of each flask and take sample for cell counting

22.
Seal and incubate flasks for 24 hr, replace medium with fresh medium at standard serum concentration and volume

Variations. *Gradient medium.* If serum and regular culture medium are used, then it is possible to culture cells directly from the eluate. Fetal bovine serum causes less reaggregation than calf or horse serum. If serum is found to be unsuitable, gradients can be formed from bovine serum albumin [Catsimpoulas et al., 1978] or Ficoll (Pharmacia).

Aggregation. Aggregation can be reduced by enclosing the separation chamber in a cooling jacket and running the whole process at 4°C. Mixers M_1, M_2, and M_3 must all be kept cold also. Water-driven magnetic stirrers (Calbiochem) may be used at S_1 and S_2 to minimize overheating of gradient.

Pump. Gravity is used in this example as the cheapest and simplest method for generating the flow of gradient medium, but if desired, a peristaltic pump (pulse free) may be inserted at V_3.

Sedimentation of cells at unit gravity is a simple low technology method of separating cells. It works well for many cell types, e.g., brain [Cohen et al., 1978], hemopoietic cells [McCool et al., 1970; Petersen and Evans, 1967], and HeLa/fibroblast mixtures (Fig. 14.4) and can be performed in regular physiological media. The cells must be singly suspended, however, and there is a practical limit of about 10_8 cells that may be separated.

Pretlow and others [Pretlow, 1971; Pretlow et al., 1978; Hemstreet, et al., 1980] have used specially formed gradients of Ficoll to separate cells by sedimentation velocity at higher g forces on a zonal rotor. The

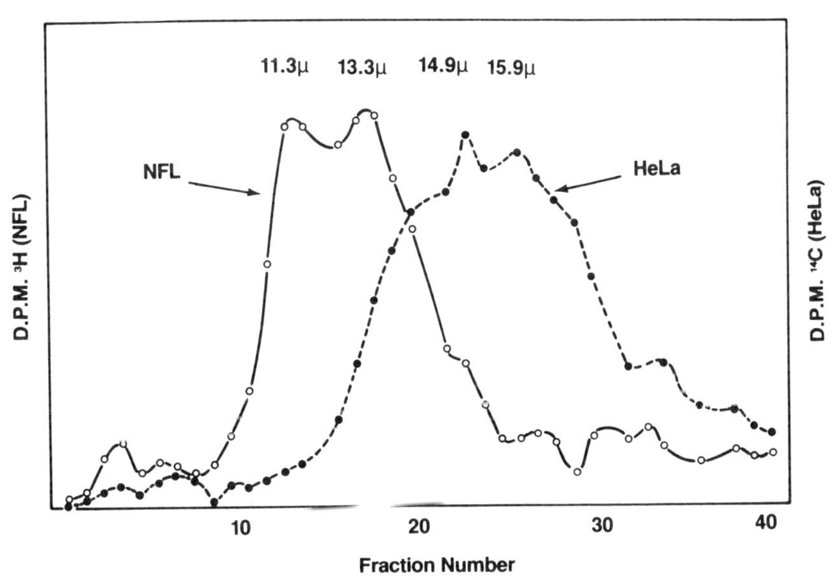

Fig. 14.4. *Elution profiles of artificial mixtures of HeLa and NFL (normal human fetal lung fibroblasts) after sedimentation at 1g for 3 hr. Cultures were prelabeled with ^{3}H-leucine or ^{14}C-thymidine and the distribution of cells in the eluate was determined by dual-isotope scintillation counting. Open circles and solid line, NFL, solid circles and broken line, Hela. The numbers above each peak are the calculated cell diameters derived from the sedimentation velocity by equation 14.1. The values obtained for HeLa were confirmed by micrometry, but the values for NFL did not agree with the micrometer readings which were around 16–18 µm [after Freshney et al., 1982a, reproduced with permission from the publisher].*

gradients are shallow and of relatively low density to minimize the effect of cell density on sedimentation rate. Cells of many different types have been separated by this method, and it appears to have a wide application.

Centrifugal Elutriation

The centrifugal elutriator (Beckman) is a device for increasing the sedimentation rate and improving the yield and resolution by performing the separation in a specially designed centrifuge and rotor (Fig. 14.5). Cells in the suspending medium are pumped into the separation chamber in the rotor while the rotor is turning. While the cells are in the chamber, centrifugal force will tend to force the cells to the outer edge of the rotor (Fig. 14.6). Meanwhile the suspending medium is pumped through the chamber such that the centripetal flow rate approximates to the sedimentation rate of the cells. If the cells were uniform, they would remain stationary, but since they vary in size, density, and cell surface configuration, they tend to sediment at different rates.

As the sedimentation chamber is tapered, the flow rate increases toward the edge of the rotor and a continuous range of flow rates is generated. Cells of differing sedimentation rates will, therefore, reach equilibrium at different positions in the chamber. The sedimentation chamber is illuminated by a stroboscopic light and can be observed through a viewing port. When the cells are seen to reach equilibrium, the flow rate is increased and the cells are pumped out into receiving vessels. The separation can be performed in complete medium and the cells cultured directly afterward.

Equilibrium is reached in a few minutes and the whole run may take 30 min. On each run 10^8 cells may be separated and the run may be repeated as often as necessary. The apparatus is, however, fairly expensive and a considerable amount of experience is required before effective separations may be made. A number of cell types have been separated by this method [Greenleaf et al., 1979; Schengrund and Repman, 1979; Meistrich et al., 1977a], as have cells of different phases of the cell cycle [Meistrich et al., 1977b].

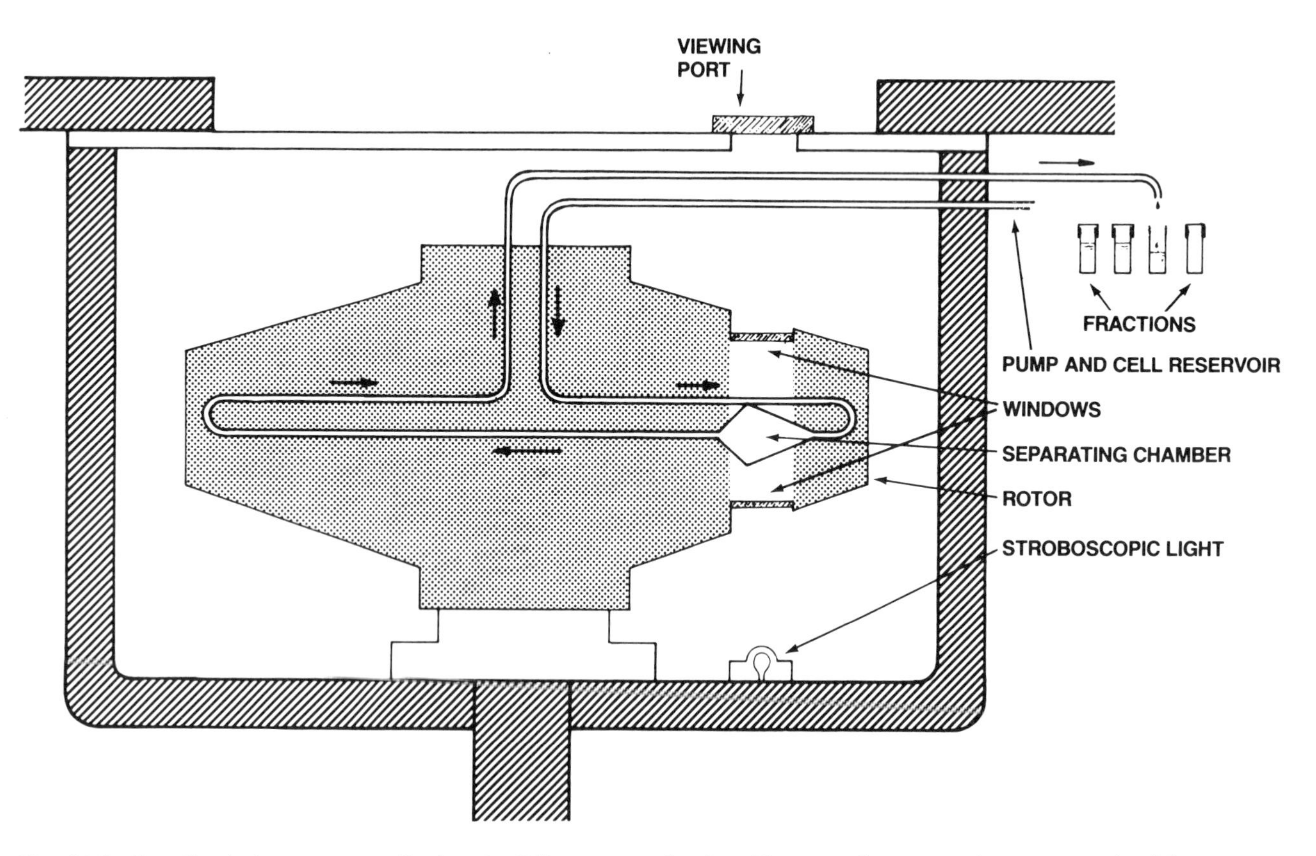

Fig. 14.5. *Centrifugal elutriator rotor (Beckman). Cell suspension and carrier liquid enter at the center of the rotor and are pumped to the periphery to enter the outer end of the separation chamber. The return loop is via the opposite side of the rotor to maintain balance.*

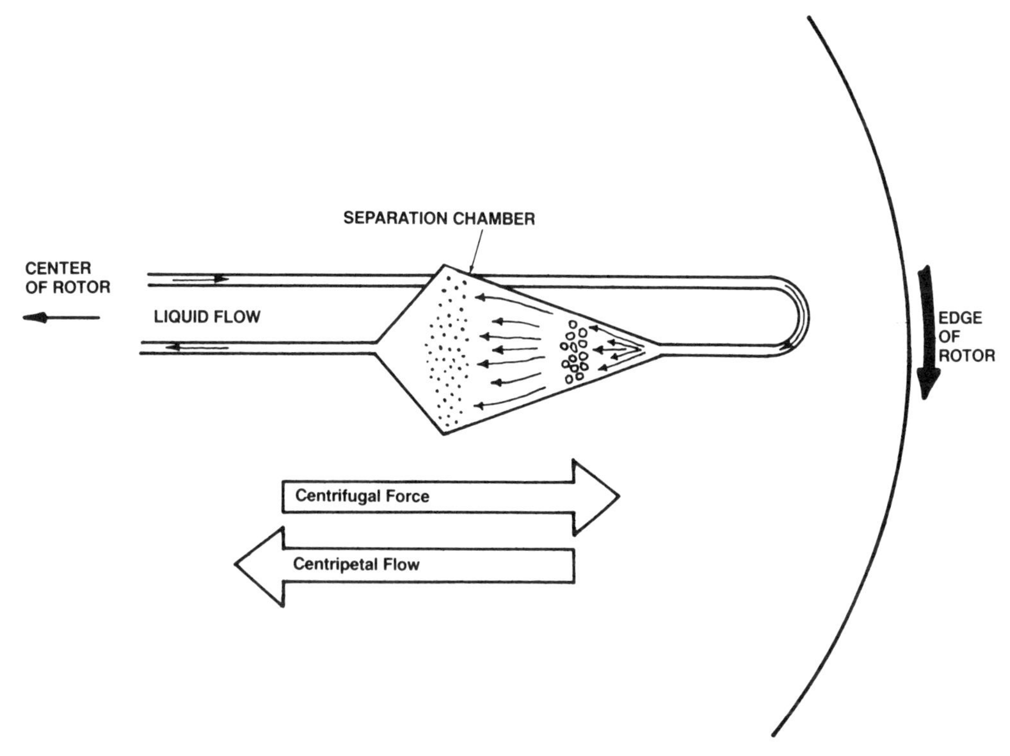

Fig. 14.6. *Separation chamber of elutriator rotor.*

CELL DENSITY AND ISOPYCNIC SEDIMENTATION

Separation of cells by density can be performed at low or high *g* using conventional equipment. The cells sediment in a density gradient to an equilibrium position equivalent to their own density (isopycnic sedimentation). Physiological media must be used and the osmotic strength carefully monitored. The density medium should be nontoxic, nonviscous at high densities (1.10 g/ml), and exert little osmotic pressure in solution. Serum albumin [Turner et al., 1967], dextran [Schulman, 1968], Ficoll (Pharmacia) [Sykes et al., 1970], metrizamide (Nygaard) [Munthe Kaas and Seglen, 1974], and Percoll (Pharmacia) [Pertoft, 1968; Pertoft and Laurent, 1977; Wolff and Pertoft, 1972] have all been used successfully; Percoll (colloidal silica) is one of the more effective media currently available. (For isolation of lymphocytes on Ficoll/Metrizoate, see Chapter 21.)

Outline

Form gradient (1) by layering different densities of Percoll (2) by high-speed spin or (3) with special gradient former. Centrifuge cells through Percoll gradient (or allow to sediment at unit gravity), collect fractions, and culture directly (Fig. 14.7).

Materials

culture medium (sterile)
medium + 20% Percoll (sterile)
25-ml centrifuge tubes (sterile)
PBSA (sterile)
0.25% trypsin (sterile)
syringe or gradient harvester (sterile)
24-well plates or microtitration plates (sterile)
refractometer or density meter
hemocytometer
cell counter

Protocol

1. Prepare gradient: (1) Prepare two media, one regular culture medium and one with 20% Percoll. (2) Adjust the density of the Percoll solution to 1.10 g/ml and its osmotic strength to 290 mOsm/kg. (3) Mix the two media in varying proportions to give the desired density range (e.g., 1.020–1.100 g/ml) in 10 or twenty steps. (4) Layering one step over another, build up a stepwise density gradient in a 25-ml centrifuge tube (see Fig. 14.8a). Gradients may be used immediately or left overnight.

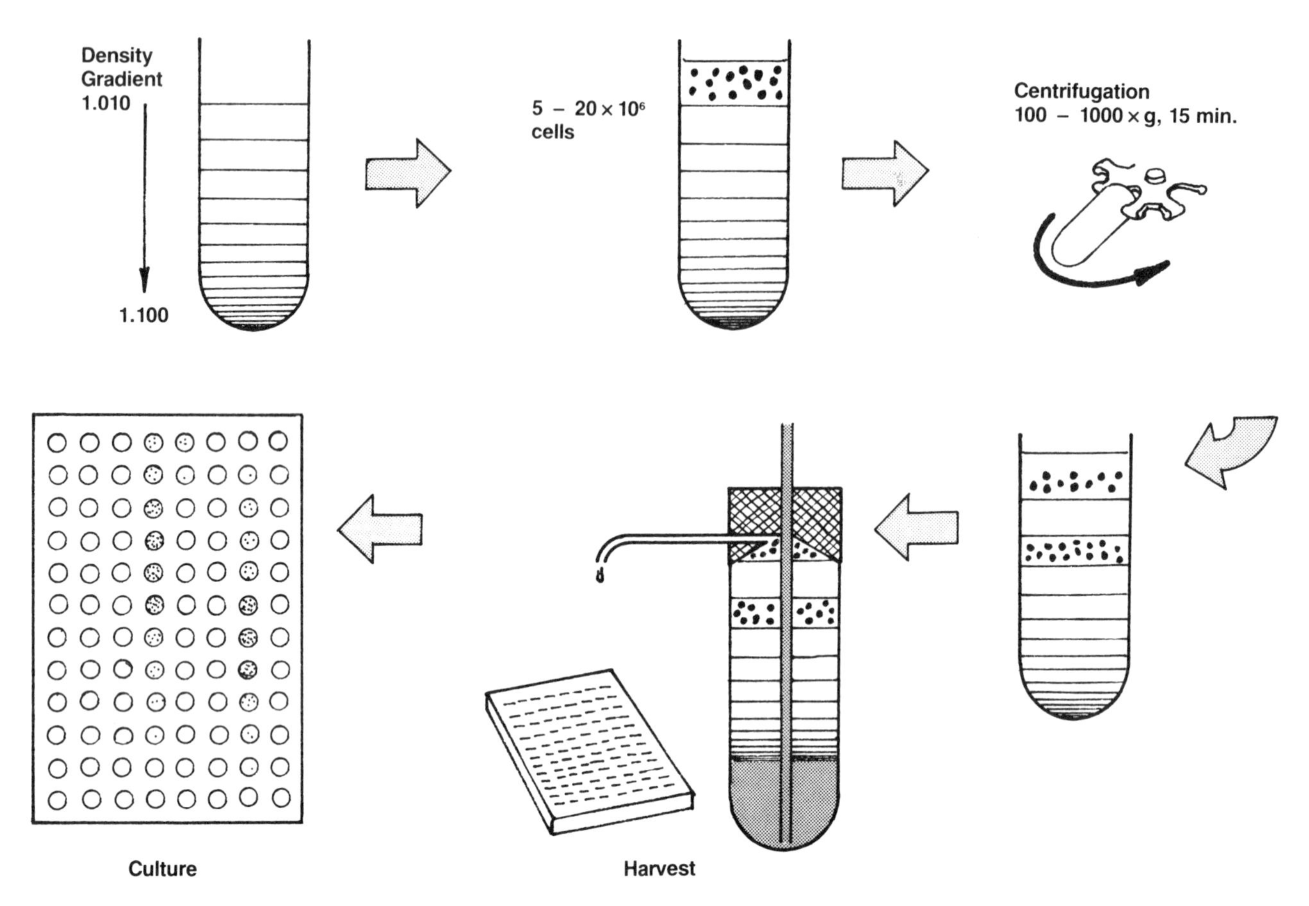

Fig. 14.7. *Cell separation by isopycnic centrifugation.*

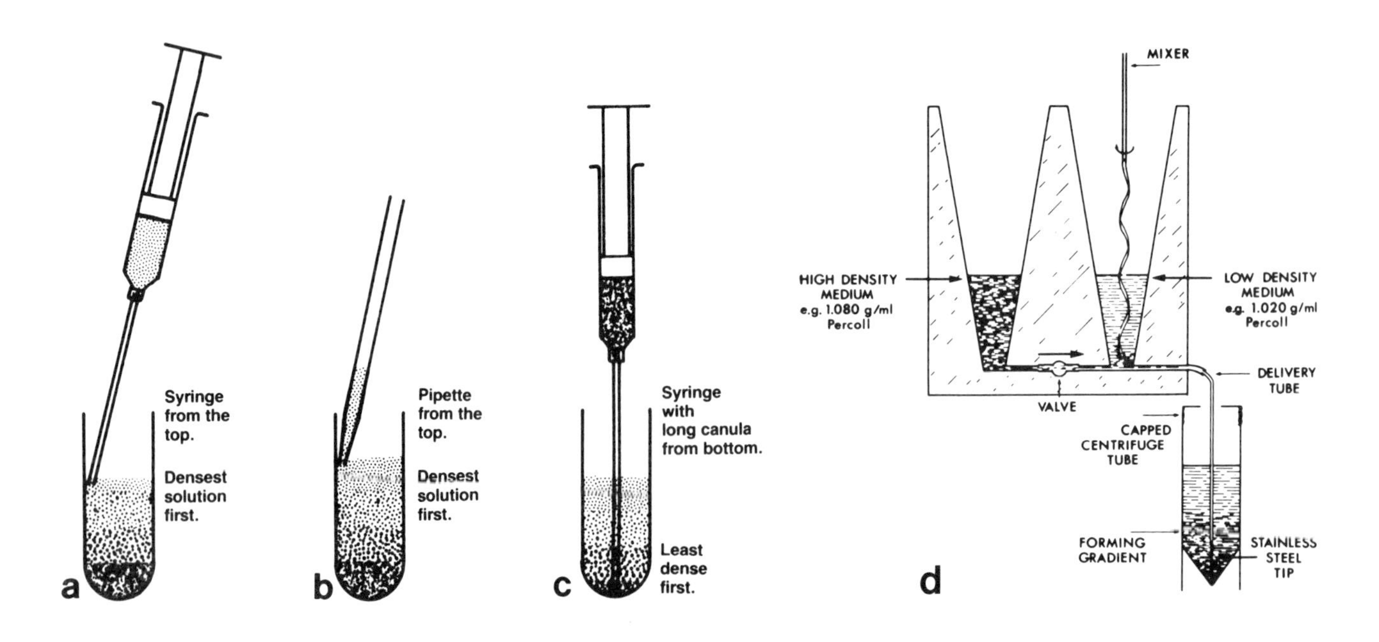

Fig. 14.8. *Layering density gradients (a) by syringe from the top, (b) by pipette from the top, (c) by syringe from the bottom, (d) by gradient mixing device (Buchler).*

Alternatively, place medium containing Percoll of density 1.085 g/ml in a tube and centrifuge at 20,000*g* for 1 hr. This generates a sigmoid gradient (Fig. 14.9), the shape of which is determined by the starting concentration of Percoll, the duration and centrifugal force of the centrifugation, the shape of the tube, and the type of rotor.

A continuous linear gradient may be produced by mixing, for example, 1.020 g/ml with 1.08 g/ml Percoll in a gradient-forming device (Fig. 14.8b)(MSE/Fisons, Pharmacia)

2.

Trypsinize cells and resuspend in medium plus serum. Check that they are singly suspended

3.

Layer up to 2 × 10^7 cells in 2 ml medium on top of the gradient

4.

The tube may be allowed to stand on the bench for 4 hr or centrifuged for 20 min at between 100 and 1,000 *g*

5.

Collect fractions using a syringe, or a gradient harvester (Fig. 14.10)(MSE/Fisons). Fractions of 1 ml may be collected into a 24-well plate or 0.1 ml into microtitration plates. Samples should be taken at intervals for cell counting and determination of the density (ρ) of the gradient medium. Density may be measured on a refractometer (Hilger) or density meter

6.

Change the medium to remove the Percoll after 24–48 hr incubation

Variations. Cells may be incorporated into the gradient during formation by centrifugation. Only one spin is required although spinning the cells at such a high *g* force may damage them.

Other media. Ficoll is one of the most popular media as it, like Percoll, can be autoclaved. It is a little more viscous at high densities and may cause agglutination of some cells. Metrizamide (Nygaard), a nonionic derivative of metrizoate, which is a radioopaque-iodinated substance used in radiography (Isopaque, Hypaque, Renografin) and in lymphocyte purification (e.g., Lymphoprep) (see Chapter 21), is less viscous at high densities [Rickwood and Birnie, 1975] but may be incorporated into some cells (Fig. 14.11) as is Isopaque [Splinter et al., 1978]. Where such media are used, cells should always be layered on top of the gradient and not mixed in during formation.

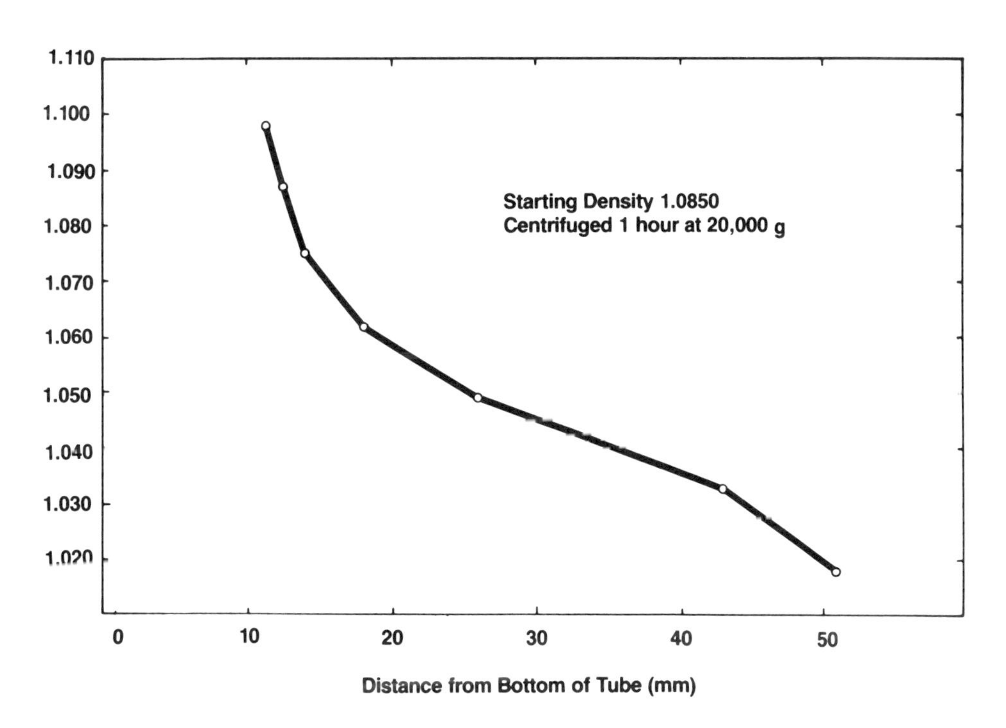

Fig. 14.9. *Gradient generated by spinning Percoll at 20,000g* for 1 hr.

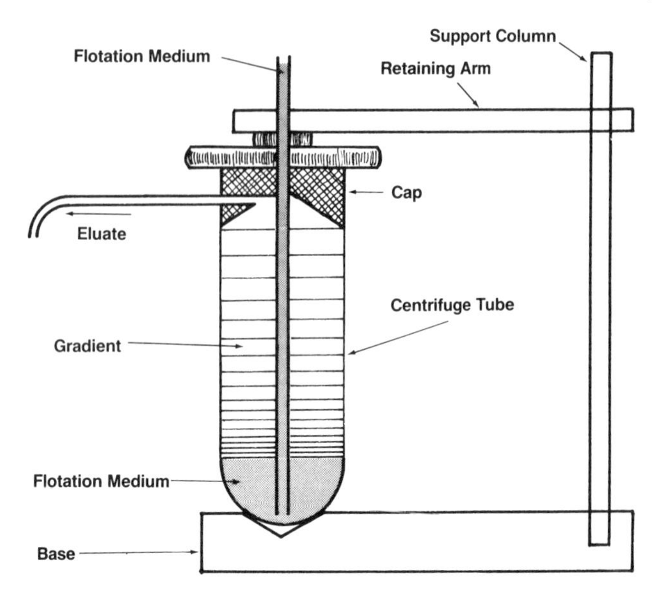

Fig. 14.10. *Gradient harvester (Fisons/M.S.E.). Fluorechemical FC43 is pumped down the inlet tube to the bottom of the gradient and displaces the gradient and cells upward and out through the delivery tube. (After an original design by Dr. G.D. Birnie.)*

Marker beads. Pharmacia manufactures colored marker beads of standard densities which may be used to determine the density of regions of the gradient.

Isopycnic sedimentation is quicker than velocity sedimentation at unit gravity and gives a higher yield of cells for a given gradient volume. It is ideal where clear differences in density exist between cells. Cell density may be affected by the gradient medium (e.g., metrizamide, Fig. 14.11), by the position of the cells in the growth cycle, and by serum (Fig. 14.12).

This type of separation can be done on any centrifuge, as high *g* forces are not required, and can even be performed at 1*g*.

FLUORESCENCE-ACTIVATED CELL SORTING

This technique [Herzenberg et al., 1976; Kreth and Herzenberg, 1974] operates by projecting a single cell stream through a laser beam in such a way that the light scattered from the cells is detected by a photomultiplier and recorded (Figs. 14.13, 14.14). If the

TABLE 14.1. Cell Separation Methods

Method	Basis for separation	Comments	Reference
Sedimentation velocity at 1 *g*	Cell size	Simple technique	Miller and Phillips [1969]
By centrifugation "isokinetic gradient"	Cell size, density, and surface configuration	Computer-designed gradient in zonal rotor	Pretlow [1971]
Centrifugal elutriation	Cell size, density, and surface configuration	Rapid; high cell yield	Meistrich et al. [1977a,b]
Isopycnic sedimentation	Cell density	Simple and rapid	Pertoft and Laurent [1977]
Flow cytophotometry (fluorescence-activated cell sorting)	Cell surface area fluorescent markers, fluorogenic enzyme substrates, multiparameter	Complex technology and expensive Very effective, high resolution but low yield	Kreth and Herzenberg [1974]
Affinity chromatography	Cell surface antigens, cell surface carbohydrate	Elution of cells from columns difficult, better in free suspension	Edelman [1973]
Counter current distribution	Affinity of cell surface constituents for solvent phase	Poor viability after separation	Walter [1977]
Electrophoresis in gradient curtain	Surface charge		Platsoucas et al. [1979] Kreisberg et al. [1977]

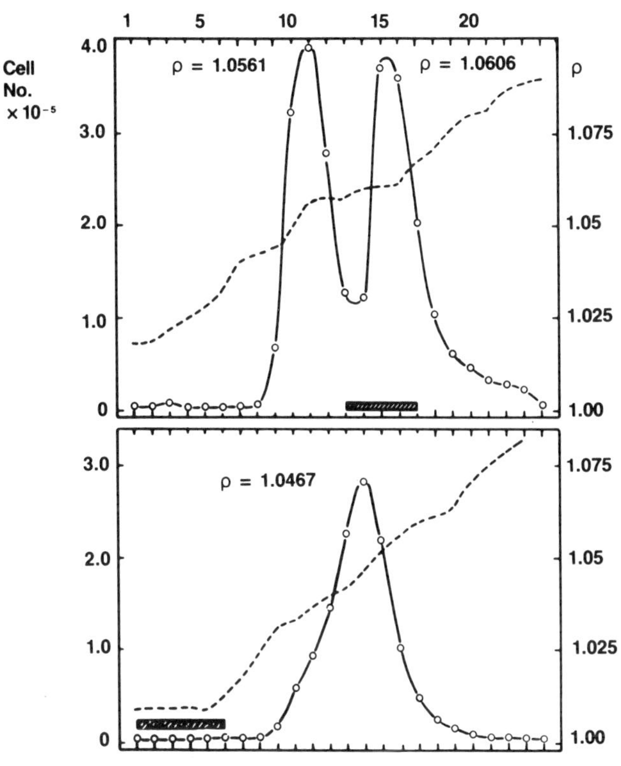

Fig. 14.11. *Incorporation of metrizamide (Nygaard) into cells during isopycnic centrifugation. MRC-5 cells (human diploid embryonic lung fibroblasts) were layered in metrizamide-containing medium of the appropriate density in the center of the gradient or in medium alone at the top of the gradient. After centrifugation, the cells were eluted and counted, and samples were taken from each fraction to determine the density [reproduced from Freshney, 1976a, with permission of the publisher]. Slashed bars, position of cells at start.*

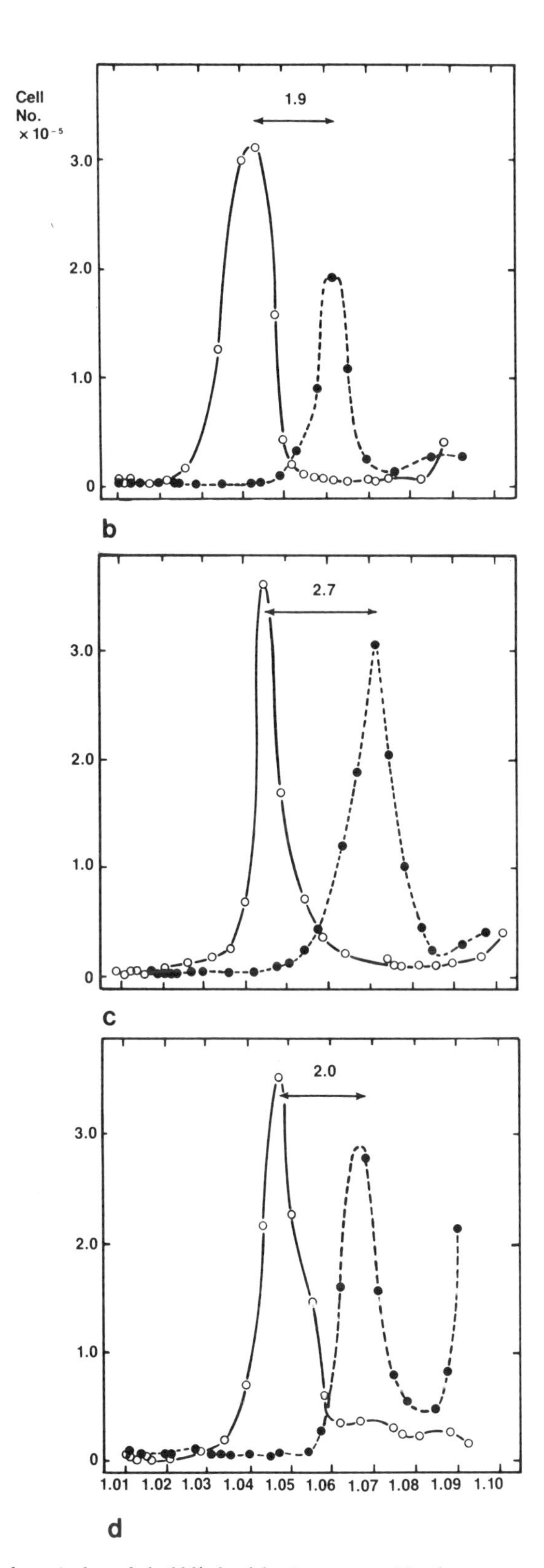

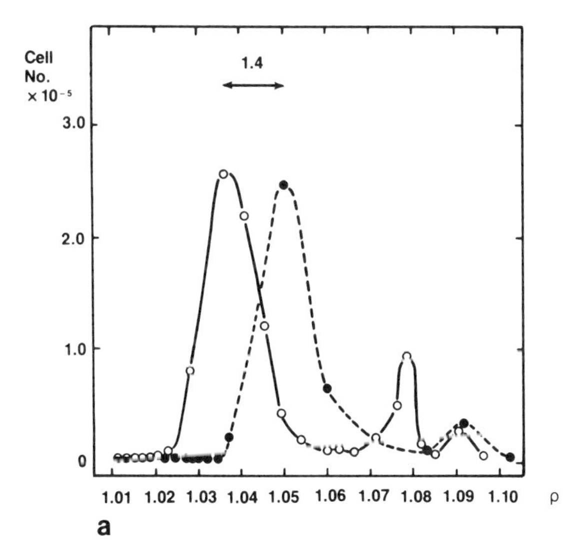

Fig. 14.12. *Sedimentation profiles of HeLa and MRC-5 cells, centrifuged to equilibrium in gradients of metrizamide in culture medium. a,b. Cells taken from log phase of growth. c,d. Cells taken from plateau phase. Gradients in a and c contained no serum; those in b and d, 10% fetal bovine serum. Numbers over the arrows are the differences in density between the peaks, multiplied by 100. Solid circles, HeLa; open circles, MRC-5 [From Freshney, 1976a, with permission of the publisher.]*

Fig. 14.13. *Fluorescence-Activated Cell Sorter (FACS). This is the Becton Dickinson version of the flow cytophotometer. a. Flow chamber panel on left and computer/readout on right. b. Close-up of flow chamber compartment (see also Fig. 14.14).*

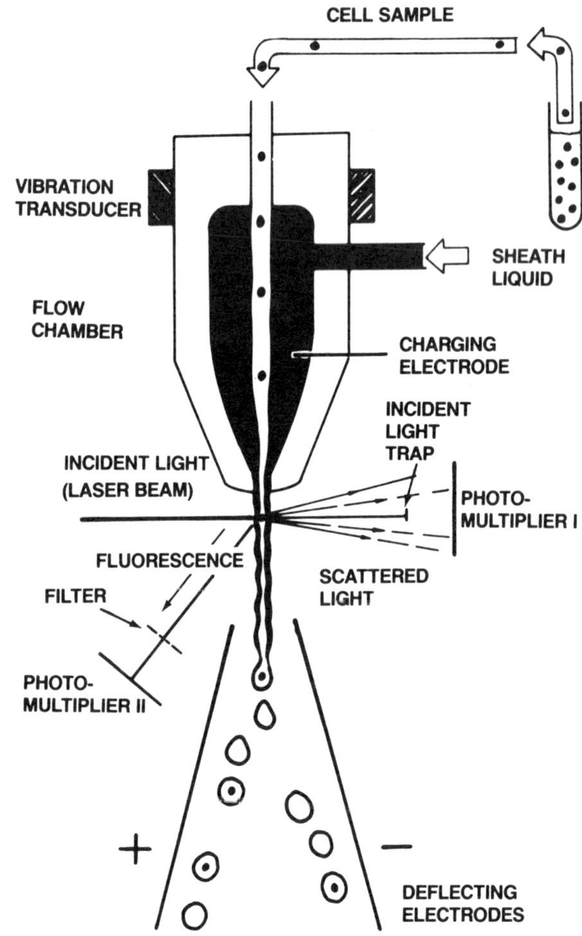

Fig. 14.14. *Principle of operation of flow cytophotometer (see text). [Reproduced from Freshney et al., 1982a, with permission of the publisher.]*

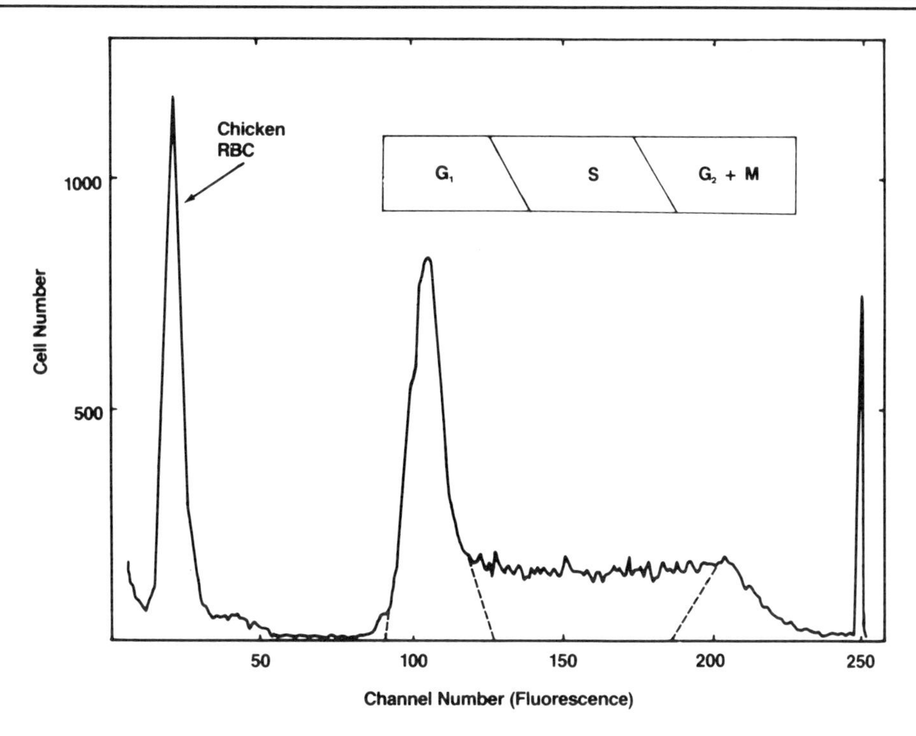

Fig. 14.15. *Printout from FACS II. Friend (Murine erythroleukemia) cells were fixed in methanol as a single cell suspension and stained with Chromomycin A_3. The cell suspension was then mixed with chicken erythrocytes, fixed and stained in the same way, and run through the FACS II. The printout plots cell number on the vertical axis and channel number (fluorescence) on the horizontal axis. Since fluorescence is directly proportional to the DNA per cell, the trace gives a distribution analysis of the cell population by DNA content. The lowest DNA content is found in the chicken erythrocytes, included as a standard. The major peak around channel 100 represents those cells as the G1 phase of the cell cycle. Cells around channel 200 are, therefore, G2 and metaphase cells (double the amount of DNA per cell) and cells with intermediate values are in S or the DNA synthetic phase (see also Chapters 15 and 22). Cells accumulated in channel 250 are those where DNA value is off scale or cells which have formed aggregates (by courtesy of Dr. B.D. Young).*

cells are pretreated with a fluorescent stain (e.g., Chromomycin A_3 for DNA) or fluorescent antibody, the fluorescence emission excited by the laser is detected by a second photomultiplier tube. This information is processed and displayed as a two- (Fig. 14.15) or three-dimensional graph on an oscilloscope. If specific coordinates are then set to delineate sections of the display, the cell sorter will divert cells with the properties that would place them within these coordinates (e.g., high or low light scatter, high or low fluorescence) into a receiver tube placed below the cell stream. The cell stream is deflected by applying a charge to it as it passes between two oppositely charged plates. The charge is applied briefly and at a set time after the cell has cut the laser beam such that only one cell is deflected into the receiver. A low cell concentration in the cell stream is required such that the gap between cells is sufficient to prevent two cells being deflected together.

All cells having similar properties will be collected into the same tube. A second set of coordinates may be set and a second group of cells collected simultaneously into a second tube by changing the polarity of the cell stream and deflecting the cells in the opposite direction. All remaining cells will be collected in a central reservoir.

This method may be used to separate cells with any differences that may be detected by light scatter (e.g., cell size) or fluorescence (e.g., DNA, RNA, protein, enzyme activity, specific antigens). It is an extremely powerful tool but limited by cell yield (about 10^7 cells is a reasonable maximum) and the very high cost of the instrument. It also requires a full-time, skilled operator.

OTHER TECHNIQUES

The many other techniques which have been used successfully to separate cells are too numerous to describe in detail. They are summarized below and listed in Table 14.1.

Electrophoresis either in a Ficoll gradient [Platsoucas et al., 1979] or by curtain electrophoresis: The second technique is probably more effective and has been used to separate kidney tubular epithelium [Kreisberg et al., 1977].

Affinity chromatography on antibody [Varon and Manthorpe, 1980; Au and Varon, 1979] or plant lectins [Pereira and Kabat, 1979] bound to nylon fiber [Edelman, 1973] or Sephadex (Pharmacia). These techniques appear to be useful for fresh blood cells but less so for cultured cells.

Counter current distribution [Walter, 1975, 1977] has been used to purify murine ascites tumor cells, but the viability may be too low for subsequent culture.

As so many techniques exist, it is difficult for the novice to know where to start. Like so many other areas of investigation, it is best to start with a simple technique such as velocity sedimentation at unit gravity or isopycnic centrifugation. Density gradient analysis may also be used in conjunction with velocity sedimentation in a two-stage fractionation. If there are still problems of resolution or yield, then it may be necessary to employ high-technology methods such as centrifugal elutriation, curtain electrophoresis, or fluorescence-activated cell sorting.

Chapter 15 Characterization

When cultured cells are used for a specific purpose, e.g., production of virus, induction of an enzyme, etc., the property that is being studied may act as a major criterion for the identification of the line. But the question of cell identity is such a crucial one that at least two other criteria should be employed. Apart from the functional significance of the characteristic, there are four other main reasons why characterization of the cell line is important.

(1) Correlation of culture with tissue of origin: (a) identification of the lineage to which the cell belongs, and (b) position of the cell within that lineage, i.e., the precursor status of the cell (see Chapter 2).

(2) Monitoring a cell line for instability and variation (see Chapter 18).

(3) Checking for cross contamination (see Chapter 17).

(4) Identifying selected sublines or hybrid cell lines.

MORPHOLOGY

This is the simplest and most direct technique used to identify cells. It has, however, certain shortcomings which should be recognized. Most of these are related to the plasticity of cellular morphology in response to different culture conditions, e.g., epithelial cells growing in the center of a confluent sheet are usually regular, polygonal, and with a clearly defined edge, while the same cells growing at the edge of a patch may be more irregular, distended, and, if transformed, may break away from the patch and become fibroblastoid in shape. Fibroblasts from hamster kidney or human lung or skin assume multipolar or bipolar shapes and are well spread on the culture surface, but at confluence they are bipolar and less well spread. They also form characteristic parallel arrays and whorls which are visible to the naked eye. Mouse 3T3 cells and human glial cells grow like multipolar fibroblasts at low cell density but become epithelial-like at confluence (Fig. 15.1). Alterations in the substrate [Gospodarowicz, 1978; Freshney, 1980], and the constitution of the medium [Coon and Cahn, 1966] can also affect cellular morphology. Hence, comparative observations should always be made at the same stage of growth and cell density in the same medium, and growing on the same substrate (see Fig. 15.1).

The terms "fibroblastic" and "epithelial" are used rather loosely in tissue culture and often describe the appearance rather than the origin of the cells. Thus a bipolar or multipolar migratory cell, the length of which is usually more than twice its width, would be called "fibroblastic," while a monolayer cell which is polygonal, with more regular dimensions, and which grows in a discrete patch along with other cells, is usually regarded as "epithelial." However, where the identity of the cells has not been confirmed, the terms "fibroblast-like" or "fibroblastoid" and "epithelial-like" or "epithelioid" should be used (see Fig. 12.1).

Frequent brief observations of living cultures, preferably with phase-contrast optics, are more valuable than infrequent stained preparations studied at length. They will give a more general impression of the cell's morphology and its plasticity and will also reveal differences in granularity and vacuolation which bear on the health of the culture. Unhealthy cells often become granular and then display vacuolation around the nucleus.

It is useful to keep a set of photographs for each cell line as a record in case a morphological change is suspected. This record can be supplemented with photographs of stained preparations.

Staining

A polychromatic blood stain such as Giemsa provides a convenient method of preparing a stained culture. The recommended procedure is as follows.

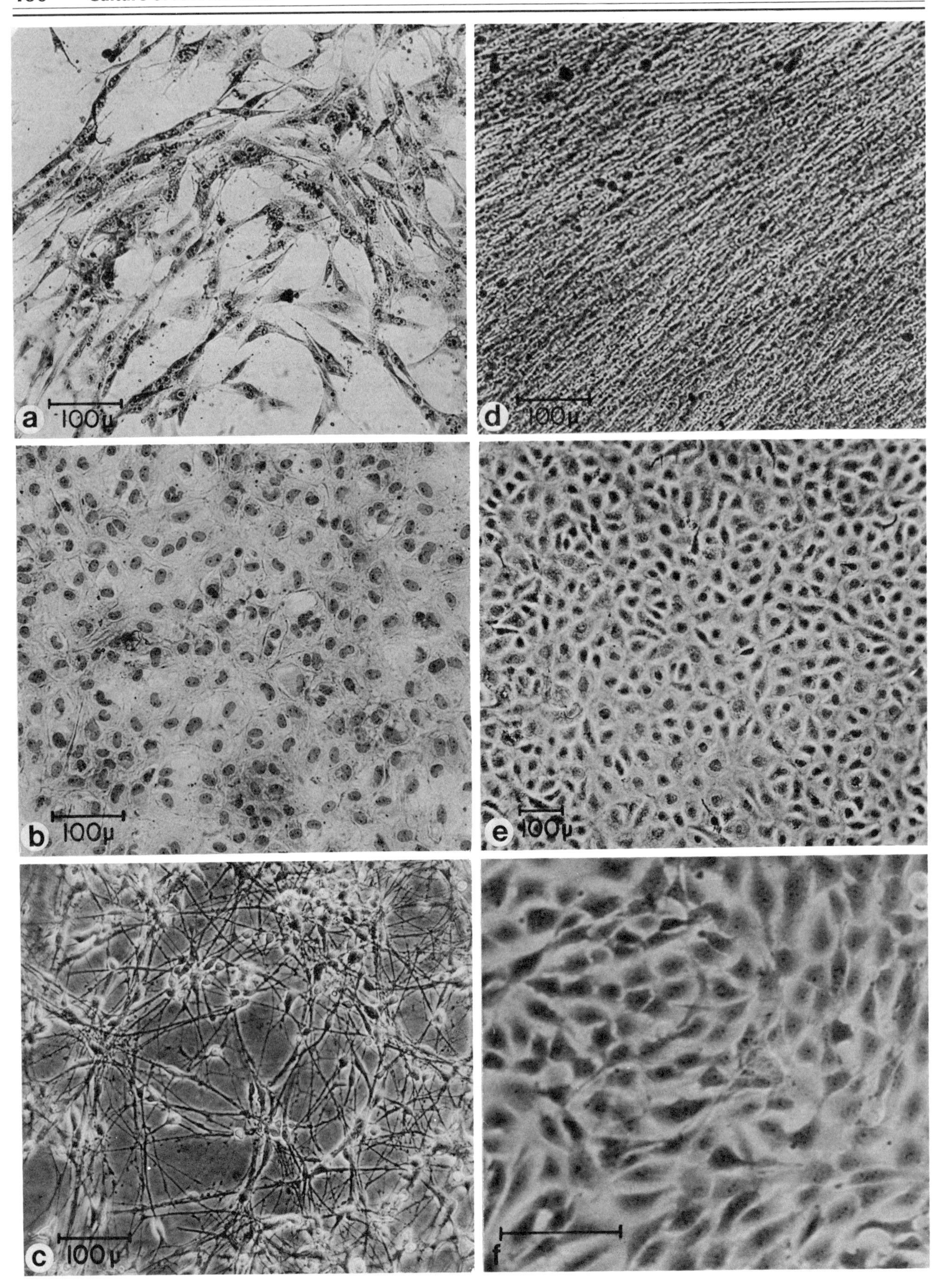
a
100μ
b
100μ
c
100μ
d
100μ
e
100μ
f

Outline

Fix the culture in methanol and stain directly with Giemsa. Wash and examine wet.

Materials

BSS, undiluted Giemsa stain
methanol
deionized water

Protocol

1. Remove medium and discard
2. Rinse monolayer with BSS, and discard rinse
3. Add fresh BSS, 5 ml per 25 cm^2
4. Add methanol slowly to the BSS with mixing. Tilt the bottle so that the BSS lies on the side opposite the monolayer, add 1 ml methanol, mix, and run the mixture over the cells. Tilt the bottle back and add another 1 ml, rinse monolayer, and continue until 50% methanol is reached
5. Discard 50% methanol/BSS mixture and replace with fresh methanol. Leave for 10 min
6. Discard methanol and replace with fresh anhydrous methanol, rinse monolayer, and discard methanol
7. At this point, the flask may be dried and stored or stained directly. It is important that staining should be done directly from fresh methanol or with a dry flask. If the methanol is poured off and the flask is left for some time, water will be absorbed by the residual methanol and will inhibit subsequent staining
8. Add neat Giemsa stain, 2 ml per 25 cm^2, making sure the entire monolayer is covered and remains covered
9. After 2 min, dilute stain with 8 ml water and agitate gently for a further 2 min
10. Discard stain and wash vigorously in running tap water until any pink cloudy background stain (precipitate) is removed but stain is not leached out of cells
11. Pour off water, rinse in deionized water, and examine on microscope while monolayer is still wet. Store dry and rewet to examine

Note. Giemsa staining is a simple procedure giving a good high-contrast stain but precipitated stain may give a spotted appearance to the cells. This occurs (1) due to oxidation at the surface of the stain forming a scum, and (2) throughout the solution particularly on the surface of the slide when water is added. Washing off stain by replacement rather than pouring off or removing slides is designed to prevent slides coming in contact with scum. Vigorous washing at the end is designed to remove precipitate left on the slide.

Culture Vessels for Cytology — Monolayer Cultures

(1) Regular 25-cm^2 flasks or 50-mm petri dishes.

(2) Coverslips (glass or Thermanox (Lux) in multiwell dishes (see Fig. 8.4), petri dishes, or Leighton tubes (Bellco, Costar; see Fig. 8.7).

(3) Microscope slides in 90-mm petri dishes or with attached multiwell chambers (Lab-Tek, Bellco).

(4) Petriperm dishes (Heracus), cellulose acetate or polycarbonate filters, Melinex, Thermanox, and Teflon-coated coverslips have all been used for E.M. cytology studies. Some pretreatment of filters or Telfon may be required (e.g., gelatin, collagen, fibronectin, or serum coating; see Chapter 8).

(5) The Gabridge chamber/dish (Bionique), which can be used with a variety of different plastic membranes or standard glass or plastic coverslips, allowing culture directly on a thin, optically clear, surface for high-resolution microscopy (Fig. 15.2).

Fig. 15.1. *Examples of variations in cell morphology. a. BHK-21 (baby hamster kidney fibroblasts), clone 13, in log growth. The culture is not confluent, and the cells are well spread and randomly oriented (although some orientation is beginning to appear). b. Cells of an epithelial-like morphology from fetal human intestine (FHI). c. Astrocytes from human astrocytoma. This pattern is quite characteristic but is lost as the cells are passaged, and a morphology not unlike b, c, and f develops. d. Plateau phase BHK-21 C13 cells. The cells are smaller, more highly condensed, and have assumed a parallel orientation with each other. e. Bovine aortic endothelium. Similar regular appearance to b, though cells more closely packed. f. Again a similar pavementlike appearance as found in b and c but now produced by 3T3 cells, mouse fibroblasts. With experience, these cell types may be distinguished, but their similarity underlines the need for criteria for identification other than morphology. a, b, and d, Giemsa stained. c,e, and f, phase contrast. e, by courtesy of Dr. P. Del Vecchio.*

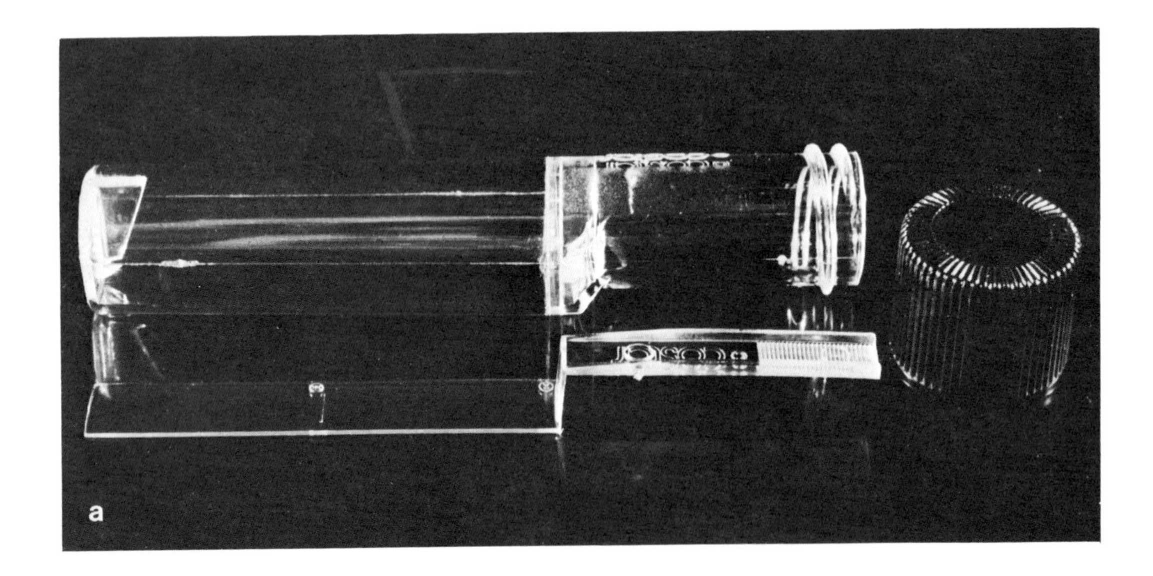

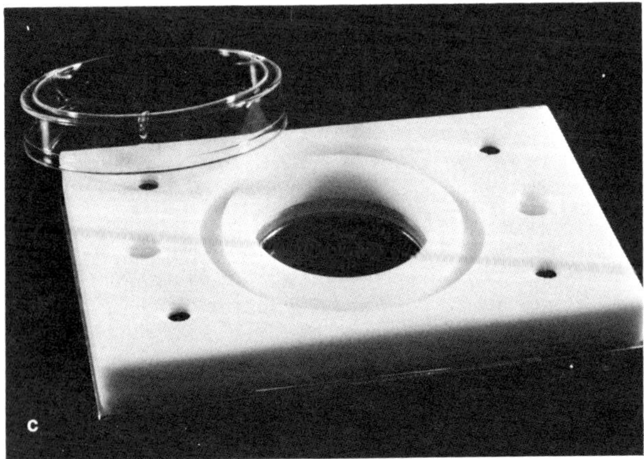

Fig. 15.2. *Culture vessels designed for cytological observation. a. Costar disposable Leighton tube with coverslip with handle for easy retrieval. b. Lab-Tek plastic chambers on regular microscope slide. One, two, four, and eight chambers per slide are available. c. Reusable chamber to take a regular glass coverslip (Bionique) (courtesy of Dr. M. Gabridge).*

Suspension Culture

The following are four ways of preparing cytological specimens from suspension cultures.

Smear. (as used in the preparation of blood films).

Materials

Concentrated cell suspension
serum
microscope slides
methanol

Protocol

1. Place a drop of cell suspension 10^6 cells/ml or more, in 50–100% serum, on one end of a slide
2. Dip the end of a second slide into the drop and move it up the first slide, distributing a thin film of cells on the slide (Fig. 15.3)
3. Dry off quickly and fix in methanol

Centrifugation. The Cytospin (Shandon) (Fig. 15.4) is a centrifuge with sample compartments specially designed to spin cells down onto a microscope slide.

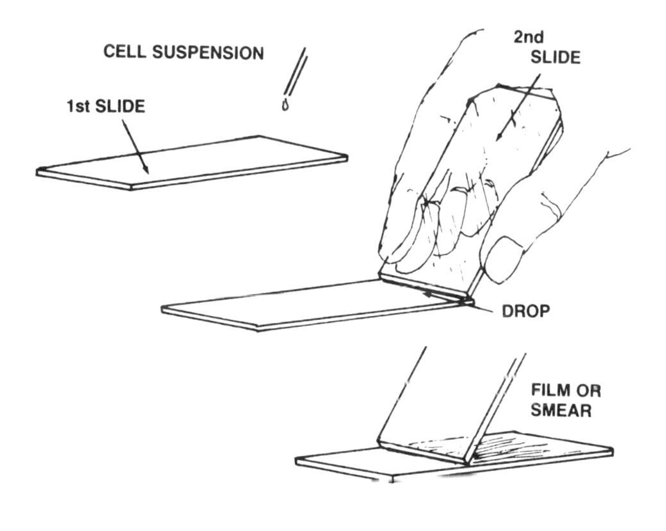

Fig. 15.3. *Preparing a cell smear. Top: A drop of cell suspension in serum or serum-containing medium is placed on a slide. Center and bottom: A second slide is used to spread the drop.*

Materials

Same as for smear preparation.

Protocol

1. Place approximately 100,000 cells in 250 μl of medium in the sample block
2. Switch on and spin the cells down onto the slide at 100g for 5 min
3. Dry off slide quickly and fix in MeOH

Note. Some centrifuges (e.g., IEC) have centrifuge buckets designed for preparing cytological preparations (Fig. 15.5).

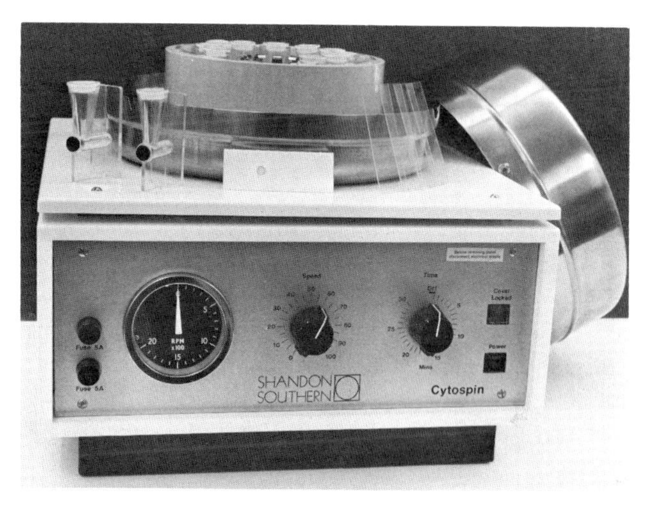

Fig. 15.4. *Shandon Cytospin. Centrifuge for making slide preparations from cell monolayers (photograph by courtesy of Shandon Scientific).*

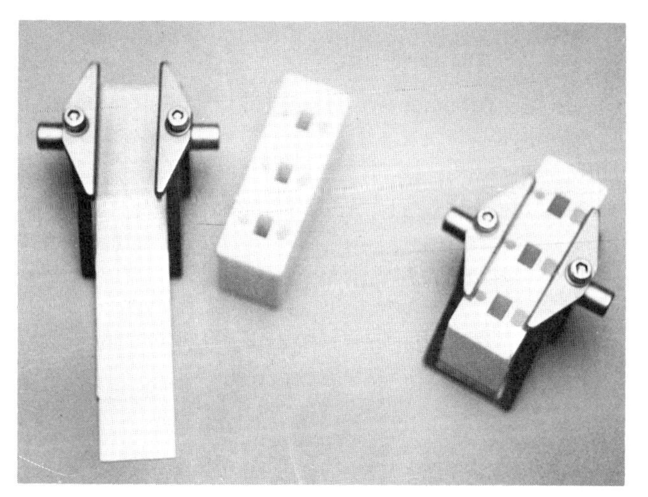

Fig. 15.5. *Centrifuge carriers for spinning cells onto a slide (IEC).*

Drop Technique. Same as for chromosome preparation (see below, this chapter) but omit Colcemid and hypotonic treatment.

Filtration. This technique is used in exfoliative cytology (see manufacturers' instructions for further details: Gelman, Millipore, Nucleopore).

Materials

Filters (e.g., 25-mm Nucleopore, 0.5-μm porosity)
filter holder (Gelman, Nucleopore, Millipore) stand
cell suspension (~10^6 cells in 5–10-ml medium with 20% serum)
20 ml BSS
50 ml methanol
vacuum pump or tap siphon
Giemsa stain
mountant (DePex or Permount)

Protocol

1. Set up filter assembly (Fig. 15.6) with 25-mm diameter, 0.5-μm-porosity polycarbonate filter
2. Draw cell suspension on to filter using a vacuum pump. Do not let all the medium run through
3. Add 10 ml BSS gently when cell suspension is down to 2 ml
4. Repeat when BSS is down to 2 ml
5. Add 10 ml methanol to BSS and repeat until pure methanol is being drawn through filter
6. Switch off vacuum before all the methanol runs through
7. Lift out filter and air dry
8. Stain filter in Giemsa and dry
9. Mount on a slide in DePex or Permount by pressing the coverslip down to flatten the filter

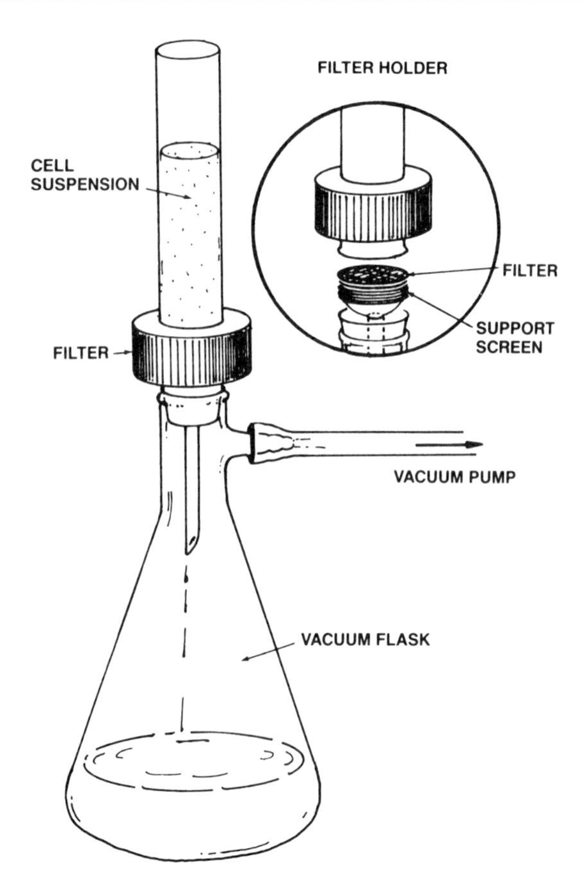

Fig. 15.6. *Filter assembly for cytological preparations (see text).*

Photography

There are two major frame sizes you may wish to consider; 35 mm, best for routine color transparencies and high-volume black and white, and 3½ × 4½ in. (12 cm × 9 cm) Polaroid with positive/negative film (type 665) for low-volume black and white. With Polaroid, you obtain an instant result and know that you have the record without having to develop and print a film. The unit cost is high, but there is a considerable saving in time.

Specific instructions are supplied with microscope cameras, but a few general guidelines may be useful.

1. Choose the film and set the exposure meter before bringing out the culture
2. Make sure the culture is free of debris, e.g., change the medium on a primary culture before photography
3. Choose the appropriate field quickly, avoiding imperfections or marks on the flask (always label on the side of the flask or dish and not on the top or

bottom). Rinse medium over the upper inside surface if condensation has formed and let it drain down before attempting to photograph

4. Focus carefully first the microscope and then the camera eyepiece if there is one

5. Turn up the light, check the focus and exposure, and expose; then turn down the light immediately to avoid overheating the culture

Note. An infrared filter may be incorporated to minimize overheating.

6. If Polaroid, check exposure on finished print and repeat if necessary at a different setting. If 35 mm, bracket the exposure by rephotographing at half and double the exposure

7. Return the culture to the incubator

8. Label Polaroid prints immediately. If 35 mm, keep a record against the frame number; otherwise the prints will be difficult to identify

9. File photographs in a readily accessible way, e.g., albums or filing cabinet sheets with transparent pockets

CELL GROWTH KINETICS

The quantitative aspects of cell growth kinetics are dealt with in more detail in Chapter 19, and the growth cycle has been described in Chapter 12.

Growth Cycle

Analysis of the growth cycle (see Chapter 12) gives the population doubling time and saturation density of the culture; these should remain constant under standard culture conditions. Variations imply altered culture conditions (medium, serum, etc.), contamination, cross contamination, transformation, or senescence.

Labeling Index

The labeling index (see Chapter 19) with [^{3}H]thymidine indicates how many cells are in cycle at any time. If cells are labeled continuously for a period of time equivalent to two cell cycles, the labeling index will express the *growth fraction,* i.e., the proportion of cells capable of division. This is characteristically low in normal cultures and high in transformed cultures (see next chapter) if measured when the culture is at saturation density.

Cell Cycle Time

The time taken for a cell to progress through the cell cycle and the length of the constituent phases of the cycle are often characteristic of the cell strain (see Chapters 19 and 22). In practice this technique is too complex for routine use, and the population doubling time derived from a growth curve is usually the figure quoted; e.g., under optimal conditions CHO cells double every 12 hr, PyY, 15 hr; HeLa, 24 hr; and many finite cell lines every 30–40 hr, although these times will vary in different media.

CHROMOSOME CONTENT

Chromosome content is one of the most characteristic and well-defined criteria for identifying cell lines and relating them to the species and sex from which they were derived. See the Committee on Standardized Genetic Nomenclature for Mice [1972] for mouse karyotype; Committee for a Standardized Karyotype of *Rattus norvegicus* [1973] for rat, and Paris Conference [1975], or *An International System for Human Cytogenetic Nomenclature* [1978] for human. Chromosome analysis can also distinguish between normal and malignant cells as the chromosome number is more stable in normal cells (except in mice where the chromosome complement of normal cells can change quite rapidly after explantation into culture).

Chromosome Preparations [Rothvels and Siminovitch, 1958; Worton and Duff, 1979]

Outline

Cells arrested in metaphase and swollen in hypotonic medium are fixed and dropped on a slide, stained, and examined (Fig. 15.7).

Materials

Cultures of cells in log phase
10^{-5} M colcemid in BSS
PBSA
0.25% crude trypsin
centrifuge tubes
centrifuge
hypotonic solution:
  0.04 M KCl
  0.025 M sodium citrate

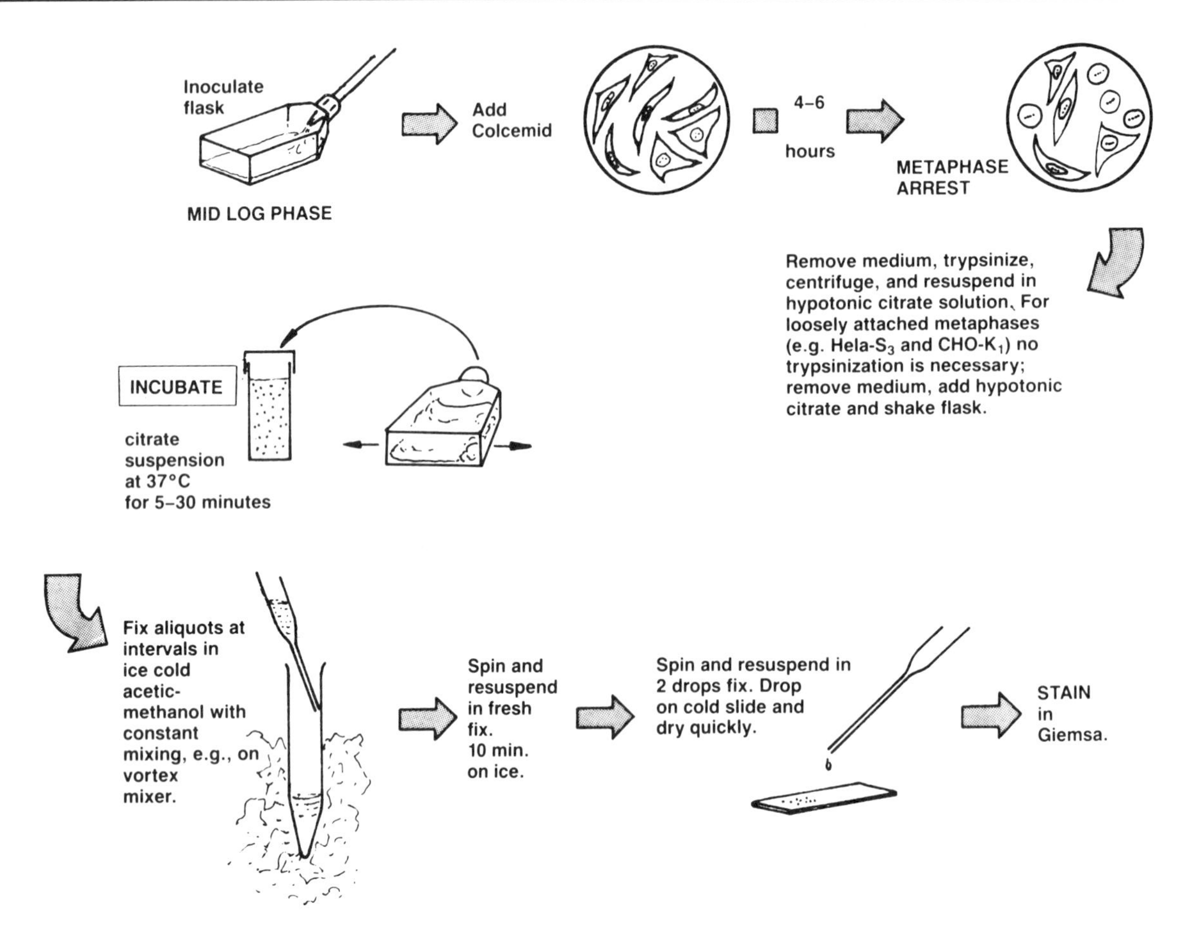

Fig. 15.7. *Chromosome preparation.*

fixative:

1 part glacial acetic acid plus 3 parts anhydrous methanol or ethanol, made up fresh and kept on ice

vortex mixer
Pasteur pipettes
ice
slides
Giemsa stain
slide dishes
00 coverslips
DPX or Permount

Protocol

1.
Set up 75-cm^2 flask culture at between 2 and 5 × 10^4 cells/ml (4 × 10^3–10^4 cells/cm^2) in 20 ml

2.
Approximately 5 d later when cells are in the log phase of growth, add colcemid (10^{-7} M, final) to the medium already in the flask

3.
After 4–6 hr, remove medium gently, add 5 ml 0.25% trypsin, and incubate 10 min

4.
Centrifuge cells in trypsin and discard supernatant trypsin

5.
Resuspend the cells in 5 ml of hypotonic solution and leave for 20 min at 36.5°C

6.
Add equal volume of freshly prepared ice-cold acetic methanol, mixing constantly, and then centrifuge at 100*g* for 2 min

7.
Discard the supernatant mixture, "buzz" the pellet on a vortex mixer, and slowly add fresh acetic methanol with constant mixing

8.
Leave 10 min on ice

9.
Centrifuge for 2 min at 100*g*

10.
Discard supernatant acetic methanol and resuspend pellet with "buzzing" in 0.2 ml acetic methanol, to give a finally dispersed cell suspension
11.
Draw one drop into the tip of a Pasteur pipette and drop on a cold slide. Let the drop run down the slide as it spreads
12.
Dry off rapidly over a beaker of boiling water and examine on microscope. If cells are evenly spread and not touching, prepare more slides at same cell concentration. If piled up and overlapping, dilute two- to fourfold and make a further drop preparation. If satisfactory, prepare more slides. If not, dilute further and repeat
13.
Stain with Giemsa: (a) Immerse slides in neat stain for 2 min. (b) Place dish in sink and add approximately 10 *V* water, allowing surplus stain to overflow from top of slide dish. (c) Leave for a further 2 min. (d) Displace remaining stain with running water and finish by running slides individually under tap to remove precipitated stain (pink, cloudy appearance on slide). (e) Check staining under microscope. If satisfactory, dry slide thoroughly and mount 00 coverslip in DePex or Permount

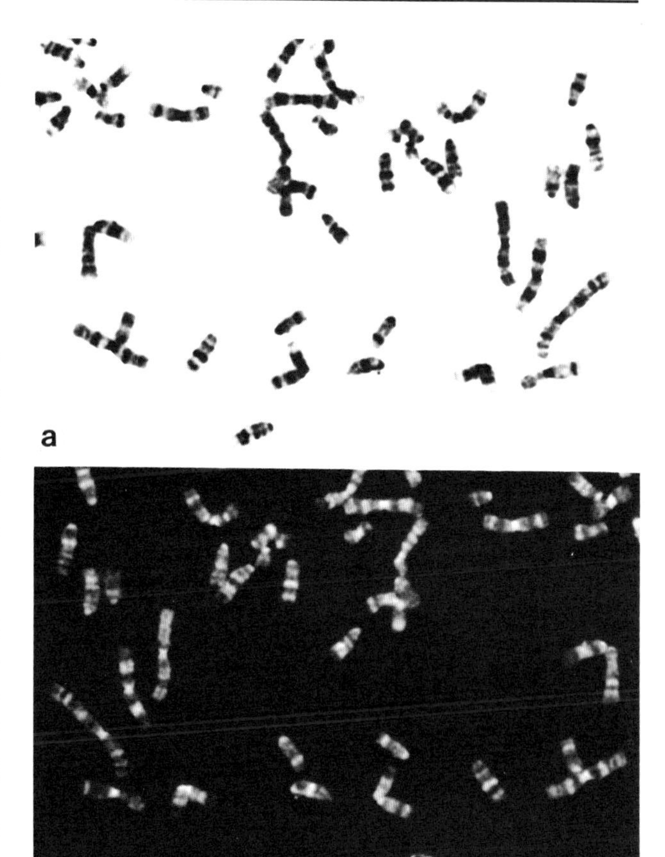

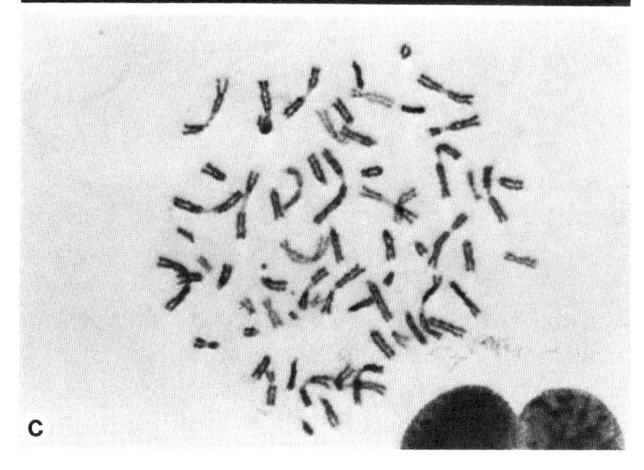

Fig. 15.8. *Chromosome staining. a. Human chromosomes banded by standard trypsin-Giemsa technique. b. Same preparation as a stained with Hoechst 33258. c. Human/mouse hybrid stained with Giemsa at pH 11. Human chromosomes are less intensely stained than mouse. Several human/mouse chromosomal translocations can be seen (By courtesy of Dr. R.L. Church).*

Chromosome Banding

This group of techniques [see Yunis, 1974] was devised to enable individual chromosome pairs to be identified where there is little morphological difference between them [Wang and Fedoroff, 1972, 1973]. The chromosomal proteins are partially digested by crude tryrpsin, producing a banded appearance on subsequent staining. The banding pattern is characteristic for each chromosome pair (Fig. 15.8).

Outline

Fixed chromosome spreads are treated briefly with trypsin, stained, and mounted.

Materials

As for chromosome preparation
0.05% trypsin
crude tissue culture grade, in PBSA or CMF
about 20 slides with well-spread chromosomes
95% methanol, 2% Giemsa in 0.01 M phosphate buffer, pH 7.0

Protocol

1.
Reduce Colcemid treatment to 1–2 hr. Fixed, unstained preparations should be left for about 48 hr to 1 wk at 4°C over desiccant or placed at 36.5°C for 1 hr
2.
Place five slides in the trypsin solution at room temperature and remove after 15, 30 s, 1, 2, and 5 min, rinse in PBS, 95% methanol, and dry
3.
Stain for 5 min in Giemsa
4.
Wash slides thoroughly in tap water, rinse in deionized water, and dry
5.
Examine unmounted on microscope: (a) If understained, return to stain for a further 1–5 min. (b) Select trypsin treatment giving clearest banding. If no banding is apparent, increase trypsin treatment (duration, temperature, or concentration). If chromosomes are fuzzy or all of the protein is digested out leaving "ghosts," reduce trypsin treatment. (3) Trypsinize and stain remaining slides. (Preparations may be destained by soaking in methanol for 10 min, dried, and reexposed to trypsin if necessary)

Analysis. 1. Count chromosome number per spread for between 50 and 100 spreads (need not be banded). Closed-circuit television or a camera lucida attachment may help. You should attempt to count all the mitoses that you see and classify them (a) by chromosome number. (b) If counting is impossible, as "near dipolid uncountable" or "polyploid uncountable." Plot the results as a histogram (see Fig. 2.2)

2. Prepare karyotype. Photograph about 10 or 20 good spreads of banded chromosomes and print on 20 × 25 cm high-contrast paper. Cut out the chromosomes, sort into sequence, and stick down on paper (see Figs. 15.9,15.10).

Variations. ***Metaphase block.*** (1) Vinblastine, 10^{-6} M, may be used instead of colcemid; (2) duration of the metaphase block may be increased to give more metaphases for chromosome counting, but chromosome condensation will increase, making banding very difficult.

Collection of mitosis by "shake-off" technique. Some cells, e.g. CHO and HeLa, detach readily when in metaphase. This allows trypsinization for collection of metaphases to be eliminated. (1) Add colcemid. (2) Remove carefully and replace with hypotonic citrate/KCl. (3) Shake the flask to dislodge cells in metaphase either before or after incubation in hypotonic medium. (4) Fix as before.

Hypotonic treatment. Substitute 0.075 M KCl alone or HBSS diluted to 50% with distilled water. Duration of hypotonic treatment may be varied from 5 min to 30 min to reduce lysis or increase spreading.

Spreading. There are perhaps more variations at this stage than any other, designed to improve the degree and flatness of the spread. They include: (1) dropping cells onto slide from a greater height. Clamp the pipette and mark the position for the slide using a trial run with fixative alone. (2) Flame drying. Dry slide after dropping cells by heating over a flame or actually burn off the fixative by igniting the drop on the slide as it spreads (this may make banding more difficult later). (3) Ultracold slide. Chill slide on solid CO_2 before dropping on cells. (4) Refrigerate fixed cell suspension overnight before dropping. (5) Drop cells on a chilled slide (e.g., steep in cold alcohol and dry off), then place over a beaker of boiling water.

Banding. (1) Use trypsin + EDTA rather than trypsin alone. (2) *Q-banding* [Caspersson et al., 1968]. Quinacrine mustard or dihydrochloride stains the interband regions unstained in the Giemsa technique. No trypsinization is required and the fixed preparations are stained in 5% (w/v) quinacrine dihydrochloride in 45% acetic acid, rinsed and mounted in deionized water at pH 4.5 [Lin and Uchida, 1973; Uchida and Lin, 1974]. The preparation is examined by uv fluorescence. (3) *C-banding.* This technique emphasizes the centromeric regions. The fixed preparations are pretreated for 15 min with 0.2 N HCl, 2 min with 0.07 N NaOH, and then treated overnight with SSC (either 0.03 M sodium citrate, 0.3 M NaCl, or 0.09 M sodium citrate, 0.9 M NaCl) before staining with Giemsa stain [Arrighi and Hsu, 1974].

Techniques have been developed for discriminating between human and mouse chromosomes, principally to aid the karyotypic analysis of human and mouse hybrids. These include fluorescent staining with Hoechst 33258 which causes mouse centromeres to fluoresce more brightly than human [Lin et al., 1974; Hilwig and Gropp, 1972] and alkaline staining with Giemsa ("Giemsa-11") [Bobrow et al., 1972; Friend et al., 1976].

Chromosome counting and karyotyping will allow species identification of the cells and, when banding is

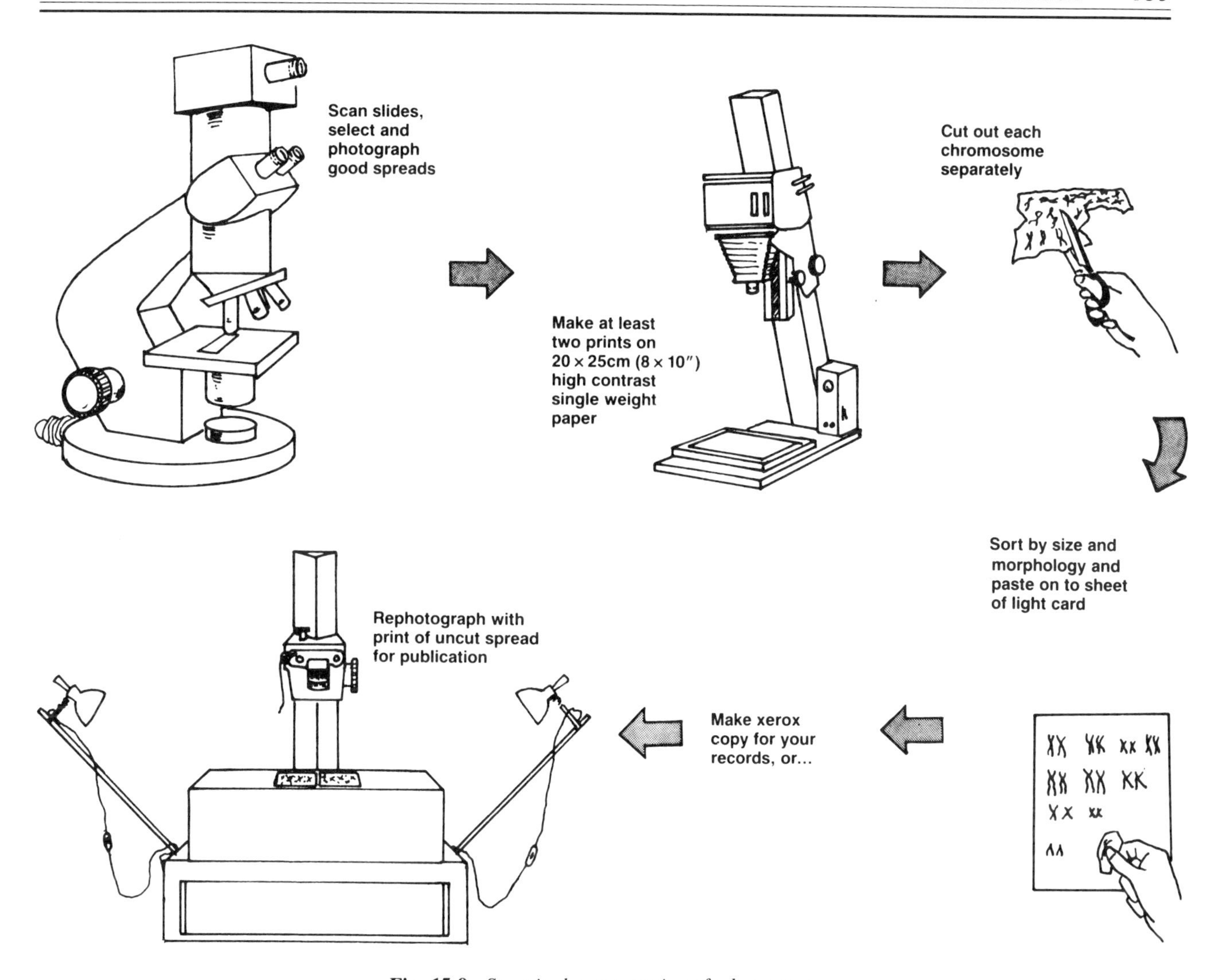

Fig. 15.9. *Steps in the preparation of a karyotype.*

used, will distinguish cell line variation and marker chromosomes. Banding and karyotyping is time consuming and chromosome counting with a quick check on gross chromosome morphology may be sufficient to confirm or exclude a suspected cross contamination.

DNA CONTENT

The amount of DNA per cell is relatively stable in normal cell lines such as human fibroblasts and glia, chick and hamster fibroblasts, but varies in cell lines from the mouse and from many neoplasms. DNA can be measured by microdensitometry of Feulgen-stained cells [Pearse, 1968] or by ethidium bromide fluorescence and microphotometry. The advent of flow cytophotometry (flow cytofluorimetry, fluorescence-activated cells sorting; see Chapter 14) has made the assay of DNA per cell much more quantitative and reproducible (see Fig. 14.15).

RNA AND PROTEIN

Histochemical reactions and flow cytophotometry also enable measurement of RNA and protein per cell. These are prone to considerable fluctuations, but in some cases, the ratio of RNA: DNA or protein:DNA may be found characteristic of the cell type if measured under standard culture and assay conditions.

Qualitative analysis of total cell protein will reveal differences between cells when whole cell, or cell membrane extracts are run on two-dimensional gels [O'Farrell, 1975]. This produces a characteristic "fingerprint" similar to polypeptide maps of protein hydrolysates, but contains so much information that interpretation can be difficult. Labeling the cells with [^{32}P], [^{35}S]methionine, or a combination of ^{14}C-labeled amino acids, followed by autoradiography, may make analysis easier, but it is not a technique suitable

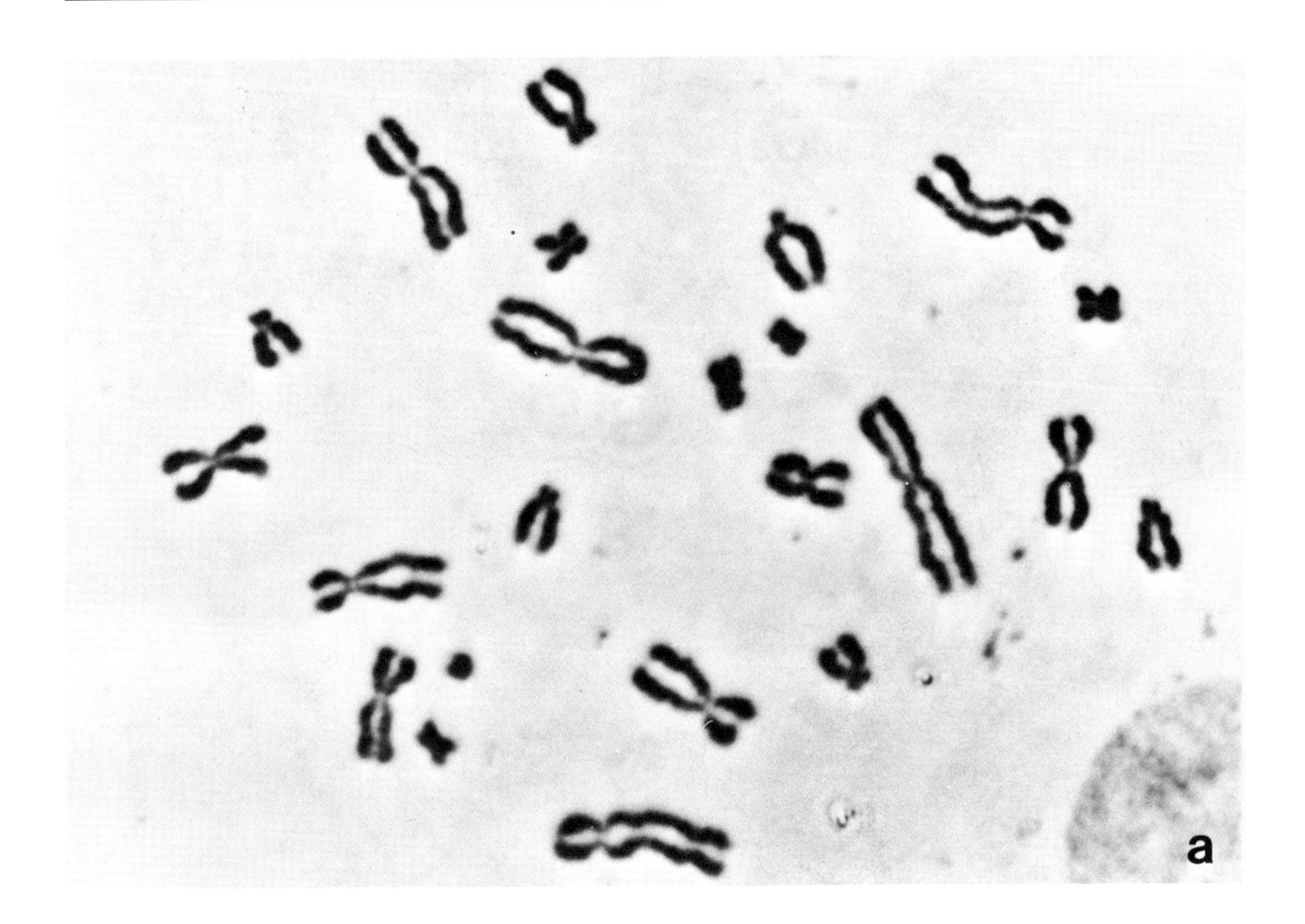

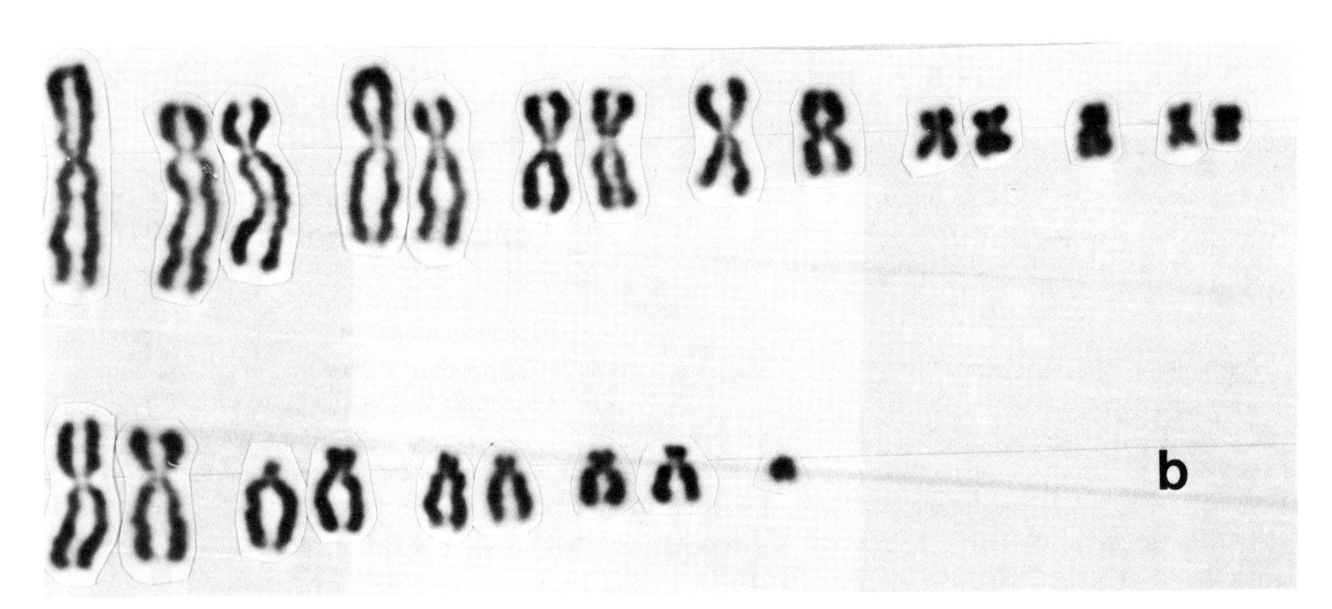

Fig. 15.10. *Example of karyotype. Chinese hamster cells recloned from the Y-5 strain of Yerganian and Leonard [1961] (Acetic-orcein stained).*

for routine use unless the technology for preparing 2D gels is currently in use in the laboratory.

ENZYME ACTIVITY

Specialized functions *in vivo* are often expressed in the activity of specific enzymes, e.g., urea cycle enzymes in liver, alkaline phosphatase in endothelium. Unfortunately, many enzyme activities are lost *in vitro* for the reasons discussed in Chapter 2 and are no longer available as markers of tissue specificity. Liver parenchyma loses arginase activity within a few days and cell lines from endothelium lack high alkaline phosphatase activity. However, some cell lines do express specific enzymes such as tyrosine aminotransferase in the rat hepatoma HTC cell lines [Granner et al., 1968]. When looking for specific marker enzymes, the constitutive (uninduced) level and the induced level should be measured and compared with a number of control cell lines. Glutamyl synthetase activity, for instance, characteristic of astroglia in brain, is increased severalfold when the cells are cultured in the presence of glutamate instead of glutamine.

Induction of enzyme activity will require specialized conditions for each enzyme and these may be obtained from the literature. Common inducers are glucocorticoid hormones such as dexamethasone, polypeptide hormones such as insulin and glucagon, or alteration in substrate or product concentrations in the medium, as in the example above with glutamyl synthetase.

Isoenzymes

Enzyme activities can also be compared qualitatively between cell strains, as many enzyme activities are expressed by different, though related, species of molecules. These so-called isoenzymes or isozymes may be separated chromatographically or electrophoretically and the distribution patterns (zymograms) found to be characteristic of species or tissue. Paul and Fottrel [1961] demonstrated differences in esterase zymograms between normal and malignant mouse cells, and human cells (Fig. 15.11), and O'Brien and Kleiner [1977] have described a number of very useful isozymic markers for human cell lines after electrophoresis of cell extracts [see also Macy, 1978]. Electrophoresis media include agarose, cellulose acetate, starch, and polyacrylamide. In each case, a crude enzyme extract is applied to one point in the gel and a potential difference applied across the gel. The different isozymes migrate at different rates and can be detected later by staining with chromogenic substrates. Stained gels may be read directly by eye and photographed, or scanned with a densitometer.

Isozyme analaysis has proved to be of great value in determining the chromosomal constitution of somatic cell hybrids (see Chapter 22) [Harris and Hopkinson, 1976; Nichols and Ruddle, 1973; Meera Khan, 1971; Van Someren et al., 1974].

ANTIGENICITY

If immunological expertise is available or if you are prepared to take time to establish it, labeling cells with specific antibodies is one of the best methods of characterization currently available [e.g., Raff et al., 1979; Barnstable, 1980]. The major controlling factor is the specificity of the antibody, and this must be clearly demonstrated with appropriate controls. The development of monoclonal antibodies (see Chapter 22) has made a number of highly specific antibodies available, though their specificity for cell type must still be demonstrated.

Antibodies are commonly raised in rabbits, goats, sheep, swine, or mice and purified by absorption with crude liver extract and, if necessary, extracts of other cell lines. After the binding of the antibody to its target cell, it may be detected with a second antibody raised against the immunoglobulin of the animal in which the first was raised, e.g., if the first antibody was raised in rabbit, the second might be swine antirabbit globulin. The second antibody is bound covalently to fluorescein isothiocyanate or rhodamine and will fluoresce on examination with the appropriate wavelength of light [Coons and Kaplan, 1950; Kawamura, 1969]. Rhodamine fades more slowly than fluorescein, but most fluorescent labels eventually fade under the wavelength of light used to excite the fluorescence. For this reason, secondary antibodies linked to horesradish peroxidase are often used [Taylor, 1978; Avrameas, 1970]. When the preparation is incubated with diaminobenzidine, an insoluble polymer is formed giving a yellow or brown color localized to the area where the antibody is bound. These preparations are permanent and usually more sensitive but are prone to more variation and background staining than fluorescence and can be difficult to interpret if the staining is light.

Specific cell surface antigens are usually stained in living cells (at 4°C in the presence of sodium azide to inhibit pinocytosis) while intracellular antigens are stained in fixed cells, sometimes requiring light trypsinization to permit access of the antibody to the antigen.

HLA and blood group antigens can be demonstrated on many human cell lines, and serve as useful charac-

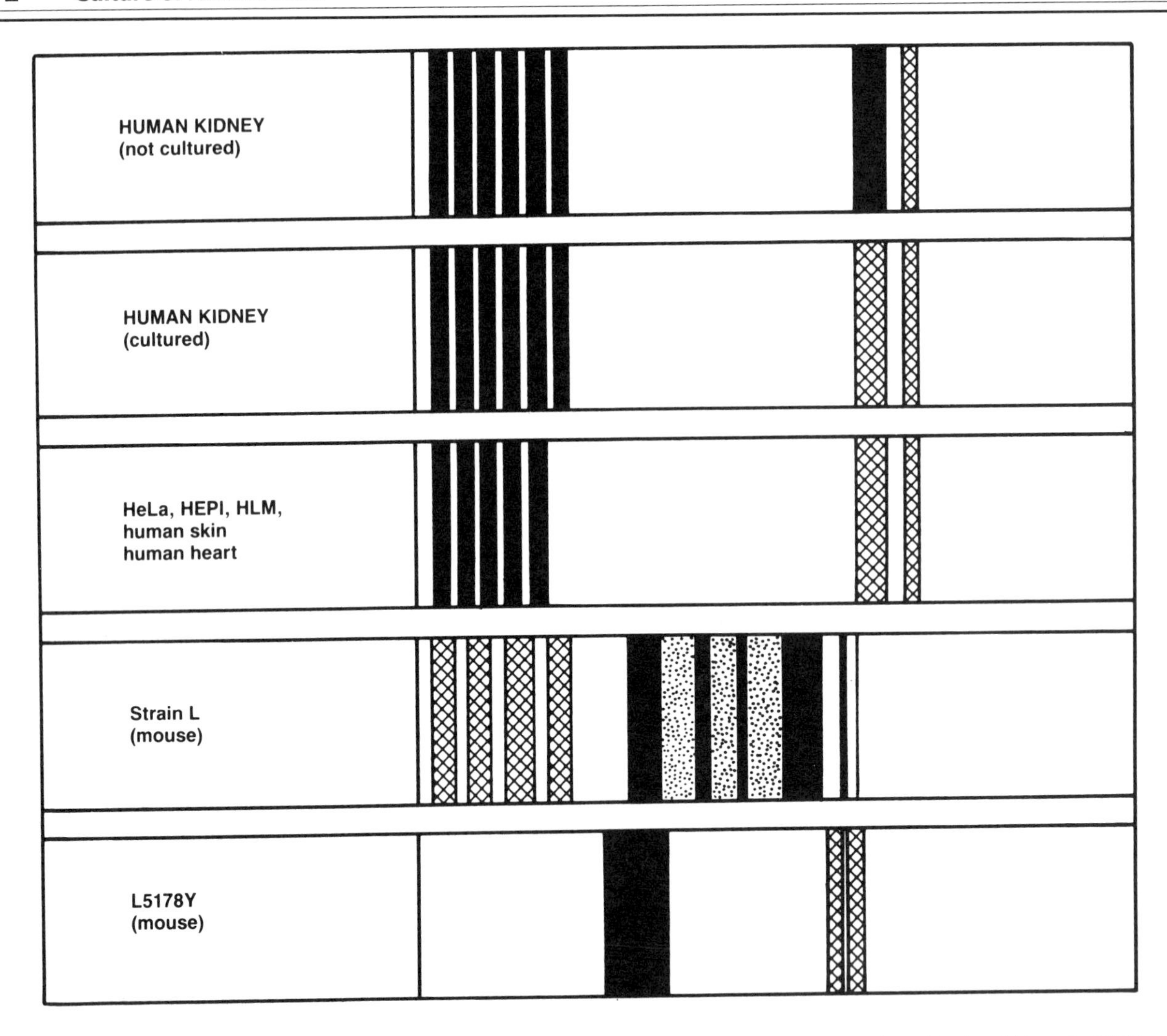

Fig. 15.11. *Isozyme patterns revealed by electrophoresis. Starch gel zymogram for esterase [from Paul and Fottrell, 1961].*

terization tools, especially where the donor patient profile is known [Espmark et al., 1978; Pollack et al., 1981; Stoner et al., 1980].

DIFFERENTIATION

Many of the characteristics described under antigenic markers or enzyme activities may also be regarded as markers of differentiation and as such they can help to correlate cell lines with their tissue of origin. Other examples of differentiation, and as such highly specific markers of cell line identity, are given in Table 2.2, and the appropriate assays for these properties may be derived from the references cited (see also Chapter 20).

While much of the interest in characterization is related to the study of specialized functions and their relationship to the cells behavior *in vivo*, these techniques are also important in confirming the identity of a cell line and excluding the possibility of cross contamination.

In Vitro Transformation and Its Relationship to Malignancy *In Vivo*

Chapter 16

The growth of a tumor in an animal represents a breakdown in the regulation of cell proliferation at that site. There are a number of events consequent upon this, such as abnormal differentiation, loss of normal tissue histology, or deformation of surrounding tissues, which are typical of the tumor but not causal. In order that a tumor be considered malignant, it must invade, progressively, the surrounding tissue, or spread to local or distant secondary sites (metastasis), or both. Hence the term "malignancy" has a fairly clear meaning *in vivo* but is difficult to translate into cell culture conditions. Tumor cells express a number of properties, including cell division and invasion as well as proteolytic activity, stimulation of angiogenesis, and glycolytic metabolism, all of which are examples of the inappropriate expression of normal cellular properties, rather than *de novo* expression of tumor-specific properties. The search for characteristics peculiar to tumor cells has been singularly unsuccessful with the possible exception of the development of aneuploidy.

"Transformation" has been used to describe the transition of a nontumor-forming (nonneoplastic) into a tumor-forming (neoplastic) cell, induced by a virus, a chemical carcinogen, or resulting from a spontaneous event or series of events [Pastan, 1979]. This is more properly known as "neoplastic transformation" [Schaeffer, 1979]. However, the relationship between "transformation" *in vitro* and "malignancy" *in vivo* is not clear. A normal cell can transform *in vitro* and the derived cell line may or may not produce a tumor when inoculated back into the host. That tumor may be (but is not necessarily) malignant. On the other hand, a cell culture derived from a malignant tumor may yet undergo transformation *in vitro*, exhibiting an increased growth rate, reduced anchorage dependence, more pronounced aneuploidy, and so on.

Transformation and malignancy, in addition to being nonsynonymous, are not necessarily equivalent. The acquisition of malignant potential by a cell may be the result of a series of events, some, but not all, causally related. The event(s) occurring during transformation *in vitro* may be similar to those in carcinogenesis but not identical.

Since malignancy, as such, cannot be demonstrated *in vitro*, we are obliged to use a number of properties associated with cells from malignant tumors grown *in vitro*. In the present discussion, I shall describe markers which might be used in cell identification and do not imply a causal relationship between these properties and the expression of malignancy *in vivo*, but clearly many of the properties discussed have an obvious functional relationship to malignancy.

Two approaches have been used to explore malignancy-associated properties. (1) Cells have been cultured from malignant tumors and characterized. (2) Transformation *in vitro* with a virus or a chemical carcinogen has produced cells which were tumorigenic and which could be compared with the untransformed cells. The second system provides transformed cells which can be shown to be malignant and they can be compared with untransformed cells which are not. Unfortunately, many of the characteristics of cells transformed *in vitro* have not been found in cells derived from spontaneous tumors. Ideally, tumor cells and equivalent normal cells should be isolated and characterized. Unfortunately, there have been relatively few instances where this has been possible, and even then, although the cells may belong to the same lineage, their position in that lineage is not always clear (see Fig. 2.3), and comparison not strictly justified.

At best, there are a number of generally accepted properties which can be recognized in many tumor cells *in vitro*. Many occur in normal cells, confirming the conclusion that malignancy is not the expression of abnormal characteristics *de novo* but rather the inappropriate expression of normal properties, and none is common to all neoplastic cell lines.

ANCHORAGE INDEPENDENCE

Many of the properties associated with neoplastic transformation *in vitro* are the result of cell surface modifications [Hynes, 1974; Nicolson, 1976], e.g., changes in the binding of plant lectins [Ambrose et al., 1961; Aub et al., 1963; Willingham and Pastan, 1975; Reddy et al., 1979] and in cell surface glycoproteins [Hynes, 1976; Lloyd et al., 1979; Van Beek, 1979; Warren et al., 1978], which may be correlated with the development of invasion and metastasis *in vivo*. Fibronectin (LETS protein, large extracellular transformation sensitive) is lost from the surface of transformed fibroblasts [Hynes, 1973; Vaheri et al., 1976]. This may contribute to a decrease in cell-cell and cell-substrate adhesion [Easty et al., 1960] and to a decreased requirement for attachment and spreading for the cells to proliferate [MacPherson and Montagnier, 1964]. In addition, loss of cell-cell recognition, a product of reduced adhesion, leads to a disorganized growth pattern and loss of density limitation of growth (see below). This results in the ability of cells to grow detached from the substrate, either in stirred suspension culture or suspended in semisolid media such as agar or Methocel. There is an obvious analogy with detachment from the tissue in which a tumor arises and the formation of metastases in foreign sites, but how valid this analogy is, is not clear.

Suspension Cloning

MacPherson and Montagnier [1964] were able to demonstrate that polyoma-transformed BHK21 cells could be grown preferentially in soft agar while untransformed cells cloned very poorly. Subsequently, it has been shown that colony formation in suspension is frequently enhanced following viral transformation. The position regarding spontaneous tumors is less clear, however, in spite of the fact that Freedman and Shin [1974; Kahn and Shin, 1979] demonstrated a close correlation between tumorigenicity and suspension cloning in Methocel. Although Hamburger and Salmon [1977] have shown that many human tumors contain a small percentage of cells (< 1.0%) which are clonogenic in agar, we [Freshney and Hart, 1982] and others [Laug et al., 1980] have shown that a number of normal cells will also clone in suspension (see Fig. 13.11) with equivalent efficiency. Since normal fibroblasts are among these cells which will clone in suspension, the value of this technique for assaying for the presence of tumor cells in short-term cultures from human tumors is in some doubt. It remains a valuable technique for assaying neoplastic transformation *in vitro* by tumor viruses and has been used extensively by Styles [1977] to assay for carcinogenesis.

The technique of cloning in suspension is described in Chapter 13. Variations with particular relevance to the assay of neoplastic cells are in the choice of suspending medium. It has been suggested [Neugut and Weinstein, 1979] that agar may only allow the most highly transformed cells to clone while agarose (lacking sulphated polysaccharides) is less selective. Montagnier [1968] was able to show that untransformed BHK21 cells, which would grow in agarose but not in agar, could be prevented from growing in agarose by the addition of dextran sulphate.

Contact Inhibition and Density Limitation of Growth

The loss of contact inhibition (see Chapter 12) may be detected morphologically by the formation of a disoriented monolayer of cells or rounded cells in foci within the regular pattern of normal surrounding cells. This is illustrated in Figure 16.1 where 3T3 cells transformed by bovine papilloma virus DNA are compared with spontaneous transformants. Cultures of human glioma show a disorganized growth pattern and exhibit reduced density limitation of growth by growing to a higher saturation density than normal glial cell lines [Freshney et al., 1980a,b]. As variations in cell size will influence the saturation density, the increase in the labeling index with [^{3}H]-thymidine at saturation density is a better measurement of reduced density limitation of growth. Human glioma, labeled for 24 hr at saturation density with [^{3}H]-thymidine, gave a labeling index of 8% while normal glial cells gave 2%.

Outline

The culture is grown to saturation density in nonlimiting medium conditions and the percentage of cells labeling with [^{3}H]-thymidine determined autoradiographically.

Materials

Culture of cells ready for subculture
PBSA
0.25% trypsin
24-well plates containing 13-mm

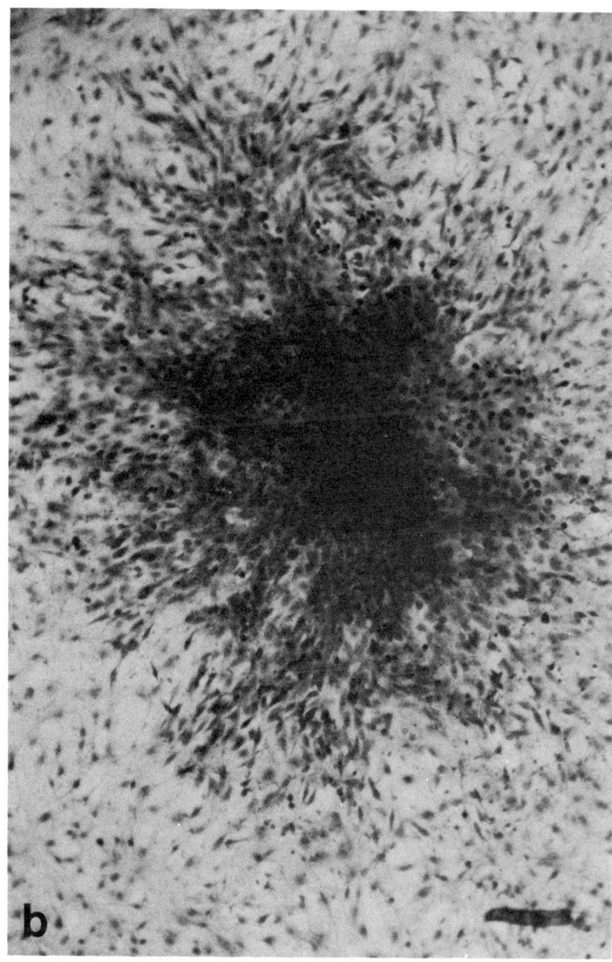

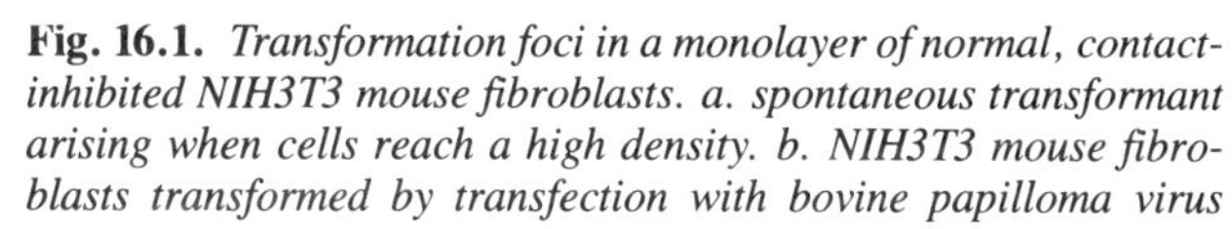

Fig. 16.1. *Transformation foci in a monolayer of normal, contact-inhibited NIH3T3 mouse fibroblasts. a. spontaneous transformant arising when cells reach a high density. b. NIH3T3 mouse fibroblasts transformed by transfection with bovine papilloma virus DNA cloned in bacterial plasmid pAT-153, coprecipitated with $Ca_3(PO_4)_2$. By courtesy of D. Spandidos, photographs by M. Freshney.*

coverslips
medium
9-cm petri dishes (one per coverslip)
medium containing 0.1 μCi/ml [^{3}H]-thymidine (2 Ci/mmol)

Protocol

1.
Trypsinize cells and seed 10^5 cells/ml into 24-well plate, 1 ml/well, each well containing a 13-mm-diameter coverslip

2.
Incubate in humidified CO_2 incubator for 1–3 d

3.
Transfer the coverslips to 9-cm petri dishes containing 20 ml medium and return to CO_2 incubator

4.
Continue culturing, changing medium every 2 d after cells become confluent on the coverslips. Trypsinize and count cells from two coverslips every 3–4 d. As cells become denser on the coverslip, it may be necessary to add 200–500 units/ml crude collagenase to the trypsin to achieve complete dissociation of the cells for counting

5.
When cell growth ceases, i.e., two sequential counts show no significant increase, change to fresh medium containing 0.1 μCi/ml [^{3}H]-thymidine (2 Ci/mmol), and incubate for 24 h

6.
Transfer coverslips back to a 24-well plate and trypsinize the cells for autoradiography (see Chap-

ter 21). They may be fixed in suspension and dropped on a slide as for chromosome preparations (without the hypotonic treatment), centrifuged onto a slide using the Shandon Cytospin, or trapped on Millipore or Nucleopore filters by vacuum filtration (see Chapter 15)

Note. It is necessary to trypsinize high-density cultures for autoradiography because of their thickness and the weak penetration of β-emission from ^3H (mean path length in water is approximately 1 μm). Labeled cells in the underlying layers would not be detected by the radiosensitive emulsion due to the absorption of the β-particles by the overlying cells.

Analysis. Count the number of labeled cells as a percentage of the total. Scan the autoradiographs under the microscope and count the total number of cells, and the proportion labeled in representative parts of the slide. A suggested scanning pattern for a circular array of cells (such as would be produced by the drop technique or the Cytospin) is given in Figure 19.6.

Growth on Confluent Monolayers

Aaronson et al. [1970], showed that transformed mouse 3T3 cells and human fibrosarcoma cells were able to form colonies on confluent monolayers of normal cells, while normal fibroblasts were not. In the author's laboratory, cells cultured from both normal and malignant breast tissue formed colonies on contact-inhibited monolayers of human fetal intestinal cells (FHI cells) (Fig. 13.9) as did normal epidermal cells (Fig. 16.2). Rheinwald and Green [1975] had previously shown that normal epidermal cells grew on 3T3 monolayers. We have since found that human glioma will grow on confluent feeder layers, but cells derived from normal glia will not (see Fig. 13.10). In some cases, therefore, where an appropriate control cell can be tested, this technique may discriminate between malignant and normal cells, but this is not universally true.

Outline

Prepare cells as for cloning and seed onto confluent monolayers of contact-inhibited cells (see Fig. 13.4).

Materials

14 × 25-cm^2 flasks
medium
PBSA
0.25% trypsin
cultures of tumor cells
tubes for dilutions
hemocytometer or cell counter

Protocol

1.
Seed fourteen, 25-cm^2 flasks with feeder cells (3T3 or other cell line which will become contact inhibited after reaching confluence)

2.
Feed cultures until all available substrate is covered with cells, and mitosis is greatly reduced. If a cell line is used which does not arrest at confluence, treat cells during exponential growth with mitomycin-C or irradiate (see Chapter 13) and reseed sufficient cells to produce confluent monolayers directly with no spaces between cells

3.
Trypsinize putative tumor cells and suitable control cells and dilute to 10^4, 10^3, and 10^2 cells/ml in 20 ml and replace medium in each pair of feeder layer flasks with 5 ml of cell suspension, and with medium alone in two flasks

4.
Seed two flasks at each concentration with each cell type without feeder layers

5.
Incubate for 2–3 wk with medium changes every 2 d

6.
Wash, fix, and stain (see Chapter 15)

Analysis. Count foci of tumor cells (usually morphologically distinguishable) and express as percentage of number of cells seeded. Compare with flasks with no feeder layers.

Variations. It is possible to have proliferation of cells seeded onto confluent monolayers without the formation of discrete colonies. In our experience, human fibrosarcoma and glioma seeded onto confluent monolayers of fetal human intestine do not form colonies, but still proliferate, infiltrating the monolayer as they do so. In order to quantify this type of growth, DNA synthesis is first inhibited in the monolayer (see Chapter 13) by treatment with mitomycin-C or by irradiation. Growth of cells seeded onto this layer can then be monitored by measuring the incorporation of [^{3}H]-thymidine.

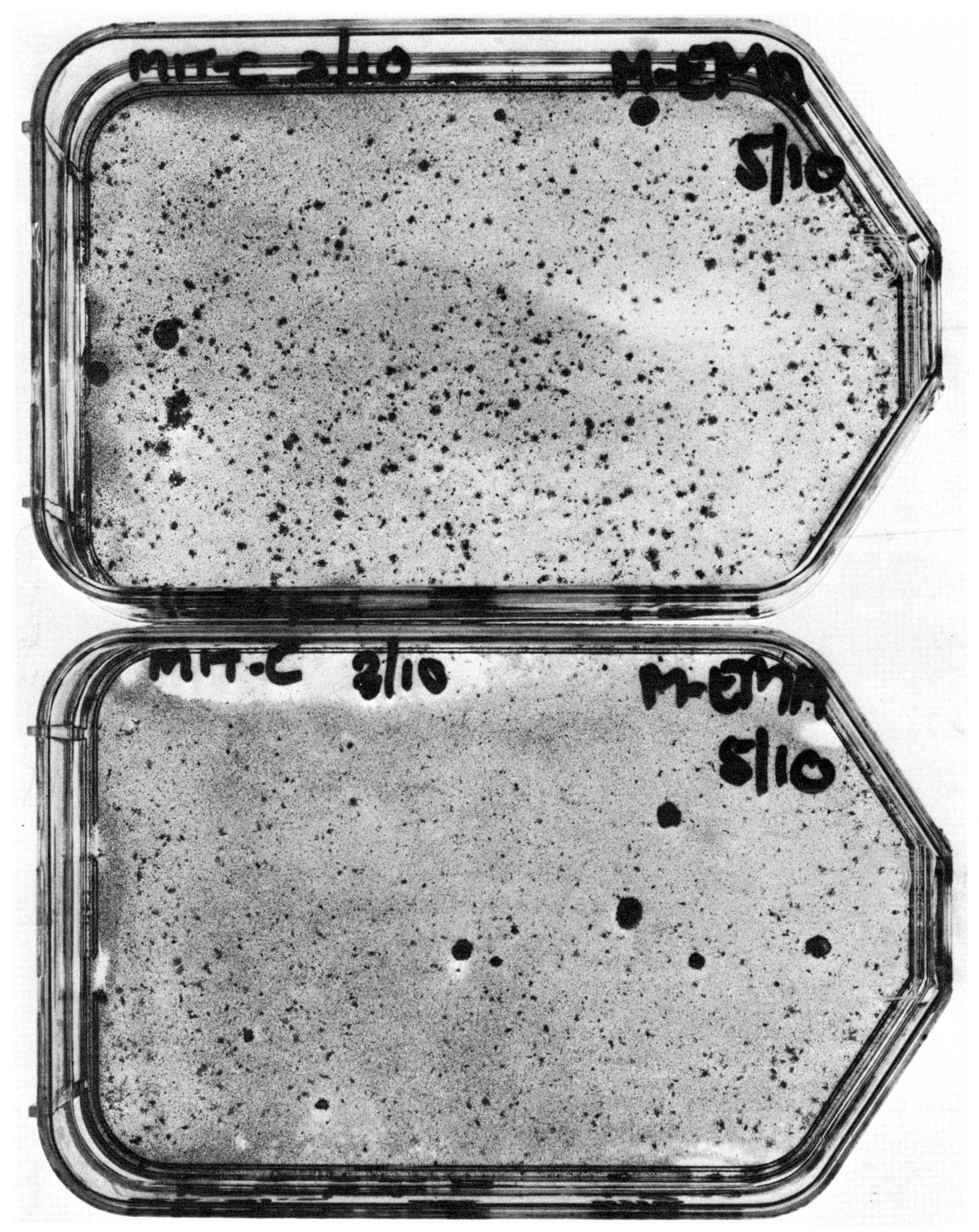

Fig. 16.2. *Epidermal cells from a skin biopsy of a benign naevus growing on a confluent monolayer of fetal human intestinal epithelial cells [Owens et al., 1974]. The large, dark-staining, circular colonies resemble keratinocytes. The smaller colonies may have been melanocytes (fibroblasts do not normally form colonies readily in this system), but no further characterization was done on them. The lower flask was treated with 1* mM *dibutyryl cyclic AMP inhibiting the small colony type but not the larger (putative keratinocytes). Cyclic AMP stimulates growth of keratinocytes (see Chapter 20) and may have done so here (top flask, four colonies; bottom flask, eight colonies).*

GENETIC ABNORMALITIES

Whether mutation is the prime cause of neoplastic transformation or not, chromosomal abnormalities and variations in DNA content per cell are found frequently in cells derived from malignant tumors.

Chromosomal Aberrations

Both changes in ploidy and increases in the frequency of individual chromosomal aberrations can be found [Biedler, 1976]. Figure 16.3 demonstrates variations in chromosome number found in human glioma and melanoma in culture. Chromosome analysis is described in Chapter 15 [see also Sandberg, 1980].

DNA content. Microdensitometry of Feulgen-stained preparations [Wright and Dendy, 1976] and flow cytofluorimetry [Traganos et al., 1977] show that the DNA content of tumor cells may vary from the normal. DNA analysis does not substitute for chromosome analysis, however, as cells with an apparently normal DNA content can yet have an aneuploid karyotype. Deletions and polysomy may cancel out, or translocations may occur without net loss of DNA.

CELL PRODUCTS AND SERUM DEPENDENCE

Transformed cells have a lower serum dependence than their normal counterparts [Temin, 1966; Eagle et al., 1970]. Lindgren et al. [1975] showed that a short exposure to serum would trigger glioma cells into cycle while normal glial cells required serum to be present throughout G1.

A possible explanation for the low serum dependence of tumor cells lies in the demonstration of Todaro and de Larco [1978] and others, that tumor cells may secrete their own growth factors. While the production of these polypeptides is not assayed as easily as some of the foregoing properties, the increasing availability of specific antibodies against them could bring this approach within the reach of any laboratory with the appropriate immunological expertise. Westall et al. [1978] have suggested some homology between a FGF-like peptide from brain and myelin basic protein, once claimed as a tumor marker.

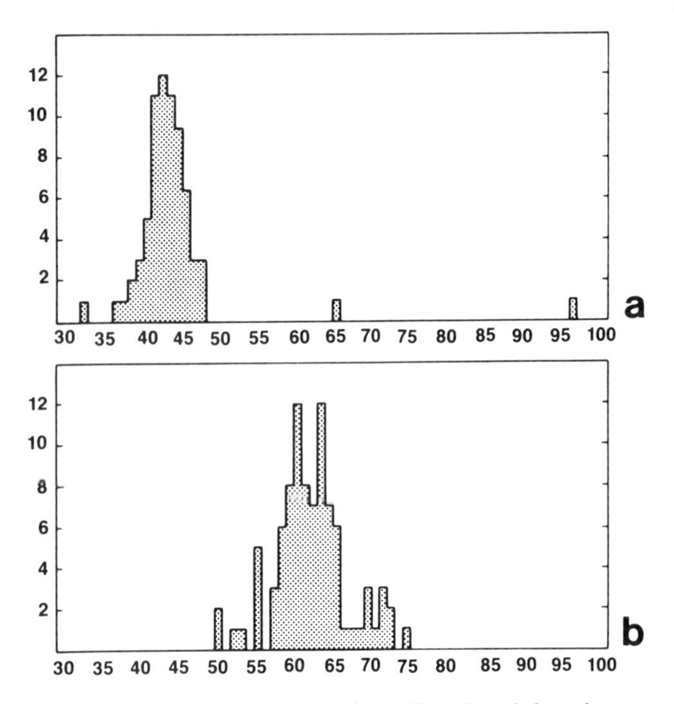

Fig. 16.3. *Chromosome counts for cells cultured from human anaplastic astrocytoma (a) and human metastatic melanoma (b).*

Tumor Angiogenesis Factor (TAF)

Among the growth factors released by tumor cells is one capable of inducing blood vessel proliferation [Phillips et al., 1976]. Fragments of tumor, pellets of cultured cells, or cell extracts, implanted on the surface of the chorioallantoic membrane (CAM) of the hen's egg, promote an increase in vascularization which is apparent to the naked eye some days later (Fig. 16.4). Since this assay is not readily quantified, stimulation of mitogenesis in monolayer cultures of vascular endothelium may provide the basis for a more quantitative assay [Freshney and Frame, 1982].

Plasminogen Activator

Other cell products which can be recognized are proteolytic enzymes [Mahdavi and Hynes, 1979], long since associated with theories of invasive growth [Ossowski et al., 1979]. Since proteolytic activity may be associated with the cell surface of many normal cells and is absent on some tumor cells, an equivalent normal cell must be used as a control when using this criterion [e.g., Hince and Roscoe, 1978]. Frame [personal communication] has found plasminogen activator is higher in some cultures from human glioma than in cultures from normal brain (Fig. 16.5), and others have shown previously that plasminogen activator is associated with many different tumors [Rifkin et al., 1974; Nagy et al., 1977]. Plasminogen activator may be measured by clarification of a fibrin clot or release of free soluble ^{125}I from [^{125}I] fibrin [Strickland and Beers, 1976; Unkeless et al., 1974]. A simple chromogenic assay has also been developed by Whur et al. [1980].

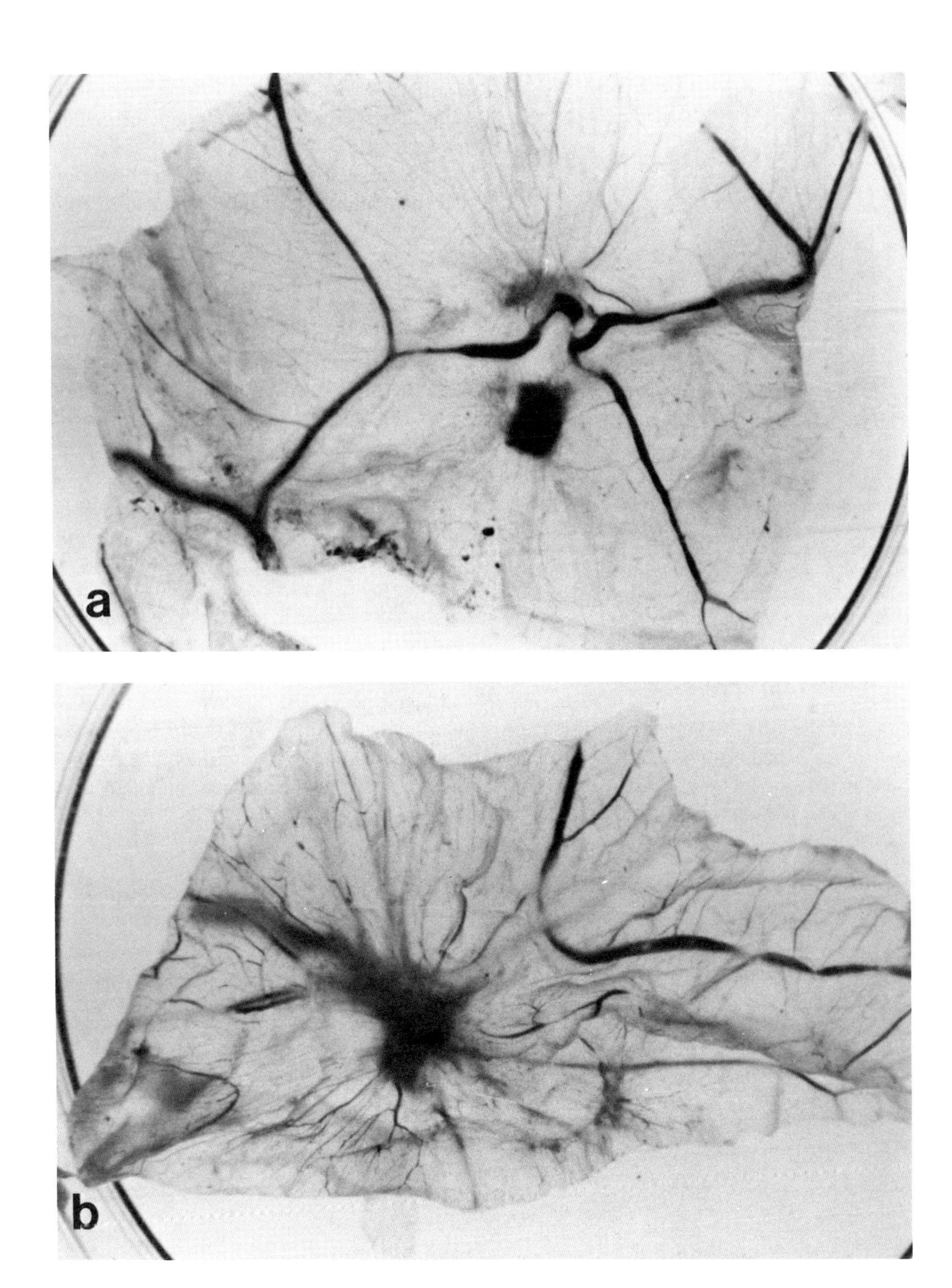

Fig. 16.4. *Induction of angiogenesis in chick chorioallantoic membrane by tumor cell extract. A crude extract of Walker 256 carcinoma cells absorbed into sterile filter paper was placed on the chorioallantoic membrane at 10 d incubation, and the membrane was removed 2 wk later. a. Control. b. Walker 256 extract. By courtesy of Mrs. Margaret Frame.*

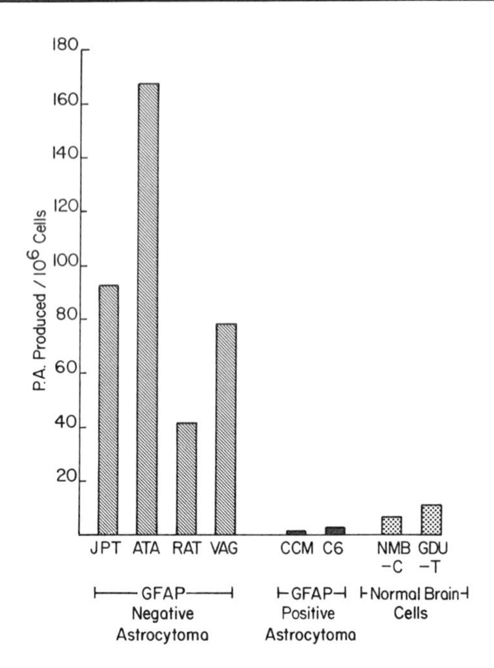

Fig. 16.5. *Plasminogen activator produced by tumor cells* in vitro *(arbitrary units). The four gliomas, JPT, ATA, RAT, and VAG were all higher than cells cultured from normal brain (NMB-C, GDU-T). It was also found that the only cells to produce the differentiated glial marker glial fibrillary acidic protein, CCM and C6, had the lowest P.A. of all.*

INVASIVENESS

An advantage of the CAM assay is that subsequent histology may reveal whether tumor cells have penetrated the underlying membrane. Easty and Easty [1974] showed that invasion of the CAM could be demonstrated in organ culture, and others [Hart and Fidler, 1978] have attempted, with some limited success, to construct a chamber capable of quantitating the penetration of tumor cells across the CAM. Mareel et al. [1979] developed an *in vitro* model for invasion, using chick embryo heart fragments cocultured with reaggregated clusters of tumor cells. Invasion appears to be correlated with the malignant origin of the cells, and the application of this technique to human tumor cells is now being explored.

TUMORIGENESIS

The only definition of malignancy that is generally accepted is the demonstration of the formation of invasive or metastasizing tumors *in vivo*. Transplantable tumor cells (10^6 or less) injected into isogeneic hosts will produce invasive tumors in a high proportion of cases while 10^6 normal cells of similar origin will not. Where spontaneous tumors, particularly from humans, are being studied, it is necessary to use immune-suppressed or immune-deficient host animals. The genetically athymic "nude" mouse [Giovanella et al., 1974] and thymectomized irradiated mice [Selby et al., 1980; Bradley et al., 1978] have both been used extensively as hosts for xenografts of human and other tumors. The take rate varies, however, and many clearly defined tumor cell lines and tumor biopsies have failed to produce tumors as xenografts, and frequently those which do fail to metastasize, although they may be invasive locally.

Chapter 17
Contamination

TYPES OF MICROBIAL CONTAMINATION

Bacteria, yeasts, fungi, molds and mycoplasmas all appear as contaminants in tissue culture, and where protozoology is carried on in the same laboratory, some protozoa can infect cell lines. Usually the species or type of infection is not important unless it becomes a frequent occurrence. It is only necesary to note the type (bacterial rods or cocci, yeast, etc.), how detected, the location where the culture was last handled, and the operator's name. If a particular type of infection recurs frequently, it may be beneficial to have it identified to help to find the origin. [For more detailed screening procedures for microbial contamination, see McGarrity, 1979; Fogh, 1973; Cour et al., 1979; Hay et al.,1979].

Characteristic features of microbial contamination are as follows: (1) Sudden change in pH; usually a decrease with most bacterial infections, very little change with yeast until contamination is heavy, and sometimes an increase in pH with fungal contamination. (2) Cloudiness in medium, sometimes with a slight film or scum on the surface or spots on the growth surface which dissipate when the flask is moved. (3) When examined on a low-power microscope (~ × 100), spaces between cells will appear granular and may shimmer with bacterial contamination (Fig. 17.1c). Yeasts appear as separate round or ovoid particles which may bud off smaller particles (Fig. 17.1a). Fungi produce thin filamentous mycelia (Fig. 17.1b), and sometimes denser clumps of spores. With toxic infection, some deterioration of the cells will be apparent. (4) On high-power microscopy (~ × 400), it may be possible to resolve individual bacteria and distinguish between "rods" and cocci. At this magnification, the shimmering seen in some infections will be seen to be due to mobility of the bacteria. Some bacteria form clumps or associate with the cultured cells. (5) If a slide preparation is made, the morphology of bacteria can be resolved more clearly at × 1,000, but this is not usually necessary. Microbial infection may be confused with precipitates of media constituents, particularly protein, or with cell debris, but can be distinguished by their regular particle morphology. Precipitates may be crystalline or globular and irregular but are not usually as uniform. If in doubt, plate out a sample of medium on nutrient agar (see Chapter 10). (6) Mycoplasmal infections (Fig. 17.1d–f) cannot be detected by naked eye other than by signs of deterioration in the culture. The culture must be tested specially by fluorescent staining, Orcein or Giemsa staining, autoradiography, or microbiological assay (see below). Fluorescent staining of DNA by Hoechst 33258 [Chen, 1977] is the easiest and most reliable method (see below) and reveals mycoplasmal infections as fine particulate or filamentous staining over the cytoplasm at × 500 magnification (Fig. 17.2). The nuclei of the cultured cells are also brightly stained by this method, as are any other microbial contaminations.

It is important to appreciate that mycoplasmas do not always reveal their presence with macroscopic alterations of the cells or media. Many mycoplasma contaminants grow slowly and do not destroy host cells. However, they can alter the metabolism of the culture in subtle ways. As mycoplasmas take up thymidine from the medium, infected cultures show abnormal labeling with ^{3}H-thymidine. Immunological studies can also be totally frustrated by mycoplasmal infections as attempts to produce antibodies against the cell surface may raise antimycoplasma antibodies. Mycoplasmas can alter cell behavior and metabolism in many other ways, so there is an absolute require-

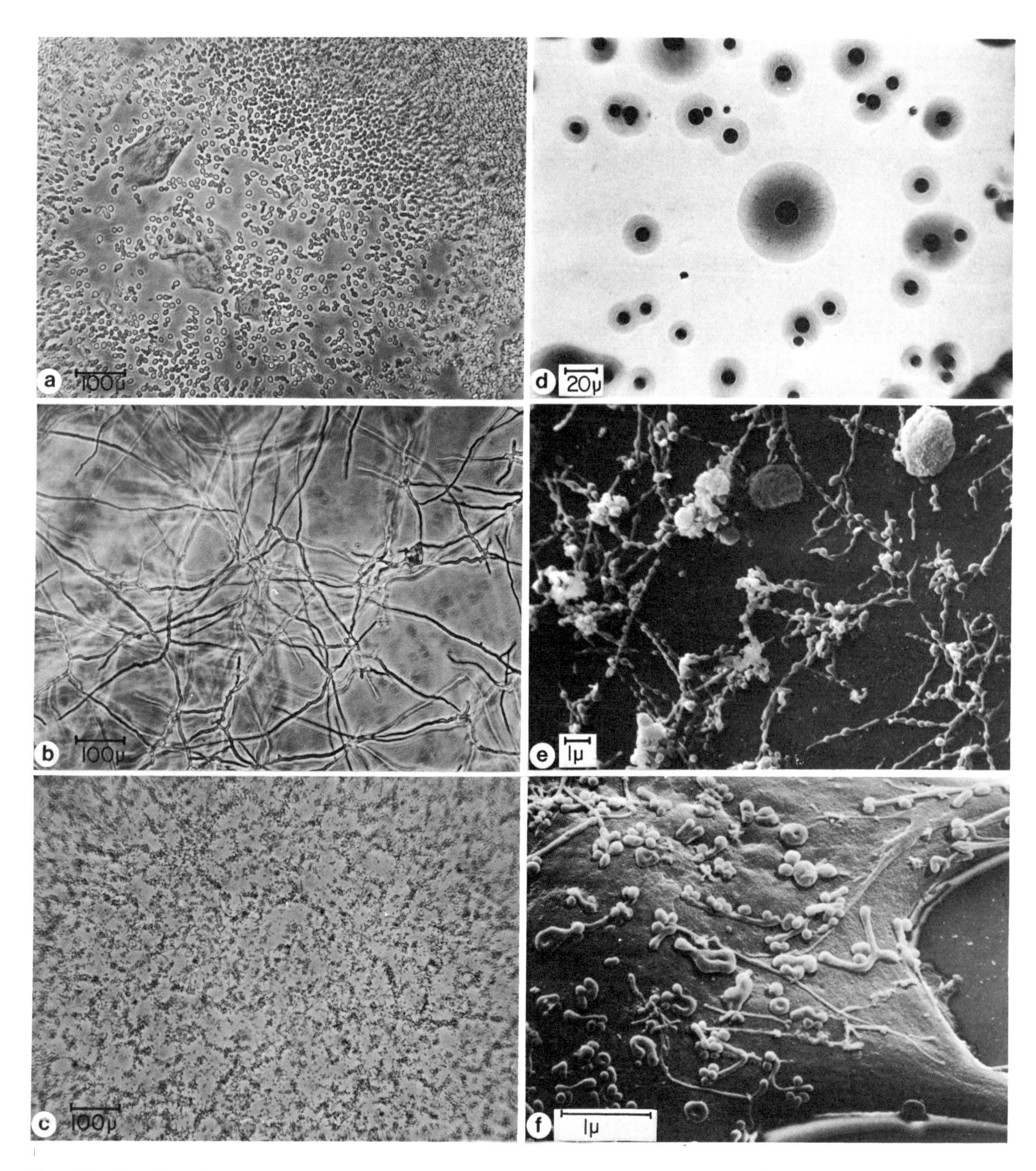

Fig. 17.1. *Examples of microorganisms found as contaminants of cell cultures. a. Yeast. b. Mold. c. Bacteria. d. Mycoplasma colonies growing on special nutrient agar (not as seen in cell culture — see Fig. 17.2). e, f. Scanning electron micrograph of mycoplasma growing on the surface of cultured cells (d–f, by courtesy of Dr. M. Gabridge).*

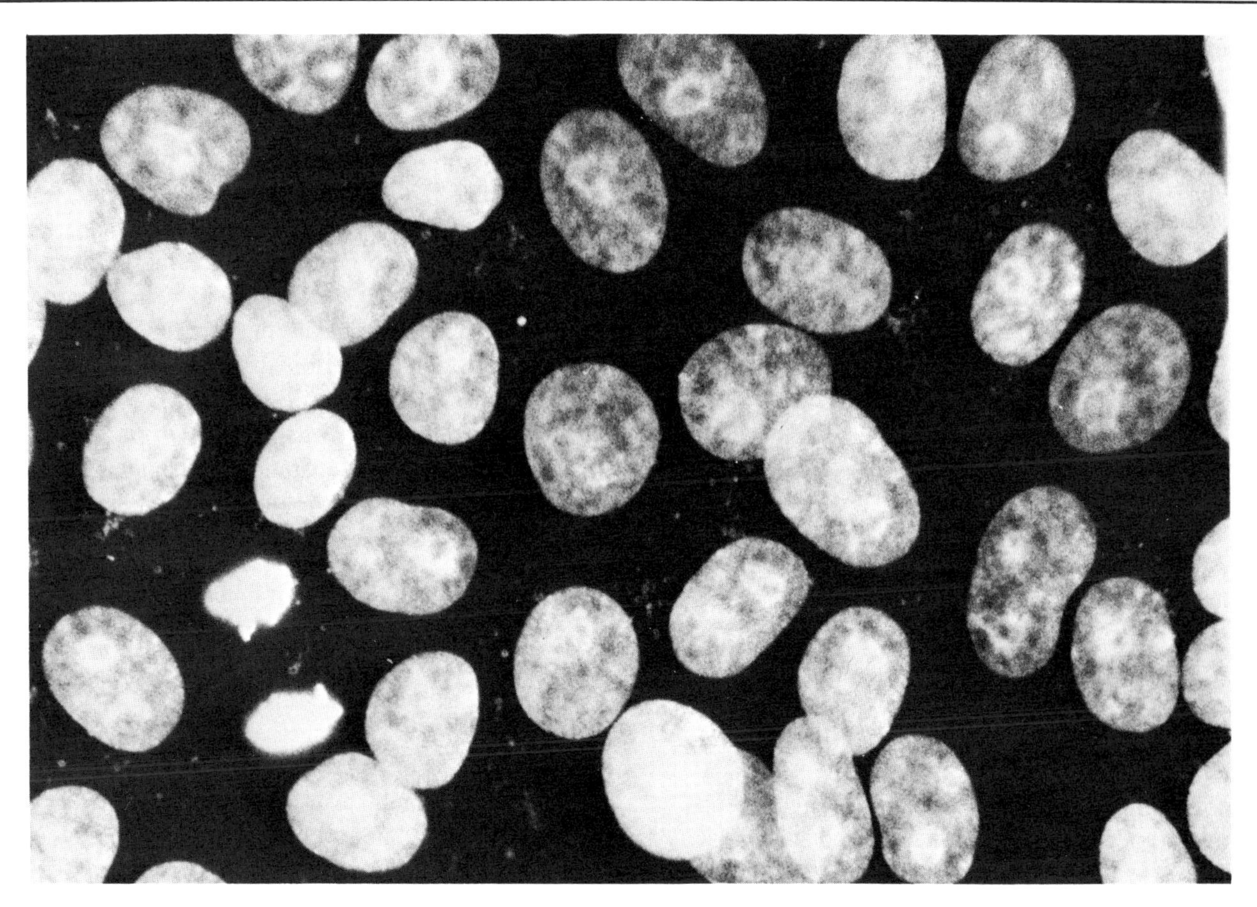

Fig. 17.2. *Mycoplasmal contamination of Vero cells. Cells fixed in acetic methanol and stained with Hoechst 33258 as described in text. Nuclei fluoresce brightly; mycoplasma visible as small points of fluorescence between nuclei.*

ment for routine, periodic assays for possible covert contamination of all cell cultures, particularly continuous or established cell lines.

Monitoring Cultures for Mycoplasmas

Superficial signs of chronic mycoplasmal infection include reduced rate of cell proliferation, reduced saturation density [Stanbridge and Doerson, 1978], and agglutination during growth in suspension. Acute infection causes total deterioration with perhaps a few resistant colonies. "Resistant" colonies are not necessarily free of contamination and may carry a chronic infection.

Fluorescent Technique for Detecting Mycoplasmas

Principle

The cultures are stained with Hoechst 33258, a fluorescent dye, which binds specifically to DNA [Chen, 1977]. Since mycoplasmas contain DNA, they can be detected readily by their characteristic particulate or filamentous pattern of fluorescence on the cell surface and, if the contamination is heavy, in surrounding areas.

Outline

Fix and stain subconfluent cultures or smears and look for fluorescence other than over the nucleus.

Monolayer Cultures

Materials

Hoechst 33258 stain
50 ng/ml in BSS without phenol red
PBSA
deionized water
fresh acetic methanol (1:3, cold)
mountant: glycerine in citrate buffer (see Appendix)

Protocol

1.
Seed culture at regular passage density (2×10^4–10^5 cells/ml, 4×10^3–2.5×10^4 cells/cm^2) and

incubate at 36.5°C until they reach 20–50% confluence. Allowing cultures to reach confluence will impair subsequent resolution of mycoplasma. Cultures may be grown on coverslips in a multiwell plate, or in a 50-mm petri dish or 25-cm^2 flask without coverslips (see Chapter 15)

2. Remove medium and discard
3. Rinse monolayer with BSS without phenol red and discard rinse
4. Add fresh BSS diluted 50:50 with acetic methanol, rinse monolayer, and discard rinse
5. Add pure acetic methanol, rinse, and discard acetic methanol
6. Add fresh acetic methanol and leave for 10 min
7. Remove acetic methanol and discard
8. Dry monolayer completely if to be stored (Samples may be accumulated at this stage and stained later)
9. If proceeding directly, wash off acetic methanol with deionized water and discard wash
10. Add Hoechst 33258 in BSS without phenol red and leave 10 min at room temperature
11. Remove stain and discard
12. Rinse monolayer with water and discard rinse
13. Mount a coverslip in a drop of mountant, blot off surplus from edges of coverslip, and seal with nail varnish or Correctine
14. Examine by epifluorescence with BG-3 (or equivalent) excitation filter, and Zeiss 53/44 (or equivalent) barrier filter

Suspension Cultures

Materials

Cytospin or Cytobuckets
slides
(other materials as for monolayer culture)

Protocol

1. Seed at regular passage concentration (2×10^4–10^5/ml) and harvest at mid-log phase (usually between 2–5 $\times 10^5$/ml)
2. Prepare slide (see Chapter 15 for more complete description): (a) (i) Cytospin: add 10^5 cells in 200 μl to sample chamber and centrifuge cells onto microscope slide, (ii) air-dry slide, fix in acetic methanol for 10 min, (iii) dry off fixative at fan. Proceed as from step 9 above. (b) Drop technique: See Chromosome Preparations (Chapter 15) from after trypsinization stage onward. Cells should be well spread and spaced out

Analysis. Check for extranuclear fluorescence. Mycoplasmas give pin points or filaments of fluorescence over the cytoplasm and sometimes in intercellular spaces. The pin points are close to the limits of resolution with a $\times$ 50 objective and are usually regular in size. In drop or Cytospin preparations, mycoplasmas can accumulate at the periphery of the cell spot. Furthermore, not all of the cells will necessarily be infected. Most of the preparation should be scanned, therefore, before declaring the culture uninfected.

Fluorescence outside the nucleus can be observed in the uninfected cultures where there is evidence of cell damage, e.g., primary cultures or cells recently recovered from frozen storage. Usually the fluorescent particles in this case are irregular in size and shape and disappear following subculture.

If there is any doubt regarding the interpretation of this test, it should be repeated, allowing about 1 wk for any low-level infection to increase. However, it is possible to have a low-level cryptic contamination which does not increase and which is consequently very difficult to detect. Such contaminations may be revealed by the microbiological culture method (see below), or by coculturing the suspected cells with two or three alternative host cell lines, such as 3T6, which may allow the infection to become overt and readily detectable by fluorescence.

Alternative Methods

Several other methods have been reported for the detection of mycoplasmal infections such as the detection of mycoplasma-specific enzymes like arginine deiminase or nucleoside phosphorylase [see Levine and Becker, 1978; Schneider and Stanbridge, 1975], but the DNA-fluorescence method is simpler and, though not specific for mycoplasmas, will detect any DNA-containing infection, which is, after all, the prime objective. Three other methods have been reported, however, which are of general importance. The first depends on microbiological culture of the organism although it is best not attempted unless you have the necessary expertise as these organisms are quite fastidious. The cultured cells are seeded into mycoplasma broth [Taylor-Robinson, 1978], grown for 6 days and plated out onto special nutrient agar. Colonies form in about 8 days and can be recognized by their size (~ 200 μm diameter) and their characteristic "fried egg" morphology—dense center with lighter periphery (Fig. 17.1d).

It is necessary to grow known mycoplasmal cultures as a positive control to confirm that the culture conditions are adequate, an element of the microbiological method which is in itself a disadvantage to most tissue culture laboratories, which should be kept clear of mycoplasmas at all times.

While the use of selective culture conditions and examination of colony morphology enables the species of mycoplasma to be identified, the microbiological culture method is much slower and more difficult to perform than the fluorescence technique. Commercial screening for mycoplasma is available (e.g., Flow Laboratories, Microbiological Associates), using microbiological culture.

Other methods use staining with aceto-orcein or Giemsa stain (see under Staining, Chapter 15). In both cases, particulate cytoplasmic staining is regarded as indicative of mycoplasmal infection but both are more difficult to interpret than the fluorescent method and can give false positives due to nonspecific precipitation of stain.

One other method which has been used quite successfully is autoradiography with [^{3}H]thymidine [Nardone et al., 1965]. The culture is incubated overnight with 0.1 μCi/ml high specific activity [^{3}H]thymidine and an autoradiograph prepared (see Chapter 21). Grains over the cytoplasm are indicative of infection (see Fig. 21.15) and this can be accompanied by a lack of nuclear labeling due to the thymidine being trapped at the cell surface by the mycoplasma.

DETECTION OF MICROBIAL CONTAMINATION

Potential sources of contamination are listed in Table 17.1 along with the precautions that should be taken to avoid them. Even in the best laboratories, however, contaminations do arise, so the following procedure is recommended.

1. Check for contamination at each handling by eye and on microscope. Every 2–3 months check for mycoplasmas (2–3 wk if primary cultures are used)
2. If a contamination is suspected but not obvious and cannot be confirmed *in situ,* clear the hood or bench of all but your suspected culture and one can of Pasteur pipettes. Because of the potential risk to other cultures, this is best done after all your other culture work is finished. Remove a sample from the culture and place on a microscope slide (Kova-slides are convenient for this as they do not require a coverslip). Check on microscope, preferably by phase contrast. If contamination is confirmed, discard pipettes, swab hood or bench with 70% alcohol, and do not use for at least 30 min
3. Note the nature of contamination, etc., on record sheet
4. If new contamination (not a repeat and not widespread,) discard (a) culture, (b) medium bottle used to feed it, and (c) any other bottle, i.e., tryspin, which has been used in conjunction with this culture. Discard into disinfectant, preferably in a fume hood, and outside the tissue culture area
5. If new and widespread (i.e., in at least two different cultures), discard all media and stock solutions, trypsin, etc
6. If similar contamination is repeated, check stock solutions for contaminations by (a) incubation alone or in nutrient broth (see Chapter 10), (b) plating out on nutrient agar (Oxoid, Difco) (see Chapter 10). (c) If a and b fail and contamination is still suspected, incubate 100 ml, filter through 0.2 μm filter, and plate out filter on nutrient agar
7. If contaminataion is widespread, multispecific, and repeated, check sterilization procedures, e.g., temperature of ovens and autoclaves particularly in the center of load, times of sterilization, packaging, storage (e.g., unsealed glassware (see Chapter 10) should be resterilized every 24 hr), and integrity of aseptic room and laminar flow hood filters

TABLE 17.1. Routes of Contamination

Source	Route or cause	Prevention
Manipulations, pipetting, dispensing, etc.	Nonsterile surfaces and equipment. Spillage on necks and outside of bottles and on work surface. Touching or holding pipettes too low down, touching necks of bottles, screw caps. Splash-back from waste beaker. Sedimentary dust or particles of skin settling on culture or bottle.	Clear work area of items not in immediate use. Swab regularly with 70% alcohol. Do not pour if it can be avoided. If you must: (1) do so in one smooth movement, (2) discard the bottle that you pour from, (3) wipe up any spillage with sterile swab moistened with 70% alcohol. Dispense or transfer by pipette, autodispenser or transfer device (see Chapter 6). Discard into beaker with funnel or, preferably, by drawing off into reservoir with vacuum pump (Figs. 4.5, 5.4). Do not work over (vertical laminar flow and open bench) or behind and over (horizontal laminar flow) an open bottle or dish.
Solutions	Nonsterile reagents and media.	Filter or autoclave before use. Monitor performance of autoclave with recording thermometer or sterility indicator (see Trade Index). Check integrity of filters after use (bubble point). Test all solutions after sterilization.
Glassware and screw caps	Dust and spores from storage. Ineffective sterilization.	Dry-heat sterilize or autoclave before use. Do not store unsealed for more than 24 hr. Check oven and autoclave regularly and monitor each load.
Tools, instruments, pipettes	Contact with nonsterile surface or material. Invasion by insects, mites or dust. Ineffective sterilization.	Sterilize by dry heat before use. Monitor performance of oven. Resterilize instruments (70% alcohol, burn and cool off) during use. Do not grasp part of instrument or pipette which will later pass into culture vessel. Do not store for more than 24 hr, unless sealed with tape.
Culture flasks, media bottles in use	Dust and spores from incubator or refrigerator. Media under cap spreading to outside of bottle.	Use screw caps in preference to stoppers. Wipe flasks and bottles with 70% alcohol before using. Flame necks and caps (without opening) before placing in laminar flow hood. Cover cap and neck with aluminum foil during storage or incubation.
Room air	Draughts, eddies, turbulence, dust, aerosols.	Clean filtered air. Reduce traffic and extraneous activity. Wipe floor and work surfaces regularly.

Table 17.1 continued on following page.

TABLE 17.1. (continued)

Source	Route or cause	Prevention
Work surface	Dust, spillage.	Swab before, after, and during work with 70% alcohol. Mop up spillage immediately.
Operator, hair, hands, breath, clothing	Dust from skin, hair, or clothing dropped or blown into culture. Aerosols from talking, coughing sneezing, etc.	Wash hands thoroughly. Bare arms. Keep talking to a minimum and face away from work if you do. Avoid working with a cold or throat infection or wear a mask. Tie back long hair, or wear a cap. Do not wear the same lab coat as in general lab area or animal house. Change to gown or apron.
Hoods	Perforated filter. Spillages, particularly in crevices or below work surface.	Check filters regularly for holes and leaks. Clear round and below work surface regularly.
Tissue samples	Infected at source or during dissection.	Do not bring animals into tissue culture lab. Incorporate antibiotics in dissection fluid (see Chapter 11).
Incoming cell lines	Contaminated at source or during transit.	Handle alone, after all other sterile work is finished, swab down bench or hood carefully after use, and do not use until next morning. Check for contamination by growing for two weeks without antibiotics (keep duplicate in antibiotics at first subculture). Check for contamination visually, by phase contrast microscopy and Hoechst stain for mycoplasma.
Mites, insects, and other infestation in wooden furniture, benches, incubators and on mice, etc., taken from animal house.	Entry of mites, etc., into sterile packages.	Seal all sterile packs. Avoid wooden furniture if possible, use plastic laminate, one piece, or stainless steel bench tops. If wooden furniture is used, seal with polyurethane varnish or wax polish and wash regularly with disinfectant. Keep animals out of tissue culture lab.
Anhydric incubators	Growth of molds and bacteria on spillages.	Wipe up any spillage with 70% alcohol on a swab. Clean out regularly.
CO_2, humidified incubators	Growth of molds and bacteria in humid atmosphere on walls and shelves. Spores, etc., carried on forced air circulation.	Clean out weekly with detergent and 70% alcohol. Enclose open dishes in plastic boxes with tight-fitting lids (but do not seal). Swab with 70% alcohol before opening. Fungicide or bacteriocide in humidifying water (but check first for toxicity).

8. Do not attempt to decontaminate cultures unless they are irreplaceable. If necessary, decontaminate by (a) washing five times in BSS containing a higher than normal concentration of antibiotics (see DBSS in Appendix), and (b) adding antibiotics (as in DBSS) to medium for 48 hr. If possible, the infection should be tested for sensitivity to a range of individual antibiotics. (c) Remove antibiotics and culture without for 48 hr. (d) Recycle b and c twice. (e) Culture for 2 wk without antibiotics to check that contamination has been eliminated. Check by phase-contrast microscopy and Hoechst staining (see above)

Tumor tissue can sometimes be decontaminated by animal passage [Van Diggelen et al., 1977]. Cytotoxic antibodies may be effective [Pollock and Kenny, 1963], particularly against mycoplasmas.

Polyanethol sulphonate [Mardh, 1975], 5-bromouracil in combination with Hoechst 33258 and uv light [Marcus et al., 1980], and antibiotics such as Tylosin [Friend et al., 1966], Kannamycin, and Gentamycin, may be effective in removing mycoplasmas. Schimmelpfeng et al. [1968] were able to eliminate mycoplasmal infections by coculturing with macrophages.

The general rule should be, however, that contaminated cultures are discarded, and that decontamination is not attempted unless it is absolutely vital to retain the cell strain. Complete decontamination, especially with mycoplasmas, is difficult to achieve and attempts to do so may produce hardier, antibiotic-resistant strains of the contaminant.

CROSS CONTAMINATION

During the history of tissue culture, a number of cell strains have evolved with very short doubling times and high plating efficiencies. Although these properties make such cell lines valuable experimental material, they also make them potentially hazardous for cross-infecting other cell lines. The extensive cross contamination of many cell lines with HeLa cells [Gartler, 1967; Nelson-Rees and Flandermeyer, 1977] is now well known, but many operators are still unaware of the seriousness of the risk. To avoid cross contamination:

(1) Either obtain cell lines from a reputable cell bank where appropriate characterization has been performed (see Appendix) or perform the necessary characterization yourself as soon as possible (see Chapter 15).

(2) Handle rapidly growing lines, such as HeLa, on their own and after other cultures.

(3) Never use the same bottle of medium, trypsin, etc., for different cell lines.

(4) Wherever possible, do not put a pipette back into a bottle of medium, trypsin, etc., after it has been in a culture flask containing cells. Add medium and any other reagents to the flask first and then add the cells last.

(5) Check the characteristics of the culture regularly and suspect any sudden change in morphology, growth rate, etc. Confirmation of cross contamination may be obtained by chromosome [Nelson-Rees and Flandermeyer, 1977] or isoenzyme [O'Brien and Kleiner, 1977] analysis (see Chapter 15 for details of chromosome analysis).

CONCLUSIONS

Check living cultures regularly for contaminations using normal and phase-contrast microscopy, and for mycoplasmas by fluorescent staining of fixed preparations.

Do not maintain all cultures routinely in antibiotics. Grow at least one set of cultures of each cell line in the absence of antibiotics for a minimum of 2 wk at a time to allow cryptic contaminations to become overt.

Do not attempt to decontaminate a culture unless it is irreplaceable and then do so under strict quarantine.

Quarantine all new lines that come into your laboratory until you are sure that they are uncontaminated.

Do not share media or other solutions among cell lines or among operators.

Check cell line characteristics (see Chapter 15) periodically to guard against cross contamination.

It cannot be overemphasized that cross contaminations can and do occur. It is essential that the above precautions be taken and regular checks of cell strain characteristics be made.

Chapter 18
Instability, Variation, and Preservation

It is a fundamental property of living organisms that they diversify to provide sufficient variation for selection to be a useful mechanism in the adaptation of the species to its environment. It is, therefore, not surprising that cells in culture behave in a similar fashion. Spontaneous variation and selection occur as with any microorganism.

During the evolution of a cell line from a primary culture and during subsequent maintenance as a cell line or purified cell strain, there is evidence of both phenotypic and genotypic instability. This arises as a result of variations in culture conditions, selective overgrowth of constituents of the cell population, and genomic variations.

Since the constitution of a culture may vary from time to time it is important (1) to standardize the culture conditions, (2) to select a period in the life history of the cell line where variation is at a minimum, (3) to select a pure characterized cell strain if possible, and (4) to preserve a seed stock to recall into culture at intervals, to maintain consistency.

ENVIRONMENT

Environment has been discussed already in Chapters 8 and 9. Once the appropriate conditions have been adopted, they should be adhered to, as alterations in media, serum, substrates, etc., will alter phenotypic expression. Test batches of serum, select one that has the required properties, and reserve enough to be sufficient for 6 months to 1 yr. Repeat the process before changing to a new batch (see also Chapter 9 and below).

SELECTIVE OVERGROWTH, TRANSFORMATION, AND SENESCENCE

Following isolation of a primary culture, the predominant phenotype may change as cells with a higher proliferative capacity will tend to overgrow the more slowly dividing and nondividing cells (see Chapter 2). As an example, consider a culture taken from carcinoma of the colon. Following explantation, these cultures can be predominantly epithelial; but after the first subculture, the epithelial cells steadily lose ground, and the fibroblasts take over. Preserving the epithelial component during this phase is a major problem and requires various selective culture and separation techniques (see Chapters 13 and 14). Occasionally, however, a transformed line may appear without using selective conditions. This may be a minority component of the original culture which has undergone transformation and now has increased growth capacity (shorter doubling time, infinite survival), and will ultimately outgrow the fibroblasts which grow more slowly and have a finite life-span in culture.

To counter cell line variation, finite cell lines should be used between certain generation limits; e.g., human diploid fibroblasts gain a fair degree of uniformity by the tenth generation and remain fairly stable up to about the 30th. Beyond 40 generations, senescence can be anticipated, although some lines may survive longer.

The culture should be grown through the first five or six generations until sufficient cells have been produced to freeze down a seed stock (see below).

GENETIC INSTABILITY

Evidence of genetic rearrangement can be seen in chromosome counts (see Fig. 2.2) and karyotype analysis. While the mouse karyotype is made up exclusively of small telocentric chromosomes, several metacentrics are apparent in many continuous murine cell lines. Furthermore, while virtually every cell in the animal has the normal diploid set, this is more variable in culture. In extreme cases, e.g., continuous cell strains such as HeLa-S3, less than half of the cells will have exactly the same karyotype, i.e., they are *heteroploid* .

Most continuous cell strains, even after cloning, contain a range of genotypes which are constantly changing. As transformation often involves chromosomal rearrangement, it is possible that it can only occur in cells with the capacity for chromosomal alterations. Alternatively, transformation may cause genetic instablility to arise in a previously stable genotype. Hence, transformed or continuous lines retain a capacity for genetic variation which is not apparent *in vivo* or in many finite cell lines.

There are two main causes of genetic variation: (1) the spontaneous mutation rate appears to be higher *in vitro*, associated, perhaps, with the high rate of cell proliferation, and (2) mutant cells are not eliminated unless their growth capacity is impaired. It is not surprising that phenotypic variation will arise as a result of this genetic variation. Minimal deviation rat hepatoma cells, grown in culture, express tyrosine aminotransferase activity constitutively and may be induced further by dexamethasone [Granner et al., 1968], but, as can be seen from Figure 18.1, subclones of a cloned strain of H4-II-E-C3 [Pitot et al., 1964] differed both in the constitutive level of the enzyme and in its capacity to be induced by dexamethasone.

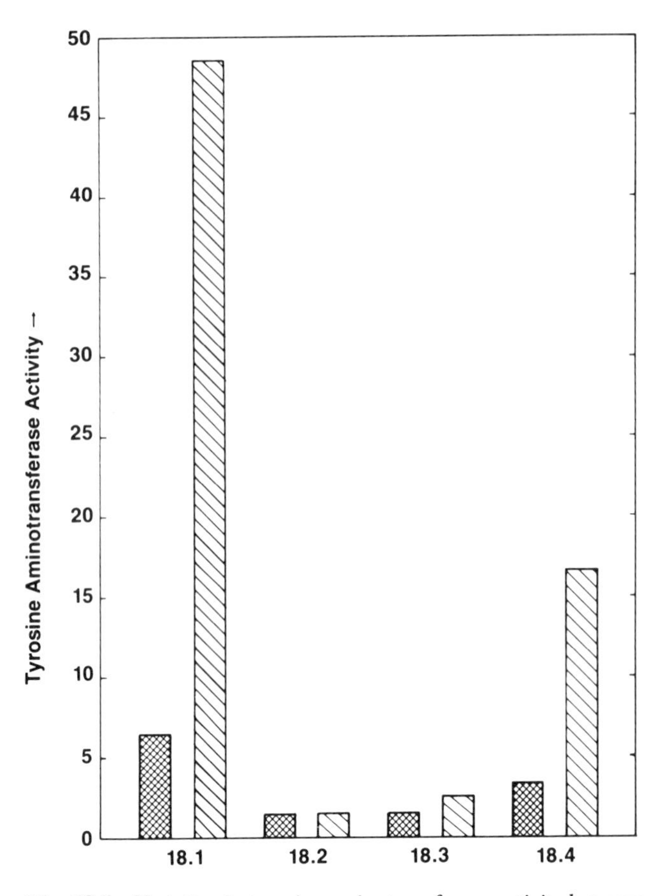

Fig. 18.1. *Variation in tyrosine aminotransferase activity between four subclones of clone 18 of a rat minimal deviation hepatoma cell strain. H-4-II-E-C3 cells were cloned; clone 18 was isolated, grown up, and recloned, and the second-generation clones were assayed for tyrosine aminotransferase activity with and without pretreatment of the culture with dexamethasone. Crosshatching, basal level; hatching, induced level. Data provided by J. Somerville.*

PRESERVATION

In order to minimize genetic drift in continuous cell lines, to avoid senescence in noncontinuous cell lines, and to guard against accidental loss by contamination, it is now common practice to freeze aliquots of cells to be thawed out at intervals as required.

Selection of Cell Line

A cell line is selected with the required properties. If it is a finite cell line, it is grown to between the fifth and tenth population doubling to create sufficient bulk of cells for freezing. Continuous cell lines should be cloned (see Chapter 13) and an appropriate clone selected and grown up to sufficient bulk to freeze. Prior to freezing, the cells must be characterized (see Chapter 15) and checked for contamination (see Chapter 17), particularly cross contamination.

Continuous cell lines have the advantage that they will survive indefinitely, grow more rapidly, and can be cloned more easily; but they may be less stable genetically. Finite cell lines are usually diploid or close to it, stable between certain passage levels, but are harder to clone, grow more slowly, and will eventually die out or transform (Table 18.1).

Standardization of Medium and Serum

The type of medium used will influence the selection of different cell types and regulate their phenotypic expression (see Chapters 2 and 9). Consequently, once a medium has been selected, standardize on that medium and, preferably, on one supplier if it is being purchased ready-made.

Variation in serum. Considerable variation [See Olmsted, 1967; Honn et al., 1975] may be anticipated between batches of serum resulting from differing methods of preparation and sterilization, different ages and storage conditions, and variations in animal stocks from which the serum was derived, including strain

TABLE 18.1. Characteristics of Finite and Continuous Cell Lines

	Finite cell lines	Continuous cell lines
Advantages	Genotypically and phenotypically closer to cell type *in vivo*. Usually euploid.	Rapidly growing. Infinite life-span. Easily cloned. May be propagated in suspension. Hardy. Simple medium requirements. Low serum requirement.
Disadvantages	Finite life-span. Slow growing. Lower plating efficiencies.	Genetically unstable. Transformed. Often aneuploid.

differences, pasture, climate, and so on. It is important to select a batch, use it for as long as possible, and replace it, eventually, with one as similar as possible (see Chapter 9).

Cell Freezing

When a cell line has been produced, or a cloned cell strain selected, with the desired characteristics and free of contamination, then a "seed stock" should be stored frozen. This will protect the cell line from change by genetic drift, from risk of contamination, and from technical problems such as incubator failure.

Storage in liquid nitrogen (Fig. 18.2) is currently the most satisfactory method of preserving cultured cells. The cell suspension, preferably at a high concentration, should be frozen slowly, at 1°C per min [Leibo and Mazur, 1971; Harris and Griffiths, 1977] in the presence of a preservative such as glycerol or dimethyl sulphoxide [Lovelock and Bishop, 1959]. The frozen cells are transferred rapidly to liquid nitrogen when they reach -50°C or below and are stored immersed in the liquid nitrogen or in the gas phase above the liquid. When required, the cells are thawed rapidly and reseeded at a relatively high concentration to optimize recovery.

If liquid nitrogen storage is not available, cells may be stored in a conventional freezer. The temperature should be as low as possible, but significant deterioration may yet occur even at -70°C. Little deterioration is found at -196°C [Greene et al., 1967].

Outline

The culture is grown to late log phase and a high cell density suspension is prepared and frozen slowly with a preservative (Fig. 18.3). When required, aliquots are thawed rapidly and reseeded at high cell density.

Materials

- cultures to be frozen
- if monolayer: PBSA and 0.25% crude trypsin
- medium
- hemocytometer or cell counter
- dimethylsulphoxide or glycerol
- glass or plastic ampules—if glass, ampule sealer with gas/O_2 burner
- canes or racks for storage
- cotton wool
- polystyrene box
- forceps
- protective gloves

To thaw:

- protective gloves and face mask
- bucket of water at 37°C with lid
- forceps
- 70% alcohol
- swab
- culture flask
- centrifuge tube
- culture medium

Protocol

1. Check culture for (a) healthy growth, (b) freedom from contamination, and (c) presence of specific characteristics required of the line for subsequent use (viral propagation, differentiation, antigenic constitution, etc.)
2. Grow up to late log phase and, if monolayer, trypsinize—if suspension, centrifuge
3. Resuspend at approximately 5×10^6–2×10^7 cells/ml in culture medium containing serum and a preservative such as dimethyl sulphoxide (DMSO) or

Fig. 18.2. *Liquid nitrogen freezers. a. Narrow-necked freezer with storage on canes in cannisters, b. c. Wide-necked freezer, storage in cannisters or in drawers, d; see also Fig. 18.6.*

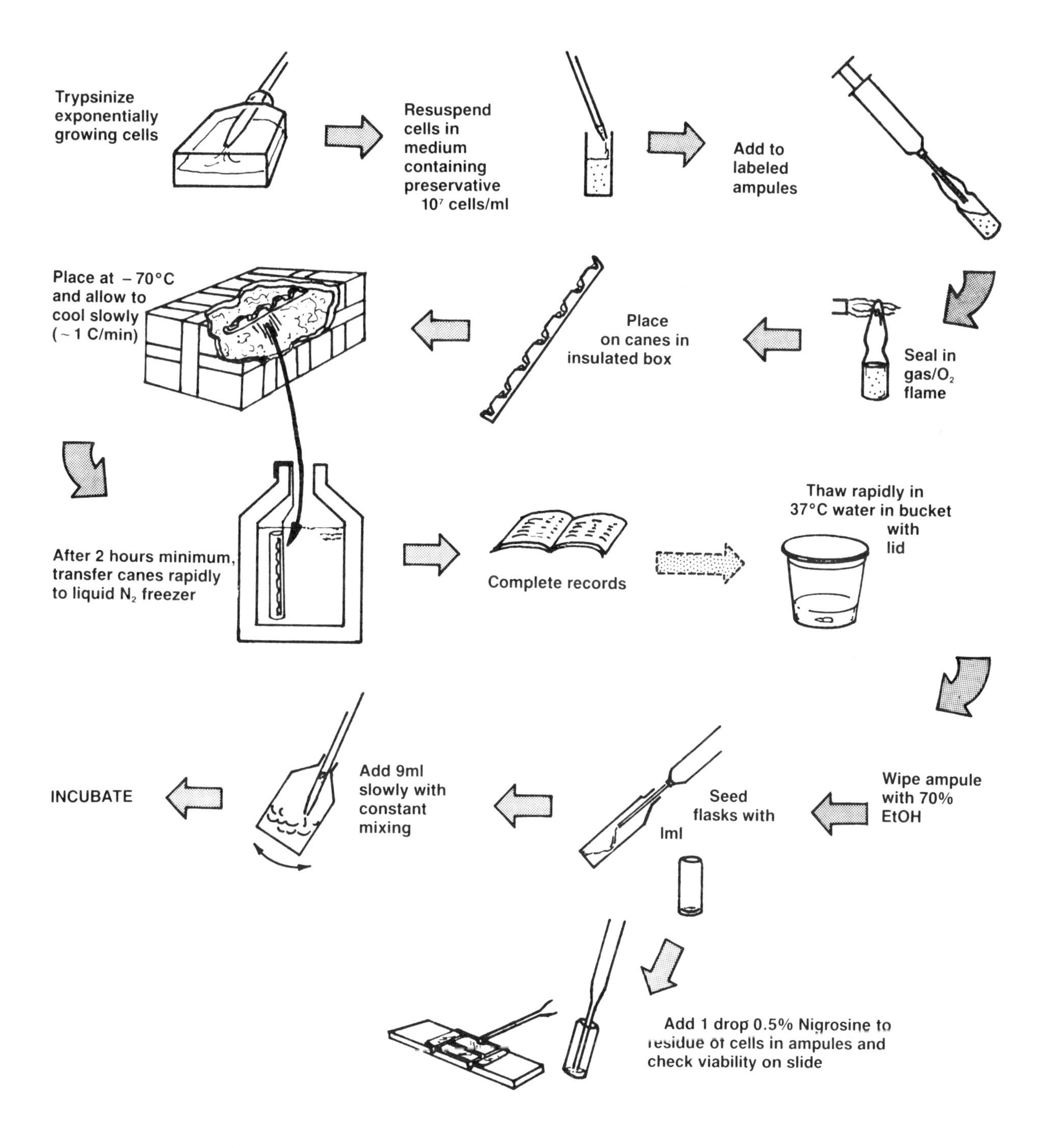

Fig. 18.3. *Preparation of cells for freezing and subsequent recovery after storage.*

glycerol at a final concentration of 10%. Make sure the preservative is pure and free from contamination. DMSO should be colorless. Glycerol should be not more than 1-yr old as it may become toxic after prolonged storage.

It is *not* advisable to place ampules on ice in an attempt to minimize deterioration of the cells. A delay of up to 30 min at room temperature is not harmful when using DMSO and is beneficial when using glycerol

Note. DMSO is a powerful solvent. It will leach impurities out of rubber and some plastics and should, therefore, be kept in its original stock bottle or in glass tubes with glass stoppers. It can also penetrate many synthetic and natural membranes including *skin* and rubber gloves [Horita and Weber, 1964]. Consequently, any potentially harmful substances in regular use (e.g., carcinogens) may well be carried into the circulation through the skin and even through rubber gloves. DMSO should always be handled with caution because of its known hazardous potential, particularly in the presence of any toxic substance.

4.
Dispense cell suspensions into 1- or 2-ml prelabeled glass or plastic ampules and seal (Figs. 18.4, 18.5)

Note. Glass ampules are still most commonly used but must be perfectly and quickly sealed. If sealing takes too long, the cells will heat up and die, and the air in the ampule will expand and blow a hole in the top (Fig. 18.5). If the ampule is not perfectly sealed (a) it may inspire liquid nitrogen during freezing and storage in the liquid phase of the nitrogen freezer, and will subsequently explode (violently) on thawing, or (b) it may become infected.

It is possible to check for leakage by placing ampules in a dish of 1% methylene blue in 70% alcohol at 4°C for 10 min. If the ampules are not properly sealed, the methylene blue will be drawn in and the ampule should be discarded. The ampules may need to be relabeled after this procedure. This step is inconvenient and time consuming and may only be necessary when sealing glass ampules for the first few times. When experience

Fig. 18.4 *Semiautomatic ampule sealer (Kahlenburg-Globe). Ampules are placed between the rollers and the flat plate by the operator's left hand. As the jig carrying the burner is moved forward, the ampule rotates in the flame and is finally ejected against the plate at the bottom of the picture.*

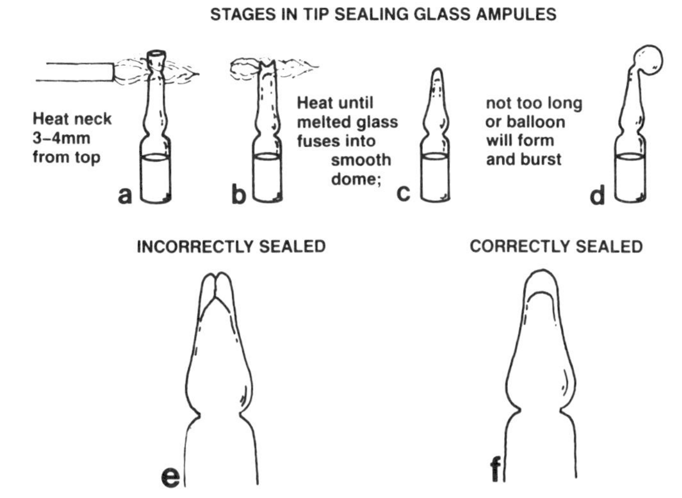

Fig. 18.5. *Appearance of ampules during and after sealing. a. Tip melting. b. Molten glass folds inward. c. Sides coalesce to form dome. d. Overheating can cause air in the ampule to blow out the glass forming a balloon, or bursting it completely. e. Incorrectly sealed ampule; fine capillary hole left. f. Correctly sealed ampule. Glass fused evenly inside and outside.*

is obtained, a well-sealed ampule may be recognized by the appearance of the tip (Fig. 18.5f)

Plastic ampules are unbreakable but must be sealed with the correct torsion on the screw cap or they may leak. They are of a larger diameter and taller than equivalent glass ampules. Check that they will fit in the canes or racks used for storage. (Special canes for plastic ampules are available.)

5.

Place ampules on canes for cannister storage or leave loose for drawer storage. Lay on cotton wool in a polystyrene foam box with a wall thickness of ~ 15mm. This, plus the cotton wool, should provide sufficient insulation such that the ampules will cool at 1°C/min when the box is placed at -70°C or -90°C in a regular deep freeze or insulated container with solid CO_2. Most cultured cells survive best at this cooling rate; but if recovery is low, try a faster rate (i.e., less insulation) [Leibo and Mazur, 1971]

Note. Programmable coolers are available to control the cooling rate (Fig. 18.6), usually by sensing the temperature of the ampule, and running in liquid N_2 at the correct rate, to achieve a preprogrammed cooling rate. They are, however, very expensive, and have few advantages unless you wish to vary the cooling rate [e.g., Foreman and Pegg, 1979].

Union Carbide also markets a special neck plug which carries ampules and places them in the neck of the freezer at the desired level to obtain a set cooling rate (Fig. 18.7).

6.

When the ampules have reached -70°C (1½–2hr after placing at -70°C if starting from 20°C ambient), transfer to liquid N_2 freezer. This must be done quickly as the cells will deteriorate rapidly if the temperature rises above -50°C. Protective gloves and face mask should be used when handling liquid nitrogen

Note. Storage in the liquid phase (i.e., freezer filled with liquid N_2) (Fig. 18.8) has the advantage that the holding time (time for total evaporation without replenishment) is usually around 3 wk, whereas storage in the gas phase (above the liquid N_2 where volume is then reduced to about one-fifth to one-tenth of the freezer volume) means that the holding time will be reduced to 7 to 10 d. Storage in the gas phase, however, avoids subsequent problems of explosion during thawing of glass ampules which have not been perfectly sealed. If the vial is potentially hazardous, it *must* be stored in the gas phase.

7.

When ampules are safely located in freezer, make sure that the appropriate entries are made in the freezer index. Records should contain (a) freezer index showing what is in each part of the freezer and (b) a cell strain index, describing the cell line, its designation, what its special characteristics are, and where it is located. The ampules should also carry a label with the cell strain designation and preferably the date and user's initials, although this is not always feasible on the available space.

Remember, cell cultures stored in liquid nitrogen may well outlive you! They can easily outlive your stay in a particular laboratory. The record should be, therefore, readily interpreted by others and sufficiently comprehensive so that the cells may be use to others

Thawing

Protocol

1.

When you wish to recover cells from the freezer, check the index, label your culture flask, then retrieve the ampule from the freezer, check that it

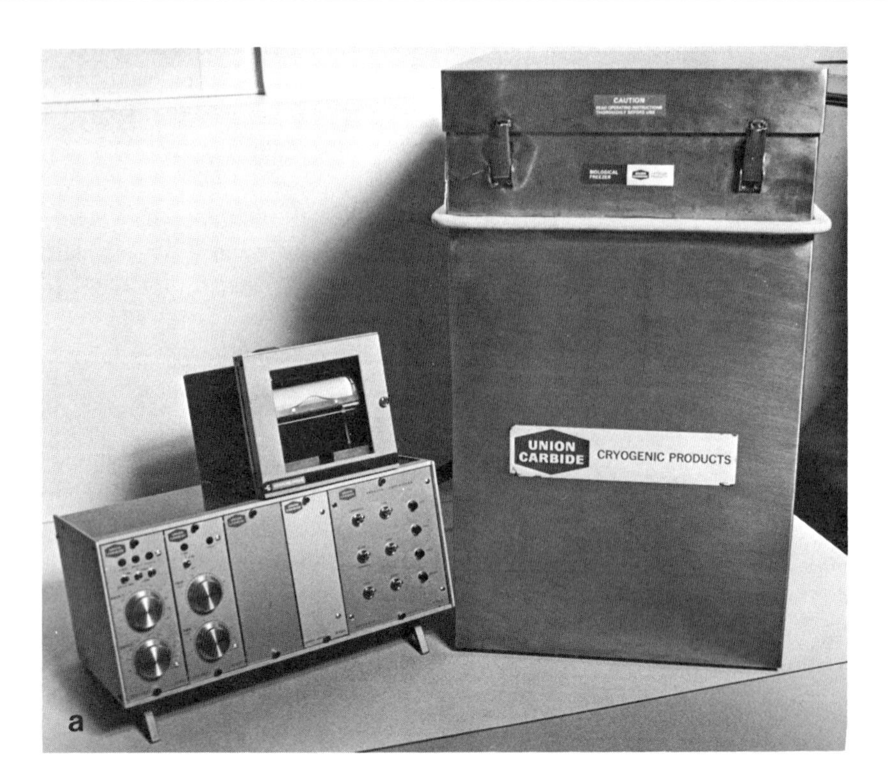

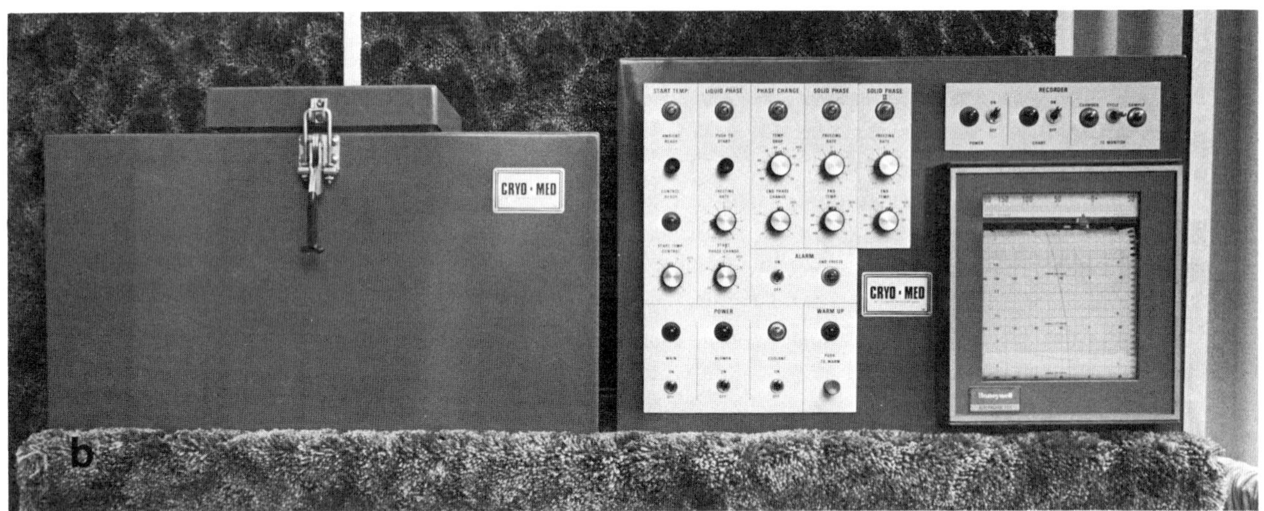

Fig. 18.6. *Programmable coolers capable of regulating the cooling rate before and after freezing. Ampules are placed in an insulated box and the cooling rate regulated by injecting liquid N_2 into the box at a rate determined by a sensor in the box and a preset program in the consol unit (Union Carbide, Cryo-med).*

is the correct one, and place it in 10 cm of water at 37°C, in a 5-l bucket with a lid

Note. If ampule has been stored in the gas phase, a lid is not necessary; but ampules stored under liquid nitrogen may inspire the liquid, and on thawing, will explode violently. A plastic bucket with a lid is, therefore, essential in this case to contain any explosion.

2.
When thawed, swab the ampule with 70% EtOH and break open. (Prescored ampules are available; these are much easier to open.) Transfer the contents to a culture flask. The dregs in the ampule may be stained with nigrosine to determine viability (see Chapter 19).

Add medium slowly to the cell suspension: 10

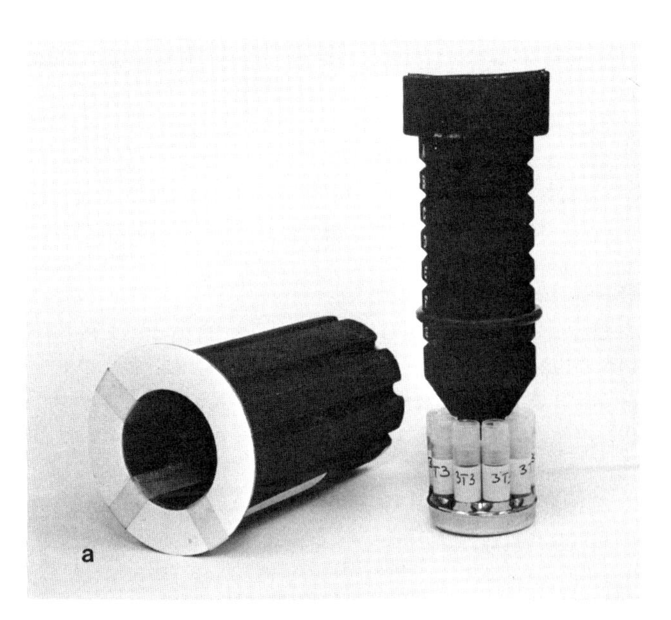

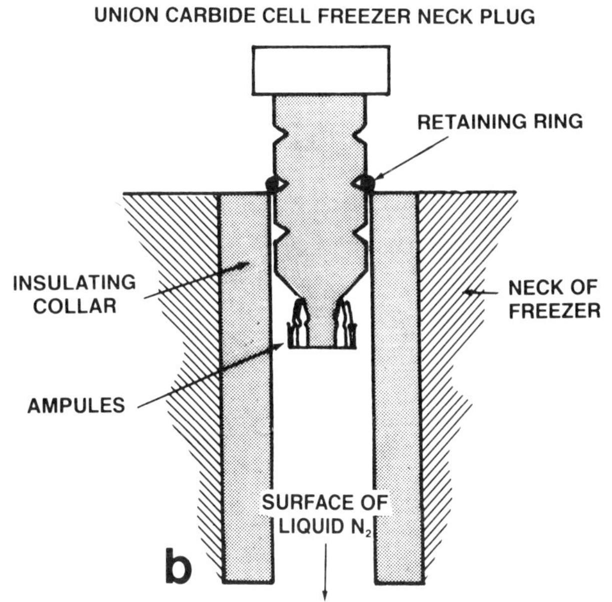

Fig. 18.7. *Modified neck plug for narrow-necked freezers allowing controlled cooling at different rates. The "O" ring is used to set the height of the ampules within the neck of the freezer. The lower the height, the faster the cooling.*

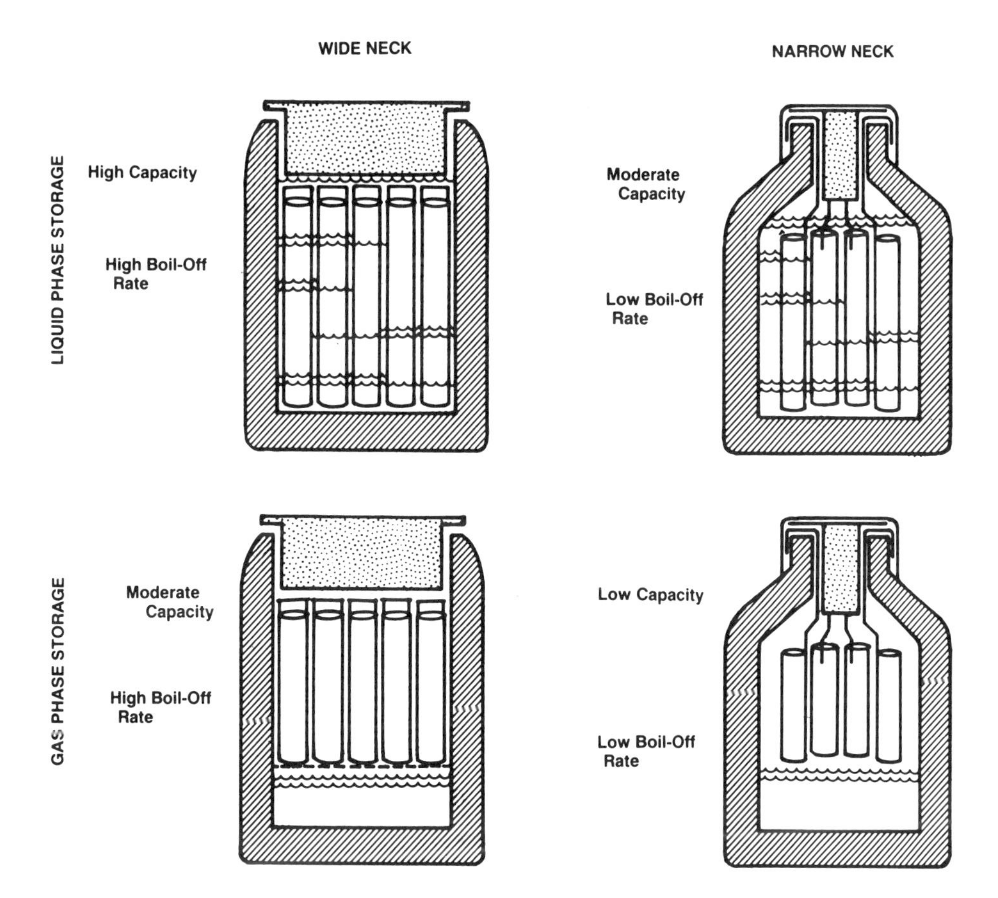

Fig. 18.8. *Variations in freezer design and usage.*

ml over about 2 min added dropwise at the start and then a little faster, gradually diluting the cells and preservative. This is particularly important with DMSO where sudden dilution can cause severe osmotic damage and reduce survival by half

In some cases, e.g., L5178Y lymphoma with DMSO, the preservative may be too toxic for the cells to survive on thawing, [M. Freshney, personal communication]. In these cases, dilute the cells slowly, as above, and centrifuge for 2 min at 100*g*, discard the supernatant medium with preservative, and resuspend the cells in fresh medium for culture.

The number of cells frozen should be sufficient to allow for 1:10 or 1:20 dilution on thawing to dilute out the preservative but still keep the cell concentration higher than at normal passage, e.g., for cells subcultured normally at 10^5 ml, 10^7 should be frozen in 1 ml, and the whole seeded, after thawing into 20 ml, giving 5×10^5/ml ($5 \times$ the normal seeding density) and diluting the preservative from 10% to 0.5%, at which concentration it is less likely to be toxic. Residual preservative may be diluted out as soon as the cells start to grow (for suspension cultures) or the medium changed as soon as the cells have attached (for monolayers).

The dye exclusion viability and the approximate take (e.g., proportion attached versus those still floating after 24 hr) should be recorded on the appropriate record card to assist in future thawing. One ampule should be thawed from each batch as it is frozen, to check that the operation was successful.

During a prolonged series of experiments, lasting more than 6 months, stock cultures should be replaced from the freezer at regular intervals, say every 3 months, thereby preserving the properties of the cell line.

Although variability cannot be eradicated entirely from cultured cell lines, it may be minimized by adopting the procedures described above. In summary, (1) select finite lines around the fifth generation, or clone, isolate, and grow up a continuous cell strain, (2) confirm characteristics of the strain, (3) check for microbial and cross contamination, (4) freeze down a large number of ampules, (5) thaw one ampule to test survival, and (6) replace stock regularly from freezer.

Chapter 19
Quantitation and Experimental Design

CELL GROWTH KINETICS

Much of the analysis of cell behavior *in vitro* depends on a proper understanding of cell proliferation, the analysis of which utilizes cell counting, cell sizing, plating efficiency determination by colony counting, and autoradiography of DNA synthesis.

Cell Counting

While estimates can be made of the stage of growth of a culture from its appearance under the microscope, standardization of culture conditions and proper quantitative experiments are difficult unless the cells are counted before and after, and preferably during, each experiment. Cell behavior changes considerably from one phase of the growth cycle to another such that reproducible measurements cannot be made unless the state of growth is known.

Hemocytometer. The concentration of a cell suspension may be determined by placing the cells in an optically flat chamber under a microscope (Fig. 19.1). The cell number within a defined area is counted and the cell concentration derived from the count.

Materials

Hemocytometer (Improved Neubauer)
PBSA
0.25% crude trypsin
medium
tally counter
Pasteur pipette
microscope

Protocol

1.
Prepare the slide: (a) Clean the surface of the slide with 70% alcohol taking care not to scratch the semi-silvered surface. (b) Clean the coverslip and, wetting the edges very slightly, press it down over the grooves and semi-silvered counting area (see Fig. 19.1). The appearance of interference patterns ("Newton's rings"—rainbow colors between coverslip and slide like those formed by oil on water) indicates that the coverslip is properly attached, thereby determining the depth of the counting chamber

2.
Trypsinize monolayer or collect sample from suspension culture. Approximately 10^5 cells/ml minimum are required for this method so the suspension may need to be concentrated by centrifuging (100*g* for 2 min) and resuspending in a smaller volume

3.
Mix the sample thoroughly and collect about 20 μl into the tip of a Pasteur pipette or micropipette. Do not let the fluid rise in a Pasteur pipette or cells will be lost in the upper part of the stem

4.
Transfer the cell suspension immediately to the edge of the hemocytometer chamber and let the suspension run out of the pipette and be drawn under the coverslip by capillarity. Do not overfill or underfill the chamber or its dimensions may change due to surface tension; the fluid should run to the edges of the grooves only. Reload the pipette and fill the second chamber if there is one

5.
Blot off any surplus fluid (without drawing from under the coverslip) and transfer the slide to the microscope stage

6.
Select × 10 objective and focus on grid lines in chamber (see Fig. 19.1). If focusing is difficult because of poor contrast, close down the field iris

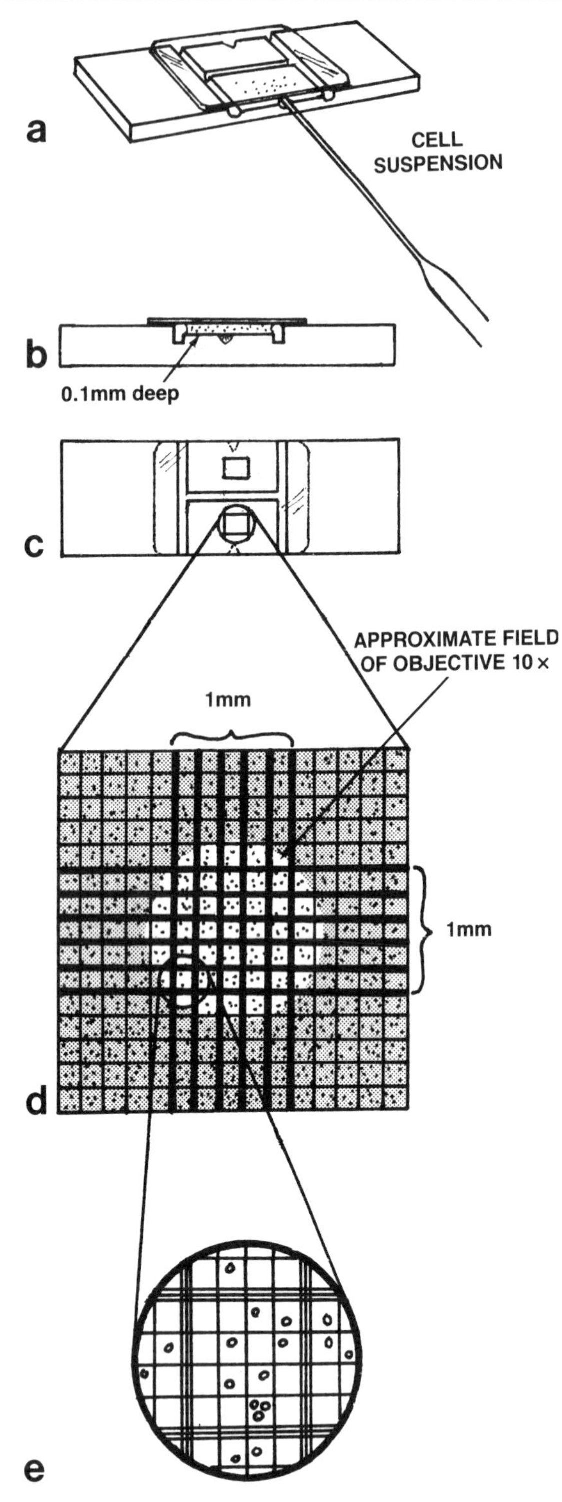

Fig. 19.1. *Hemocytometer slide (Improved Neubauer). a. Adding cell suspension to the assembled slide. b. Longitudinal section of slide showing position of cell sample in 0.1-mm-deep chamber. c. Top view of slide. d. Magnified view of total area of grid. Light central area is that area which would be covered by the average × 10 objective (depending on field of view of eye piece). This covers approximately the central 1 mm² of the grid. e. Magnified view of one of the 25 smaller squares bounded by triple parallel lines, making up the 1-mm² central area. This is subdivided by single grid lines into 16 small squares to aid counting.*

or make the lighting slightly oblique by tilting the mirror or offsetting the condenser

7. Move the slide so that the field that you see is the central area of the grid and is the largest area that you can see bounded by three parallel lines. This area is 1 mm^2. Either it will almost fill the field, or the corners will be slightly outside the field, depending on the field of view

8. Count the cells lying within this 1-mm^2 area, using the subdivisions (also bounded by three parallel lines) and single grid lines as an aid to counting. Count cells which lie on the top and left-hand lines of each square but not those on the bottom or right-hand lines to avoid counting the same cell twice. For routine subculture, attempt to count between 100 and 300 cells per mm^2; the more cells that are counted, the more accurate the count becomes. For more precise quantitative experiments, 500–1,000 cells should be counted

9. If there are very few cells ($< 100/mm^2$), count one or more additional squares (each 1 mm^2) surrounding the central square

10. If there are too many cells ($> 1,000/mm^2$), count only five small squares (each bounded by three parallel lines) across the diagonal of the larger (1-mm) square

11. If the slide has two chambers, move to the second chamber and do a second count. If not, rinse the slide and repeat the count with a fresh sample

Analysis. Calculate the average of the two counts, and derive the concentration of your sample as follows:

$$c = \frac{n}{v}$$

where c = cell concentration (cells/ml), n = number of cells counted, and v = volume counted (ml). For the Improved Neubauer slide, the depth of the chamber is 0.1 mm, and assuming only the central 1 mm^2 is used, v is 0.1 mm^3 or 10^{-4} ml. The formula becomes: $c = n \times 10^4$.

Hemocytometer counting is cheap and gives you the opportunity to see what you are counting. If the cells are mixed previously with an equal volume of a viabil-

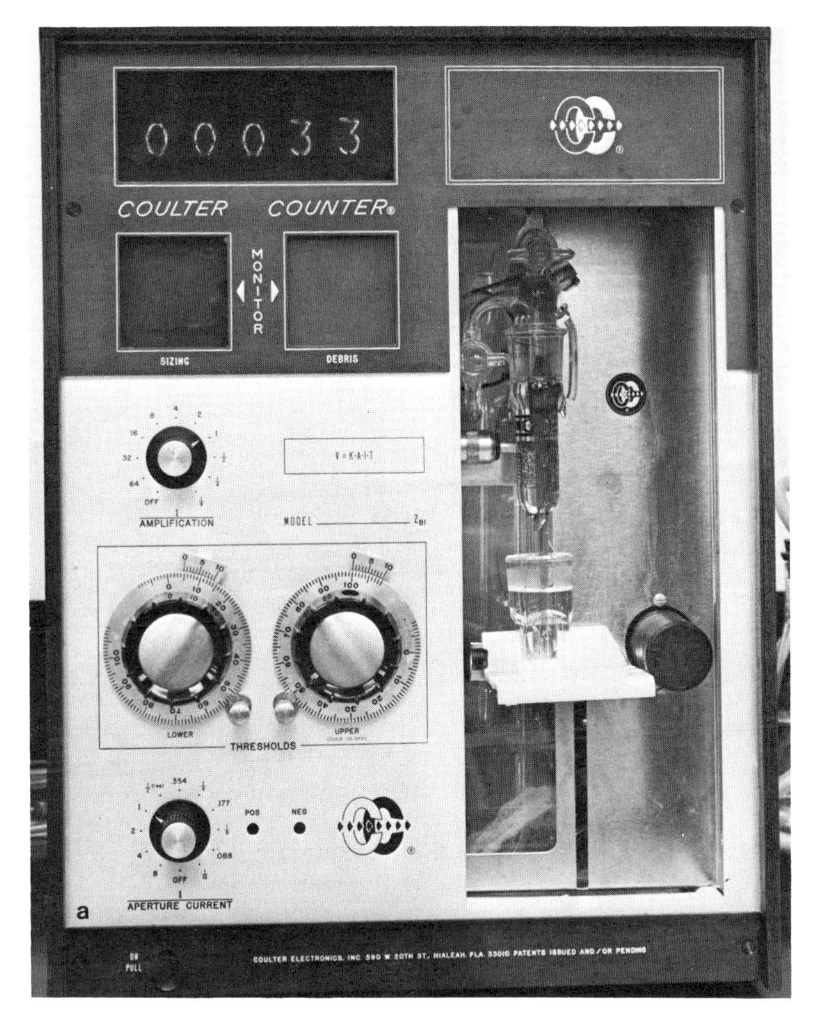

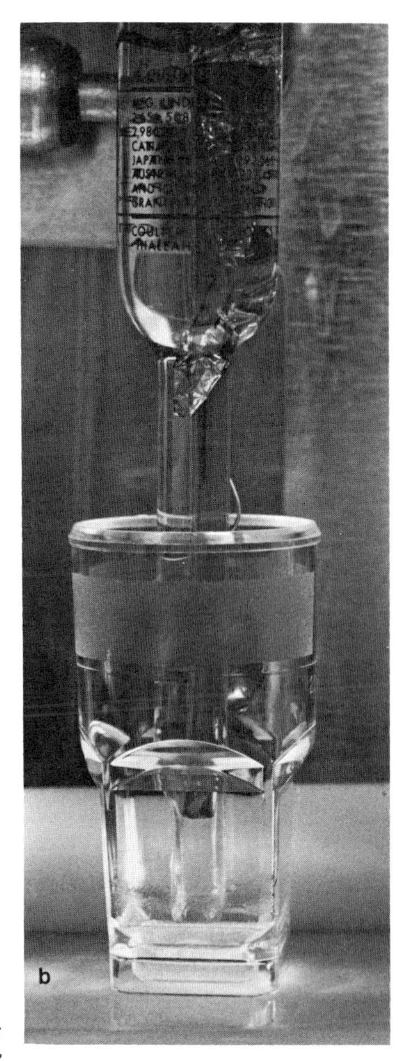

Fig. 19.2. *Coulter counter. This version is capable of counting and sizing cells; cheaper models are available for counting only. a. Front view of counter. b. Orifice tube and electrode in sample cup.*

ity stain (see below), a viability determination may be performed at the same time. The procedure is, however, rather slow and prone to error both in the method of sampling and the size of samples and requires a minimum of 10^5 cells/ml.

Most of the errors occur by incorrect sampling and transfer of cells to the chamber. Make sure the cell suspension is properly mixed before you take a sample, and do not allow the cells time to settle or adhere in the tip of the pipette before transferring to the chamber. Ensure also that you have a single cell suspension as aggregates make counting inaccurate. Larger aggregates may enter the chamber more slowly or not at all.

If aggregation cannot be eliminated during preparation of the cell suspension (see Table 12.1), lyse the cells in 0.1 M citric acid containing 0.1% crystal violet at 37°C for 1 hr and count the nuclei [Sanford et al., 1951].

Electronic particle counting. Although a number of different automatic methods have been developed for the counting of cells in suspension, the system devised originally by Coulter Electronics is the most common. Briefly, cells drawn through a fine orifice change the current flow through the orifice, producing a series of pulses which are sorted and counted.

Coulter counter. (Fig. 19.2). There are two main components of the system (Fig. 19.3): (1) an orifice tube connected to a pump and a mercury manometer by a two-way valve (Fig. 19.3a); and (2) an amplifier, pulse height analyzer, and scaler connected to two electrodes—one in the orifice tube and one in the sam-

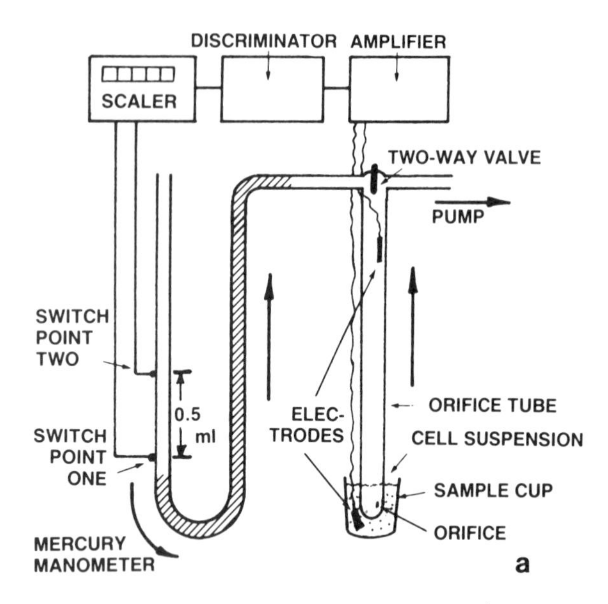

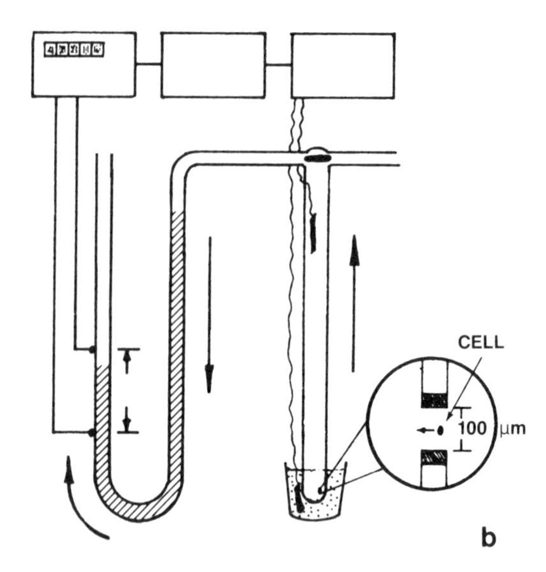

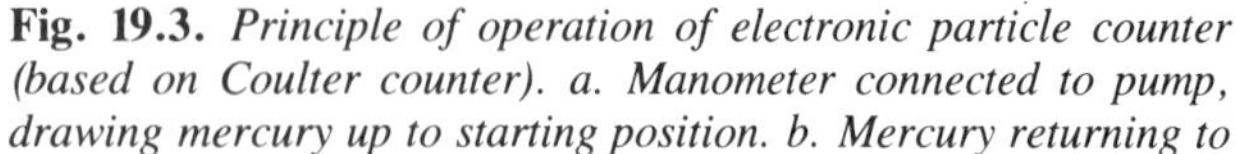

Fig. 19.3. *Principle of operation of electronic particle counter (based on Coulter counter). a. Manometer connected to pump, drawing mercury up to starting position. b. Mercury returning to equilibrium, drawing sample though orifice, and activating count cycle. Inset: magnified view of orifice in section.*

ple beaker—the current to them controlled by switch points on the mercury manometer.

When the two-way valve is turned vertically, the mercury manometer and orifice tube are connected to the pump (Fig. 19.3a). While liquid is drawn through the orifice generating a signal on the cathode ray oscilloscope, the mercury is drawn up the manometer to a preset level determined by the negative pressure generated by the pump. When the valve is restored to the horizontal position (Fig. 19.3b), the pump is disconnected, and the mercury monometer is connected directly to the orifice tube. As the mercury returns to equilibrium, fluid carrying the cell suspension is drawn through the orifice.

As the mercury travels along the tube, it passes two switch points; the first starts the count cycle, the second stops it. The mercury displacement between the two switches is 0.5 ml, hence 0.5 ml of cell suspension is drawn in through the orifice during the count cycle.

As each cell passes through the orifice, it changes the resistance to the current flowing through the orifice by an amount proportional to the volume of the cell. This generates a pulse (amps^{-1}) which is amplified and counted. Since the size of the pulse is proportional to the volume of the cell or particle passing through the orifice, a series of signals of varying pulse height are generated. A threshold control is set on the front panel to eliminate electronic noise and fine particulate debris but to retain pulses derived from cells (see below). This setting controls a pulse height analyzer circuit between the amplifier and the scaler which only allows pulses to pass to the scaler above the preset threshold.

Operation of Cell Counter

Outline

A sample of cells is diluted in electrolyte (physiological saline or PBS) and placed under the orifice tube and counted by drawing 0.5 ml of the sample through the counter.

Materials

Culture
PBSA
0.25% crude trypsin
medium
counting cup
counting fluid (see Appendix)

Protocol

1\. Trypsinize cell monolayer or collect sample from

suspension culture. The cells must be well mixed and singly suspended

2.
Dilute the sample of cell suspension 1:50 in 20 ml counting fluid in a 25-ml beaker or disposable sample cup. An automatic dispenser will speed up this dilution and improve reproducibility

3.
Mix well and place under tip of orifice tube, ensuring that the orifice is covered and that the external electrode lies submerged in the counting fluid in the sample beaker

4.
Set the two-way valve vertically until red light comes on and then extinguishes (mercury has passed both switch points)

5.
Clear the display

6.
Restore two-way valve to the horizontal position and allow count to proceed (red light comes on)

7.
When the red light extinguishes, the count is finished. Note the count and replace the sample with fresh counting fluid

Analysis. The counter takes 0.5 ml of the 1:50 dilution, so multiply the final count on the read-out by 100 to give the concentration of the cell suspension.

Problems. 1. Count stops, will not start, or counts slowly (i.e., takes longer than 25 sec.):

Orifice clogged. Free with tip of finger or fine brush.

2. Count lower than expected, orifice blocks frequently. Cell suspension aggregated:

Disperse cells by pipetting original sample vigorously, redilute, and proceed.

3. High electrical activity on oscilloscope screen but will not count:

Electrode out of beaker or disconnected.

4. Red light comes on but will not go out, or will not come on at all:

(a) Blocked orifice as in 1, (b) insufficient negative pressure. Pump has failed or leakage in tubing. Check pump and connections.

5. Gurgling sound from pump:

Counting fluid from waste reservoir has been drawn into pump:

Switch off, disconnect, and dry out pump. Relubricate, reconnect, and start again. The level in the waste reservoir should be checked regularly so that it may be emptied before the pump becomes contaminated.

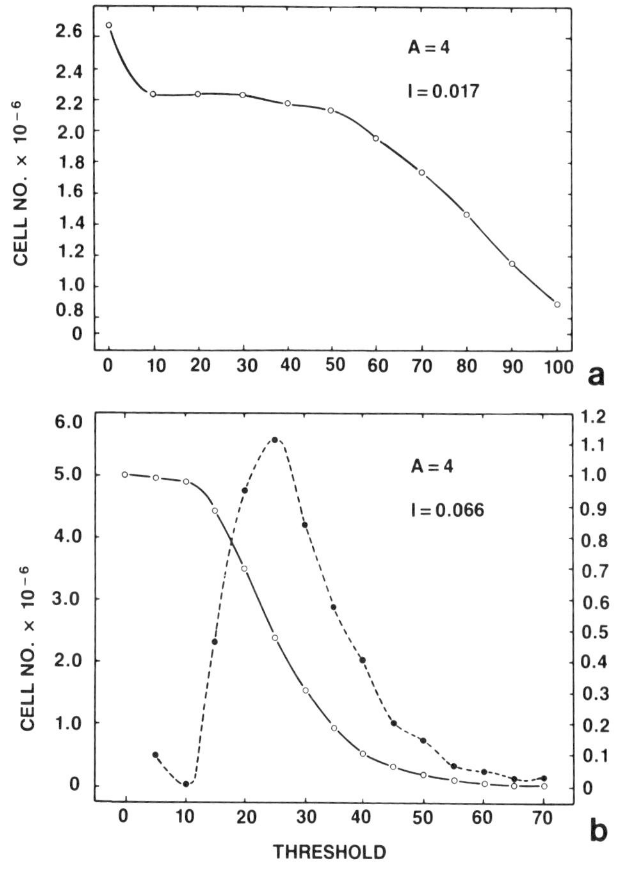

Fig. 19.4. *Calibration of cell counter for a specific cell type. a. A sample of LS cells was counted at a range of threshold values. (Counters with automatic simultaneous pulse height analysis will plot this curve automatically.) The correct threshold setting is that equivalent to the center of the plateau, i.e., 20. b. Repeat counts (higher cell concentration) with aperture current reset to 0.066. Plateau is now between zero and ten (unsuitable for routine counting) and cell count falls to near zero by threshold setting 70. By differentiating the first curve (solid line) a cell size analysis is obtained (dotted line). A = attenuation. I = aperture current.*

6. High background (a) line or radio interference, usually from electrical equipment (motors, fluorescent lights, incubators):

Check and eliminate by fitting suppressors to equipment. Line filters are available but are not always effective. Check grounding (earthing) of counter, particularly the case.

(b) Particulate matter in counting fluid:

Filter through disposable Millex or equivalent filter.

Calibration. To set the threshold in the correct position, perform a series of counts on the cell suspension, moving the threshold up in increments of 5 or 10 units from zero to 100. Plotting the counts against the threshold settings should produce a curve with a plateau (Fig. 19.4a). If the plateau is too short, decrease the attenuation and repeat. If the curve is too drawn out and does not reach a plateau, increase the attenuation.

Cell sizing. If the amplification is set such that the curve of cell counts versus threshold reaches a value at 100, which is <5% of the count at 5, then the bulk of the cell population has been analyzed and differentiating the curve (subtracting each count from the previous count at a lower threshold setting) will give a plot of relative size distribution (Fig. 19.4b). If standard latex particles are treated in the same way, a standard curve is obtained and the absolute cell size may be derived from the formula

$$v = KAIT$$

where v = cell volume, A = Attenuation setting (inverse of amplification), I = current setting, T = threshold setting, and K = constant. Derive K from standard latex particles and substitute in above equation to derive v for your cell sample.

More sophisticated counters are available with automatic cell sizing capabilities. They operate on the same principle but instead of controlling the threshold manually, the pulses from the amplifier are fed into a more sophisticated pulse height analyzer and sorter (Channelyser) which gives an instant readout on an oscilloscope and will print the size distribution histogram on an X-Y recorder. A counter with this facility will cost about three or four times as much as the simple version described above but makes cell sizing faster, easier, and more accurate.

Electronic cell counting is rapid and has a low inherent error due to the high number of cells counted. It is prone to misinterpretation, however, as cell aggregates, dead cells, and particles of debris of the correct size will all be counted indiscriminately. The cell suspension should be examined carefully before dilution and counting.

Electronic particle counters are expensive, but if used correctly, they are very convenient and give greater speed and accuracy to cell counting. There are now several such instruments available, but the Coulter D (Industrial) remains one of the cheapest.

Stained monolayers. There are occasions when cells cannot be harvested for counting or are too few to count in suspension. This situation is encountered with some multiwell plates or Terasaki plates. In these cases, the cells may be fixed and stained *in situ* and counted by eye on a microscope. Since this procedure is tedious and subject to high operator error, isotopic labeling or estimation of total DNA or protein is preferable though these measurements may not correlate directly with cell number, e.g. if the ploidy of the cell varies. A rough estimate of cell number per well can also be obtained by staining the cells with crystal violet and measuring absorption on a densitometer. This method has also been used to calculate the number of cells per colony in clonal growth assays [McKeehan et al., 1977].

Growth Cycle

As described in Chapter 12, following subculture, cells will progress through a characteristic growth pattern of lag phase, exponential or "log" phase, and stationary or "plateau" phase (see Fig. 12.3). The log and plateau phases give vital information about the cell line: the population doubling time during log growth and the maximum cell density achieved in plateau (saturation density). Measurement of the population doubling time is used to quantify the response of the cells to different inhibitory or stimulatory culture conditions such as variations in nutrient concentration, hormonal effects, or toxic drugs. It is also a good monitor of the culture during serial passage and enables the calculation of cell yields, and the dilution factor required at subculture.

It must be emphasized that the population doubling time is an average figure and describes the net result of a wide range of division rates, including zero, within the culture. Clonal growth analysis is required for the examination of individual growth rates within the population (see below). Hence, the population doubling time is distinct from the cell cycle time or generation time and the relationship between them will depend on the growth fraction (see below). A growth cycle is performed each time the culture is passaged and can be analyzed in more detail as described in Chapter 12.

Plating Efficiency

When cells are plated out as a single cell suspension at low cell densities (2–50 cells/cm^2), they will grow as discrete colonies (see Chapter 13). When these are counted the results are expressed as the plating efficiency:

$$\frac{\text{No. of colonies formed}}{\text{No. of cells seeded}} \times 100 = \text{plating efficiency}$$

If it can be confirmed that each colony grew from a single cell, this term becomes the *cloning efficiency*.

Strictly according to the definition, plating efficiency measurements are derived from counting colonies over a certain size (usually 16–50 cells) growing from a low inoculum of cells, and the term should not be used for the recovery of adherent cells after seeding at higher cell densities (e.g., 2×10^4 cells/cm^2). This is more properly called the *seeding efficiency*:.

$$\frac{\text{No. of cells recovered}}{\text{No. of cells seeded}} \times 100 = \text{seeding efficiency}$$

It should be measured at a time when the maximum number of cells has attached but before mitosis starts. This provides a crude measurement of recovery in, for example, routine cell freezing or primary culture.

Clonal growth assay by dilution cloning.

The protocol for dilution cloning [Puck and Marcus, 1955] is given in Chapter 14. When colonies have formed, remove medium, rinse carefully in BSS, and fix and stain colonies (see Chapter 15).

Analysis. 1. Count colonies and calculate plating efficiency. Magnifying viewers (e.g., Fig. 19.5) help to make counting easier.

2. The size distribution of the colonies may also be determined (e.g., to assay the growth promoting ability of a test medium or serum (see Chapter 9) by counting the number of cells per colony or by densitometry. Fix in 1% glutaraldehyde, stain the colonies with crystal violet, and measure absorption on a densitometer [McKeehan et al., 1977].

Automatic colony counting. If the colonies are uniform in shape and quite discrete, they may be counted on an automatic *colony counter* (e.g., New Brunswick, Artek, Micromeasurements Ltd.) which scans the plate with a conventional tv camera and analyzes the image to give an instantaneous readout of colony number. A size discriminator gives size analysis based on colony diameter (not always proportional to cell number, as cells may pile up in the center of the colony).

Though expensive, these instruments can save a great deal of time and make colony counting more objective. They will not cope well with colonies which overlap or have irregular outlines.

Labeling Index

If a culture is labeled with [^{3}H]-thymidine, cells that are synthesizing DNA will incorporate the isotope.

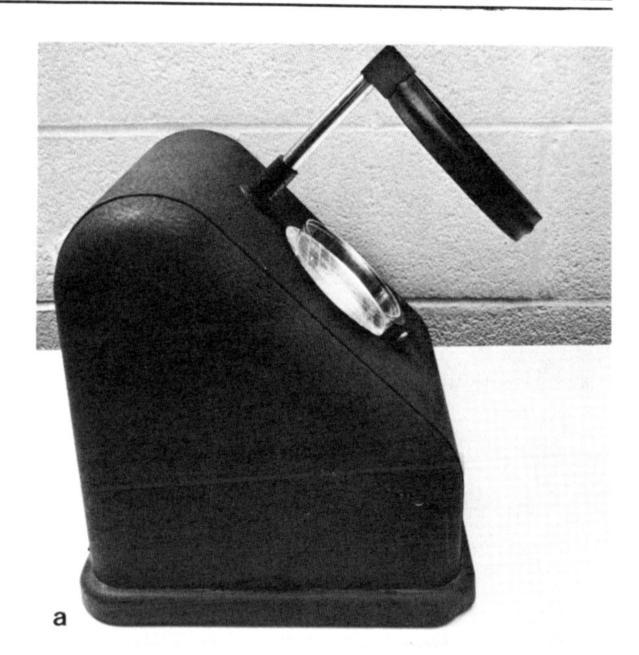

Fig. 19.5. *a. Simple magnifying colony counter. Versions are available with an electronically activated marking pen which records the count automatically. b. Bellco projection viewer. Magnifies the plate and allows more discrimination in scoring colonies.*

The percentage of labeled cells, determined by autoradiography [e.g., Westermark, 1974; Maciera-Coelho, 1973] (see Chapter 21 and Fig. 21.15a), is known as the labeling index (L.I.). Measurement after a 30-min exposure to [^{3}H]-thymidine shows a large difference between exponentially growing cells (L.I. = 10–20%)

and plateau cells (L.I. ⩽ 1%). Since the L.I. is very low in plateau, exposure times may have to be increased to 24 hr. With this length of label normal cells can be shown to have a lower labeling index with thymidine than neoplastic cells (see Chapter 16).

Outline

Grow cells to appropriate density, label with [^{3}H]-thymidine for 30 min, wash, fix, remove unincorporated precursor, and prepare autoradiographs.

Materials

Culture of cells
multiwell plate(s) containing 13-mm Thermanox coverslips
PBSA
0.25% crude trypsin
growth medium
hemocytometer or cell counter
[^{3}H]-thymidine (2 Ci/mmol)
HBSS
acetic methanol (1 part glacial acetic acid to 3 parts methanol) icecold, freshly prepared microscope slides
DPX
10% trichloroacetic acid ice-cold
deionized water
methanol

Protocol

1. Set up cultures at 2×10^4/ml–5×10^4/ml in 24-well plates containing coverslips
2. Allow cells to attach, start to proliferate (48–72 hr) and grow to desired cell density
3. Add [^{3}H]-thymidine to medium, 5 μCi/ml (2 Ci/mmol) and incubate for 30 min

   Note. Some media, e.g., Ham's F10 and F12, contain thymidine. In these cases, the concentration of radioactive thymidine must be increased to give the same specific activity in the medium. In prolonged exposure to high specific activity, [^{3}H]-thymidine causes radiolysis of the DNA. This can be reduced by using [^{14}C]-thymidine or [^{3}H]-thymidine of a lower specific activity.
4. Remove labeled medium and discard (care, radioactive!). Wash coverslip three times with BSS. Lift the coverslip off the bottom (but not right out of the well) at each wash to remove isotope from underneath
5. Add 1:1 BSS: acetic methanol, 1 ml per well and remove immediately
6. Add 1 ml acetic methanol at 4°C and leave for 10 min
7. Remove coverslips and dry at fan
8. Mount coverslip cells uppermost on a microscope slide
9. When mountant is dry (overnight), place slides in 10% trichloroacetic acid at 4°C in a staining dish and leave 10 min. Replace trichloroacetic acid twice during this extraction
10. Rinse in deionized water, then in methanol, and then dry
11. Prepare autoradiograph (see Chapter 21)

Analysis. Count percentage labeled cells. To cover a representative area, follow the scanning pattern illustrated in Figure 19.6.

Growth Fraction

If cells are labeled with [^{3}H]-thymidine for varying lengths of time up to 48 hr, the plot of labeling index against time increases rapidly over the first few hr and then flattens out to a very low gradient, almost a plateau (Fig. 19.7). The level of this plateau, read against the vertical axis, is the growth fraction of the culture, i.e., the proportion of cells in cycle at the time of labeling.

Outline

Label culture continuously for 48 hr, sampling at intervals for autoradiography.

Protocol

As for labeling index except that at step 3, incubation should be carried out for 15 min, 30 min, 1, 2, 4, 8, 24, and 48 hr.

Analysis. Count labeled cells as a percentage of the total, using scanning pattern from previous protocol. Plot labeling index against time.

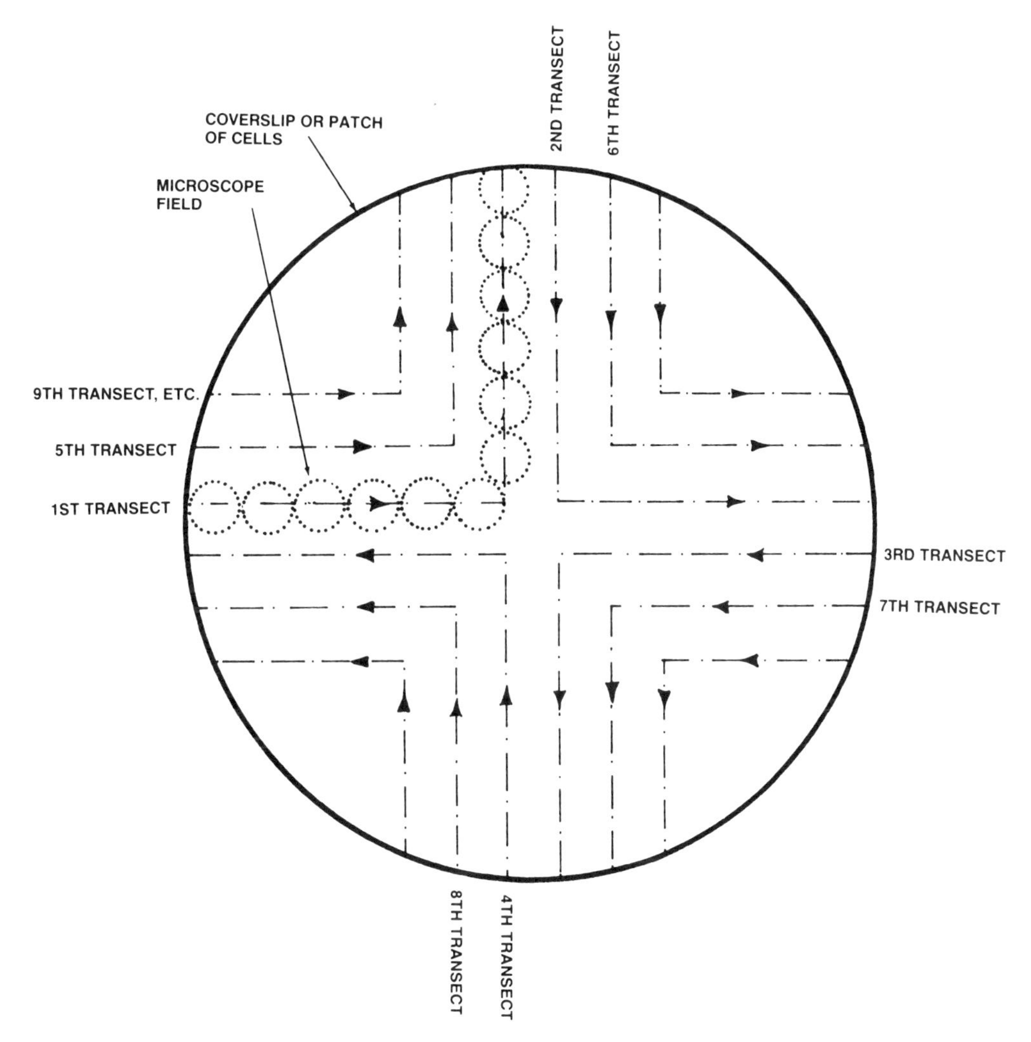

Fig. 19.6. *Scanning pattern for analysis of cytological preparations. Each dotted circle represents one microscope field and the whole, greater, circle, the extent of the specimen (e.g., a coverslip, culture dish or well, or spot of cells on a slide). Guide lines can be drawn with a nylon tipped pen.*

Note. Autoradiographs with ^{3}H can only be prepared where the cells remain as a monolayer. If they form a multilayer, they must be trypsinized after labeling and slides prepared by the drop technique (see Chapter 15), as the energy of β-emission from ^{3}H is too low to penetrate an overlying layer of cells.

Mitotic Index

This is the fraction or percentage of cells in mitosis, determined by counting mitoses in stained cultures as a proportion of the whole population.

Cell Cycle Time (Generation Time)

To determine the length of the cell cycle and its constituent phases, cells are labeled with [^{3}H]-thymidine for 30 min, the label is removed, and the appearance of the label in mitotic cells is determined autoradiographically at 30-min intervals up to 48 hr. The plot of the percentage of labeled mitoses against time takes the form of Figure 19.8, from which the cell cycle time and the length of its constituent phases may be derived.

CYTOTOXICITY AND VIABILITY

It is often necessary to determine the proportion of viable cells in a population, either as a step in controlling the standardization of the culture before it is used or in assaying the response of a culture to an experimental variable—e.g., in the preparation of a primary culture, or following treatment of a cell line with a cytotoxic drug or antibody. A wide range of tech-

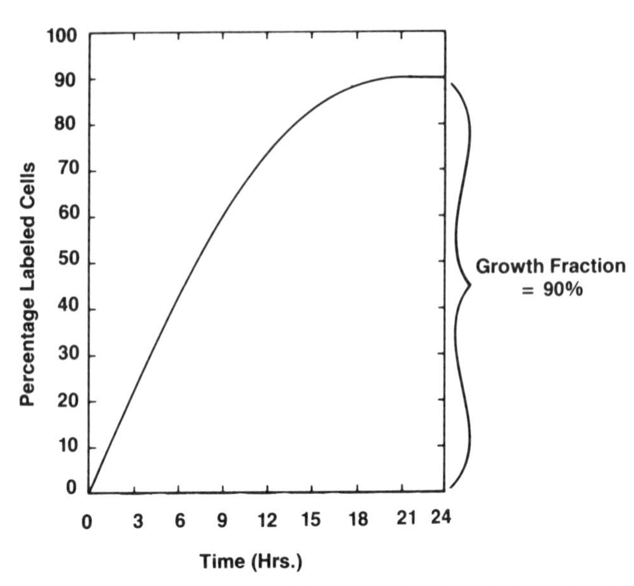

Fig. 19.7. *Determination of the growth fraction. Cells are labeled continuously with [^{3}H]-thymidine and the percentage labeled cells determined at intervals afterward by autoradiography (see text).*

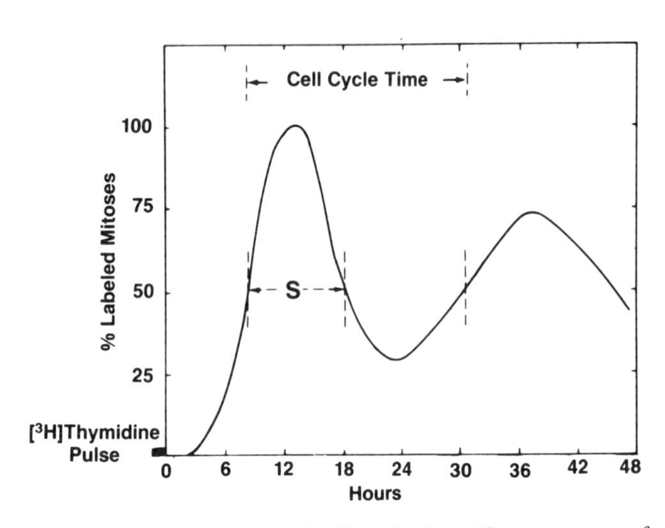

Fig. 19.8. *Determination of cell cycle time. The percentage of labeled mitosis relative to the total number of mitoses is plotted against time after removal of the thymidine. The time between the midpoint of the first ascending curve and the second rising curve is the cell cycle time [after Van't Hof, 1973; Maciera-Coelho, 1973; for further discussion, see Quastler, 1963; and Van't Hof, 1968].*

niques have been used [Malinin and Perry, 1967], too many to cover here, but some examples in common use will be mentioned.

Short-Term Tests—Viability

Many tests rely on the controlled permeability of the plasma membrane of living cells.

Dye exclusion

Viable cells are impermeable to nigrosine (naphthalene black), trypan blue, and a number of other dyes [Kaltenbach et al., 1958].

Outline

A cell suspension is mixed with stain and examined by low-power microscopy.

Materials

Cells
PBSA
0.25% trypsin
growth medium
hemocytometer
viability stain
Pasteur pipettes
microscope
tally counter

Protocol

1. Prepare cell suspension at a high concentration (~ 10^6 cells/ml) by trypsinization or centrifugation and resuspension
2. Take a clean hemocytometer slide and fix the coverslip in place (see above)
3. Add one drop of cell suspension to one drop of stain on the open surface of the slide, mix, transfer to the edge of the coverslip, and allow to run into the counting chamber
4. Leave 1–2 min (do not leave too long or viable cells will deteriorate and take up stain)
5. Place on microscope under a × 10 objective
6. Count the number of stained cells and the total number of cells
7. Wash hemocytometer and return to box

Analysis. Calculate the percentage of unstained cells. This is the percentage viability by this method. If the volumes of cell suspension and stain are measured accurately at step 3, this method of viability determination can be incorporated into the hemocytometer cell counting protocol.

Dye exclusion viability tends to overestimate viability; e.g., 90% of cells thawed from liquid nitrogen may exclude trypan blue but only 60% prove to be capable of attachment 24 hr later

Dye uptake. Viable cells take up diacetyl fluorescein and hydrolyse it to fluorescein to which the cell membrane of live cells is impermeable [Rotman and Papermaster, 1966]. Live cells fluoresce green; dead cells do not. Nonviable cells may be stained with ethidium bromide and will fluoresce red. Viability is expressed as the percentage of cells fluorescing green.

Chromium release. Reduced $^{51}Cr^{3+}$ is taken up by viable cells and oxidized to $^{51}Cr^{2+}$ to which the membrane of viable cells is impermeable [Holden et al., 1973; Zawydiwski and Duncan, 1978]. Dead cells release the $^{51}Cr^{2+}$ into the medium. A reduction in viability is detected by γ-counting aliquots of medium from cultures labeled previously with $Na_2{}^{51}CrO_4$ for released ^{51}Cr. The test works well over a few hours but over longer periods spontaneous release of ^{51}Cr may be a problem.

Long-Term Tests—Survival

While short-term tests are convenient and usually quick and easy to perform, they only reveal cells which are dead (i.e., permeable) at the time of the assay. Frequently, cells subjected to toxic influences, e.g., antineoplastic drugs, will only show an effect several hours or even days later. The nature of the tests required to measure viability in these cases is necessarily different since, by the time the measurement is made, the dead cells may have disappeared. Long-term tests are often used to demonstrate the metabolic or proliferative capacity of cells *after* rather than during exposure to a toxic influence. The objective is to measure survival rather than short-term toxicity, which may be reversible.

Plating efficiency, as described above, is the best measure of survival and proliferative capacity, provided that the cells plate with a high enough efficiency that the colonies can be considered representative. Though not ideal, anything over 10% is usually acceptable.

Since the colony number may fall at high toxic concentrations, it is usual to compensate by seeding more cells so that approximately the same number of colonies form at each concentration. This removes the risk of cell concentration influencing survival and improves statistical reliability. In addition cells should be plated on a preformed feeder layer the density of which ($5 \times 10^3/cm^2$) greatly exceeds that of the cloning cells, where the plating efficiency of controls is $\ll$ 100%.

A typical survival curve is prepared as follows:

Outline

Treat cells with experimental agent at a range of concentrations for 3 d, changing the agent daily. Trypsinize, seed at low cell density, and incubate for 1–3 wk. Stain and count colonies.

Materials

25 cm² flasks
6- or 9-cm petri dishes
PBSA
0.25% trypsin
growth medium
hemocytometer or cell counter
agent to be tested
HBSS
methanol
Giemsa stain

Protocol

1. Prepare a series of cultures in 25-cm² flasks, three for each agent concentration, and three controls. Seed cells at 5×10^4/ml in growth medium and incubate for 48 hr, by which time the cultures will have progressed into log phase (see above)
2. Prepare 50 ml of agent to be tested at a range of concentrations in regular growth medium. Check pH and osmolality and adjust if necessary
3. Replace the medium in each group of three flasks (5 ml per flask) with control medium or one of the concentrations of the agent. Always start with the control and work from lowest to highest agent concentration to avoid any risk of carry-over
4. If the agent is slow-acting or partially reversible repeat step 3 twice; i.e., expose cultures to the agent for 3 d, replacing the agent daily by changing the medium
5. Remove medium from each group of three flasks in turn, trypsinize cells, dilute, and seed at required density for clonal growth (see Chapter 13). If toxicity is expected, increase the seeding density at

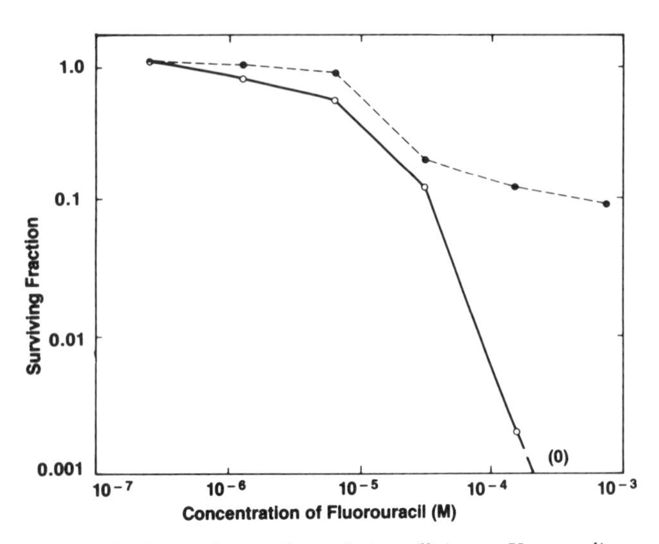

Fig. 19.9. *Survival curve from plating efficiency. Human glioma cells were plated out in the presence (dotted line) and absence (solid line) of a feeder layer after treatment with various concentrations of 5-fluorouracil. A 10% resistant fraction is apparent at 10^{-4} M drug only in the presence of a feeder layer. In the absence of the feeder layer, the small number of colonies constituting the resistant fraction were unable to survive alone.*

higher agent concentrations to keep the number of colonies forming in the same range. In addition, plate cells on to a feeder layer (see Chapter 13) of the same cells treated with mitomycin C, if plating efficiency of controls is substantially less than 100%

6.

Incubate until colonies form: fix, stain, and count colonies (see Chapter 19)

Analysis. 1. Plot relative plating efficiency (plating efficiency as a percentage of control) against drug concentration (Fig. 19.9) *(survival curve).*

2. Determine ID_{50} or ID_{90}: the concentration promoting 50% or 90% inhibition of colony formation.

3. Complex survival curves may be compared by calculating the area under the curve.

Variations. ***Concentration of agent.*** A wide concentration range, in log increments, e.g., 10^{-6} M, 10^{-5} M, 10^{-4} M, 10^{-3} M, O, should be used for the first attempt and a narrower range (log or linear), based on the indications of the first, for subsequent attempts.

Invariate agent concentrations. Some conditions tested cannot easily be varied, e.g., testing the quality of medium, water, or an insoluble plastic. In these cases, the serum concentration should be varied, as serum may have a masking effect on minor toxic effects.

Duration of exposure to agent. Some agents act rapidly; others, more slowly. Exposure to ionizing radiation, for example, need only be a matter of minutes, sufficient to achieve the required dose, while testing some antimetabolic drugs may take several days for a measurable effect.

Time of exposure to agent. Where the agent is soluble and expected to be toxic, the above procedure should be followed; but where the quality of the agent is unknown, stimulation is expected, or only a minor effect is expected (e.g., 20% inhibition rather than 100-fold or more), the agent may be incorporated during clonal growth rather than at preincubation.

Cell density. The density of the cells during exposure to an agent can alter its response, e.g., HeLa cells are less sensitive to the alkylating agent mustine at high cell densities [Freshney et al., 1975]. In this kind of experiment, the cell density should be varied in the preincubation phase, during exposure to drug.

Solvents. Some agents to be tested have low solubilities in aqueous media, and it may be necessary to use an organic solvent. Ethanol, propylene glycol, and dimethyl sulphoxide have been used for this purpose but may themselves be toxic. Use the minimum concentration of solvent to obtain solution. The agent may be made up at a high concentration in, for example, 100% ethanol, then diluted gradually with BSS, and finally diluted into medium. The final concentration of solvent should be $< 0.5\%$, and a solvent control must be included.

Take care when using organic solvents with plastics or rubber. It is better to use glass with undiluted solvents and only to use plastic where the solvent concentration is $< 10\%$.

While plating efficiency is one of the best methods for testing survival, it should only be used where the cloning efficiency is high enough for colonies that form to be representative of the whole cell population. Ideally this means that controls should plate at 100% efficiency. In practice this is seldom possible and control plating efficiencies of 20% or less are often accepted.

Plating efficiency tests are also time consuming to set up and analyze, particularly where a large number of samples is involved, and the duration of each experiment may be anything from 2 to 4 wk.

Microtitration. The introduction of multiwell plates revolutionized the approach to replicate sampling in tissue culture. They are economical to use, lend themselves to automated handling, and can be of good

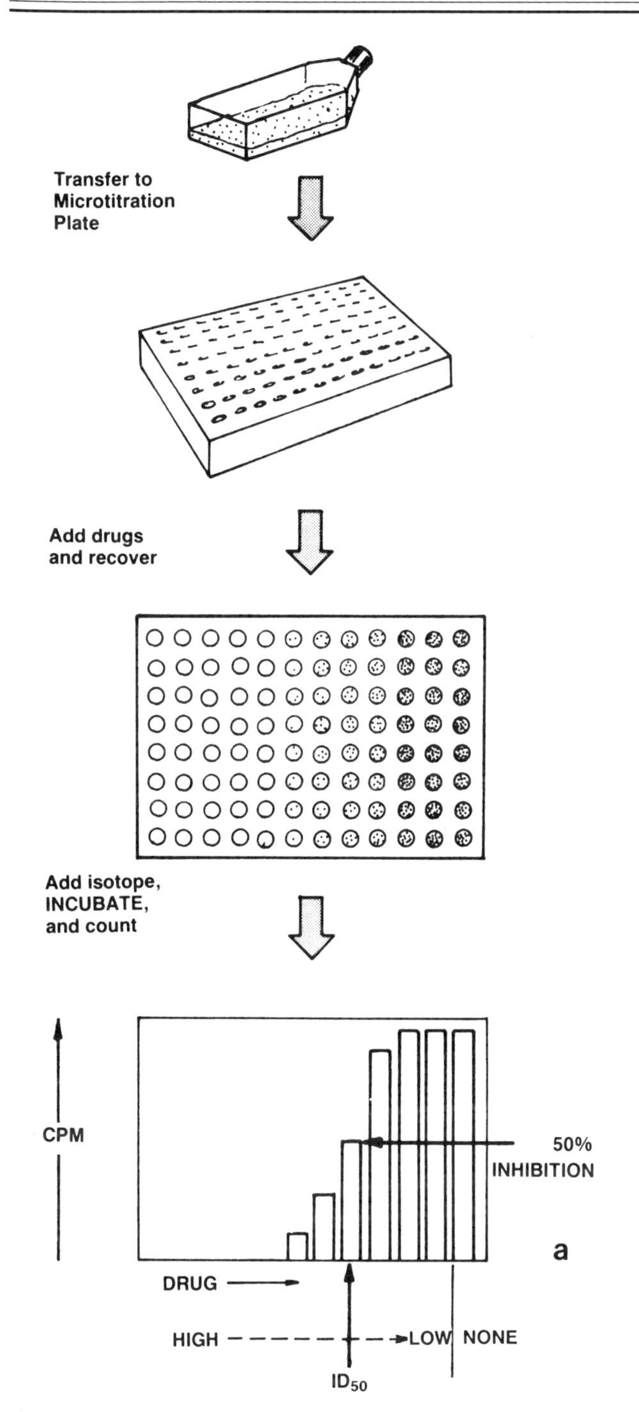

Fig. 19.10. *Microtitration assay for cytotoxicity. a. Stages of assay. b. Measurement of incorporated isotope by autofluorography.*

optical quality. The most popular is the 96 well microtitration plate, each well having 32 mm^2 growth area and capacity for 0.1 or 0.2 ml medium and up to 10^5 cells.

They may be used for cloning, for antibody, virus, and drug titration, for cytotoxicity assays of potential toxins, and for numerous other applications. The following example illustrates the use of microtitration plates in the assay of anticancer drugs but would be applicable with minor modification to any cytotoxic assay.

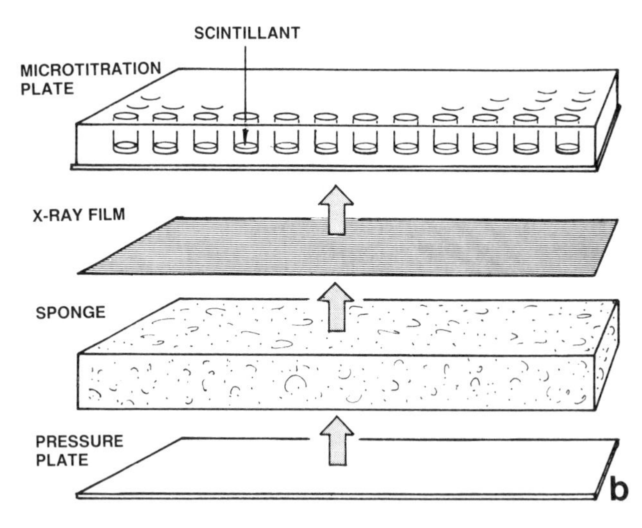

Figure 19.10. *(continued)*

Outline

Microtitration plate cultures are exposed to a range of drug concentrations during the log phase of growth and viability determined, several days after drug removal, by measuring incorporation of [^{35}S]-methionine (Fig. 19.10).

Materials

Culture of cells
PBSA
0.25% trypsin
growth medium
hemocytometer or cell counter
96-well microtitration plates
test solution
Mylar film (Flow Labs)
medium containing 0.1 μCi/ml [^{35}S]-methionine
HBSS
methanol
10% trichloroacetic acid
scintillation fluid
x-ray film (Kodak Royal)
dark box
silica gel
black light-tight bag
-70°C freezer
photographic developer and fixer

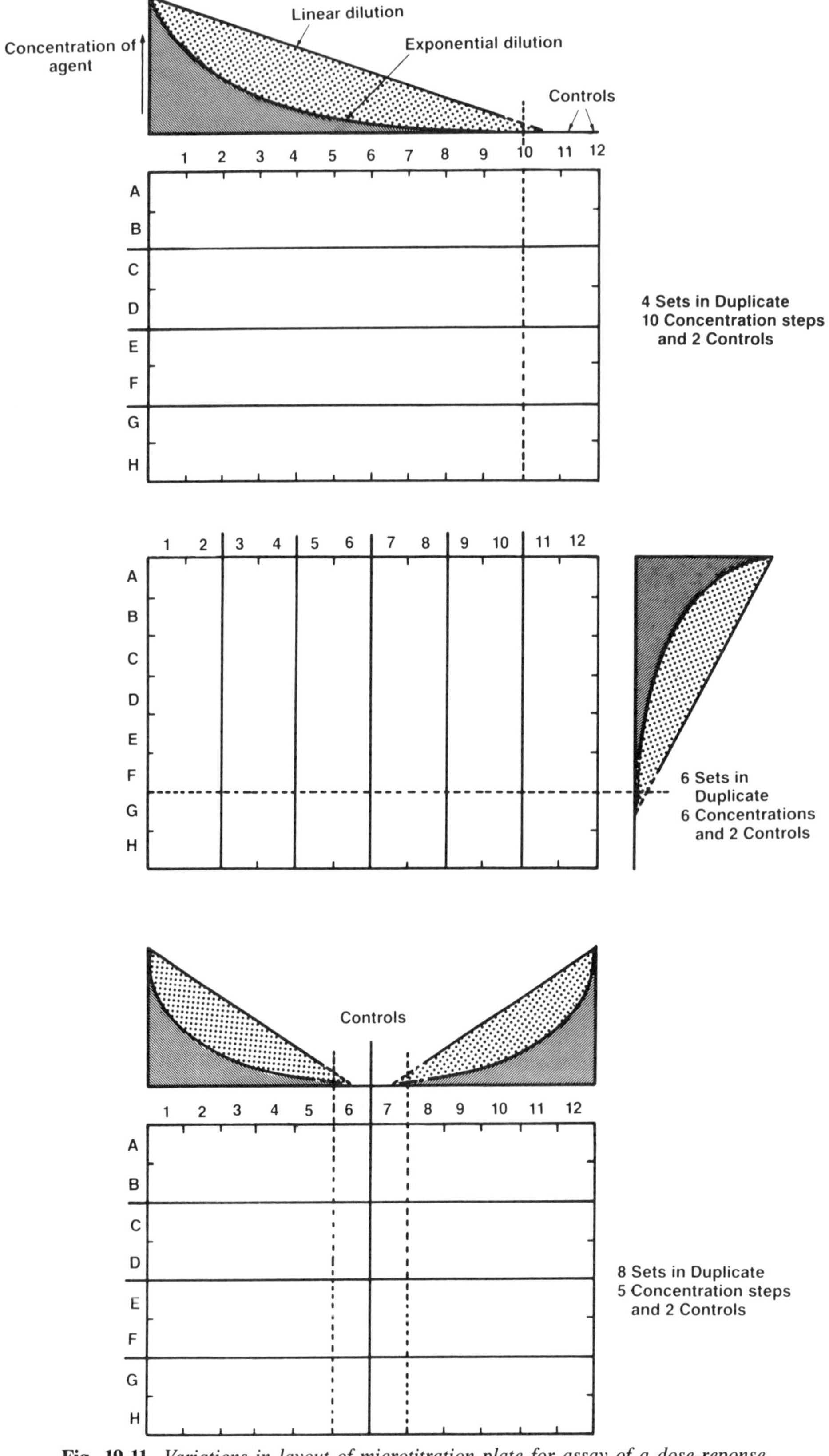

Fig. 19.11. *Variations in layout of microtitration plate for assay of a dose-reponse curve. The graphs represent agent concentrations in linear (stippled) or exponential (shaded) dilutions.*

Protocol

1.
Trypsinize cells (see Chapter 12). Seed microtitration plates at 10^3 cells/well, 0.1 ml/well (~ 3,000 cells/cm^2). Set up one duplicate plate for cell counting

2.
Place plate in CO_2 incubator with loose-fitting lid for 30 min to equilibrate with CO_2 (see Chapter 9.)

3.
Prepare drug dilutions in 1.0-ml aliquots. Dilute in complete medium and prepare similar aliquots of medium with no drug. The number of dilution steps and the volume required in each aliquot will depend on the way the plate is subdivided (Fig. 19.11). One milliliter is sufficient for duplicates, changed daily for 3 d, 0.1 ml per well

4.
Remove from incubator, Return to aseptic area, and quickly seal plate with self-adhesive Mylar film. Return plates to incubator for 48–72 hr

5.
Remove cell count plate from incubator and swab with 70% alcohol

6.
Cut Mylar film round eight wells and peel off film. Remove medium completely from these wells and add 0.1 ml of trypsin. Incubate for 15 min, at 36.5°C and then disperse the cells in the trypsin. Pool the cell suspensions from four wells, dilute to 20 ml, and count on cell counter.

If a cell counter is not available, do not trypsinize; rinse the cells in BSS and fix in methanol. At the end of the experiment the cells may be stained and counted by eye on a microscope, or an estimate of cell number made by densitometry (see above)

Note. The cells must remain in exponential growth throughout (see Chapter 12). If the cell growth curve shows that the cells are moving into stationary phase, proceed directly to 16.

7.
Change medium in remaining wells and return plate to incubator, equilibrate with CO_2 for 30 min, and reseal.

8.
Remove drug plate from incubator, swab with 70% alcohol, and gently peel off Mylar film

9.
Remove medium, two rows at a time, and add drug dilutions, the second row duplicating the first. There are a number of different ways of dividing up the plate to give different numbers of dilutions and replicates (see Fig. 19.11)

10.
Equilibrate with CO_2 in incubator for 30 min as in steps 2 and 3 above, reseal with Mylar film, and incubate for 24 hr

11.
Repeat steps 8, 9, and 10 twice more to give three complete days exposure to the drug. Do cell counts and change the medium on accompanying plate each time the drugs are renewed

12.
After drug exposure, remove drug medium and wash wells by gently adding and removing 0.1 ml medium three times ("dumping" the medium by inverting the plate can detach cells and increases the chance of contamination) and finally leave 0.1 ml of fresh medium in each well

13.
Count samples from accompanying plate and feed remaining wells

14.
Incubate for a further 5 d, changing the medium on the second or third day

15.
Change medium and count samples on cell count plate at the time the medium is changed on the drug plate and at the end

16.
Remove medium (drug plate only) and add 0.1 ml medium containing 0.1 μCi/ml [^{35}S]-methionine (the specific activity is unimportant as this is controlled by the methionine in the culture medium)

17.
Incubate for 3 hr

18.
Remove isotope and wash plate by submerging it in BSS, rubbing the wells with a gloved finger or comb to promote entry of BSS into wells. Rapid removal of medium by pouring off or flicking the plate will dislodge cells

19.
Repeat BSS wash twice, do not pour off previous wash from wells

20.
Immerse plate in 100% methanol and rub as in step 18

21.
Repeat twice in fresh methanol and leave for 10 min in final bath of methanol

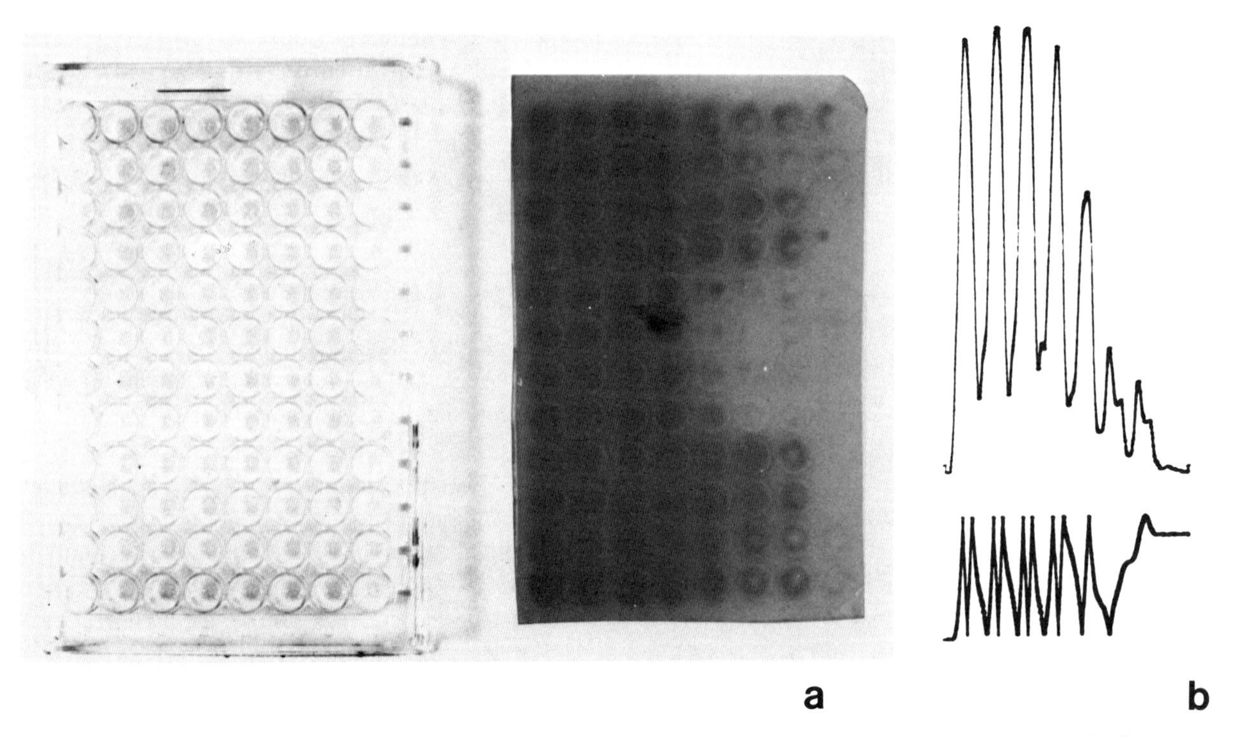

Fig. 19.12. *Autofluorograph from isotopically labeled cultures in a microtitration plate (a) and densitometer scan (b) from one row.*

22.
Pour off methanol and dry plate at fan
23.
Add ice cold 10% trichloroacetic acid (TCA) to the plate from a wash bottle. Fill the wells and stand on ice for 5 min. Remove TCA and repeat twice more with fresh TCA
24.
Wash in methanol and dry
25.
Add 50 μl scintillation fluid (e.g., Instagel) and dry down in a flat film onto the cells by centrifuging the plate for 1 hr at 20°C
26.
Bind dry plate with x-ray film (see Fig. 19.10b) (under dark-red safelight) and seal in dark box with desiccant such as silica gel
27.
After 2–14 days, open and remove film, under safelight conditions, develop for 10 min in D19, wash in tap water, fix in photographic fixer for 5 min, wash, and dry

Analysis. If the titration point is obvious, the plate may be read by eye. If not, scan plate on a densitometer, and determine ID_{50} (Fig. 19.12).

Variations. ***Duration.*** As for plating efficiency (see above), some agents may act more quickly, and the exposure period and recovery may be shortened.

Sampling. When trying the assay at first, it may be desirable to sample ([^{35}S]-methionine labeling and autofluorograph) on each day of drug exposure and recovery (Fig. 19.13). If a stable ID_{50} is reached earlier then the assay may be shortened.

End point. [^{35}S]-methionine labeling and autofluorography were chosen for speed and ease of analysis, and because active protein synthesis implies that the cells are still alive. Other alternatives are possible, however, including direct staining and cell counting *in situ* or by densitometry, fluorimetric DNA assay [Kissane and Robbins, 1958; Brunk et al. 1979], measurement of dehydrogenase activity [DiPaolo, 1965], or labeling with [^{3}H]-thymidine (DNA synthesis), [^{3}H]-uridine (RNA synthesis), or other isotopes and analysis by scintillation counting or autofluorography.

Microtitration offers a method whereby large numbers of samples may be handled simultaneously but with relatively few cells per sample. The whole population is exposed to the agent and viability determined metabolically, by staining or by counting cells from a few hours to several days later. A variety of automated handling techniques are available (autodispensers, di-

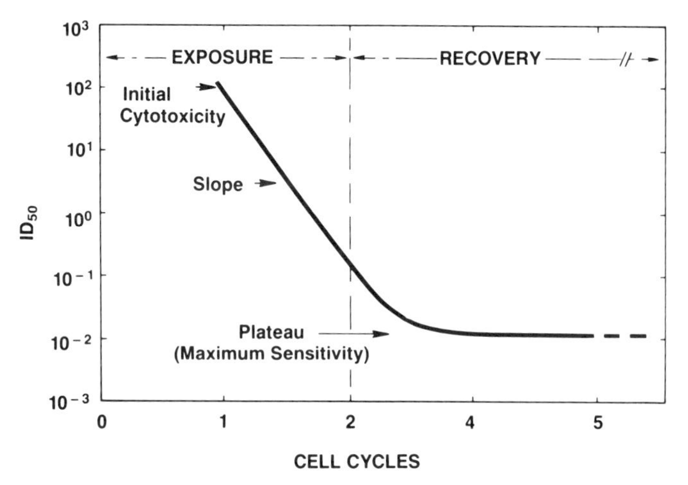

Fig. 19.13. *Time course of the fall in ID_{50}. Idealized curve for an agent with a progressive increase in cytotoxicity with time but eventually reaching a maximum effect after three cell cycles. Not all cytotoxic drugs will conform to this pattern.*

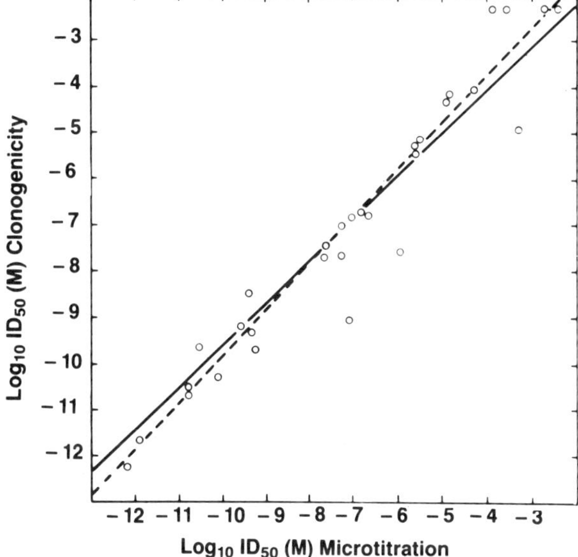

Fig. 19.14. *Correlation between microtitration and cloning in the measurement of the ID_{50} (solid line) of a group of five cell lines from human glioma and six drugs (vincristine, bleomycin, VM-26 epipodophyllotoxin, 5-fluorouracil, methyl CCNU, mithramycin.) Most of the outlying points were derived from one cell line which later proved to be a mixture of cell types. The dotted line is the regression with the data points from the heterogeneous cell line omitted [from Freshney et al., 1982].*

luters, cell harvesters, and densitometers (see Fig. 6.2), reducing the time required per sample. The volume of medium required per sample is less than one-tenth of that required for cloning, though the cell number is approximately the same.

Microtitration, however, is unable to distinguish between differential responses between cells within a population and the degree of response in each cell—e.g., a 50% inhibition could mean that 50% of the cells respond or that each cell is inhibited by 50%.

A comparison of ID_{50}'s derived by microtitration and plating efficiency assays showed a strong correlation between the two methods (Fig. 19.14) for the assay of antineoplastic drugs.

Metabolic Tests

The distinction between a metabolic test and a survival test is that a survival test examines the number of cells or amount of cellular activity remaining after prolonged exposure to an agent or short exposure and prolonged recovery in the absence of the agent. What is measured is the ability of the cell to survive, and in most cases, to continue to proliferate. In a metabolic test, on the other hand, the direct effect of the agent on one or more metabolic pathways is being measured. If, in the above protocol for microtitration assay, the labeling with ^{35}S-methionine is carried out during the first 24 hr of drug exposure, this becomes a metabolic test. There are many such tests, ranging from simple observation of the inhibition of pH depression to more specific tests depending on precursor incorporation or measurement of enzyme activity. They may be short-term (30 min or so) or long-term (several days).

In the context of cytotoxicity and viability testing, it should be kept in mind that effects on metabolism measured in the presence of an agent must be interpreted only as such. To establish an irreversible effect on cell survival, culture must be continued in the absence of the agent.

Cytometry

Histochemical and other cytological techniques (see also Chapter 15) can measure the amounts of enzyme, DNA, RNA, protein, or other cellular constituents *in situ*. It is difficult, however, to interpret the results of such techniques in a quantitative fashion. The introduction of flow cytometry [Melamad et al., 1979; Crissman et al., 1975], while losing the relationship between cytochemistry and morphology, has added a new dimension to the measurement of cellular constituents and activities [Kurtz and Wells, 1979]. The potential amount of information that may be obtained about the constituent cells of a population is so vast that the problem becomes one of intellectual interpretation rather than data collection.

For a further description of flow cytometry, see Chapter 14.

Chapter 20
Culture of Specific Cell Types

It will be apparent from the discussion in previous chapters that the expression of specialized functions in culture is controlled by the nutritional constitution of the medium, the presence of hormones and other inducer or repressor substances, and the interaction of the cells with the substrate and other cells. Each cell type may have highly specific requirements, and there is not space to review all of these here. Reviews of specialized culture techniques for specific cell types can be found in Jakoby and Pastan [1979], Kruse and Patterson [1973], Willmer [1965], Defendi [1964], Ursprung [1968], and Sato [1978, 1981].(see also Table 2.2) My main purpose here is to present an outline of some of the techniques that are available for culturing different tissue or cell types and to exemplify the diversity of cell types that can be cultured. It is useful to classify these anatomically, although there is considerable overlap in the techniques used.

EPITHELIAL CELLS

Epithelial cells are often responsible for the recognized functions of an organ, e.g., controlled absorption in kidney and gut, secretion in liver and pancreas, and gas exchange in lung. They are also of interest as models of differentiation and stem cell kinetics (e.g., epidermal keratinocytes) and are amongst the principal tissues where the common cancers arise. Consequently, culture of various epithelial cells has been a focus of attention for many years. The major problem in the culture of pure epithelium has been the overgrowth of the culture by stromal cells such as fibroblasts and vascular endothelium. Most of the variations in technique are aimed at preventing this, by nutritional manipulation of the medium or alterations in the culture substrate (Table 20.1).

Isolation of epithelial cells from donor tissue is best performed with collagenase (see Chapter 12) as this disperses the stroma but leaves the epithelial cells in small clusters, which favors their subsequent survival.

Epidermis

Rheinwald and Green [1975] showed that murine and human epidermal keratinocytes could be cultured selectively on feeder layers of irradiated 3T3 cells and could mature to form differentiated squames [Green, 1977]. Basal cell carcinoma can also be cultured by this method [Rheinwald and Beckett, 1981]. Alteration of the constituents of the medium enabled Peehl and Ham [1980; and Tsao et al., 1982] to culture keratinocytes from human foreskin selectively without feeder layers or serum and others have shown that reductions in pH [Eisinger et al., 1979], Ca^{2+} [Peehl and Ham, 1980], and temperature [Miller et al., 1980], may all contribute to improved selective growth of epidermal keratinocytes. Addition of hydrocortisone (10μg/ml) [Peehl and Ham, 1980], 10^{-10}M cholera toxin, or 10^{-6}M isoprenaline and epidermal growth factor (10 ng/ml) [Rheinwald and Green, 1977] to the medium has made continued serial subculture possible over many cell generations [Green et al., 1979]. When mouse sublingual epidermal cultures were grown on collagen rafts (see also under "liver") at the gas-liquid interface, complete histological maturation was possible [Lillie et al., 1980].

Breast

Milk [Buehring, 1972; Ceriani et al., 1979] and reduction mammoplasty are suitable sources of normal ductal epithelium from breast, the first giving purer cultures of epithelial cells. Growth on confluent feeder layers of fetal human intestine [Stampfer et al., 1980; Freshney et al., 1982] represses stromal contamination with both normal and malignant tissue (see Fig. 13.9), and optimization of the medium [Stampfer et al., 1980]

TABLE 20.1. Inhibition of Fibroblastic Overgrowth

Method	Agent	Tissue	Reference
Selective detachment	Trypsin	Fetal intestine, cardiac muscle, epidermis	Owens et al. [1974], Polinger [1970], Milo et al. [1980]
	Collagenase	Breast carcinoma	Lasfargues [1973]
Selective attachment and substrate modification	Polyacrylamide	Various tumors	Jones and Haskill [1973, 1976]
	Teflon	Transformed cells	Parenjpe et al. [1975]
	Collagen (pig skin)	Epidermis	Freeman et al. [1976]
Confluent feeder layers	Mouse 3T3 cells	Epidermis	Rheinwald and Green [1975]
	Fetal human intestine	Breast epithelium, normal	Stampfer et al. [1980]
		and malignant	Freshney et al. [1982]
		Colon carcinoma	Freshney et al. [1982]
Selective media	D-valine	Kidney	Gilbert and Migeon [1975, 1977]
	Cis-OH-Proline	Cell lines	Whei-Yang Kao [1977]
	Ethylmercurithiosalicylate	Neonatal pancreas	Braaten et al. [1979]
	Phenobarbitone	Liver	Fry and Bridges [1979]
	MCDB 151	Epidermis	Peehl and Ham [1980]
	Antimesodermal antibody	Various carcinomas	Edwards et al. [1980]

enables serial passage and cloning [Smith et al., 1981] of the epithelial cells. Cultivation in collagen gel allows three-dimensional structures to form; these correlate well with the histology of the original donor tissue [Yang et al., 1979, 1980, 1981].

As with epidermis, cholera toxin [Taylor-Papadimetriou et al., 1980] and EGF [Osborne et al., 1980] stimulate the growth of epithelioid cells from breast *in vitro*.

The hormonal picture is more complex. Many epithelial cells survive better with insulin added to the culture (1–10 IU/ml) in addition to hydrocortisone ($\sim 10^{-8}$M). The differentiation of acinar breast epithelium in organ culture requires hydrocortisone, insulin, and prolactin [Stockdale and Topper, 1966], and requirements for estrogen, progesterone, and growth hormone have been demonstrated in cell culture [Klevjer-Anderson et al., 1980].

Cervix

A modification of the epidermal culture technique can be used for the propagation of cervical epithelium [Stanley and Parkinson, 1979]. Dispase II (Boehringer, Mannheim), 1 unit/ml, allows selective detachment of the epithelial cells without dislodging the feeder layer.

Gastrointestinal Tract

Culture of normal gut lining epithelium has not been extensively reported although there are numerous reports in the literature of continuous lines from human colon carcinoma [Tom et al., 1976; Van der Bosch et al., 1981; Kim et al., 1979; Noguchi et al, 1979; Bergerat et al., 1979]. Colorectal carcinoma cells plated on confluent feeder layers of FHI (see below and Chapter 13) form colonies which apparently disappear but reappear 8–10 wk later as nodules in the monolayer (Fig. 20.1). These nodules will increase and can be subcultured with or without feeder layers and, in several cases, have given rise to continuous cell lines, some of which produce carcinoembryonic antigen CEA and sialomucin, markers of neoplasia in human colon [Freshney et al., 1982].

Owens et al. [1974] were able to culture cells from fetal human intestine as a finite cell line (FHS 74 Int). Similar results have been obtained in the author's laboratory, where more vigorous growth was observed in the epithelial cell component of these cultures and

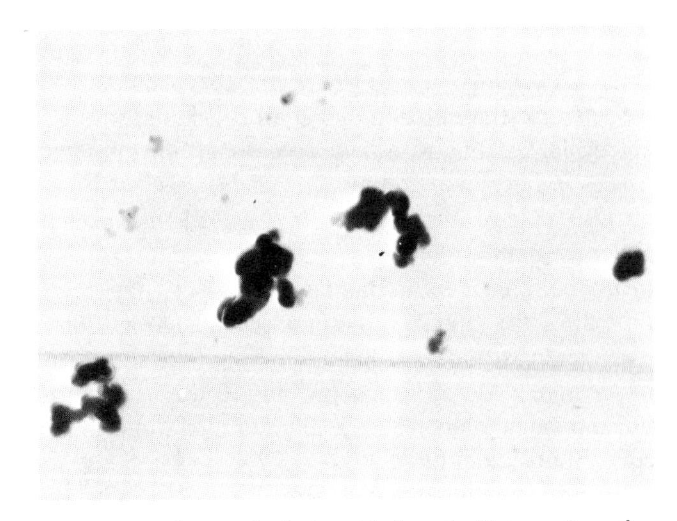

Fig. 20.1. *Colonies of colonic epithelium shed into medium after several weeks of cultivating colon carcinoma cells on a confluent feeder layer (FHI). (Courtesy of Dr. J.M. Russell.)*

fibroblastic cells were eventually diluted out, giving rise to a finite cell line, FHI.

Liver

One of the major objectives of the 1960s was the culture of cell lines of normal functional liver parenchyma. This has not yet been fully realized, but there are now many examples of epithelial cells cultured from rat liver. Although these cultures do not express all the properties of liver parenchyma, there is little doubt that the correct lineage of cells may be cultured even if the correct environment cannot be recreated for full functional expression [Guillouzo et al., 1981].

The development of the correct conditions for perfusing liver *in situ* with collagenase and hyaluronidase [Berry and Friend, 1969] (see Chapter 11) provided a technique whereby a good viable suspension of liver parenchymal cells could be plated out with high purity and form a viable monolayer, and reports such as those of Malan-Shibley and Iype [1981] suggest that these cultures may be propagated.

Some of the most useful continuous liver cell lines were derived from Reuber H35 [Pitot et al., 1964] and Morris [Granner et al., 1968] minimal deviation hepatomas of the rat. Induction of tyrosine aminotransferase in these cell lines with dexamethasone proved to be a valuable model for studying the regulation of enzyme adaptation in mammalian cells [Reel and Kenney, 1968; Granner et al., 1968].

Pitot and others have demonstrated that, as with epidermis, greater functional expression can be obtained by culturing liver parenchymal cells on free-floating collagen sheets [Michalopoulos and Pitot, 1975; Sirica et al., 1980]. The cells are seeded in medium onto a preformed collagen gel on the base of a petri dish. After the cells have attached, the gel is released from the base of the dish with a bent Pasteur pipette or spatula, allowing it to float freely in the medium. This permits access of nutrients to the cells from above and below. It is possible that diffusion of nutrients and metabolites via the collagen layer is analogous to the situation *in vivo* where epithelial cells usually lie on basement membrane, and may be important in establishing a necessary polarity in the cells.

Pancreas

There has not been the same effort expended on culture of pancreas as in liver. Pahlman et al. [1979] and Lieber et al. [1975] described neoplastic cell lines from exocrine pancreas; and Wallace and Hegre [1979] produced epithelial monolayers from fetal rat pancreas by a primary explant technique. These cultures remained free of fibroblasts for several days, and contained many endocrine cells (see also under Endocrine Cells, this chapter).

Kidney

The separation of tubular epithelium from stroma has been one of the simpler problems in the field of epithelial cell culture. As these cells have D-amino acid oxidase, they will grow in D-valine while the stromal cells do not [Gilbert and Migeon, 1975]. Kidney tubular epithelium has also proved amenable to separation from stroma by curtain electrophoresis and velocity sedimentation [Kreisberg et al., 1977]. One of the simpler approaches is to treat the finely chopped kidney with collagenase (see Chapter 11) overnight and then, following dispersal of the tissue by gentle pipetting, to allow the undisaggregated fragments of tubule and glomeruli to sediment through the more finely dispersed stroma. If the tubules are washed two or three times by repeating this differential sedimentation, a culture highly enriched for tubular epithelium and glomeruli may be obtained.

A continuous line of dog kidney epithelium has been described [e.g., MDCK, Taub and Saier, 1979; Rindler et al., 1979].

Bronchial and Tracheal Epithelium

There are a number of reports of primary culture of alveolar, bronchial, and tracheal epithelium [e.g., Fraser and Venter, 1980; Stoner et al., 1980] including the use of floating collagen [Geppert et al., 1980] and pigskin [Yoshida et al., 1980], and Steele et al. [1978] were able to produce nontumorigenic continuous cell lines by treating tracheal epithelium with a phorbol ester. More recently Lechner et al. [1981] have developed a low serum medium for clonal growth of normal lung and Carney et al. [1981] have developed a serum-free medium supplemented with hydrocortisone, insulin, transferrin, estrogen, and selenium (HITES medium) for small cell carcinoma of lung and with modifications for large cell and adenocarcinoma. These media are selective and do not support stromal cells.

MESENCHYMAL CELLS

I will include here those cells which are derived from the embryonic mesoderm, but excluding the hemopoietic system, which will be discussed below. This group includes the structural and vascular cells.

Connective Tissue

These cells are generally regarded as the weeds of the tissue culturist's garden. They survive most mechanical and enzymatic explantation techniques and may be cultured in many of the simplest media, such as Eagle's basal medium.

Although cells loosely called fibroblasts have been isolated from many different tissues and assumed to be connective tissue cells, the precise identity of cells in this class remains somewhat obscure. Fibroblast lines, e.g., 3T3 from mouse, produce types I and III collagen and release it into the medium [Goldberg, 1977]. While collagen production is not restricted to fibroblasts, synthesis of type I in relatively large amounts is characteristic of connective tissue. However, 3T3 cells can also be induced to differentiate into adipose cells [Kuriharcuch and Green, 1978]. It is possible that cells may transfer from one lineage to another under certain conditions, but such transdifferentiation has rarely been confirmed. It is more likely that mouse embryo fibroblastic cell lines are primitive mesodermal cells [Franks, personal communication] which may be induced to differentiate in more than one direction.

Human, hamster, and chick fibroblasts are morphologically distinct from mouse fibroblasts as they assume a spindle-shaped morphology at confluence, producing characteristic parallel assays of cells distinct from the pavementlike appearance of mouse fibroblasts. The spindle-shaped cell may represent a more highly committed precursor and may be more correctly termed a fibroblast. NIH3T3 cells may become spindle shaped if allowed to remain at high cell density.

It has also been suggested that fibroblastic cell lines may be derived from vascular pericytes, connective tissue–like cells in the blood vessels, but in the absence of the appropriate markers, this is difficult to confirm.

It is clearly possible to cultivate cell lines, loosely termed fibroblastic, from embryonic and adult tissues, but these should not be regarded as identical or classed as fibroblasts without confirmation with the appropriate markers. Collagen, type I, is one such marker. Thy I antigen has also been used [Raff et al., 1979] although this may also appear on some hemopoietic cells.

Adipose Tissue

Although it may be difficult to prepare cultures from mature fat cells, differentiation may be induced in cultures of mesenchymal cells (mouse 3T3) cells by maintenance of the cells at a high density for several days [Kuriharcuch and Green, 1978]. An adipogenic factor in serum appears to be responsible for the induction.

Muscle

Myoblasts from the three main categories of muscle may be grown in culture. Skeletal and cardiac myoblasts may be prepared from chick embryo as described in Chapter 11 (see also Konigsberg, 1979). Yaffe [1968] and others have described the stages of differentiation in these cultures.

Cardiac myoblast cells will also progress through differentiation *in vitro* and can be seen to contract a few days after explantation from the embryo, although they tend to lose this capacity with continued subculture. Polinger [1970] used the differential rate of attachment to reduce fibroblastic contamination of primary cultures.

Smooth muscle cells may be cultured from blood vessels following disaggregation in trypsin or collagenase [Ross, 1971; Burke and Ross, 1977]. Gospodarowicz et al., [1976] described cloned cell lines from bovine aorta derived from scraped or collagenase-treated tissue. Yasin et al. [1981] obtained cell lines from adult skeletal muscle and showed that they retained specialized markers.

Muscle cells may be identified by a number of antigenic markers including myosin and tropomyosin. Actin is not a good marker as it can be found in most cells. Creatine phosphokinase activity increases as muscle cells differentiate [Richler and Yaffe, 1970; Yaffe, 1971]. The most obvious property of all is spontaneous contraction, which is observed in both skeletal and cardiac muscle.

In common with other cells with excitable membranes (and some hemopoietic cells), muscle cells may be stained selectively with the fluorescent dye merocyanine 540) [Easton et al., 1978].

Cartilage

Coon and Cahn [1966] described a technique (see also Chapter 12) for the cultivation of cartilage-synthesizing cells from chick embryo somites. Cahn and Lasher [1967] later used this system for analysis of the involvement of DNA synthesis as a prerequisite for cartilage differentiation. Chondrocytes respond to stimulation of growth by both EFG and FGF [Gospodarowicz and Mescher, 1977] but ultimately lose their differentiated function [Benya et al., 1978].

Bone

Although bone is mechanically difficult to handle, thin slices treated with EDTA and digested in collagenase [Bard et al., 1972] give rise to cultures of osteoblasts which have some functional characteristics of the tissue. Antiserum against collagen has been used to prevent fibroblastic overgrowth without inhibiting the osteoblasts [Duksin et al., 1975]. Propagated lines have been obtained from osteosarcoma [Smith et al., 1976; Weichselbaum et al., 1976] but not from normal osteoblasts.

Endothelium

Endothelium has been successfully cultured by collagenase perfusion of bovine aorta (Fig. 15.1e) [Gospodarowicz et al., 1976, 1977, 1978; Schwartz, 1978] and human umbilical vein [Gimbrone et al., 1974], trypsinization of white matter from rat cerebral cortex [Phillips et al., 1979], and microdissection of adrenal cortex [Folkman et al., 1979]. In the author's laboratory, an endothelial cell line has been developed from a human anaplastic astrocytoma by collagenase digestion. The astrocytoma cells were overgrown during serial passage.

Endothelium can be characterized by the presence of factor VIII antigen [Booyse et al., 1975], type IV collagen [Howard et al., 1976], Weibel-Palade bodies [Weibel and Palade, 1961] and, sometimes, the formation of tight junctions, although the last feature is not always demonstrated readily in culture.

Endothelial cultures are good models for contact inhibition and density limitation of growth as cell proliferation is strongly inhibited after confluence is reached [Haudenschild et al., 1976].

Much interest has been generated in endothelial cell culture because of the potential involvement of endothelial cells in vascular disease, blood vessel repairs, and angiogenesis in cancer. Folkman and Haudenschild [1980] described, rather elegantly, the development of three-dimensional structures resembling capillary blood vessels derived from pure endothelial lines *in vitro*. Growth factors, including angiogenesis factor derived from Walker 256 cells *in vitro*, play an important part in maintaining proliferation and survival, so that secondary structures can be formed.

NEURECTODERMAL CELLS

Nerve cells appear more fastidious in their choice of substrate than most other cells [Nelson and Lieberman, 1981]. They will not survive well on untreated glass or plastic but will demonstrate neurite outgrowth in collagen [Ebendal and Jacobson, 1977; Ebendal, 1979] and poly-D-lysine [Yavin and Yavin, 1980]. Neurite outgrowth is encouraged by a polypeptide nerve growth factor (NGF) [Levi-Montalcini, 1964, 1979] and a factor secreted by glial cells [Barde et al., 1978; Lindsay, 1979] immunologically distinct from NGF.

Cell proliferation has not been found in cultures of neurons even with cells from embryonic stages where mitosis was apparent *in vivo*. Much of the work on nerve cell differentiation has, therefore, been performed on neuroblastoma cell lines [Augusti-Tocco and Sab, 1969; Lieberman and Sachs, 1978; Littauer et al., 1979] or on glial-neuronal hybrids [Minna et al., 1972; Minna and Gilman, 1973] (see Chapter 22). This remains an intriguing area with many unsolved problems.

Glia

Greater success has been obtained in culturing glial cells from avian, rodent, and human brain. Embryonic and adult brain give cultures by trypsinization [Pontén and McIntyre, 1968], collagenase digestion (see Chapter 11), and primary explant [Bornstein and Murray, 1958] which closely resemble glia. Astrocytic markers can be demonstrated for several subcultures, although there is only one report that cell lines from human adult normal brain lines express the most specific marker, glial fibrillary acidic protein (GFAP) [Gilden et al., 1976]. It is our experience that while some glial properties remain (high-affinity γ-aminobutyric acid and glutamate uptake, glutamine synthetase activity) GFAP is lost [Frame et al, 1980]. Oligodendrocytes do not readily survive subculture, but Schwann cells from optic nerve have been subcultured using cholera toxin as a mitogen [Brockes et al., 1979; Raff et al., 1978].

Cultures of human glioma can also be prepared by mechanical disaggregation, trypsinization, or collagenase digestion [Pontén, 1975; Freshney, 1980] (see Fig. 15.1c). The right temporal lobe from human males appears to be marginally better than other regions of the brain [Westermark et al., 1973], but most give a good chance of success. The glia/glioma system provides a good model for comparing normal and neoplastic cells under the same conditions.

There is good evidence that the cell lineage is the same, particularly between embryonic normal cells and tumor cells, but the position of the cells within the lineage, as with many cultured cells, is still debatable.

A number of gliomas have been cultured from rodents among which the C_6 deserves special mention [Benda et al., 1968]. This cell line expresses the astrocytic marker, glial fibrillary acidic protein, in up to 98% of cells [Freshney et al., 1980a] but still carries the enzymes glycerol phosphate dehydrogenase and 2′3 ′ cyclic nucleotide phosphorylase [Breen and de Vellis, 1974], both of which are oligodendrocytic markers. This appears to be an interesting example of a precursor cell tumor which can mature along two distinct phenotypic routes simultaneously.

Linser and Moscona [1980] separated the Müller cells of the neural retina from pigmented retina and neurons and demonstrated that full functional development could not be achieved unless the Müller cells (astroglia) were recombined with neurons from the retina. Neurons from other regions of the brain were ineffective.

Pigment cells were cultured successfully by Coon [Coon and Cahn, 1966] from chick pigmented retina and propagated over many generations. As with the chick embryo cartilage cells, a fraction derived from embryo extract was required for the function differentiation of these cells.

Other normal pigment cells have proven difficult to culture although cultures have been obtained from human uveal melanocytes [Meyskens et al., 1980]. Pigment cells from skin do not survive readily although cultures can be obtained from melanomas with a reasonable degree of success [Creasey et al, 1979; Mather and Sato, 1979a,b]. Primary melanomas are often contaminated with fibroblasts, but since they can be cloned on confluent feeder layers of normal cells (see Chapter 13; Fig. 16.3) [Freshney et al., 1982; Creasey et al., 1979] purification may be possible. In general, however, greater success is obtained with secondary growth from lymph nodes, or from distant metastatic recurrences. Sato [1979] has described conditions for serum-free culture of cell lines from human and murine melanoma.

Endocrine Cells

The problems of culturing endocrine cells [O'Hare et al., 1978] are similar to the culture of any other specialized cell but accentuated because the relative number of secretory cells may be quite small. Sato and colleagues [Sato and Yasamura, 1966; Buonassisi et al., 1962] cultured functional adrenal and pituitary cells from rat tumors by mechanical disaggregation of the tumor [Zaroff et al., 1961] and regular monolayer culture. The functional integrity of the cells was retained by intermittent passage of the cells as tumors in rats [Buonassisi et al., 1962; Tashjian et al., 1968]. These lines are now fully adapted to culture and can be maintained without animal passage [Tashjian, 1979], in some cases in fully defined media [Hayashi and Sato, 1976].

Fibroblasts have been reduced in cultures of pancreatic islet cells by treatment with ethylmercurithiosalicylate [Braaten et al., 1974] and have also been purified by density gradient centrifugation [Prince et al., 1978]. These cells apparently produce insulin but not as propagated cell lines.

Pituitary cells, which continue to produce pituitary hormones for several subcultures, have been isolated from the mouse [DeVitry et al., 1974], but in our experience, normal human pituitary cells do not survive well and even pituitary adenoma cells gradually lose the capacity for hormone synthesis.

HEMOPOIETIC CELLS

There have been three major milestones in this area. Bradley and Metcalf [1966], Pluznik and Sachs [1965], and McCulloch and co-workers [Wu et al., 1968] developed techniques for cloning normal hemopoietic precursor cells in agar or Methocel (see also Chapter 13), in the presence of colony stimulating factor(s) [Burgess and Metcalf, 1980]. The colonies matured during growth and could not be subcloned, implying that the colony-forming unit (CFU) was a precursor cell which was not regenerated in culture. This cell, the CFU-C ("colony-forming unit-culture") is distinct from the CFU-S (spleen colony-forming unit), which is a pluripotent stem cell present in colonies forming in the spleens of sublethally irradiated mice after bone marrow reconstitution [Wu et al., 1968].

Hence, suspension colonies, which contain cells of only one lineage, survive only as primary cultures which lose repopulation efficiency and cannot be subcultured. Granulocytic colonies are the most common; but under the appropriate conditions, lymphoid [Choi and Bloom, 1970] and erythroid [Stephenson and Axelrad, 1971] colonies can be produced.

Golde and Cline [1973] obtained survival in a liquid culture system of normal and neoplastic leukocytes at high cell densities but with abundant medium by placing the cells in a small diffusion chamber immersed in medium. Dexter has also demonstrated, in a liquid culture system, that lymphoid, granulocytic, and erythroid stem cells could be propagated from bone marrow if a bone marrow culture was first prepared and allowed to form a monolayer and to act as a feeder layer for a later, second bone marrow primary culture [Dexter et al., 1977, 1979].

The third major development which occurred over several years between the earlier suspension cloning and Dexter's liquid culture system was the development of a number of functional cell lines from hemopoietic cells. Human lymphoblastic cell lines of both B and T cell lineage were developed by Moore et al. [1967] and subsequently Epstein-Barr virus has been found to be implicated in the ability of these cell lines to become permanent. A number of myeloid cell lines have also been developed from murine leukemias [Horibata and Harris, 1970] and, like some of the human lymphoblastoid lines [Collins et al., 1977], have been shown to make globulin chains, and in some cases, complete α- and γ-globulins (see Chapter 22, Production of Monoclonal Antibodies). Some of these lines can be grown in serum-free medium [Iscove and Melchers, 1978]. T-cell lines require T-cell growth factors (Interleukins) [Burgess and Metcalf, 1980; Gillis and Watson, 1981] and B-cell growth factors have also been described [Howard et al, 1981; Snedni et al., 1981].

Originally human lymphoblastoid cell lines were derived by culturing peripheral lymphocytes from blood at very high cell densities ($\sim 10^6$/ml), usually in deep culture (> 10 mm) [Moore et al., 1967]. A monolayer culture appeared in the cell pellet at the bottom of the culture tube and eventually cells were shed into suspension and started to proliferate. This could be detected by the pH drop and the cells were then subcultured. The cell concentration was kept high initially, but eventually these cells adapted to regular culture conditons and could be passaged at 10^5 cells/ml or less. More recently the development of cell lines has become easier by the use of irradiated spleen cells, antigenic stimulation (for T-cell lines), and T- and B-cell growth factors [Paul et al., 1981; Schnook et al., 1981].

Erythroid cell lines have also been cultured from the mouse. Rossi and Friend [1967] demonstrated that a mouse RNA virus (the "Friend virus") could cause splenomegaly and erythroblastosis in infected mice. Cell cultures taken from minced spleens of these animals could, in some cases, give rise to continuous cell lines of erythroleukemia cells. All of these cell lines are transformed by what is now recognized as a complex of defective and helper virus derived from Moloney sarcoma virus [Ostertag and Pragnell, 1978, 1981]. Some cell lines can produce virus which is infective *in vivo* but not *in vitro*, and the cells can also be passaged as solid tumors or ascites tumors in DBA2 or BALB-C mice.

Treatment of cultures of Friend cells with a number of agents, including DMSO, sodium butyrate, isobutyric acid, and hexamethyl-bis-acetamide, promotes erythroid differentiation [Friend et al., 1971; Leder and Leder, 1975]. Untreated cells resemble undifferentiated proerythroblasts while treated cells show nuclear condensation, reduction in cell size, and an accumulation of hemoglobin to the extent that centrifuged cell pellets are red in color. Evidence for differentiation can also be demonstrated by staining for hemoglobin with benzidine, isolating globin-specific messenger RNA, and fluorescent antibody detection of spectrin, a specific cell surface constituent of erythrocytes, on the surface of stimulated cells [Eisen et al., 1977].

Anderson et al. [1979] have shown that the human leukemic cell line K562 can also be induced to differentiate with sodium butyrate, though not with DMSO.

Macrophages

Macrophages may be isolated from many tissues by collecting the cells that attach *during* enzymatic disaggregation. The yield is rather low, however, and a number of techniques have been developed to obtain larger numbers of macrophages. Mineral oil or thioglycollate broth [Adams, 1979] may be injected into the peritoneum of a mouse, and 3 days later the peritoneal washings contain a high proportion of macrophages.

If necessary, macrophages may be purified by their ability to attach to the culture substrate in the presence of proteases, as above. They can only be subcultured with difficulty because of their insensitivity to trypsin. Methods have been developed using hydrophobic plastics, e.g., Petriperm dishes (Heraeus).

There are some reports of propagated lines of macrophages mostly from murine neoplasia [Defendi, 1976]. Normal mature macrophages do not proliferate although it may be possible to culture replicating precursor cells by the method of Dexter (see above).

GONADS

Culture of germ cells has on the whole been disappointing. Ovarian granulosa cells can be maintained and are apparently functional in primary culture [Orly

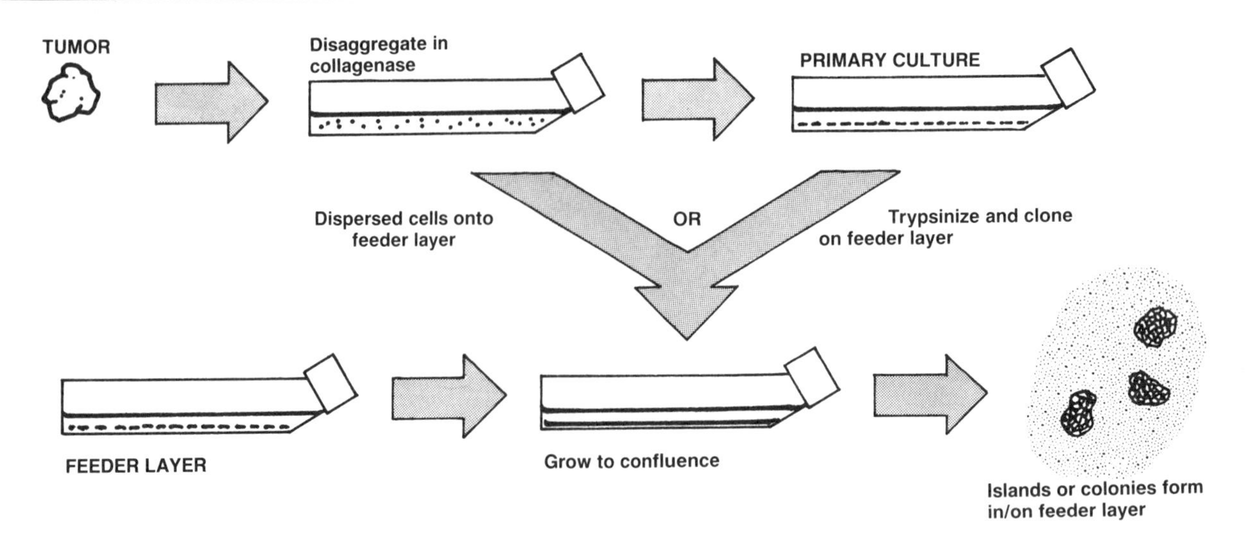

Fig. 20.2. *Selective growth on confluent feeder layers. Colonies of epithelial cells will form on confluent fetal human intestinal epithelium or on confluent normal human glia.*

et al., 1980], but specific functions are lost on subculture. A cell line started from Chinese hamster ovary (CHO-K1) [Kao and Puck, 1968] has been in culture for many years but its identity is still not confirmed. Although epithelioid at some stages of growth, it undergoes a fibroblasticlike modification when cultured in dibutyryl cyclic AMP [Ilsie and Puck, 1971].

Cellular fractions from testis have been separated by velocity sedimentation at unit gravity, but prolonged culture of these has not been reported. The TM4 is an epithelial line from mouse testis although its differentiated features have not been reported, and Sertoli cells have also been cultured from testis [Mather, 1979].

MINIMAL DEVIATION TUMORS

Several cell lines have been derived from the Reuber and Morris hepatomas of the rat [Pitot et al., 1964; Granner et al., 1968] (see above) adrenal cortex and pituitary [Sato and Yasamura, 1966] (as described above) and provide a valuable, if rare, source of continuous cell lines with differentiated properties.

TERATOMAS

When cells from an embryo are implanted into the adult, e.g., under the kidney capsule, these can give rise to tumors known as teratomas. Teratomas also arise spontaneously when groups of embryonic cells or single cells are carried over into the adult, often at an inappropriate site.

Artificially derived teratomas have been used extensively to study differentiation [Martin, 1975, 1978; Martin and Evans, 1974], as they may develop into a variety of different cell types (muscle, bone, nerve, etc.). Growth of teratoma cells on feeder layers of, for example, SC1 mouse fibroblasts, will proliferate but not differentiate, whereas when grown on gelatin without feeder layer, or in nonadherent plastic dishes, nodules form which eventually differentiate.

TUMOR TISSUE

In general, follow the techniques associated with the tissue from which the tumor was derived. Methods for the selective growth of tumor cells have followed the same course as for the suppression of fibroblastic overgrowth (Table 20.1) although methods based on anchorage independence may be more selective. Thus, growth on hydrophobic surfaces [e.g., Parenjpe et al., 1975], on polyacrylamide [Jones and Haskill, 1973, 1976] or in suspension in agar or methocel (see Chapters 13 and 16), may be selective for transformed cells. However, it is not clear whether these techniques are applicable to cultures from spontaneously arising tumors; on the whole it would seem not, although suspension cloning may work in some cases (e.g., with epithelial tumors) [see also Freshney et al., 1982].

We have found the confluent feeder layer technique (Fig. 20.2) to be effective in suppressing fibroblastic overgrowth in many epithelial tumors as reported by Stanley and Parkinson [1979] (cervix) and Rheinwald and Beckett [1981] (basal cell carcinoma), and in suppressing normal glial overgrowth in glioma (see also Chapter 13).

Chapter 21
Specialized Techniques. I

It is intended in this chapter to provide detailed information on some specialized techniques referred to in the text but so far not described. In some cases, these are not tissue culture techniques *per se*, but rather techniques associated with tissue culture and which might be used in a number of different tissue culture-based experiments, e.g., autoradiography. Other techniques involving tissue culture directly, but of a very specialized nature, e.g., monoclonal antibody production, will be described, rather more briefly, in the next chapter.

ORGAN CULTURE

Organ culture has developed into a separate discipline [Thomas, 1970] with a technology of its own. As discussed in the first chapter, organ culture favors the retention of tissue architecture, consequent cell interactions, and differentiated structure. Since organ cultures are thicker than cell cultures, they cannot rely on the low O_2 tension found at the bottom of the medium in a culture flask and they must be placed at the gas/liquid interface to maximize gas exchange [Trowell, 1959]. This is achieved by placing a small organ or tissue sample on the surface of gelled medium (plasma or agar) or on a filter and/or grid positioned at the surface of the medium (Fig. 21.1). The second technique utilizes liquid medium and makes feeding the culture simpler.

Outline

Dissect out organ or tissue, reduce to 1mm^3, or to thin membrane or rod, place on support at gas (air)/medium interface, and incubate in humid CO_2 incubator, changing the medium as required.

Materials

instruments
culture grids
sterile filters, 0.5 μm Nucleopore (sterilize by autoclaving)
medium
organ culture dishes (Falcon #3010)

Protocol

1. Prepare grids (Falcon) with sterile filters (Nucleopore) in position (see Fig. 21.7) on dishes (Falcon)
2. Add enough medium to wet the filter but not so much that it floats (~ 1.1 ml)
3. Place dishes in humid CO_2 incubator to equilibrate at 36.5°C
4. Prepare tissue or dissect out whole embryonic organs, e.g., 8-day femur or tibiotarsus of chick embryo (see Chapter 11 for dissection). Tissue must not be more than 1 mm thick, preferably less, in one dimension, e.g., 8-day embryonic tibiotarsus is perhaps 5 mm long but only 0.5–0.8 mm in diameter. A fragment of skin might be 10 mm^2 but only 200 μm thick. Tissue like liver or kidney which must be chopped down to size should be no more than 1 mm^3.

   For short dissections. HBSS is sufficient; but for longer dissections, use 50% serum in HBSS buffered with HEPES to pH 7.4
5. Take dishes from incubator and transfer tissue carefully to filters. A pipette is usually best and can be used to aspirate any surplus fluid transferred with the explant

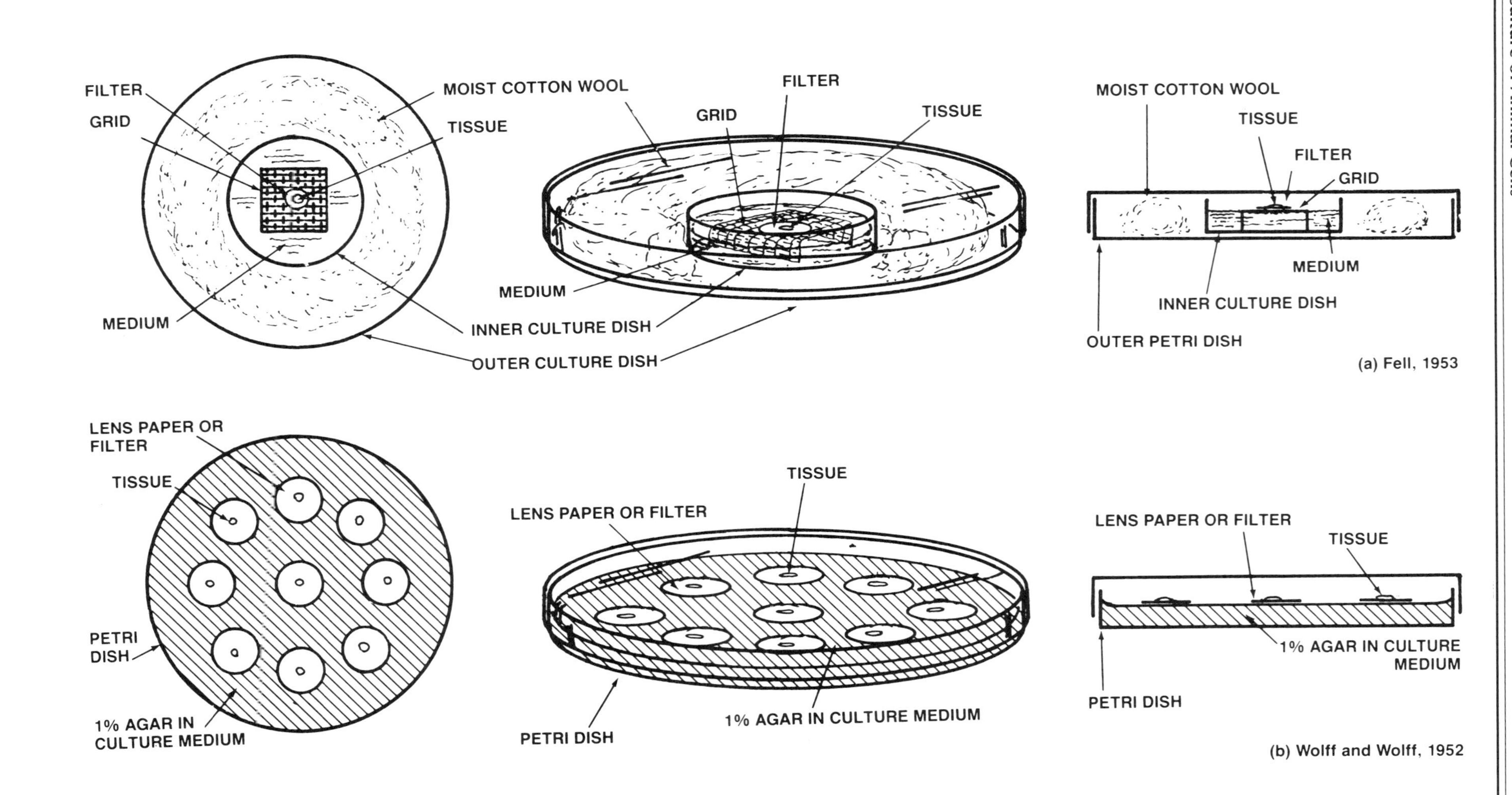

FILTER
GRID
MOIST COTTON WOOL
TISSUE
MEDIUM
INNER CULTURE DISH
OUTER CULTURE DISH
FILTER
GRID
TISSUE
MEDIUM
MOIST COTTON WOOL
TISSUE
FILTER
GRID
MEDIUM
INNER CULTURE DISH
OUTER PETRI DISH
(a) Fell, 1953
LENS PAPER OR FILTER
TISSUE
PETRI DISH
1% AGAR IN CULTURE MEDIUM
TISSUE
LENS PAPER OR FILTER
PETRI DISH
1% AGAR IN CULTURE MEDIUM
LENS PAPER OR FILTER
TISSUE
1% AGAR IN CULTURE MEDIUM
PETRI DISH
(b) Wolff and Wolff, 1952

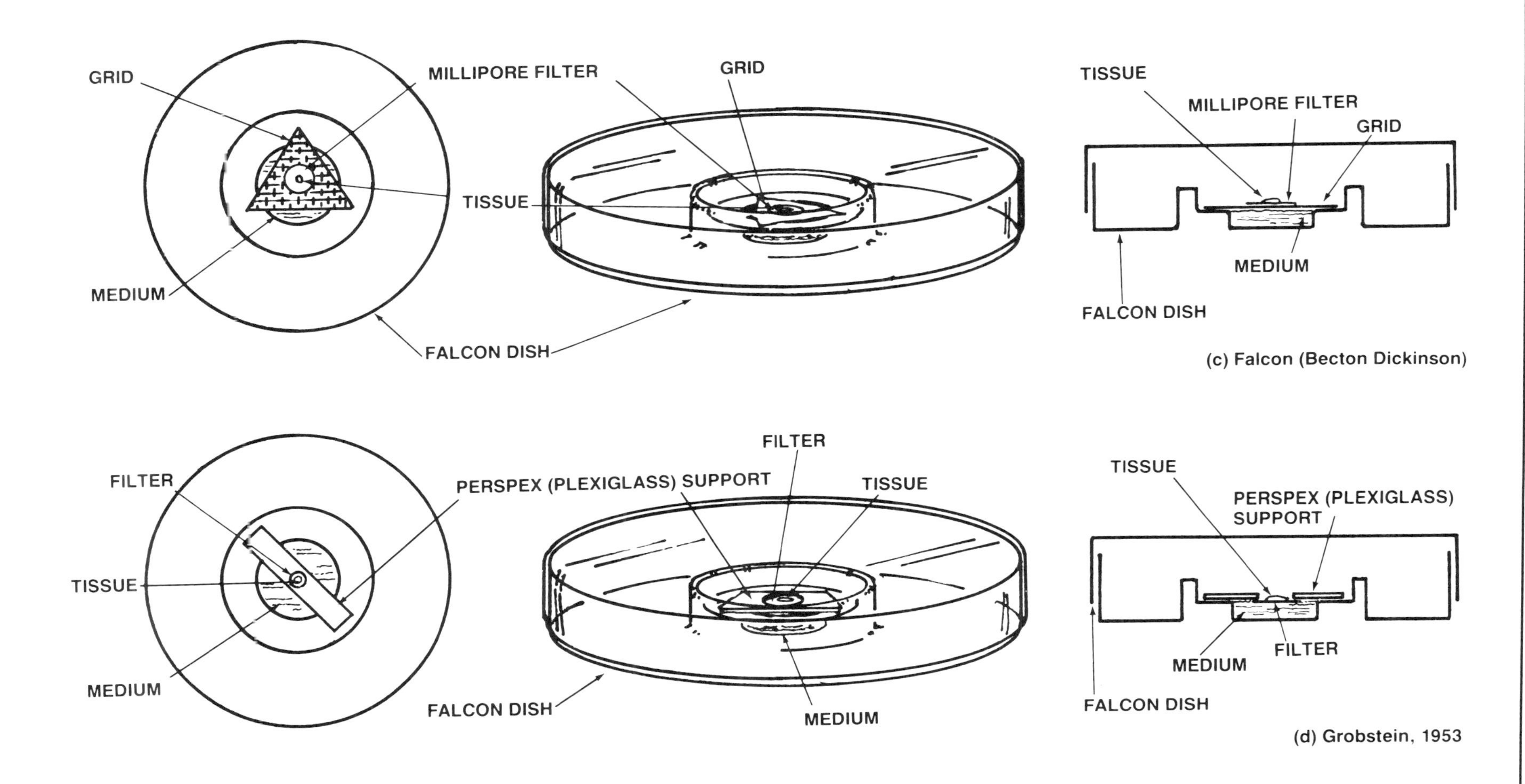

Fig. 21.1. *Types of organ culture. a. Wire gauze support in liquid medium [Fell, 1953; Trowell, 1959]. b. On filter or lens paper rafts on agar medium [Wolff and Wolff, 1952]. c. Falcon organ culture dish (#3010) with stainless steel mesh support. d. Modification for embryonic induction studies [Grobstein, 1953].*

6. Check level of medium, making sure tissue is wetted and return dishes to incubator
7. Incubate for 1–3 w, changing medium every 2 or 3 days

Analysis. Usually by histology, autoradiography, or immunocytochemistry, but assay of total amounts of cellular constituents or enzyme activity is possible, although variation between replicates will be high.

Variations. Most variations are in (1) type of medium; 199 may be used with or without serum, and BJC [Biggers et al., 1961] for cartilage or bone; (2) type of support (see Fig. 21.1); (3) O_2 tension: Embryonic cultures are usually best kept in air, but late-stage embryos, newborn, and adult tissue in 95% O_2 [Trowell, 1959; de Ridder and Mareel, 1978].

Organ cultures are useful in the demonstration of processes such as embryonic induction [e.g., Grobstein, 1953; Cooper, 1965], where the maintenance of the integrity of whole tissue is important. However, they are slow to prepare and present problems of reproducibility between samples. Growth is limited by diffusion (although growth is perhaps not necessary and may even be undesirable) and mitosis is nonrandomly distributed throughout the explant. Mitosis occurs round the periphery only, while the centers of explants frequently become necrotic (Fig. 21.2). It has been argued that this type of geometry makes organ cultures good models of tumor growth, where peripheral cell division is often accompanied by central necrosis.

HISTOTYPIC CULTURE

Various attempts have been made to regenerate tissuelike architecture from dispersed monolayer cultures. Kruse and Miedema [1965] demonstrated that perfused monolayers could grow to more than ten cells deep and organoid structures can develop in multilayered cultures if kept supplied with medium [Schneider et al., 1963; Bell et al., 1979]. Green [1978] has shown that human epidermal keratinocytes will form dematoglyphs (friction ridges) if kept for several weeks without transfer, and Folkman and Haudenschild [1980] were able to demonstrate formation of capillary tubules in cultures of vascular endothelial cells cultured in the presence of endothelial growth factor and medium conditioned by tumor cells.

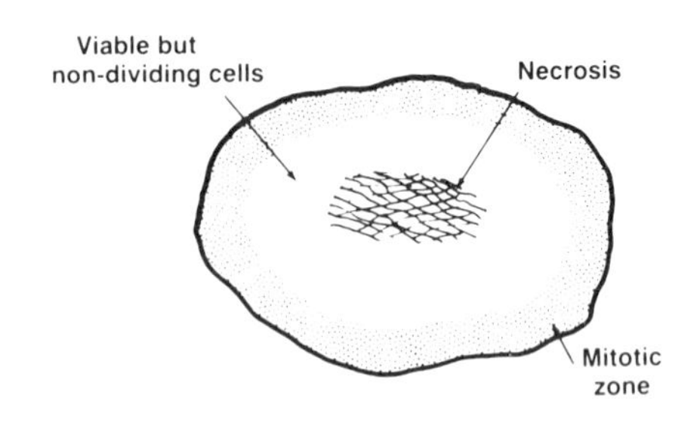

Fig. 21.2. *Diagrammatic representation of the expected distribution of mitoses (stippled area) and necrosis (shaded central area) in an organ culture explant.*

Capillary Bed Perfusion

Since medium supply and gas exchange become limiting at high cell densities, Knazek et al. [1972; Gullino and Knazek, 1979] developed a perfusion chamber from a bed of plastic capillary fibers. This is now available commercially from Amicon. The fibers are gas- and nutrient-permeable and support cell growth on their outer surfaces. Medium, saturated with 5% CO_2 in air, is pumped through the centers of the capillaries, and cells are added to the outer chamber surrounding the bundle (Fig. 21.3; see also Fig. 8.3). The cells attach and grow on the outside of the capillary fibers fed by diffusion from the perfusate and can reach tissue-like cell densities. There is an option between two types of plastic and different ultrafiltration properties giving molecular weight cut-off points at 10,000, 50,000 or 100,000 daltons, regulating the diffusion of macromolecules from the medium to the cells.

It is claimed that cells in this type of high-density culture behave as they would *in vivo.* Choriocarcinoma cells release more human chorionic gonadotrophin [Knazek et al., 1974] and colonic carcinoma cells produce elevated levels of CEA [Rutzky et al., 1979; Quarles et al., 1980]. There are considerable technical difficulties in setting up the chambers, however, and they are costly.

Sampling cells from these chambers and determination of the cell concentration are also difficult. However, they appear to present an ideal system for studying the synthesis and release of biologically generated compounds.

Reaggregation and Spheroids

When dissociated cells are cultured in a gyratory shaker, they may reassociate into clusters. Dispersed cells from embryonic tissues will sort during reaggre-

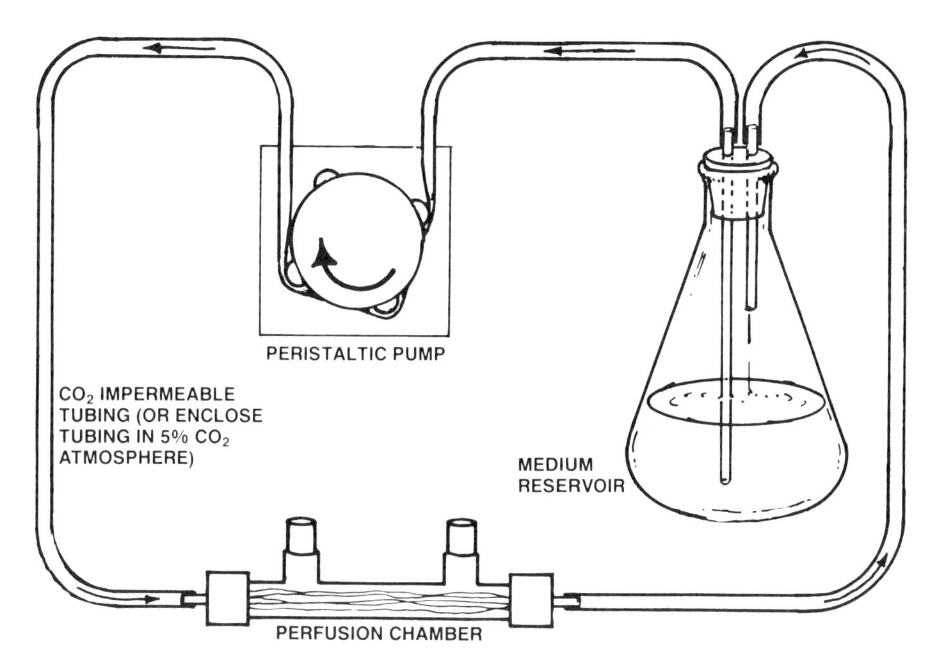

Fig. 21.3. *Vitafiber (Amicon) apparatus for perfused culture on capillary bundles of permeable plastic. Medium is circulated from a reservoir to the culture chamber by a peristaltic pump. As silicone tubing is gas permeable, the apparatus should be enclosed in an atmosphere of 5% CO_2. Alternatively, the length of tubing between the pump and the culture chamber may be enclosed in a polyethylene bag purged with CO_2.*

gation in a highly specific fashion [Linser and Moscona, 1980]—e.g., Müller cells of the chick embryo retina reaggregated with neuronal cells from the retina were inducible for glutamine synthetase; but those reaggregated with neurons from other parts of the brain were not. Cells in these heterotypic aggregates appear to be capable of sorting themselves into groups and forming tissue-like structures. This property is less easily demonstrated in adult cells, although some results suggest that it may be possible for adult cells to form organoid structures [Douglas et al., 1976, 1980; Bell et al., 1979].

Homotypic reaggregation also occurs failry readily, and spheroids generated in gyratory shakers or by growth on agar have been used as models for chemotherapy *in vitro* [Twentyman, 1980] and for the characterization of malignant invasion [Mareel et al., 1980]. As with organ cultures, growth is limited by diffusion and a steady state may be reached where cell proliferation in the outer layers is balanced by central necrosis.

MASS CULTURE TECHNIQUES

While organ culture and perfusion may be directed at induction or retention of specific differentiated functions, there is also a requirement for the mass production of cells whose differentiated properties are either unimportant (e.g., as virus substrate) or expressed spontaneously (e.g., immunoglobulins in lymphoblastoid cell lines). Mass culture might be defined as encompassing from 10^9 cells to semi-industrial pilot plant (10^{11} or 10^{12} cells). The method employed depends on whether the cells proliferate in suspension or require to be anchored to the substrate.

Suspension Culture

Increasing the bulk of suspension cultures is relatively simple since only the volume need be increased. Above about 5-mm-deep agitation of the medium is necessary, and above 10 cm sparging with CO_2 and air is required to maintain adequate gas exchange (Fig. 21.4). Stirring of such cultures is best done slowly with a large surface area paddle or large diameter magnetic stirrer bar. The stirring speed should be between 30 and 100 rpm, sufficient to prevent cell sedimentation but not so fast as to grind or shear the cells. If a bar is used, it must be kept off the base of the culture vessel with a collar or be suspended from above. Antifoam (Dow Chemical Co.) must be included where the serum concentration is above 2% particularly if the medium is sparged. In the absence of serum, it may be necessary to increase the viscosity

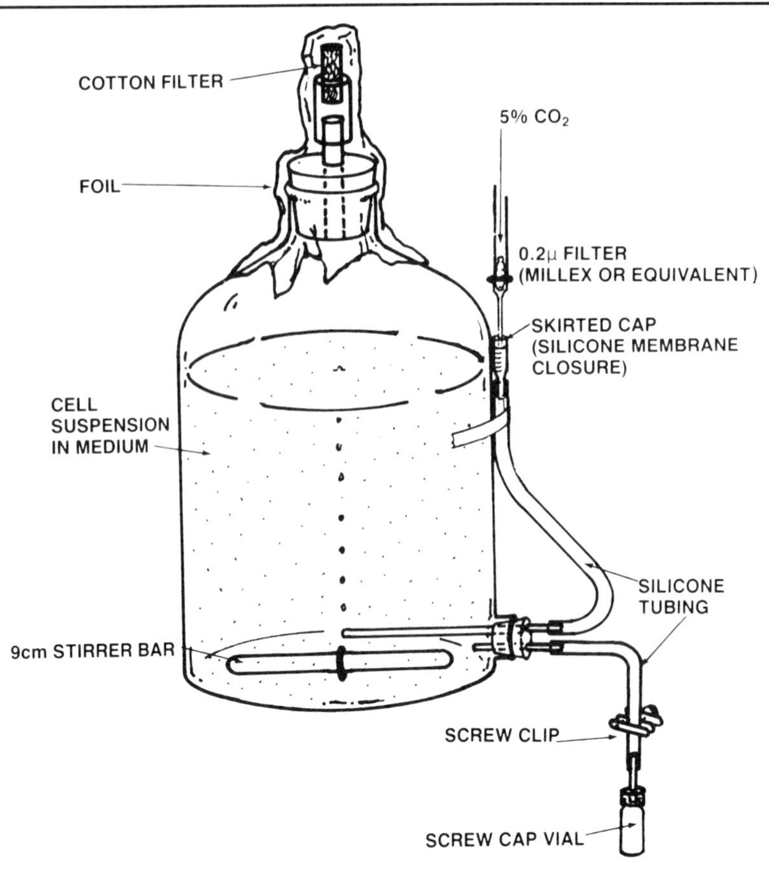

Fig. 21.4. *Bulk culture of cells in suspension. Standard 5- or 10-l aspirators may be modified as illustrated. Optimum mixing is achieved by using a large stirrer bar, with a central collar, and a slow stirring speed (~ 60 rpm). (Apparatus developed by the staff of the Beatson Institute for Cancer Research, Glasgow, Scotland.)*

of the medium with (1–2%) carboxymethyl cellulose (molecular weight $\sim 10^5$).

The procedure for setting up a 4-l culture of suspended cells is as follows:

Outline

Grow pilot culture of cells and add to prewarmed, pregassed aspirator of medium. Stir slowly with sparging until required cell concentration is reached and harvest.

Materials

- medium with antifoam
- pilot culture
- prepared aspirator (Fig. 21.4)
- magnetic stirrer
- supply of 5% CO_2
- bunsen
- gas lighter
- counting fluid
- cell counter or hemocytometer

Protocol

1.

Prepare a standard 5- or 10-l aspirator as in Figure 21.4, with a one-holed silicon rubber stopper at the top and a two-holed stopper at the bottom. The top stopper carries a glass tube with a cotton plug and the bottom stopper has (a) a glass tube and silicone rubber connection to an inlet port closed with a silicone membrane closure ("skirted cap") and (b) a screw cap fixed to a tube leading from the stopper with a screw cap vial inserted in the cap.

A Teflon-coated bar magnet is placed in the aspirator. It should be as large as possible, while still able to turn freely in the bottom of the aspirator (~ 9–12 cm) and have a central collar to raise it up from the base of the aspirator to avoid grinding cells below the bar.

Sterilize by autoclaving 100 kPa (15 lb/in^2) for 20 min

2.

Set up "starter culture," using a standard screw-capped 1-l reagent bottle, or equivalent, with a Teflon-coated bar magnet in the bottom (see Fig. 12.2). Add 400 ml medium and seed with cells at 5×10^4–10^5/ml. Place on magnetic stirrer rotating at 60 rpm and incubate until 5×10^5–10^6 cells/ml is reached

3. Add 4-l medium and 0.4 ml antifoam to sterile aspirator. The antifoam should be added directly to the aspirator using a disposable pipette or syringe. Place on magnetic stirrer in 36.5°C room or incubator. Connect 5% CO_2 air line via fresh, sterile, 25 mm, 0.2 μm Millex filter to sterile disposable hypodermic needle and insert needle into skirted cap. Turn on gas at a flow rate of approximately 10–15 ml/min and stir at 60 rpm. Incubate for about 2 hr to allow temperature and CO_2 tension to equilibrate

4. Bring aspirator and starter culture back to laminar flow hood, remove top stopper, keeping aluminum foil in place, and starter bottle cap, taking care to keep stopper sterile. Flame neck of aspirator and starter culture bottle and pour starter culture into aspirator

5. Replace stopper in aspirator and return to incubator or hot-room. Reconnect 5% CO_2 line and restart stirrer at 60 rpm. Adjust gas flow to 10–15 ml/min

6. Incubate for 4–7 d, sampling (see below) every day to check cell growth. When cell concentration reaches desired level, disconnect aspirator, run off cells into centrifuge bottles, and centrifuge at 100 *g* for 10 min

Sampling. Open screw clip and run 5–10 ml into vial to remove cells and medium which have been stagnant in the delivery line. Discard vial and contents and replace with fresh vial. Collect second 5–10 ml, perform a cell count, and check viability by dye exclusion.

Analysis. For best results, cells should not show a lag period of more than 24 hr and should still be in exponential growth when harvested. Plot cell counts daily and harvest at approximately 10^6 cells/ml.

Variations. ***Continuous culture—"Biostat."*** If it is required that the cells be maintained at a set concentration, e.g., at mid-log phase, cells may be removed and medium added daily or cells may be run off and medium added continuously using the skirted cap entry port to add medium and the screw cap outlet to collect into a larger reservoir (Fig. 21.5). The volume of medium that must be added daily may be calculated from the growth rate; a volume of medium equivalent to the total volume of the aspirator culture must be added every time the cells double, i.e., for a steady state culture $R = V/D$ where R is the flow rate of medium into the aspirator (and the flow rate of cells leaving the aspirator), V = total volume of culture in ml, and D = doubling time of culture in minutes. When this is achieved, the cell concentration in the culture (and the effluent) should remain constant. Thus, a 4-l culture with a doubling time of 24 hr will need 4 l of fresh medium per day or 2.8 ml/min.

Production of cells in bulk is best done by the "batch" method outlined first. The "steady-state" method is required for monitoring metabolic changes related to cell density but is more expensive in medium and is more likely to lead to contamination.

Suspension cultures can also be grown in bottles rotating on a special rack as for monolayer cultures (see below).

Monolayer cells. Anchorage-dependent cells cannot be grown in liquid suspension except on microcarriers (see below), but transformed cells, e.g., virally transformed or spontaneously transformed continuous cell lines, can. Because these cells are still capable of attachment, the culture vessels will require to be coated with a water-repellent silicone (e.g., Repelcote) and the calcium concentration may need to be reduced. MEMS medium is a variation of Eagle's MEM with no calcium in the formulation, which has been used for the culture of HeLa-S_3 and other cells in suspension.

Monolayer Culture

For anchorage-dependent monolayer cultures, it is necessary to increase the surface area of the substrate in proportion to the cell number and volume of medium. This requirement has prompted a variety of different strategies, some simple, others complex.

Nunclon cell factory. The simplest system for scaling up monolayer cultures is the Nunclon Cell Factory (Figs. 21.6, 21.7) (see Table 8.1). This is made up of ten rectangular petri dish–like units, total surface area 6,500 cm^2, interconnected at two adjacent corners by vertical tubes. Because of the positions of the apertures in the vertical tubes, medium can only flow between

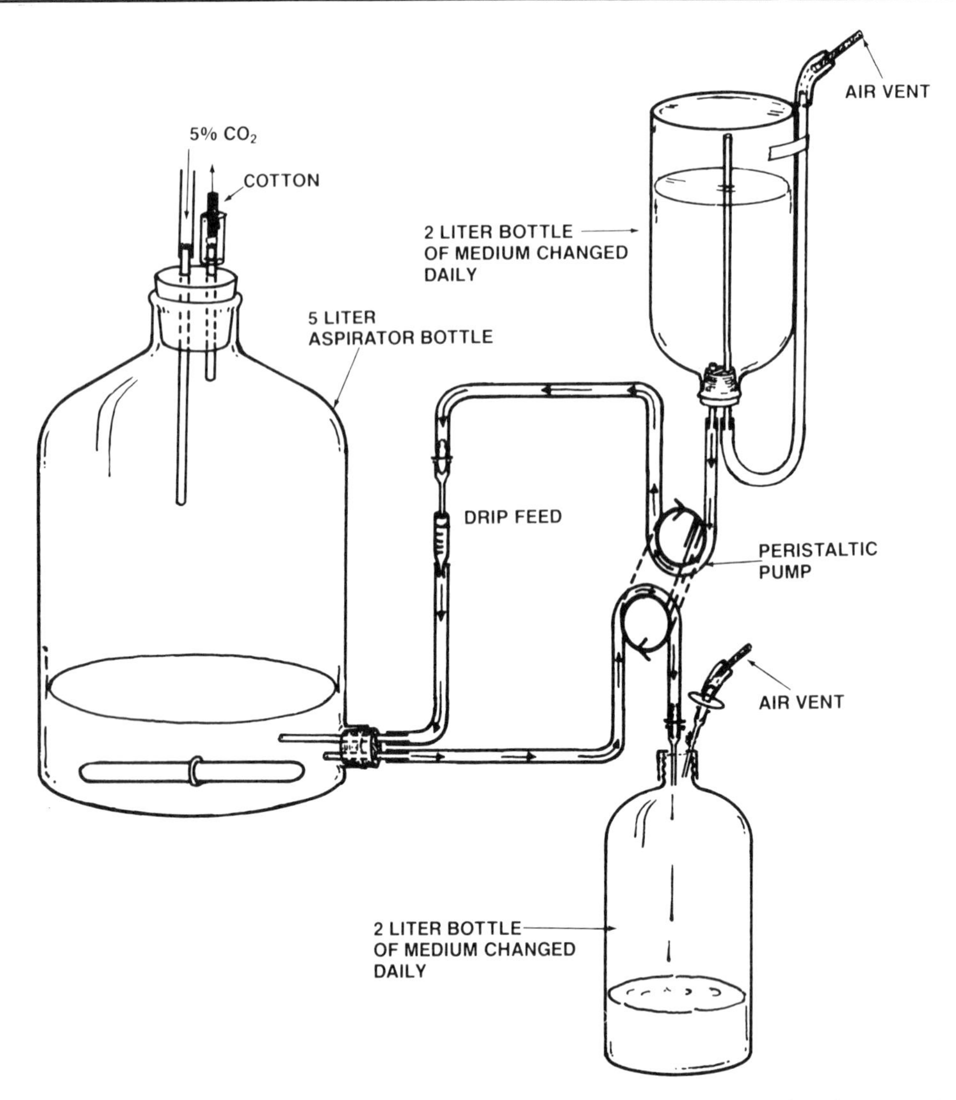

Fig. 21.5. *"Biostat." A modification of the suspension culture vessel of Figure 21.4, with continuous matched input of fresh medium and output of cell suspension. The objective is to keep the culture conditions constant rather than to produce large numbers of cells. Bulk culture,* per se, *is best performed in batches in the apparatus in Figure 21.4.*

compartments when the unit is placed on end. When the unit is rotated and laid flat, the liquid in each compartment is isolated, although the apertures in the interconnecting tubes still allow connection of the gas phase. The cell factory has the advantage that it is not different in the geometry or the nature of its substrate from a conventional flask or petri dish. The recommended method of use is as follows:

Outline

Prepare a cell suspension in medium and run into the chambers of the unit. Lay the unit flat and gas with CO_2. Seal and incubate.

Materials

monolayer cells
medium
0.25% crude trypsin
PBS
hemocytometer or cell counter and counting fluid
culture chamber
silicone tubing and connectors

Protocol

1. Trypsinize cells (Chapter 12), resuspend, and dilute to 2×10^4 cells/ml in 1,500 ml medium

2.
Place chamber on long edge with supply tube to the bottom (see Fig. 21.7) and run cells and medium in through supply tube. Medium in all chambers will reach the same level
3.
Clamp off supply tube and disconnect from medium reservoir
4.
Rotate unit through 90° in the plane of the monolayer so that it lies on a short edge with the supply tube at the top

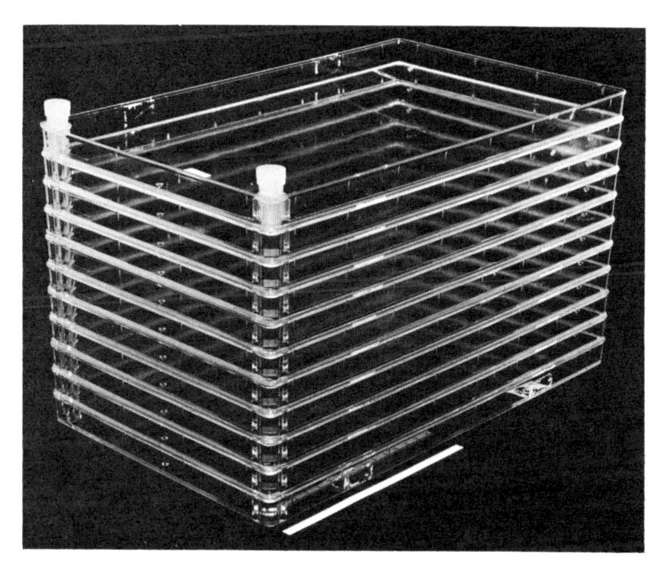

Fig. 21.6. *Nunclon "Cell Factory."*

5.
Rotate unit through 90° perpendicular to the plane of the monolayer so that it now lies flat on its base with the culture surfaces horizontal. To transport to incubator, tip medium away from supply port
6.
If it is necessary to gas the culture, loosen clamp on supply line and purge unit with 5% CO_2 in air for 5 min, then clamp off both supply and outlet. The unit may be gassed continuously if desired
7.
To change medium (or collect medium), reverse step 5 and then 4, flame clamped line, open clamp, and drain off medium
8.
Replace medium as in steps 2–6
9.
To collect cells, remove medium as in step 7, add 500 ml PBS, and remove. Add 500 ml trypsin at 4°C and remove after 30 s. Incubate, add medium after 15 min, and shake to resuspend cells. Run off cells as in step 7
10.
The residue may be used to seed the next culture, although this does make it difficult to control the seeding density. It is better to discard the chamber and start fresh

Analysis. Following growth in these chambers is difficult, so a single tray or chamber is supplied to act as a pilot culture. It is assumed the single tray will behave as the multichamber unit.

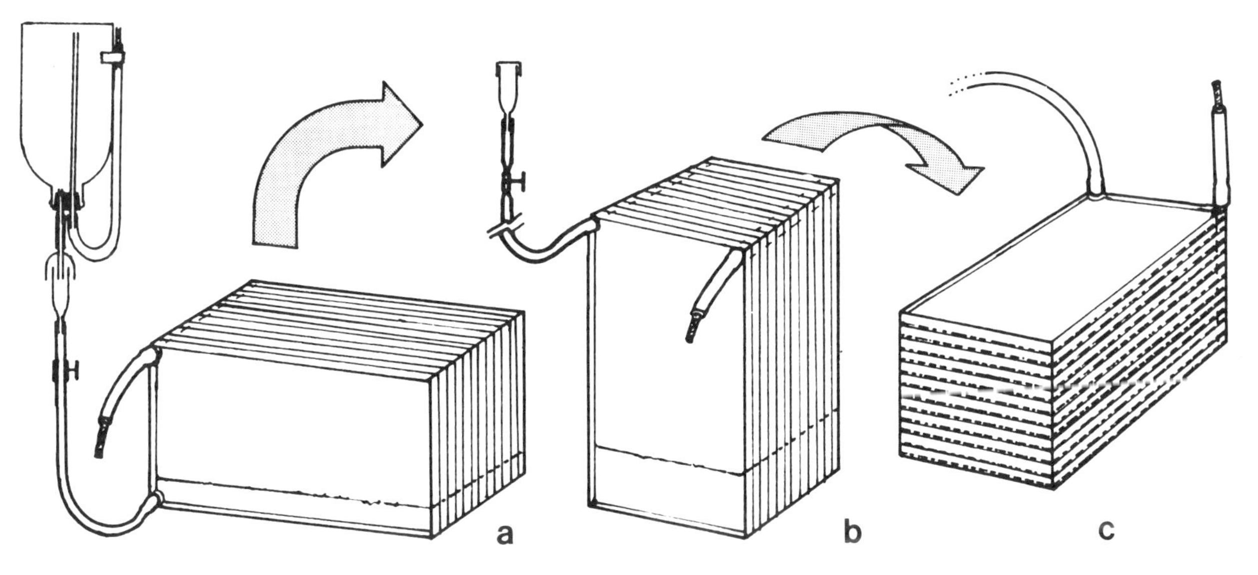

Fig. 21.7. *Filling Nunclon Cell Factory. a. Run medium in. b. Rotate onto short side away from inlet. c. Lay down flat, seal inlet or connect to 5% CO_2 line.*

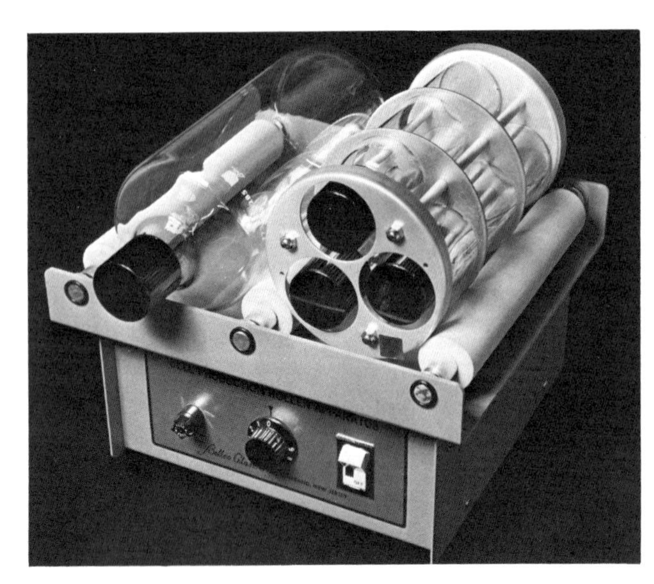

Fig. 21.8. *Roller culture bottles on rack. Small, bench top rack (Bellco).*

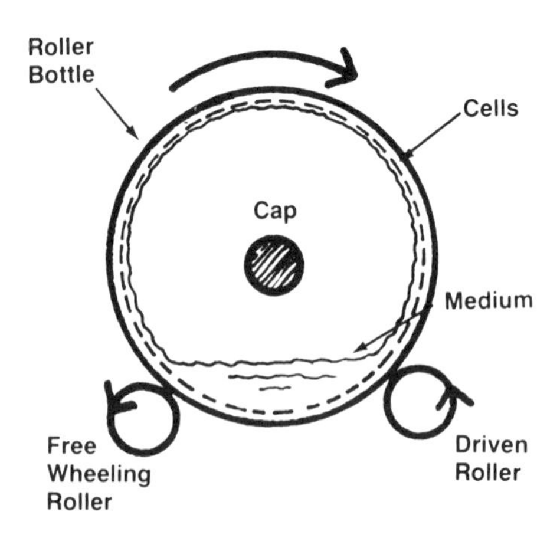

Fig. 21.9. *Roller bottle culture. Cell monolayer (dotted line) is constantly bathed in liquid but only submerged for about one-fourth of the cycle, enabling free gas exchange.*

The supernatant medium can be collected repeatedly for virus or cell product purification. Collection of cells for analysis depends on the efficiency of trypsinization.

This technique has the advantage of simplicity but can be expensive if the unit is discarded each time cells are collected. It was designed primarily for harvesting supernatant medium but is a good method for producing large numbers of cells (3×10^8 – 3×10^9) for a pilot run, or on an intermittent basis.

Roller culture. If cells are seeded into a round bottle or tube which is then rolled around its long axis, the medium carrying the cells runs around the inside of the bottle (Figs. 21.8, 21.9). If the cells are nonadhesive, they will be agitated by the rolling action but remain in the medium. If the cells are adhesive they will gradually attach to the inner surface of the bottle and grow to form a monolayer. This system has three major advantages over static monolayer culture: (1) the increase in surface area, (2) the constant, but gentle, agitation of the medium, and (3) the increased ratio of medium surface area to volume, allowing gas exchange to take place at an increased rate through the thin film of medium over cells not actually submerged in the deep part of the medium.

Outline

Seed cell suspension in medium into round bottle and rotate slowly on a roller rack.

Materials

- medium and medium dispenser
- PBSA
- 0.25% crude trypsin
- monolayer cultures
- hemocytometer or cell counter and counting fluid
- roller bottles
- supply of 5% CO_2
- roller apparatus

Protocol

1. Trypsinize cells and seed at usual density

   Note. The gas phase is large in a roller bottle so it may be necessary to blow a little 5% CO_2 into the bottle (e.g., 2 s at 10 l/min). If medium is CO_2/HCO_3^- buffered, then the gas phase should be purged with 5% CO_2 (30 s to 1 min at 20 l/min depending on the size of the bottle; see Chapter 12).

2. Rotate bottle slowly around its axis at 20 rev/hr until cells attach (24–48 hr)

3. Increase rotational speed to 60–80 rev/hr as cell density increases

4. To feed or harvest medium, take bottles to sterile work area and draw off medium as usual and replace with fresh medium. A transfusion device

Fig. 21.10. *Examples of roller culture bottles. Center and left, disposable plastic (Falcon, Corning); right, glass.*

Fig. 21.11. *Roller drum apparatus for roller culture of large numbers of small bottles or tubes (New Brunswick Scientific).*

(see Fig. 6.1) is useful for adding fresh medium, provided that the volume is not critical. If the volume of medium is critical, it may be dispensed by pipette or metered by a peristaltic pump (Camlab, Jencons) (see Chapter 6)

5.

To harvest cells, remove medium, rinse with 50–100 ml PBSA and discard PBSA, add 50–100 ml trypsin at 4°C, roll for 15 s, by hand or on rack at 20 rpm. Draw off trypsin, incubate the bottle for 5–15 min, add medium, shake, and/or wash off cells by pipetting

Analysis. Monitoring cells in roller bottles can be difficult, but it is usually possible to see cells on an inverted microscope. With some microscopes, the condensor needs to be removed, and with others, the bottle may not fit on the stage. Choose a microscope with sufficient stage accommodation.

For repeated harvesting of large numbers of cells or for repeated collection of supernatant medium, the roller bottle system is probably most economical, although it is labor intensive and requires investment in a bottle rolling unit ("roller rack"; Fig. 21.8,9).

Variations. *Aggregation.* Some cells may tend to aggregate before they attach. This is difficult to overcome but may be improved by reducing the initial rotational speed to 5 or even 2 rev/hr or trying a different type or batch of serum.

Size. A range of bottles, both disposable and reusable, are available (see Table 8.1, Figs. 21.8, 21.10).

Volume. Medium volume may be varied. A low volume will give better gas exchange and may be better for untransformed cells. Transformed cells, which are more anaerobic and produce more lactic acid, may be better in a larger volume. The volumes given in Table 8.1 are mean values and may be halved or doubled as appropriate.

Mechanics. The system where bottles are supported on rollers (see Fig. 21.8) is now the most popular because it is most economical in space. Roller drums (see Fig. 21.11), once popular, demand more space. They are still used for smaller bottles or tubes.

Spiral Propagator (Sterilin). [House et al., 1972; see also House, 1973] (Fig. 21.12) The sterile propagator comprises a spiral coil of tissue culture-treated polystyrene sheet inside a plastic container. Cells adhere to and grow on both surfaces of the sheet giving a total area of 8,500 cm^2. Mixing of the medium and adequate gas exchange is achieved by sparging with 5% CO_2 in air.

Outline

Cells are suspended in medium, and run into the chamber, which is first rolled around its long axis to allow the cells to adhere and then placed upright and sparged for the remainder of the culture period.

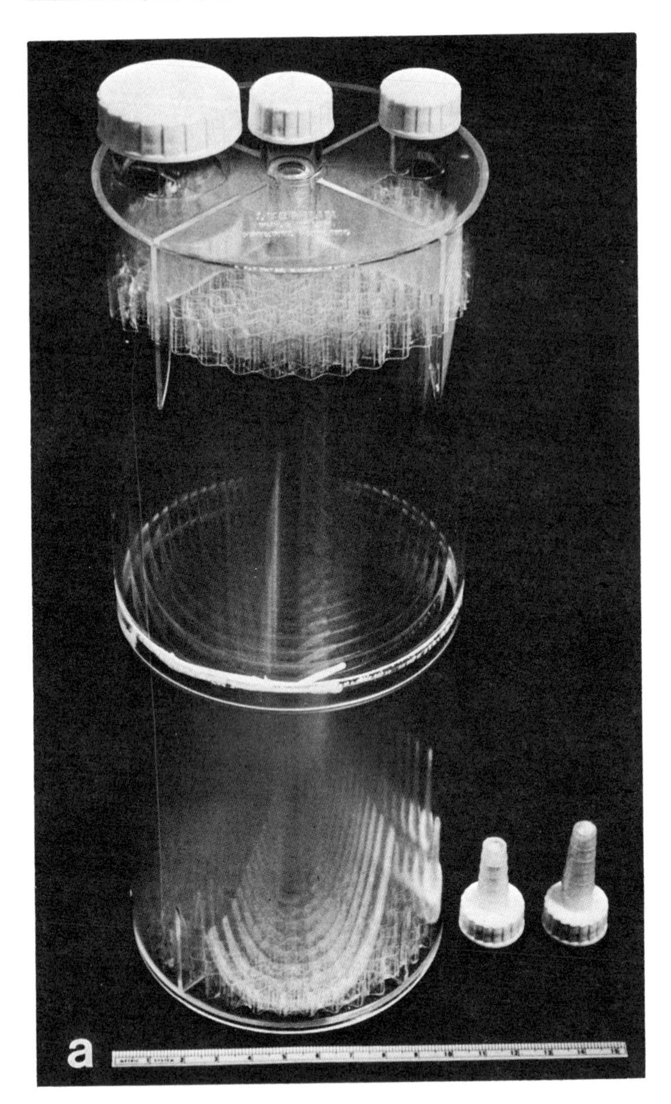

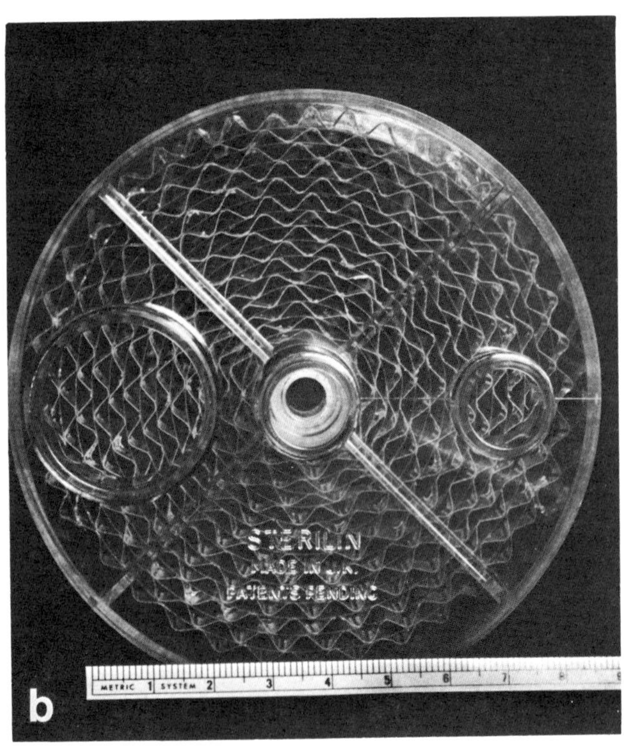

Fig. 21.12. *Sterilin spiral culture vessel, developed in collaboration with the Imperial Cancer Research Fund Laboratories, Lincoln's Inn Fields, London. The vessel is filled with cell suspension in medium and rolled slowly (2–5 rev/hr) overnight to allow cells to attach to both sides of spiral film. It is then incubated vertically and 5% CO_2 pumped through slowly to mix and gas the medium a. Side view with gassing connections. b. Top view showing spiral.*

Materials

- medium
- dispenser
- PBSA
- 0.25% crude trypsin
- monolayer culture
- hemocytometer or cell counter and counting fluid
- Sterilin spiral culture vessel and connectors
- roller rack capable of 4 rev/hr (it may be necessary to make rings to go round the culture vessel if the distance between the rollers is too great)

Protocol

1. Trypsinize cells and dilute to required concentration in 1,500 ml CO_2-buffered medium with antifoam (e.g., 2×10^4/ml, 3×10^7 cells total for 3T3)
2. Run medium plus cells into chamber with bottle and dispenser device (see Fig. 6.1)
3. Place screw caps in place and tighten
4. Rotate slowly (4 rev/hr) around long axis for 24 hr
5. Stand flask on end and connect center inlet to 5% CO_2 line and cotton-plugged outlet to outer narrow opening. Bubble gas through at 20–50 ml/min (~2 bubbles per second)
6. Replacement of medium is not recommended; but if necessary, pour off from outer narrow outlet and refill as in step 2 above

7.
To harvest cells, pour off medium, add 500 ml trypsin and run around inside of coils, checking by eye that all surface is covered
8.
Incubate on roller for 10–15 min
9.
Pour off cells and centrifuge to remove trypsin
10.
To reseed, rinse out with 1 l of BSS, pour off and repeat from step 1. Reseeding may increase the risk of contamination

The Sterilin spiral can give a higher yield of cells than the Nunclon Cell Factory because of its greater surface area, but it is limited in the types of cells which will grow in it satisfactorily. Human diploid fibroblasts, for example, tend to form aggregates before they attach and grow slowly from the aggregates.

Microcarriers. Monolayer cells may be grown on plastic microbeads of approximately 100 μm diameter, made of polystyrene (Biosilon, Nunc), Sephadex (Supabeads, Flow and Cytodex, Pharmacia), or polyacrylamide (Biorad). Culturing monolayer cells on microbeads gives maximum ratio of surface area to medium volume and has the additional advantage that the cells may be treated as a suspension. While the Nunclon Cell Factory gives an increase in scale with conventional geometry, microcarriers require a significant departure from usual substrate design. This has relatively little effect at the microscopic level as the cells are still growing on a smooth surface at the solid liquid interface. The major difference created by microcarrier systems is in the mechanics of handling [Thilly and Levine, 1979; Clark et al., 1979]. Efficient stirring without grinding the beads is essential and a paddle system (Fig. 21.13) rotating at 30 rpm appears best. Technical literature is available from microcarrier suppliers to assist in setting up satisfactory cultures.

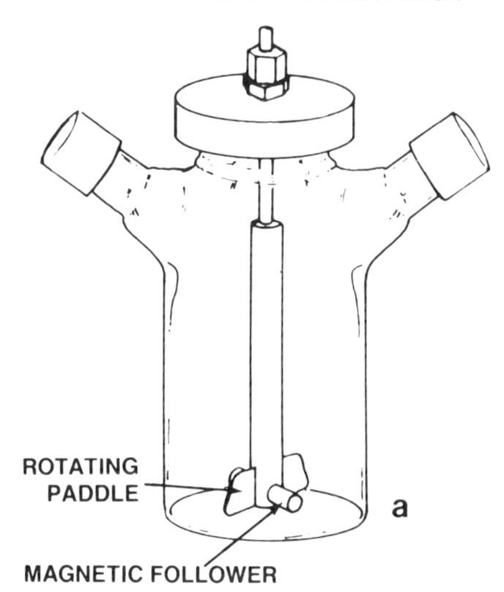

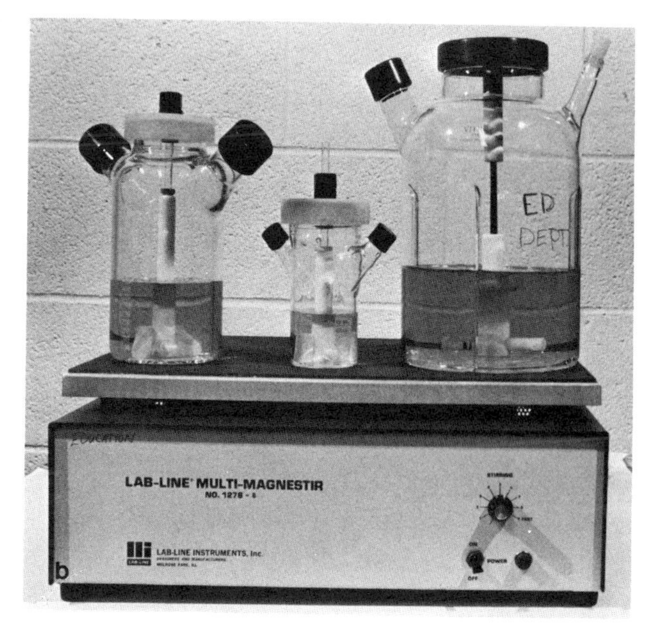

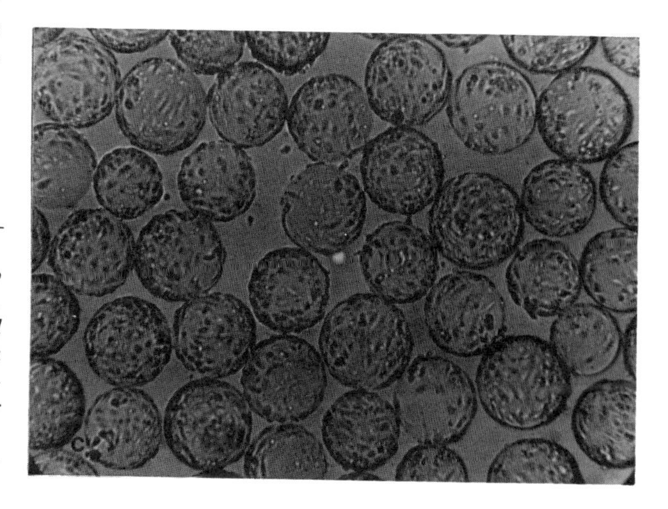

Fig. 21.13. *Microcarrier culture. a. The apparatus is similar to that used for suspension culture as the principle is the same. Modifications include a larger surface area for stirring, provided by a paddle, to enable slower stirring speeds to be used, and a suspended stirrer bar. Both modifications are designed to minimize shearing and grinding of the cells. b. A range of microcarrier culture vessels (Bellco) on a multiplace stirrer rack (Lab-Line). c. Vero cells growing on microcarriers (courtesy of Flow Laboratories, Irvine, Scotland).*

Many other mass culture techniques exist [McLimans, 1979] but they are of such specialized application that they will not be described in detail here. Linbro produced a multiplate system, similar to the Sterilin Chamber, but with plates at right angles to the long axis of the chamber. This resembled the multiplate system of Schleicher [1973] but was smaller. Amicon also manufactures larger perfusion chambers in a similar style to the Vitafiber system (see above). The potential of these systems for large-scale high-density culture has yet to be explored, but they may be valuable in recreating high tissue-like cell densities both for production of natural substances and for synthesizing large numbers of cells in a tissue-like matrix.

Lymphocyte Preparation

There is a variety of methods for the preparation of lymphocytes, but flotation on a combination of Ficoll and sodium metrizoate (e.g., Hypaque) is the most common [Boyum, 1968a,b; Perper et al. 1968].

Outline

Whole citrated blood or plasma depleted in red cells is layered on top of a dense layer of Ficoll and sodium metrizoate. After centrifugation most of the lymphocytes are found at the interphase between the Ficoll/metrizoate and the plasma.

Materials

- blood sample
- clear centrifuge tubes or universal containers
- Dextraven 110 (Fisons)
- PBSA
- Lymphoprep (Flow) (Ficoll/metrizoate, adjusted to 1.077 g/cc (Pharmacia, Nygaard))
- centrifuge
- syringe or Pasteur pipette
- serum-free medium
- hemocytometer or cell counter

Protocol

1. Add Dextraven 110 to blood sample to final concentration of 10% and incubate at 36.5°C for 30 min to allow most of the erythrocytes to sediment
2. Collect supernatant plasma, dilute 1:1 with PBSA and layer 9 ml onto 6 ml Lymphoprep or other Ficoll/sodium metrizoate mixture. This should be done in a wide transparent centrifuge tube with a cap such as the 25-ml Sterilin or Nunclon Universal Container, or the clear plastic Corning 50-ml tube, using double the above volumes
3. Centrifuge for 15 min at 400 *g* (measured at center of interface)
4. Carefully remove plasma/PBSA without disturbing the interphase
5. Collect the interface with a syringe or Pasteur pipette and dilute to 20 ml in serum-free medium (e.g., RPMI 1640 [Moore et al., 1967])
6. Centrifuge at 70 *g* for 10 min
7. Discard supernatant fluid and resuspend pellet in 2 ml serum-free medium. If several washes are required, e.g., to remove serum factors, resuspend cells in 20-ml serum-free medium and centrifuge two or three times more, and finally resuspend pellet in 2 ml
8. Count cells on hemocytometer (count only nucleated cells) or on electronic counter

Lymphocytes will be concentrated in the interface, along with some platelets and monocytes. Granulocytes will be found mostly in the Ficoll/metrizoate and erythrocytes will pellet at the bottom of the tube. Removal of monocytes and residual granulocytes and red cells may not be necessary or even desirable depending on subsequent use of the lymphocytes. If purer preparations are required, fractionation on density gradients of metrizamide (Nygaard) or Percoll (Pharmacia) or by centrifugal elutriation (see Chapter 14) may be attempted. Alternatively, specific subpopulations of lymphocytes may be purified on antibody or lectin-bound affinity columns (Pharmacia).

Blast transformation. [Hume and Weidemann, 1980] Lymphocytes in purified preparations, or in whole blood, may be stimulated with mitogens such as phytohemagglutinin (PHA), pokeweed mitogen (PWM), or antigen [Berger, 1979]. The resultant response may be used to quantify the immunocompetence of the cells. PHA stimulation is also used to produce mitosis for chomosomal analysis of peripheral blood [Kinlough and Robson, 1961; Rothvells and Siminovitch, 1958].

Materials

medium + 10% FBS or autologous serum
phytohemaglutinin (PHA), 5 μg/ml
test tubes or universal containers
microscope slides
Colcemid, 0.01 μg/ml in BSS
0.075 M KCl

Protocol

1. Using the washed interface fraction from step 7 above, incubate 2 × 10^6 cells/ml in medium, 1.5–2.0 cm deep, in HEPES or CO_2-buffered DMEM, CMRL 1066, or RPMI 1640 supplemented with 10% autologous serum or fetal bovine serum
2. Add PHA, 5 μg/ml, to stimulate mitosis from 24 to 72 hr later
3. Collect samples at 24, 36, 48, 60, and 72 hr and prepare smears or cytocentrifuge slides to determine optimum incubation time (peak mitotic index)
4. Add 0.001 μg/ml (final concentration) Colcemid for 2 hr when peak of mitosis is anticipated [Berger, 1979]
5. Centrifuge cells after Colcemid treatment, resuspend in 0.075 M KCl for hypotonic swelling, and proceed as for chromosome preparation in Chapter 15

AUTORADIOGRAPHY

The following description is intended to cover autoradiography of any small molecular precursor into a cold acid–insoluble macromolecule such as DNA, RNA, or protein. Other variations may be derived from this or found in the literature [Rogers, 1979; Stein and Yanishevsky, 1979].

Isotopes suitable for autoradiography are listed in Table 21.1. A low energy emitter, e.g., ^3H or ^{55}Fe, in combination with a thin emulsion, gives high intracellular resolution. Slightly higher energy emitters, e.g., ^{14}C and ^{35}S, give localization at the cellular level. Still higher energy isotopes, e.g., ^{131}I, ^{59}Fe, and ^{32}P, give poor resolution at the microscopic level but are used for autoradiographs of chromatograms and electropherograms where self-absorption of low energy emitters limits detection. Low concentrations of higher energy isotopes (^{14}C and above) used in conjuction with thick nuclear emulsions produce tracks useful in locating a few highly labeled particles, e.g., virus particles infecting a cell.

TABLE 21.1. Isotopes Suitable for Autoradiography

Isotope	Emission	Energy (mV) (mean)
^3H	β^-	0.018
^{55}Fe	x-rays	0.0065
^{125}I	x-rays	0.035
		0.033
^{14}C	β^-	0.155
^{35}S	β^-	0.167
^{45}Ca	β^-	0.254

Tritium is used most frequently for autoradiography at the cellular level because the β-particles released have a mean range of about 1 μm, giving very good resolution. Tritium-labeled compounds are usually less expensive than the ^{14}C- or ^{35}S-labeled equivalents and have a long half-life. Because of the low energy of emission, however, it is important that the radiosensitive emulsion is positioned in close proximity to the specimen, with nothing between the cell and the emulsion. Even in this situation only the top 1 μm of the specimen will irradiate the emulsion.

β-particles entering the emulsion produce a latent image in the silver halide crystal lattice within the emulsion at the point where they stop. The image may be visualized as metallic silver grains by treatment with an alkaline reducing agent (developer) with subsequent removal of the remaining unexposed silver halide by an acid fixer.

The latent image is more stable at low temperature and in anhydrous conditions, so sensitivity may be improved by exposing in a refrigerator or freezer over desiccant. This will reduce background grain formation by thermal activity.

Outline

Cultured cells are incubated with the appropriate isotopically labeled precursor (e.g., [^3H]thymidine to label DNA), washed, fixed, and dried (Fig. 21.14). Any extractions necessary, e.g., to remove unincorporated precursors, are performed, and the specimen is coated with emulsion in the dark and left to expose. When subsequently developed in photographic developer, silver grains can be seen overlying areas where radioisotope was incorporated (Fig. 21.15).

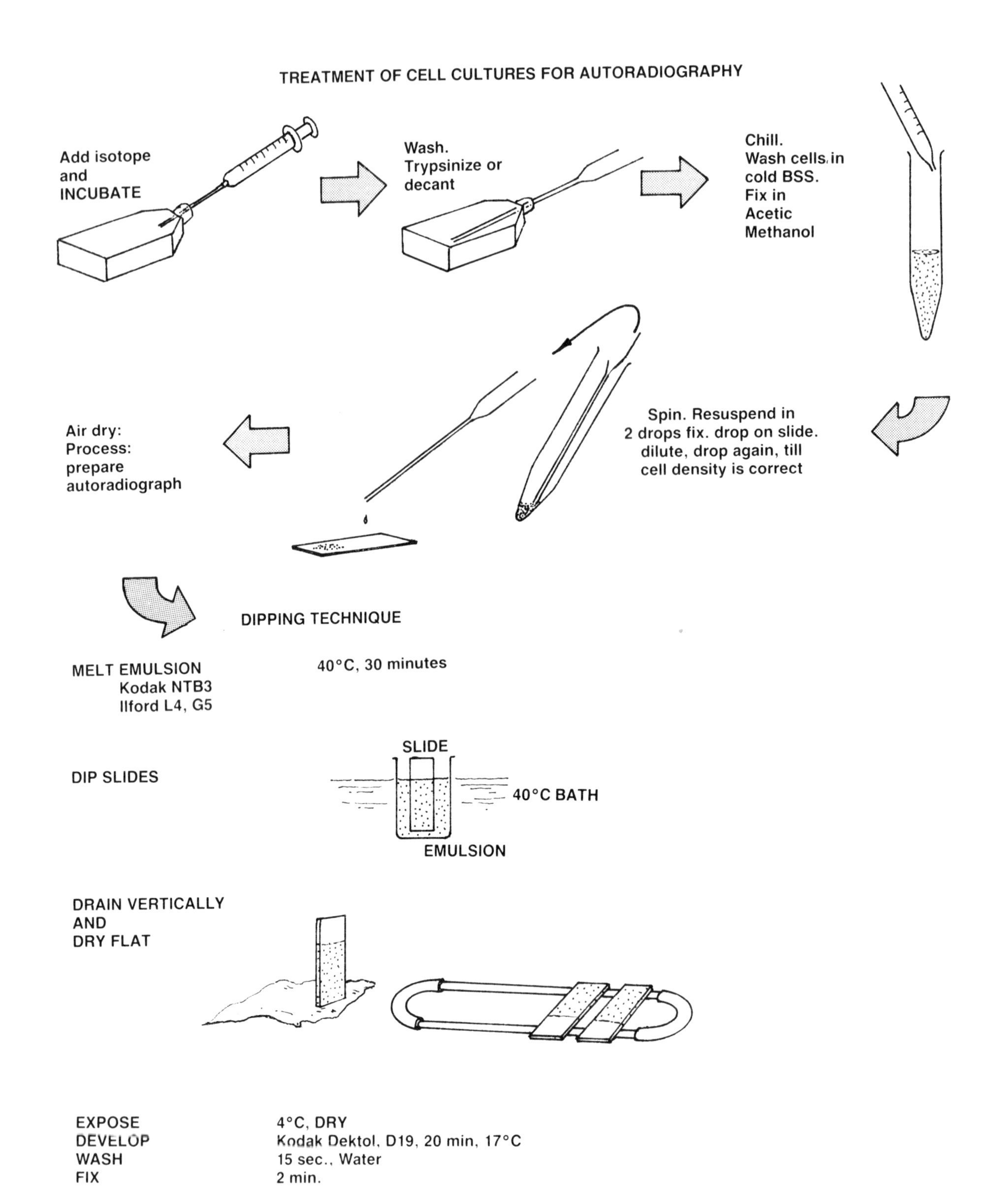

EXPOSE 4°C, DRY
DEVELOP Kodak Dektol, D19, 20 min, 17°C
WASH 15 sec., Water
FIX 2 min.

WASH, DEHYDRATE, STAIN AND MOUNT COVERSLIP

Fig. 21.14. *Steps in preparing an autoradiograph from a cell culture.*

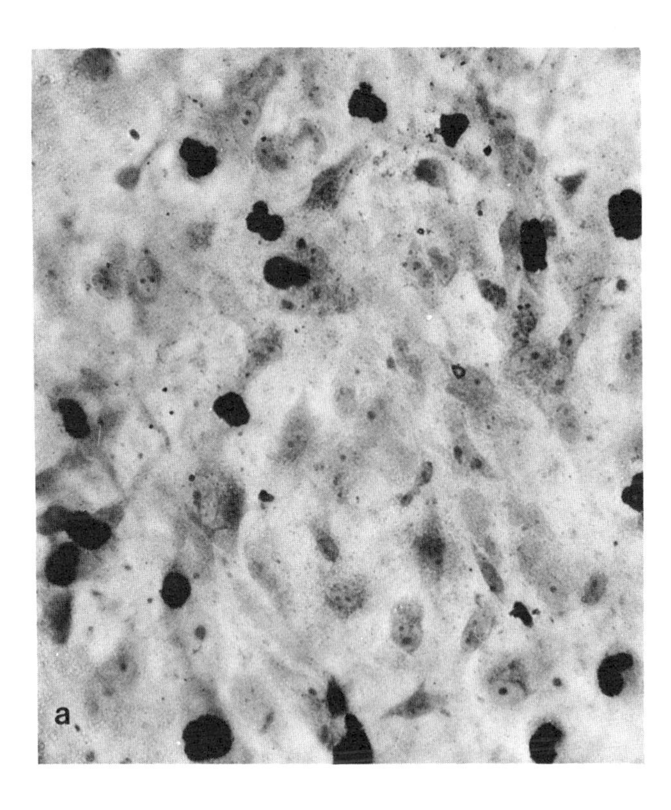

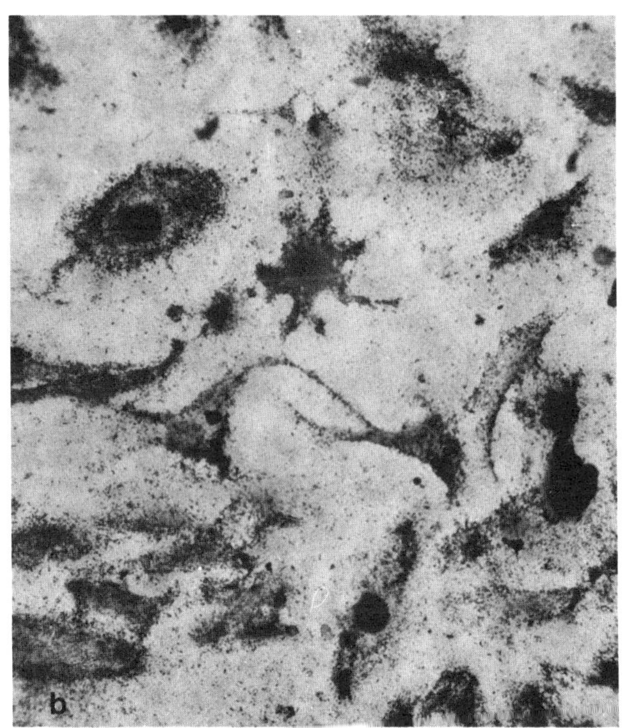

Fig. 21.15. *Autoradiograph. This is an example of [^{3}H]thymidine incorporation into a cell monolayer. Normal glial cells were incubated with 0.1 μCi/ml (200 Ci/mMol) [^{3}H]thymidine for 24 hr, washed, and processed as in text. a. Typical densely-labeled nuclei, suitable for determining labeling index (see Chapter 19). b. Similar culture infected with mycoplasma; cytoplasm is now labeled also.*

Materials

Setting up culture:
- cells
- PBSA
- trypsin
- medium
- hemocytometer or cell counter
- counting fluid if using automatic cell counter
- coverslips or slides and petri dishes (may be non-tissue culture grade if coverslips or slides are used) or plastic bottles

Labeling with isotope and setting up autoradiographs:
- isotope
- HBSS
- protective gloves
- containers for disposal of radioactive pipettes
- container for radioactive liquid waste
- acetic methanol (1:3, ice cold, freshly prepared)
- DPX
- 10% TCA
- emulsion (Kodak NTB2, Ilford G5), diluted 1 + 2 in distilled or deionized water
- light-tight microscope boxes
- silica gel
- dark vinyl tape
- black paper or polyethylene

Processing:
- D19 developer (Kodak)
- photographic fixer (Kodak, Ilford)
- hypoclearing agent (Kodak)
- coverslips (00)
- Giemsa stain
- 0.01 M phosphate buffer (pH 6.5)

Protocol

1.

Prepare culture. Monolayer cells may be grown on coverslips (Thermanox), slides (Lab-Tek, Bellco), or in conventional plastic bottles or petri dishes

Note. Follow local rules for handling radioisotopes. Since these vary, no special recommendation will be made here.

2.

Add isotope (usually in the range 0.1–10 μCi/ml,

100 Ci/mmol) for 0.5–48 hr as appropriate

3.

Remove medium containing isotope, wash cells carefully in BSS, discarding medium and washes (Radioactive!)

Note. All glassware must be carefully washed and free of isotopic contamination. Plastic coverslips should be used in preference to glass to minimize radioactive background. Be particularly careful with spillages; mop up right away. Wear gloves and change regularly, e.g., when you move from incubation (high level of isotope) to handling washed, fixed slides (low level of isotope)

4.

Fix cells in ice-cold acetic methanol. Coverslips should be mounted on a slide with DPX or Permount, cells uppermost. Cell suspensions may be centrifuged onto a slide (Cytospin) or drop preparations made (see Chapter 15). Prepare several extra control slides for use later to determine correct durations of exposure. All preparations will be referred to as "slides" from now on

5.

Extract acid-soluble precursors (when labeling DNA, RNA, or protein) with ice-cold 10% TCA (3 × 10 min), and perform any other control extractions, e.g., with lipid solvent or enzymatic digestion

6.

Wash slide in distilled water and methanol and dry

7.

Take to darkroom [see also Kopriwa, 1963] and under dark red safelight, melt emulsion, in water bath at 40°C and dilute with two parts deionized distilled water. It is convenient to place aliquots of the diluted emulsion in containers suitable for the number of slides to be handled at one time. If sealed in a dark box, these may be stored at 4°C until required

8.

Still under safelight, dip slides in emulsion, making sure that the cells are completely immersed, withdraw, blot the end of the slide, and allow to dry flat

9.

When dry (~30 min), transfer to light-tight microscope slide boxes (Clay Adams, Raven Scientific) with a desiccant, such as silica gel, and seal with dark vinyl tape (e.g., electrical insulation tape)

10.

Wrap in black paper and place in refrigerator. Make sure that this refrigerator is not used for storage of isotopes

11.

Leave at 4°C for 24 hr to 2 wk. The time required will depend on the activity of the specimen and can be determined by processing one of the extra slides at intervals

12.

To develop, return to darkroom (dark red safelight), unseal box, and allow slides to come to atmospheric temperature and humidity (~2 min)

13.

Place slides in developer (e.g., Kodak D19) for 10 min with gentle intermittent agitation

14.

Wash briefly in distilled deionized water

15.

Transfer to photographic fixer for 3–5 min

16.

Rinse in deionized water and place in hypoclearing agent (Kodak) for 1 min

17.

Wash in deionized water, five changes over 5 min

18.

Dry slides and examine on microscope. Phase contrast may be used by mounting a thin glass (00) coverslip in water. Remove coverslip when finished before water dries out or it will stick to the emulsion

19.

If staining is desired, immerse dry slide in Giemsa stain diluted 1:10 in 0.01 M, pH 6.5, phosphate buffer for 10 min. Rinse thoroughly under running tap water until color is removed from emulsion but not from cells [see also Thurston and Joftes, 1963]

Analysis. ***Qualitative.*** Determine specific localization of grains, e.g., over nuclei only, or over one cell type rather than another.

Quantitative. (1) Grain counting. Count number of grains per cell, per nucleus, etc. This requires a low grain density, about five to 20 grains per nucleus, ten to 50 grains per cell, no overlapping grains, and a low uniform background

(2) Labeling index. Count number of labeled cells as a proportion of the total. Grain density should be higher than in (1) to ease the recognition of labeled cells. If the grain density is high (e.g., ~100 grains per nucleus), set the lower threshold at, say, ten grains per nucleus or per cell; but remember that low levels of

labeling, significantly over background, may yet contain useful information.

Autoradiography is a useful tool for determining the distribution of isotope incorporation within a population, but it is less suited to total quantitation of isotope uptake or incorporation, when scintillation counting is preferable.

Variation. Autoradiographic localization of water-soluble precursors is possible with rapidly frozen specimens which have been freeze-dried or freeze-substituted to remove the water. These may be mounted dry, clamped to the radiosensitive film, or with a minimal amount of moisture (obtained by brief condensation on a cold slide) to promote adhesion of the emulsion [Novak, 1962; Hassbroek, et al., 1962]. Both processes require Kodak AR-10 stripping film (see below).

Isotopes of two different energies, e.g., ^3H and ^{14}C, may be localized in one preparation by coating the slide first with a thin layer of emulsion, coating that with gelatin alone, and finally coating the gelatin with a second layer of emulsion [Baserga, 1962; Kempner and Miller, 1962; Rogers, 1979]. The weaker β-emission from ^3H is stopped by the first emulsion and the gelatin overlay, while the higher-energy β-emission from ^{14}C, having a longer mean path length of around 20 μm, will penetrate the upper emulsion.

Soft β-emitters may also be detected in electron microscope preparations using very thin films of emulsion or silver halide sublimed directly on to the section [Salpeter, 1974; Rogers, 1979].

Adams [1980] described a method for autoradiographic preparations from petri dishes or flasks where liquid emulsion is poured directly onto fixed preparations without the necessity for trypsinization.

Radiographic emulsion may be applied to slides in the form of a film stripped from a glass plate (Kodak, AR-10)[Rogers, 1979]. The film is made up of a 5 μm layer of sensitive emulsion on a 10 μm gelatin backing. It is applied to the slide by inverting the film, peeled off the plate in 30 mm $\times$ 40 mm rectangles, after drying the prescored plate in a desiccator, on to the surface of warm water, and bringing the slide up under the film and allowing the film to drape over and around the slide. After drying, it is treated in the same way as dipping emulsion for exposure and development.

AR-10 gives a very reproducible emulsion thickness and is good for high resolution work and quantitation. As the emulsion tends to detach during processing, coat the slide before adding cells with 0.5% gelatin, 0.05% chrome alum, and avoid prolonged washing in low ionic strength.

CULTURE OF CELLS FROM POIKILOTHERMS

The approach to the culture of cells from cold-blooded animals (poikilotherms) has been similar to that employed for warm-blooded animals largely because the bulk of present-day experience has been derived from birds and mammals. Thus, the dissociation techniques for primary culture employ proteolytic enzymes such as trypsin and EDTA as a chelating agent. Fetal bovine serum appears to substitute well for homologous serum or hemolymph (and is more readily available), but modified media formulations may improve growth. A number of these media are available through commercial suppliers and the procedure is much the same as for mammalian cells—try those media and sera which are currently available, assessing for growth, plating efficiency, and specialized functions (see Chapter 9). Since the development of media for many invertebrate cell lines is in its infancy it may prove necessary to develop new formulations if an untried class of invertebrates is examined. Most of the accumulated experience so far relates to insects and molluscs.

Two reviews cover some aspects of one field [Maramorosch, 1976; Vago, 1971, 1972], but since this is a rapidly expanding area it is to be hoped that a fundamental methodological review text will be forthcoming in the near future. Culture of vertebrate cells other than birds and mammals has also followed procedures for warm-blooded vertebrates, and so far there has been insufficient interest to stimulate a major divergence in technique.

Since this is a developmental area, certain basic parameters will still need to be considered to render culture conditions optimal, and if a new species is being investigated, optimal conditions for growth may need to be established, e.g., pH, osmolality (which will vary from species to species), nutrients, and mineral concentration. Temperature may be less vital but its consistency should be maintained within the 25°–30°C range at $\pm$ 0.5°C.

Chapter 22
Specialized Techniques. II

The techniques described so far are of general importance, applicable to many different aspects of tissue culture, and are fairly easily performed. Techniques described in this chapter are either of more limited application or of a degree of technical complexity that does not allow for their full presentation here.

CELL SYNCHRONY

The percentage-labeled mitosis method for determining the duration of the stages of the cell cycle has been described in Chapter 21. In order to follow the progression of cells through the cell cycle, a number of techniques have been developed whereby a cell population may be fractionated or blocked metabolically so that on return to regular culture they will all be at the same phase.

Cell Separation

Techniques for this have been described in Chapter 19. Sedimentation at unit gravity (Figs. 14.2, 14.3) is the simplest [Shall and McLelland, 1971; Shall, 1973], but centrifugal elutriation is preferable if a large number of cells ($> 5 \times 10^7$) is required [Meistrich et al., 1977a,b] (see Figs. 14.5, 14.6). Fluorescence-activated cell sorting (see Figs. 14.13, 14.14, 14.15) can also be used in conjunction with a nontoxic, reversible DNA stain such as Hoechst 33342. The yield is lower than unit gravity sedimentation ($\sim 10^7$ cells or less) but the purity of the fractions is higher.

One of the simplest techniques for separating synchronized cells is mitotic shake off. Mitotic cells tend to round up and detach when the flask is shaken. This works well with CHO cells [Tobey et al., 1967; Petersen et al., 1968] and some sublines of HeLa-S_3. Placing the cells at 4° C for 30 min to 1 hr a few hours previously enhances the yield at "shake off" [Newton and Widly, 1959; Sinclair and Morton, 1963; Lesser and Brent, 1970; Miller et al., 1972].

Blockade

Two types of blocking have been used:

DNA synthesis inhibition (S-phase). Thymidine, hydroxyurea, cytosine arabinoside, aminopterin, etc. [Stubblefield, 1968]: The effects of these agents are variable because many are toxic. Hence, the culture will contain nonviable cells, cells blocked in S but viable, and cells which have escaped the block.

Nutritional deprivation (G1 phase). In these cases serum [Chang and Baserga, 1977] or isoleucine [Ley and Tobey, 1970] is removed from the medium for 24 hr and then restored, whereupon transit through cycle is resumed in synchrony.

A high degree of synchrony (e.g., >80%) is only achieved in the first cycle; by the second cycle it may be <60% and by the third cycle, close to random. Chemical blockade is often toxic to the cells and nutritional deprivation does not work well in many transformed cells. Physical fractionation techniques are probably most effective and do less harm to the cells.

TIME-LAPSE CINEMICROGRAPHY

This is a technique whereby living cultures may be filmed and their behavior (e.g., cell membrane ruffling, mitosis, migration) accelerated for viewing [Riddle, 1979]. A typical apparatus is depicted in Figure 22.1. It consists of (1) an inverted microscope with phase-contrast, interference-contrast or surface-interference-contrast optics (e.g., Leitz Diavert or Reichert Biovert), (2) a perfusion slide (Sterilin) (Fig. 22.2) to present the cells in optical clarity but still in optimal

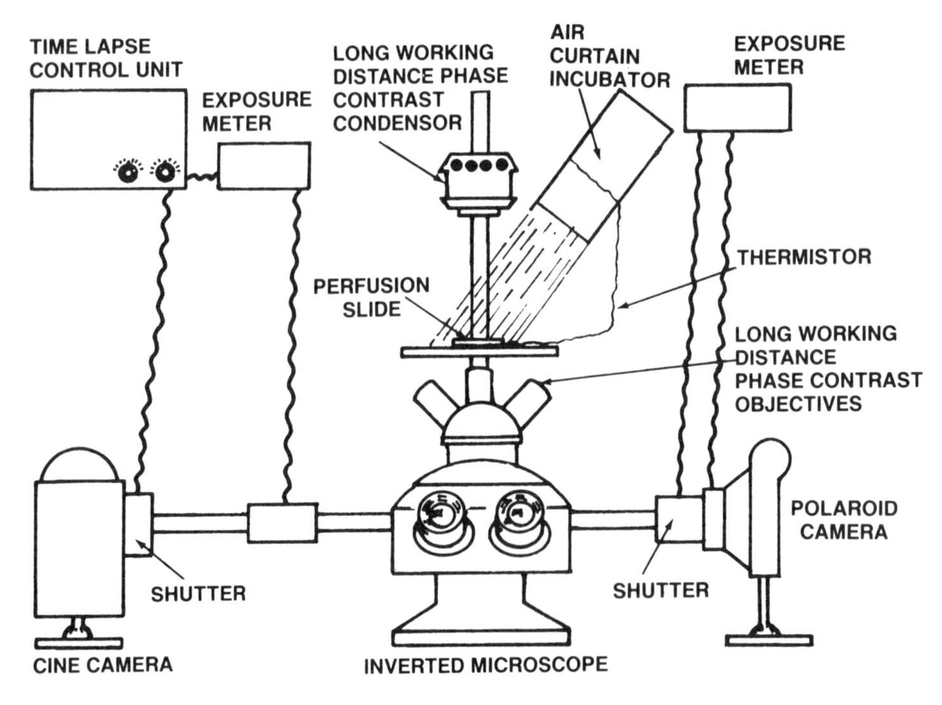

Fig. 22.1. *Suggested layout for time-lapse cinemicrography.*

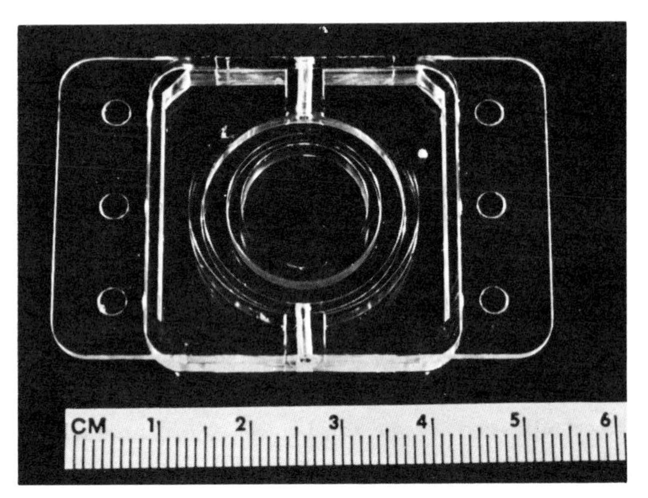

Fig. 22.2. *Perfusion slide for use in cytological observation such as time-lapse cinemicrography. Medium can be perfused in and out via the holes at top and bottom.*

physiological conditions, (3) an incubator chamber or warm air curtain (Sage) to keep the specimen at 36.5° C, (4) a time-lapse control unit (a) to determine the frequency of exposure (i.e., the lapsed time between exposures), (b) to switch on the light before each exposure and switch off the light when it is completed, and (c) to activate the camera shutter and advance film, (5) an exposure meter linked to the time-lapse control unit and camera to set the duration of each exposure, and (6) a still camera, preferably Polaroid, to take record shots at intervals without interrupting filming as this would upset the time base of the film.

Regular film may be replaced with a television camera and video recorder. This reduces the need for exposure control and gives an instant result, but poorer resolution and limited acceleration of movement.

CARCINOGENICITY AND TOXICITY TESTING

The measurement of viability by clonal growth analysis and microtitration has been described in detail in Chapter 19. One of the major applications of such tests is in the development of new anticancer agents where comparison of survival curves in clonogenicity assays with L1210, P388, Hep-2, and many other cell types can give comparative figures for relative cytotoxicity. Drug testing *in vitro* does not allow for the modification of drugs by liver metabolism en route to the target tissue, so some workers have included liver microsomal enzyme preparations in the culture medium to activate drugs such as cyclophosphamide [Sladek, 1973].

Predictive Testing

The possibility has often been considered that measurement of the chemosensitivity of cells derived from a patient's tumor might be used in designing a chemotherapeutic regime for the patient. This has never been exhaustively tested, although small scale trials have been encouraging [Limburg and Heckman, 1968; Hamburger and Salmon, 1977; Berry et al., 1975; Kauffman et al., 1980; Freshney, 1978; Dendy, 1976; Bateman et al., 1979]. What is required now is the development of reliable and reproducible culture techniques for the common tumors such as breast, lung, and colon, such that cultures of pure tumor cells capable of cell proliferation over several cell cycles may be prepared routinely. Assays might then be performed in a high proportion of cases, within 2 wk of receipt of the biopsy. So far this has not been possible, but recent developments with new defined media (see Chapter 9) may have brought this closer [Carney et al., 1981] .

Cytotoxicity Testing

Current legislation demands that new drugs, cosmetics, food additives, etc., go through extensive cytotoxicity testing before they are released [see also Berky and Sherrod, 1977]. This usually involves a large number of animal experiments which are very costly and raise considerable public concern. There is, therefore, much pressure, both emotional and economic, to perform at least part of cytotoxicity testing *in vitro*. The introduction of specialized cell lines, as well as the continued use of long-established cultures, may make this a reasonable proposition, but the tests as performed currently are limited in the same way as predictive testing. Many nontoxic substances become toxic after metabolism by the liver; and in addition, many substances, toxic *in vitro*, may be detoxified by liver enzymes. For testing *in vitro* to be accepted as an alternative to animal testing, it must be demonstrated that potential toxins reach the cells *in vitro* in the same form as they would *in vivo*. This may require additional processing by purified liver microsomal enzyme preparations as with cytotoxic drugs (see above).

The nature of the response must also be considered carefully. A toxic response *in vitro* may be measured by changes in cell survival or metabolism (Chapter 19), while the major problem *in vivo* may be a tissue response, e.g., inflammatory reactions or fibrosis. For *in vitro* testing to be more effective, construction of models of these responses will be required utilizing, perhaps, histotypic cultures reassembled from several different cell types and maintained in the appropriate hormonal milieu.

Mutagenicity

The determination of mutagenicity is a task more amenable to culture as it represents a cellular response. Metabolism by enzymes of the liver and the gastrointestinal tract may still be required, however.

Assaying for mutagenicity can be performed by a variety of standard genetic techniques, e.g., an increase in the frequency of occurence of such characterized mutants as the absence of thymidine kinase (TK^-), which makes cells resistant to bromodeoxyuridine, or hypoxanthine guanine phosphoribosyl transferase deficiency ($HGPRT^-$), making the cell resistant to 8-azaguanine [Littlefield, 1964].

As TK^- or $HGPRT^-$ cells will not clone in HAT medium (hypoxanthine, aminopterin, and thymidine) but revertant mutants will, the reversion rate of TK or HGPRT deficiencies assayed by cloning in HAT medium can also be used to assay for frame-shift mutations.

Carcinogenicity

Carcinogenesis may arise from mutation in specific target cells, e.g., proliferating precursor, or stem cells, in the intestinal crypts or basal cells of the skin. If so, carcinogenesis may be assayed by the same route as mutagenesis. In practice, much of industrial testing is performed in this way using bacteria as in the Ames test [Ames, 1980]. However, substances are known which are carcinogenic but nonmutagenic and viceversa, so carcinogenesis may not always be initiated by mutation. The growth of embryonal carcinomas, for example, carries no implication of mutagenesis, but rather a failure in cellular communication due to the retention of embryonic cells in an inappropriate site.

Current assays for carcinogenicity use such models as the induction of clonal growth of BHK21-C13 cells in soft agar [Styles, 1977]. This assay has two problems: (1) that it uses mesodermal rather than ectodermal or endodermal cells, and measures sarcomogenesis rather than carcinogenesis, whereas the common cancers arise in epithelial, pigment, and glial cells or hematopoietic cells. Only the last is mesodermal. (2) The criterion for carcinogenicity is growth in suspension, when it has been shown that a number of normal cells (glial cells, fibroblasts, and hematopoietic cells) will all clone in suspension although with a lower efficiency, but one similar to the cloning efficiency of cells from spontaneous tumors.

The development of carcinogenicity assays with epithelial lines now seems possible, but the major question is still the choice of a good criterion for transformation (see Chapter 16).

AMNIOCENTESIS

Inborn metabolic abnormalities may be identified by culturing cells collected from the amnion during early pregnancy [Valenti, 1973] (Fig. 22.3). The amniotic fluid is centrifuged, the supernatant removed (it may be used for biochemical analysis of proteins and enzyme activity), and the cells cultured as a monolayer. When the monolayer starts to proliferate, it is used for chromosome analysis as in Chapter 15.

SOMATIC CELL FUSION

For many years mammalian, and particularly human, genetic analysis was hampered by the limitations of the duration of the breeding cycle and difficulties in performing breeding experiments. The discovery by Barski et al. [1960] and Sorieul and Ephrussi [1961] that somatic cells would fuse in the presence of Sendai virus led to a burst of activity that has developed into the field of somatic cell genetics.

Briefly, somatic cells fuse if cultured with inactivated Sendai virus, or with polyethylene glycol (PEG) [Pontecorvo, 1975; Milstein, 1979]. A proportion of the cells that fuse will progress to nuclear fusion, and a proportion of these will progress through mitosis such that both sets of chromosomes replicate together and a hybrid is formed. In some interspecific hybrids, e.g., human-mouse, one set of chromosomes (the human) is gradually lost [Weiss and Green, 1967]. Thus, genetic recombination is possible *in vitro* and, in some cases, segregation as well.

Since the proportion of viable hybrids is low, selective media are required to favor the survival of the hybrids at the expense of the parental cells. TK^- and $HGPRT^-$ mutants (see above) of the two parental cell types are used, and the selection is carried out in HAT medium (Fig. 22.4) [Littlefield, 1964]. Only cells formed by the fusion of two different parental cells (heterokaryons) survive, as the parental cells, and fusion products of the same parental cell type (homokaryons) are deficient in either thymidine kinase or hypoxanthine guanine phosphoribosyl transferase. They cannot, therefore, utilize hypoxanthine or thymidine from the medium, and since aminopterin blocks endogenous synthesis of purines and pyrimidines, they are unable to synthesize DNA.

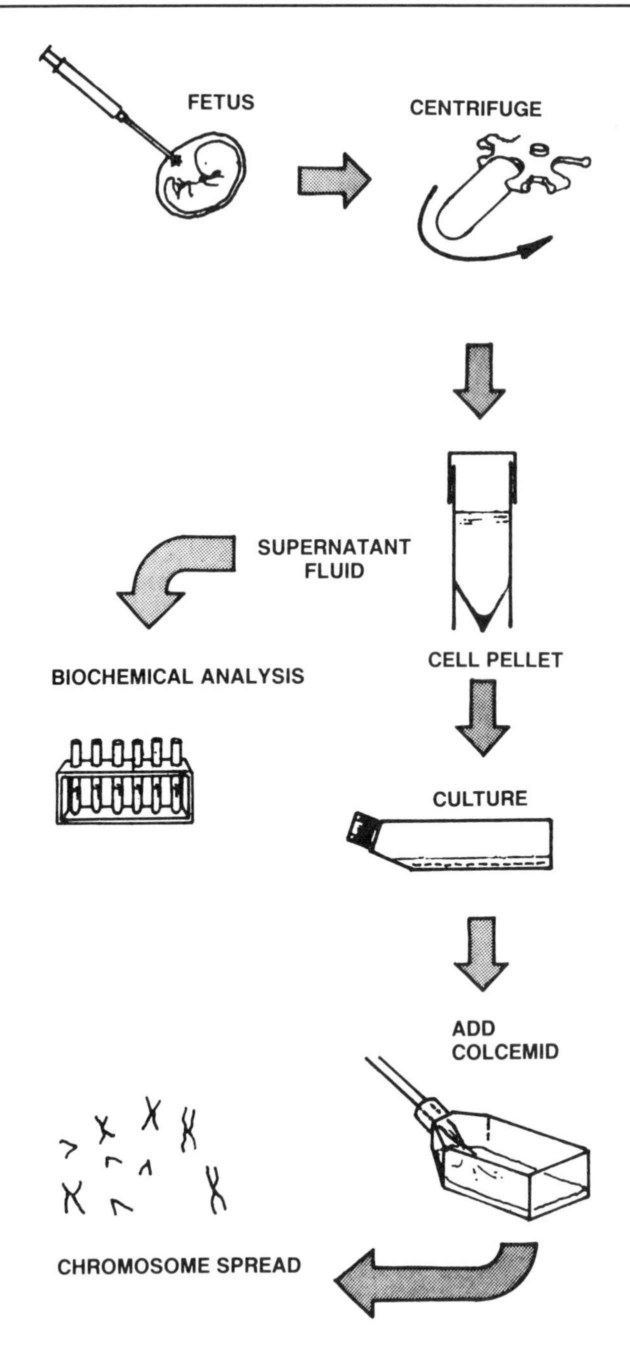

Fig. 22.3. *Culture of amniocentesis samples for detection of chromosomal abnormalities.*

Gene Transfer

In addition to fusion of whole cells, fusion of isolated nuclei, individual chromosomes, and even purified genes or gene fragments with whole cells or enucleated cytoplasts is now possible (Fig. 22.5) [Shows and Sakaguchi, 1980]. Enucleation is performed by centrifuging cytochalasin-B-treated cells such that the nuclei detach from an anchored mono-

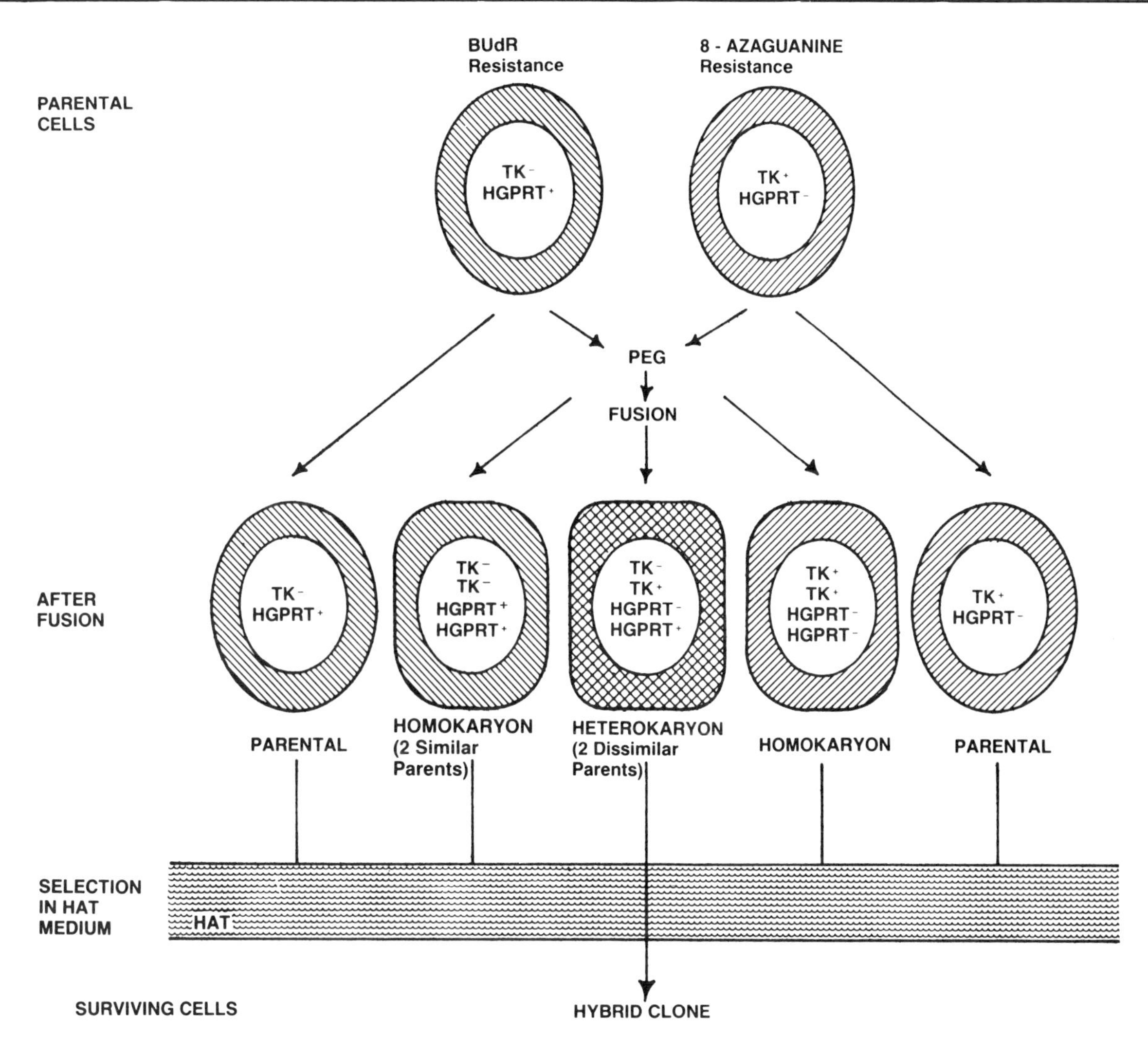

Fig. 22.4. *Somatic cell hybridization. Selection of hybrid cells after fusion (see text).*

layer and pellet at the bottom of the tube. This gives cytoplasmic residues without nuclei (cytoplasts) and nuclei with only some residual plasma membrane surrounding them (karyoplasts). Incubation of karyoplasts with cytoplasts, or whole cells, in the presence of polyethylene glycol results in fusion.

Chromosomes may be isolated from metaphase cells by hypotonic lysis and incubation of these with whole cells after coprecipitation with calcium phosphate results in their incorporation into the nucleus. The chromosomes may be fractionated by density centrifugation or flow cytophotometry and individual chromosome pairs inserted into recipient cells.

At a still higher level of resolution, DNA extracted from one cell can be incorporated into another by a technique similar to that used for whole chromosomes. The DNA may also be fragmented by restriction endonuclease, the fragments cloned in plasmids, and released by further nuclease treatment. The purified genes or fragments so produced can then be incorporated into recipient cells and their effect on gene expression determined (see Fig. 16.1b).

Production of Monoclonal Antibodies

One of the most exciting developments of somatic cell fusion arises from the demonstration that sensitized plasma cells from the spleen of an immunized mouse can be fused to continuous lines of mouse myeloma cells. Some of the resultant fusion products are capable of synthesizing immunoglobulins; if cloned, each clone produces a single specific monoclonal antibody [Milstein et al., 1979]. The steps in

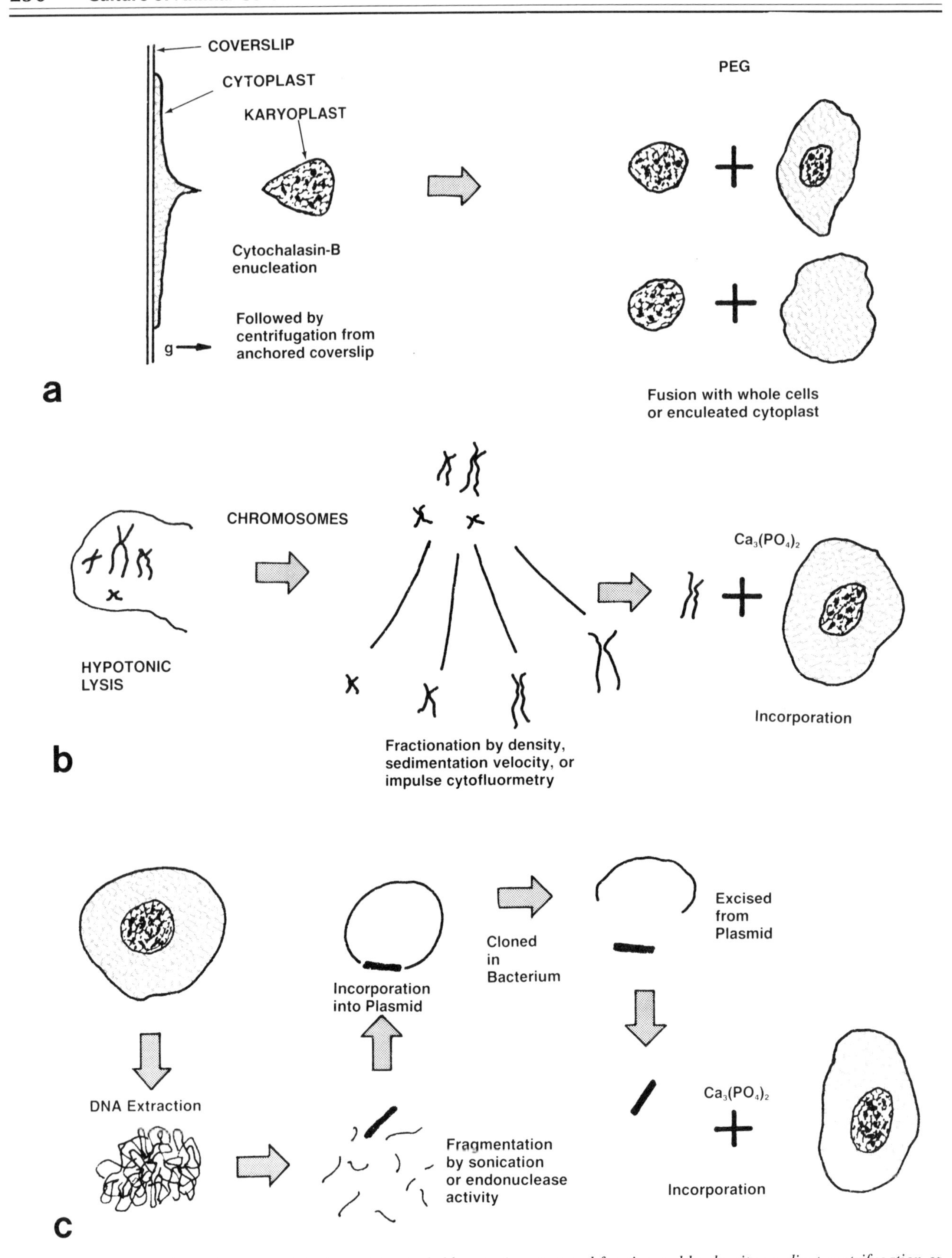

Fig. 22.5. *Gene transfer techniques. a. Whole nuclei extruded by treatment with cytochalasin B hybridized with whole cells or enucleated cytoplast. b. Chromosomes isolated from cells in mitotic arrest and fractionated by density gradient centrifugation or flow cytophotometry added to whole cells. c. Isolated DNA fragments, amplified by gene cloning techniques, added to whole cells.*

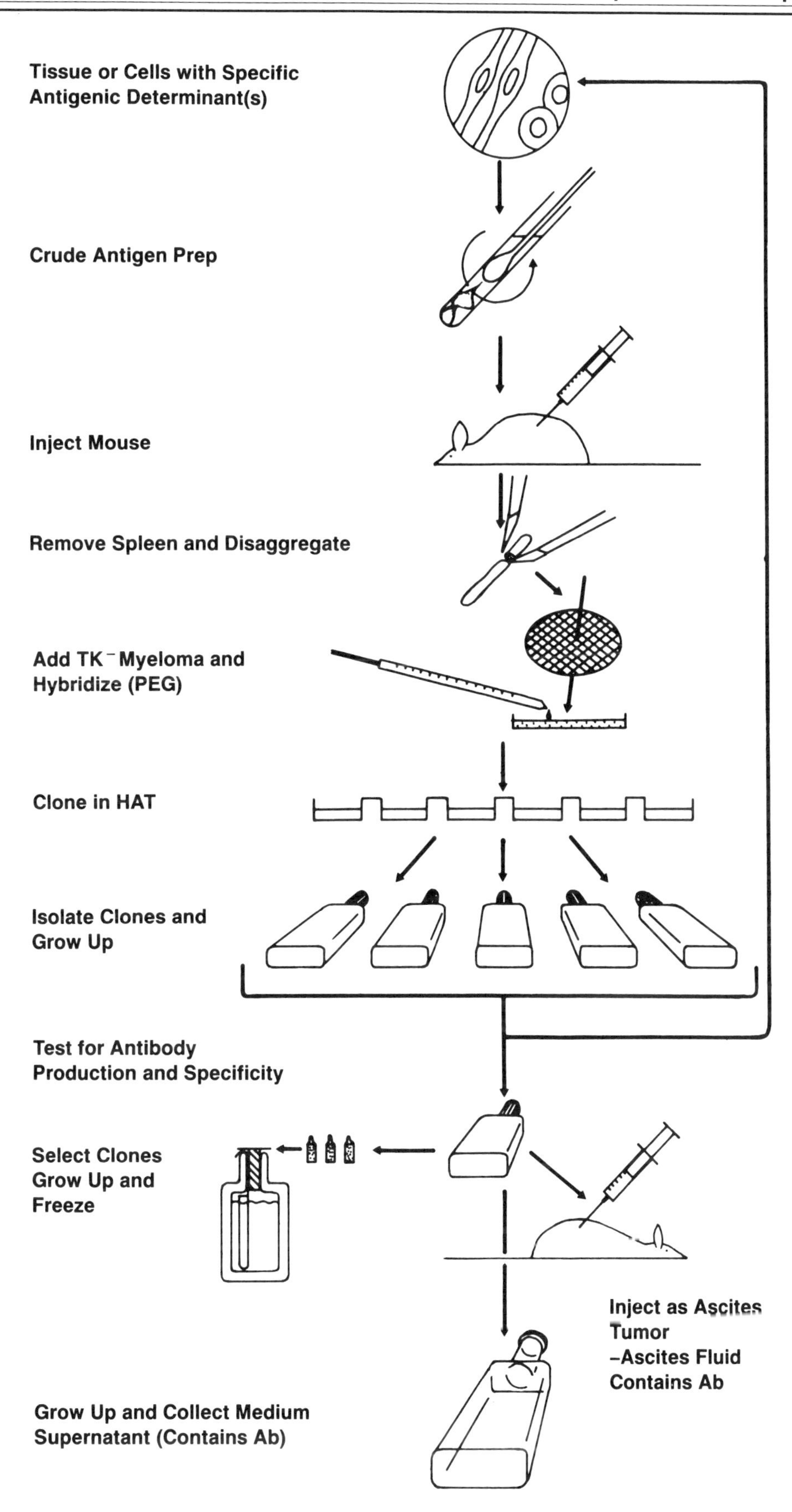

Fig. 22.6. *Schematic diagram of the production of hybridoma clones capable of secreting monoclonal antibodies. (Drawing by David Tallach.)*

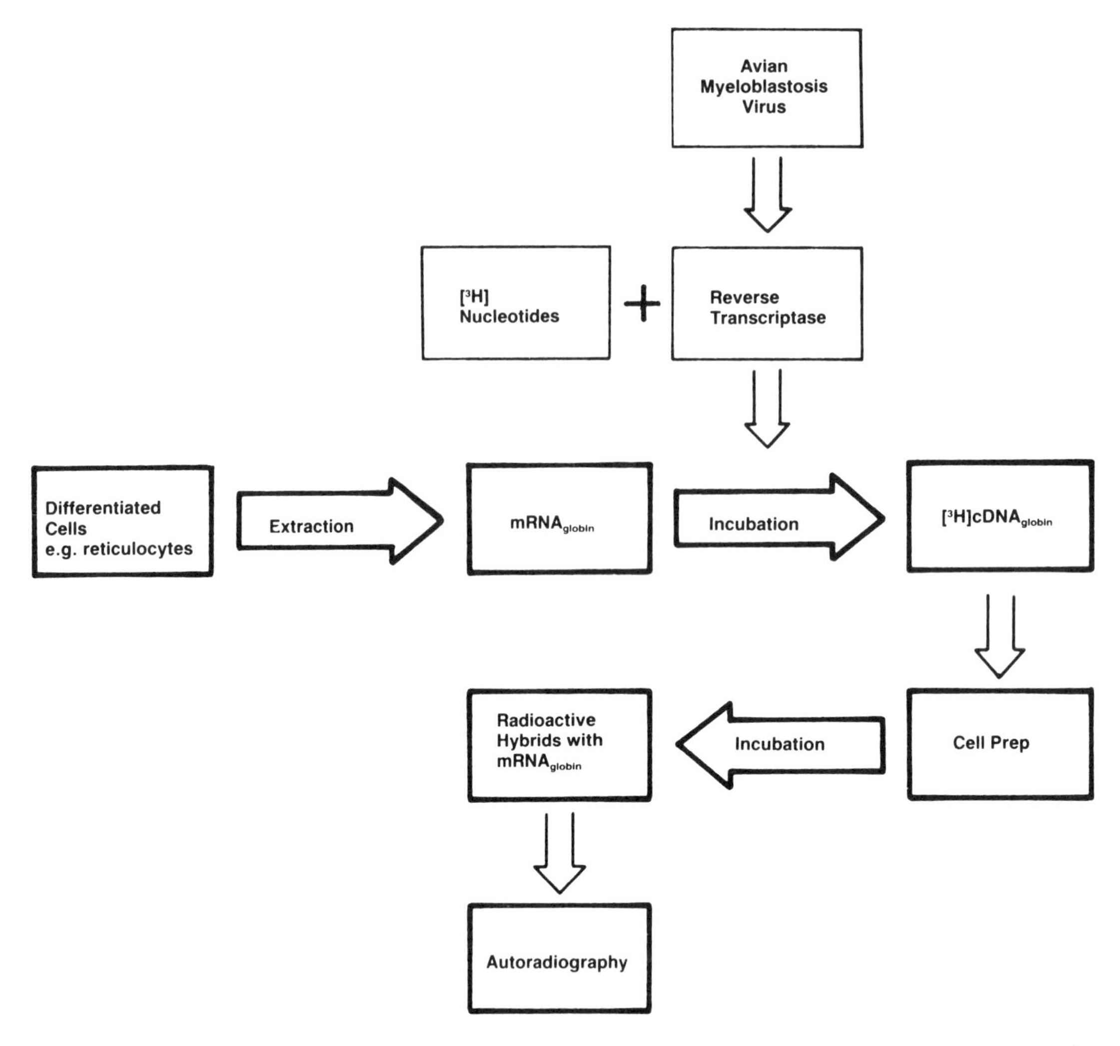

Fig. 22.7. In situ *hybridization of radioactive complementary DNA, synthesized on a messenger RNA template by the action of reverse transcriptase, with intracellular messenger RNA.*

the technique are outlined in Figure 22.6. A mouse is first immunized with crude antigen, and some days later the spleen is removed. It is then minced and placed in culture with TK^- mouse myeloma cells in the presence of polyethylene glycol. The cells fuse, and the fusion products are cloned in HAT medium. The parental cells or homokaryons will not grow in HAT medium because the myeloma is TK^- and the spleen cells are unable to grow *in vitro*. The clones are then grown and their specificity tested by radioimmune assay of their supernatant medium with appropriate target cells or antigen, anchored to microtitration plates. Where a specific antigen-antibody complex forms, it can be detected autoradiographically with ^{131}I-labeled antimouse immunoglobulin.

The required clones are recloned, retested, and then frozen. For subsequent antibody production, the cells may be propagated in suspension culture and the supernatant medium used, or they may be passaged in mice as an ascites tumor. Ascitic fluid gives very high yields of antibody although contaminated by globulins from the host animal.

The potential of this technique is enormous for the study of the immune system, production of cell type specific markers, performance of structural analysis of proteins, in the production of vaccines, and in purification of proteins by immunoaffinity chromatography.

IN SITU MOLECULAR HYBRIDIZATION

Like transformation and cloning, hybridization is a word with many meanings. In this context it implies the pairing or matching of two complementary molecules of nucleic acid, single-strand DNA with DNA, or with RNA, the complementarity induced by the base sequences in the nucleic acid polymer. In hybridization *in situ* a purified tritiated DNA molecule is used to probe for complementary RNA sequences in fixed cells, and binding is revealed by autoradiography [Wilkes et al., 1978; Conkie et al., 1974; Harrison et al., 1974] (see Chapter 21). The ^{3}H-DNA is prepared from a specific purified messenger RNA (mRNA), by incubating the mRNA with ^{3}H-nucleotides and the enzyme reverse transcriptase (Fig. 22.7). This enzyme can be extracted from avian myeloblastosis virus, and catalyzes the synthesis of DNA from nucleotides using RNA as a template. Hence the DNA that is synthesized (cDNA) is a complementary copy of the mRNA and has a high affinity for similar mRNA in the cell.

The isolation of purified proteins by monoclonal antibodies may be extended to the isolation of polysomes carrying nascent protein chains from the cytoplasm of a cell, using a specific antibody to the protein being synthesized. On dissociation, the isolated polysomes give ribosomes, nascent polypeptide chains, and the mRNA for that protein. This mRNA can then be used to make a cDNA probe as above. Hence it may be possible to make a wide range of cDNA probes for the characterization of specific cell types, or their stage in differentiation.

VIRUS PREPARATION AND ASSAY

Many of the mass propagation methods described above (Chapter 21) were developed to produce large quantities of virus for analytical and preparative purposes. The current status of our understanding of viruses and their use in the manufacture of vaccines owes much to the development of tissue culture [Habel and Salzman, 1969; Kuchler, 1974,1977; Petricciani, 1979]. The corollary is also true that the development of large-scale culture techniques and serum-free media was prompted in part by the requirements of virology.

Viral assays are of two main types: (1) cytopathic and (2) transforming. Cytopathic viruses may be assayed by their antimetabolic effects in microtitration plates or by the formation of characteristic plaques in monolayers of the appropriate host cell. A viral suspension is serially diluted and added to monolayer culture plates. The number of plaques forming at the limiting dilution is taken as equivalent to the number of infectious particles in the supernatant medium, allowing the concentration of virus in the initial sample to be calculated. Characterization of the virus may be performed with specific antisera, measuring inhibition of the cytopathic effect of the virus, or by radioimmunoassay.

Transforming viruses may be assayed by the selective growth of transformed clones in suspension [Macpherson and Montagnier, 1964], or by looking for transformation foci in monolayer cultures [Temin and Rubin, 1958] (see Chapter 16).

In conclusion, tissue culture is not limited to mammalian and avian systems, though much of modern technology and our current understanding of cellular and molecular biology have derived from these cultures. Using appropriate culture conditions, it is possible to culture cells from cold-blooded vertebrates such as reptiles, amphibia, and fish and many cell lines are now available (commercial catalogues, e.g. Flow, Gibco, and American Type Culture Collection; see Trade Index and list of Cell Banks, Repositories, and Indices).

To cover these and many other fascinating aspects of this field would take many volumes and defeat the objective of this book. It has been more my intention to provide sufficient information to set up a laboratory and prepare the necessary materials with which to perform basic tissue culture and to develop some of the more important techniques required for the characterization and understanding of your cell lines. This book may not be sufficient on its own, but with help and advice from colleagues and other laboratories, it may make your introduction to tissue culture easier and more profitable than it otherwise might have been.

Reagent Appendix

Acetic/Methanol
1 part glacial acetic
3 parts methanol
Make up fresh each time used and keep at 4°C

Agar 2.5%
2.5 g agar
100 ml water
Boil to dissolve
Sterilize by autoclaving
Store at room temperature

Amino Acids—Essential
See Eagle's MEM: "Amino Acids," Chapter 9
(Available as 50x concentrate from commercial suppliers such as Flow Laboratories and GIBCO)
Sterilize by filtration
Store at 4°C in the dark

Amino Acids—Nonessential

Ingredient	*g/liter in a 100× solution*
L-alanine	0.89
L-asparagine H_2O	1.50
L-aspartic acid	1.33
Glycine	0.75
L-glutamic acid	1.47
L-proline	1.15
L-serine	1.05

Sterilize by filtration
Store at 4°C
Use 1:100

Antibiotics
See penicillin, streptomycin, kanamycin, gentamycin, mycostatin

Antifoam (RD emulsion 9964.40)
Aliquot and autoclave to sterilize
Store at room temperature
Dilute 0.1 ml/l liter, i.e., 1:10,000

Bactopeptone, 5%
5 g Difco bactopeptone dissolved in 100 ml Hanks' BSS
Stir to dissolve
Dispense and autoclave
Store at room temperature
Dilute 1/10 for use

Balanced Salt Solutions (BSS)
See Table 9.2
Dissolve each constituent separately, adding $CaCl_2$ last and make up to 1 liter. Adjust to pH 6.5
Hanks' BSS, No phenol red: As regular recipe but omit phenol red
Sterilize by autoclaving. Mark liquid level before autoclaving. Store at room temperature. Make up to mark with sterile deionized distilled water before use, if necessary

Dissection BSS
Hanks' BSS without bicarbonate—previously sterilized by autoclaving
250 units/ml penicillin
250 μg/ml streptomycin
100 μg/ml kanamycin or 50 μmg/ml gentamycin
2.5 μg/ml Amphotericin B
(all preparations sterile, see below)
Store at −20°C

Calcium and Magnesium-Free Saline CMF
See Chapter 9, Table 9.2 for constituents and this Appendix under "Balanced Salt Solutions" for preparation

Carboxymethylcellulose (CMC)
4g CMC
90 ml Hanks' BSS
Weight out 4 g CMC and place in beaker
Add 90 ml Hanks' BSS, and bring to boil to "wet" the CMC
Allow to stand overnight at 4°C to clear
Make volume to 100 ml with Hanks' BSS
Sterilize by autoclaving. The CMC will solidify again during autoclaving but will dissolve at 4°C
For use, e.g., to increase viscosity of medium in suspension cultures, use 3 ml per 100 ml growth medium

Chick Embryo Extract [Paul, 1975]
1. Remove embryos from eggs as described in Chapter 11 and place in 9 cm petri dishes
2. Take out eyes using two pairs of sterile forceps
3. Transfer embryos to flat or round bottomed containers—two embryos to each container
4. Add an equal volume of Hanks' BSS to each
5. Using a sterile glass rod which has been previously heated and flattened at one end, mash the embryos in the BSS until they have broken up
6. Stand for 30 minutes at room temperature
7. Centrifuge for 15 minutes at 2,000*g*
8. Remove supernatant and after keeping a sample to check sterility (see Chapter 10), aliquot and store at −20°C.

Extracts of chick and other tissues may also be prepared by homogenization in a Potter homogenizer or Waring blender [Coon and Cahn, 1966].
1. Homogenize chopped embryos with an equal volume of Hanks' BSS
2. Transfer homogenate to centrifuge tubes and spin at 1000*g* for 10 min
3. Transfer the supernatant to fresh tubes and centrifuge for a further 20 min at 10,000*g*
4. Check sample for sterility (see Chapter 10), aliquot remainder and store at −20°C

0.1 M Citric Acid/0.1% Crystal Violet
21.01 g citric acid

1.0 g crystal violet
Make up to 1000 ml with deionized water
Stir to dissolve
To clarify, filter through Whatman No. 1 filter paper

CMC
See "Carboxymethylcellulose"

CMF
See Chapter 9, Table 9.2

Colcemid, 100× Concentrate
100 mg colcemid
100 ml Hanks' BSS
Stir to dissolve
Sterilize by filtration
Aliquot and store at −20°C

Collagenase
2000 units/ml in Hanks' BSS
100,000 units Worthington CLS grade Collagenase, or equivalent (Specific activity = 150–200 u/ml)
50 ml Hanks' BSS

To dissolve: stir at 36.5°C for 2 hours 4°C overnight
Sterilize by filtration
Divide into aliquots suitable for 1–2 weeks' use
Store at −20°C

Collagenase-Trypsin-Chicken Serum (CTC, Coon and Cahn, [1966])

	Volume	*Final Concentration*
Calcium and magnesium free saline [Moscona, 1952], sterile	85 ml	
Trypsin stock, 2.5%, sterile	4 ml	0.1%
Collagenase stock, 1%, sterile	10 ml	0.1%
Chick serum stock	1 ml	1.0%

Aliquot and store at −20°C

Collection Medium (for tissue biopsies)

Growth medium + antibiotics

Growth medium	500 ml
Penicillin	125,000 units
Streptomycin	125 mg
Kanamycin or	50 mg
Gentamycin	25 mg
Amphotericin	1.25 mg

Store up to 3 weeks at 4°C or at −20°C for longer periods

Counting Fluid × 10 [Paul, 1975]

Sodium chloride	700 g
Tri-sodium citrate ($2H_20$)	170 g

Distilled water up to 10 liters

Stir ingredients in water until dissolved
Filter through Whatman Filter Paper No. 1, then dispense into Pyrex bottles
Adjust pH to 7.2
Autoclave, and store at room temperature
Dilute 1:10 to use, when osmolality should be 335 mOs mols/Kg

Crystal Violet 0.1% in Water

Crystal violet	100 mg
Water	100 ml

D19 (Kodak)
See manufacturers instructions

Dexamethasone (Merck) 1 mg/ml (100×)
This comes already sterile in glass vials. To dissolve, add 5 mls of water by syringe to vial, remove, and dilute to give a concentration of 1 mg/ml
Aliquot and store at −20°C
β-methasone (Glaxo) and methylprednisolone (Sigma) may be prepared in the same way

Dissection BSS
See "Balanced Salt Solutions"

Ficoll, 20%
Sprinkle 20 g Ficoll (Pharmacia) on the surface of 80 ml water and leave overnight to settle and dissolve. Make up to 100 ml
Sterilize by autoclaving
Store at room temperature

Fixative, Photographic (Kodak or Ilford)
See manufacturer's instructions

Fixative for Tissue Culture
See "Acetic/Methanol", this Appendix

Gentamycin (Flow, GIBCO, Schering)
Dilute to 50 μg/ml for use

Giemsa in 0.01 M Phosphate Buffer pH 6.5

$NaH_2PO_4 \cdot 2H_2O$	0.01 M	1.38 g/liter
$Na_2HPO_4 \cdot 7H_2O$	0.01 M	2.68 g/liter

Combine to give pH 6.5
Dilute prepared Giemsa concentrate (Gurr, BDH, Fisher) 1:10 in 100 ml buffer. Filter through Whatman No. 1 filter paper to clarify
Use fresh each time (precipitates on storage)

Glucose, 20%
Glucose 20 g
Dissolve in Hanks' BSS and make up to 100 mls. Sterilize by autoclaving
Store at room temperature

Glutamine 200 mM

L-Glutamine	29.2 g
Hanks' BSS	1000ml

Dissolve glutamine in BSS and sterilize by filtration (see Chapter 10)
Aliquot and store at −20°C

Glutathione
Make 100× stock, i.e., 0.10M in BSS or PBSA and dilute to 1 mM for use
Sterilize by filtration
Aliquot and store at −20°C

Growth Medium
That medium normally used to maintain the cells in exponential growth with the appropriate serum added as specified

Ham's F12
See Chapter 9

Hanks' BSS
See "Balanced Salt Solutions," Chapter 9

HAT [Littlefield, 1964]

Drug	*Concentration*	*Dissolve in*	*Molarity (100× final)*
Hypoxanthine	136 mg/100 ml	0.05 N HCl	1×10^{-2} M
Aminopterin	1.76 mg/100 ml	0.1 N NaOH	4×10^{-5} M
Thymidine	38.7 mg/100 ml	BSS	1.6×10^{-3} M

For use in the HAT selective medium mix equal volumes of sterile (filtered) H, A, and T and add mixture to medium at 3%

Store H and T at 4°C, A at −20°C

HBSS

See "Hanks' BSS," Chapter 9

Hoechst 33258 [Chen, 1977]

(2- 2(4-hydroxyphenol)-6-benzimidazolyl -6-(1-methyl-4-pierpazyl)-benzimidalol-trihydrochloride

Make up 1 mg/ml stock in BSS without phenol red and store at −20°C

For use dilute 1:20,000 (1.0 μl→20 ml) in BSS without phenol red at pH 7.0

This substance may be carcinogenic. Handle with extreme care

Kanamycin Sulphate "Kannasyn" (10 mg/ml)

4 × 1 g vials

Hanks' BSS 400 mls

Add 5 ml BSS from a 400 ml bottle of BSS to each vial

Leave for a few minutes to dissolve

Remove the BSS and kanamycin from the vials and add back to the BSS bottle

Add another 5 ml BSS to each vial to rinse and return to BSS bottle

Dispense 20 ml into sterile containers and store at −20°C

Test sterility: add 2 mls to 10 mls sterile medium and incubate at 36.5 for 72 hrs

Lactalbumin Hydrolysate 5% (10×)

Lactalbumin hydrolysate	5 gms
Hanks' BSS	100 ml

Heat to dissolve

Sterilize by autoclaving

Use at 0.5%

McIlvaines Buffer pH 5.5

		To make 20 ml	*100 ml*
0.2 M Na_2HPO_4	28.4 g/l	11.37 ml	56.85 ml
0.1 M Citric acid	21.0 g/l	8.63 ml	43.15 ml

Media

The constituents of some media in common use are listed in Chapter 9 with the recommended procedure for their preparation. For those not described see Morton [1970], manufacturers catalogues (Flow, GIBCO, Microbiological Associates, etc.), or the original literature references.

MEM

See "Eagle's MEM," Chapter 9

2-Mercaptoethanol (M.W. 78) (Stock solution 5×10^{-3}M (100×))

3.9 mg in 10 ml HBSS

Sterilize by filtration

Store at −20°C or make fresh each time

Methocel

See "Methylcellulose"

Methylcellulose (1.6% in medium)

1. Add 8.0 gms of methylcellulose (4000 centipose) to a 500 ml medium storage bottle containing 250 ml of distilled water at 80–100°C and a magnetic stirrer bar
2. Shake by hand until the methylcellulose is wetted
3. Place on a magnetic stirrer and stir until cooled to room temperature
4. Remove to cold room and continue to stir overnight
5. Autoclave the methocel, when it will form an opaque solid
6. Allow to cool to room temperature and then cool to 4°C
7. Add 250 ml of 2× strength medium and stir overnight at 4°C
8. Dispense into sterile 100 ml bottles and store at −20°C. Thawed bottles may be kept at 4°C for 1 month

For use, dilute methylcellulose medium with appropriate volume of serum and add cell suspension in sufficient growth medium to give final concentration of 0.8% methylcellulose.

Mitomycin-C (Stock solution 100 μg/ml (50×))

2 mg vial

Measure 20 ml HBSS into a sterile container

Remove 2 ml by syringe and add to vial of mitomycin

Allow to dissolve, withdraw and add back to container

Store for 1 week only at 4°C in the dark

(Cover container with aluminum foil.)

For longer period store at −20°C

Dilute to 2 μg/10^6 cells for use

Mycostatin (Nystatin) (2 mg/ml, 100×)

Mycostatin	200 mg
Hanks' BSS	100 ml

Make up by same method as kanamycin. Final concentration 20 μg/ml

Napthalene Black 1% in Hanks' BSS

Napthalene black	1 g
Hanks' BSS	100 mls

Dissolve as much as possible of the stain then filter through Whatman No. 1 filter paper

Penicillin (e.g., Crystapen Benzylpenicillin (sodium)) 1,000,000 units per vial

Use 4 vials and 400 ml Hanks' BSS

Make up as for kanamycin, final concentration 10,000 units/ml

Percoll (Pharmacia)

Made up and sterile and should be diluted with medium or HBSS until correct density is achieved

Check osmolality and adjust to 290 mOsm/Kg by adding water or less medium

Phosphate Buffered Saline PBS (Dulbecco 'A') (See Table 9.1, Chapter 9)

Oxoid Tablets, Code BR 14a, 1 tablet per 100 ml distilled water

Dispense, then autoclave

Store at room temperature, pH 7.3, osmolality 280 mOsm/Kg

PBSB contains the calcium and magnesium and should be made up and sterilized separately. Mix with PBSA, if required, immediately before use

Phosphate Buffered Saline/EDTA (PBS/EDTA) (1mM 0.372 g/liter)

Make PBSA as above

Add EDTA and stir

Dispense, autoclave and store at room temperature

Phytohemagglutinin Stock (500 μg/ml, 100×. Lyophilized (Sigma))

Dissolve powder by adding HBSS by syringe to ampule
Aliquot and store at −20°C
Dilute 1:100 for use

Sodium Citrate/Sodium Chloride (SSC)

Trisodium citrate (dihydrate)	0.03 M	8.82 g	0.09M	26.46 g
Sodium chloride	0.3 M	17.53 g	0.9 M	52.60 g
Water				1000 ml

SSC

See "Sodium Citrate/Sodium Chloride" above.

Streptomycin Sulphate (1 g per vial) (For method see Kanamycin)

Take 2 ml from a bottle containing 100 ml sterile Hanks' BSS and add to 1 g vial of streptomycin. When solution complete, return 2 ml to 98 ml Hanks'
Dilute 1/200 for use
Final concentration—50 μg/ml

Trypsin Diluent-Buffered

Sodium chloride	6.0 g
Trisodium citrate	2.9 g
Tricine (N-Tris (hydroxymethyl) methyl glycine)	1.79 g
Phenol red	0.005 g
Distilled water to	1000 ml

Stir ingredients until dissolved, adjust pH to 7.8
Filter through Whatman No. 1 filter paper
Dispense and autoclave
Osmolality—290 mOsm/Kg

Trypsin Stock 2.5% in 0.85% (0.14 M) NaCl

Trypsin solutions can be bought commercially. Alternatively, to make up a 2.5% solution in 0.85% NaCl, stir for 1 hour at room temperature or 10 hours at 4°C. If trypsin does not dissolve completely, clarify by filtration through Whatman No. 1 filter paper
Sterilize by filtration, aliquot and store at −20°C
Note. Trypsin is available as crude (e.g. Difco 1:250) or purified (e.g., Sigma 3× recrystalized) preparations. Crude preparations contain several other proteases which may be important in cell dissociation but may also be harmful to more sensitive cells. The usual practice is to use crude trypsin unless cell damage reduces viability or reduced growth is observed, when purified trypsin may be used. Pure trypsin has a higher specific activity and should therefore be used at a proportionally lower concentration, e.g., 0.05 or 0.01%.

Trypsin, Versene, Phosphate (TVP)

Trypsin (Difco 1:250)	25 mg (or 1 ml Flow or Gibco 2.5%)
Phosphate buffered saline	98 ml
Disodium EDTA ($2H_2O$)	37 mg
Chick serum (Flow)	1 ml

Mix PBS and EDTA and autoclave then add chick serum and trypsin
If using powdered trypsin, sterilize by filtration before adding chick serum. Aliquot and store at −20°C

Tryptose Phosphate Broth

10 % in Hanks' BSS

Tryptose phosphate (Difco)	100g
Hanks' BSS	1000 ml

Stir until dissolved
Aliquot and sterilize in the autoclave
Store at room temperature
Use 1:100 (final concentration 0.1%)

Viability Stain

See "Naphthalene Black"

Vitamins

See media recipes (Chapter 9)
Make up at 100× concentrated
Sterilize by filtration
Store at −20°C in the dark

Trade Index

SOURCES OF MATERIALS

Material	Source
Agar	GIBCO
	Difco
Amino acids	Sigma
Aminopterin	Sigma
Ampules—glass	Wheaton
Ampules—plastic	Nunc, *see* GIBCO
Ampule sealer	Kahlenberg-Globe
Anemometers	*See* Laminar flow cabinets
Antibiotics	*See* individual antibiotics
Antifoam RD emulsion 9964.40	Dow-Corning
	Hopkin & Williams
Automatic glassware washing machines	John Burge (Equip.) Ltd. (U.K.)
	Camlab Ltd.
	Vernitron (U.S.A.)
Automatic pipettes	Becton Dickinson
	Camlab
	Gilson
	Boehringer
	Flow
	Jencons
	Warner
Automatic pipette plugger	Bellco (A.R. Horwell, U.K.)
	Volac, *see* Camlab
Bactopeptone	Difco
Bench centrifuges	MSE
	Fisher
	DAMON/IEC
	Sorvall
	Beckman
Bovine serum albumen	Sigma
BSS, Hanks', Earle's, etc.	Flow
	GIBCO
Camera lucida	British-American Optical
	E. Leitz (Instruments) Ltd.
	Olympus
	Nikon
	Vickers Instruments
	Carl Zeiss (Oberkochen) Ltd.
Carboxymethylcellulose (CMC)	Fisher
	BDH
Cell counters	Coulter
	Grant Instruments
	GIBCO
	Particle Data
Cellulose nitrate filters	Gelman
	Millipore
	Sartorius GmbH
Centrifugal elutriator	Beckman
Centrifuges	MSE
	Fisher
	Beckman
	DAMON/IEC
Centrifuge—Cytospin	Shandon
Centrifuge tubes, tissue culture	Corning
	Falcon, *see* Becton Dickinson
Channelyzer attachment	Coulter
Chemicals	BDH
	Sigma
	Fisher
Chick plasma	GIBCO
Chicken serum	GIBCO
	Flow
Cine camera	Bolex
Cloning rings	Fisher
Closed-circuit TV	British-American Optical
	E. Leitz (Instruments) Ltd.
	Olympus
	Gallenkamp
	Nikon
	Vickers Instruments
	Carl Zeiss (Oberkochen) Ltd.
CMC	*See* Carboxymethylcellulose
CO_2 automatic change-over unit for cylinders	Distillers Co.
	Air Products
	L.I.P.
CO_2 controllers	Laboratory Impex Ltd.
	Gow Mac
	Heinicke
	Lab-Line
	Burkard Scientific (Sales) Ltd.
CO_2 incubators	Assab
	Astell Hearson
	Baird & Tatlock
	Flow
	Napco
	Gallenkamp
	Heinicke
	Heraeus (Tekmar Co. U.S.A.)
	Grant Instruments
	A.R. Horwell Ltd.
	Lab-Line (Burkard Scientific (Sales) Ltd., U.K.)
	LEEC
	New Brunswick Scientific
	Precision Scientific
	Vindon
Colcemid	Sigma
	CIBA
Collagen (Vitrogen 100)	Collagen Corporation
	Flow
Collagenase	Worthington (Millipore)

Item	Supplier
Collagenase	Sigma
Colony counter	New Brunswick Scientific
	Dynatech (Artech)
Compupet	W.R. Warner
	General Diagnostics Ltd.
Coverslips, glass, plastic	Gallenkamp
	Fisher
	Lux
	L.I.P.
	M A Bioproducts
Crystal violet (Gurr)	Searle Scientific Services
	Fisher
Cytobuckets	DAMON/IEC
Cytofluorograph	Ortho
D19	*See* Photographic
DEAE dextran	Pharmacia
	Bio-Rad
Deionizers	Fisons, *see* MSE
	Elga
	Corning
	Bellco
Densitometers	Joyce-Loebl
	Beckman Istruments
	Gilford Instruments Ltd.
	Helena (M.I. Scientific, U.K.)
	Gelman
DePex, permount	*See* stains
	Fishers
Detergents	Flow
	Decon
	Diversey
	Medical Pharmaceutical Dev. Ltd.
Dexamethasone	Sigma
Dextran	Calbiochem
	Sigma
Dextraven 110	Fisons
Diacetyl fluorescein	Fisher
	Sigma
Dialysis tubing	laboratory suppliers
Disc filter assembly for sterilization	*See* Filters
Disinfectants (Chloros (1Cl))	laboratory suppliers
Dispase	Boehringer Mannheim
DMSO	Merck
DNase	Sigma
EDTA	*See* Chemicals
Ehrlenmeyer flasks	*See* Glassware
Electronic cell counter for cell sizing	Coulter
Electronic thermometer	Comark
	Fisher
Emulsion	*See* Photographic
Epidermal growth factor	Bethesda Research Laboratories
	Uniscience
	Collaborative Research
Ethidium bromide	Sigma
Fibronectin	Uniscience
	Collaborative Research
	Bethesda Research Laboratories
Ficoll	Pharmacia
Filters	Millipore
	Sartorius
	Gelman
	Nucleopore, (Sterilin, U.K.)
	Pall (Europe) Ltd.
Filtration equipment	*See* Filters
Flow cytophotometers and flow cytofluorimeters	Becton Dickinson
	Ortho Instruments
	Coulter
Fluorescence activated cell-sorter	Becton Dickinson
Freezers −20°C	Fisher
	Gallenkamp
	New Brunswick Scientific
Freezer −70°C	Fisher
	Gallenkamp
	Revco
	Cliffco, *see* Fisons
	Kelvinator
Fungicide (Mycostatin)	GIBCO
Gas permeable caps	Camlab
Gentamicin	Schering
Giemsa stain	Searle Scientific Services
	Fisher
Gilson Pipettman	Gilson (USA)
	(*See* Anachem, U.K.)
Glassware	Corning
	Jobling
	Bellco
Glucose	Fisher
	BDH
Glutamine (L)	
solid	Sigma
200 mM	Flow
200 mM	GIBCO
Glutathione	Sigma
Glycerol	Fisher
	BDH
Gradient harvester	MSE
	Fisons
Growth factors	Collaborative Research
	Bethesda Research Laboratories
Hemocytometers	Fisher
	Shandon
Hepes	Flow (1M Solution)
	GIBCO (1M Solution)
	Sigma
Hoechst 33258 fluorescent dye	Flow (home kit)
Hyaluronidase	Worthington (*See* Millipore)
	Sigma
Hydrocortisone	Sigma
	Merck
Hypaque – Ficoll media	Flow
	Pharmacia
Hypochlorite disinfectant	laboratory suppliers
Hypoclearing agent	Kodak

Item	Supplier
	Portex, *See* Portland Plastics
	Cedanco
Patapar paper	Aseptic-thermo Indicator Company
PBS A & B	Oxoid
	Flow
	GIBCO
Penicillin	Glaxo
Percoll	Pharmacia
Petriperm dishes	Heraeus
Photograph filing	Lab-Equip
Photographic developers, fixatives and emulsions	Eastman Kodak Company
	Ilford
	Agfa-Gavaert (Braun, U.S.A.)
Phytohemagglutinin	Pharmacia
	Sigma
Pipettes	Bellco
	A.R. Horwell
	Corning
	Fisher
	Volac
	Sterilin
	Becton Dickinson
Pipette cans: round, square	Bellco
	Denley Instruments Ltd.
Pipette cylinders	local laboratory supplier
Pipette washer, pipette drier	local laboratory supplier
Plastic coverslips	Lux
	M A Bioproducts
Pleated cartridge tower	Pall
	Gelman
Polaroid camera	Polaroid
Poly-D-lysine	Sigma
Polystyrene flasks, petri dishes	*See* Tissue Culture Flasks, etc.
Polyvinyl pyrrolidone, dextran	Calbiochem
	Sigma
Pressure cooker, bench top autoclave	Astell Hearson
	Valley Forge
	Napco
Projection table	Bellco (A.R. Horwell Ltd., U.K.)
Pronase	Sigma
Pumps, peristaltic	Pharmacia
	Gilson
	LKB
	Millipore
Pumps, vacuum	Millipore
	Gelman
Quinacrine dihydrochloride (Atebrin)	Searle Scientific Services
Recording thermometer	Cambridge Instruments
	Fisher
Refractometer	Townson & Mercer Ltd.
	American Optical Co.
Refrigerators	Fisher
	Gallenkamp
	local laboratory suppliers
Repelcote	Hopkin & Williams
Reusable in line filter assembly	*See* Filters
Reverse osmosis	Millipore
	Fisons
	Elga
Roller bottles (glass)	Bellco
Roller bottles (plastic)	*See* Tissue Culture Flasks, etc.
Roller bottle rack	New Brunswick Scientific
	Bellco
Rubber bulb with inlet and outlet valve	Jencons Scientific Ltd.
Scalpels	*See* Instruments, dissecting
Scintillation fluid	Packard
	New England Nuclear
Scintillation vials, minivials	Packard
Semi-permeable nylon film	*See* Packaging
Serum	Flow Laboratories
	Microbiological Associates
	GIBCO
	Seralab
	Tissue Culture Services
Silica gel	Fisher
	BDH
Silicon tubing	ESCO
	Fisher
Slide containers for emulsion–Cyto-Mailer	Lab-Teck, *See* Fisher
Stains	Searle Scientific Services
	Fisher
Stainless steel mesh	Falcon, *See* Becton Dickinson
Streptomycin	Glaxo
Sterilizing and drying oven	Astell Hearson
	Baird & Tatlock
Sterilizing ovens	Astell Hearson
	Baird & Tatlock
	Camlab Ltd.
	Forma
	Gallenkamp Ltd.
	A.R. Horwell Ltd.
	LEEC
	New Brunswick Scientific
	Precision Scientific Co.
	Strands Scientific
Sterilizing tape (indicator)	Fisher
Sterility indicators, Themalog	Brownes Tubes
	Bennet & Co.
	Propper
Stills	Fisons
	Jencons
	Corning
Storage tank for liquid N_2	BOC Cryo Products
	Minnesota Valley Engineering
	Union Carbide Ltd.
Supplements, e.g., tryptose (lactalbumin hydrolysate, bactopeptone)	Difco
Syringes	local laboratory suppliers
Temperature controllers	*See* Thermostats
Temperature recorders	*See* Recording Thermometer

Thermostats, proportional controllers	Fisher
	Jumo
	Napco
α-Thioglycerol	Sigma
Thymidine	Sigma
Timelapse cinemicrography	British-American Optical Ltd.
	E. Leitz Ltd.
	Olympus
	The Projectina Co. Ltd.
	Vickers Instruments
	Carl Zeiss Ltd.
	Bolex
Tissue culture flasks, etc.	Nunc (GIBCO)
	Bellco
	A.R. Horwell
	Corning
	Costar (Bellco) (Northumbria Biologicals, U.K.)
	Falcon (Becton Dickinson)
	Linbro Lux (Flow)
Tissue culture media	Flow
	GIBCO
	Microbiological Associates
Trypan blue	*See* Stains
Tryptophan	*See* Amino Acids
Tryptose phosphate broth	Difco
	Oxoid
Trypsin	Flow
	GIBCO
Trypsin inhibitor (soya bean)	Sigma
Tyrosine	*See* Amino Acids
Universal containers	Sterilin
	Nunc (GIBCO)
Washing machine	Vernitron/Betterbuilt (Burge, U.K.)
Vacuum pump	*See* Pumps
Vinblastin	Sigma
Vinyl tape	3M
Vitafiber	Amicon
Vitamins, solid	Sigma
Vitamins, solution	*See* Tissue Culture Media
Vortex mixer	Gallenkamp
	Fisher
	general laboratory suppliers
X-ray film	Eastman Kodak Company
	Fuji

ADDRESSES OF COMMERCIAL SUPPLIERS

United Kingdom and Europe

ADELPHI MANUFACTURING CO.
207 Duncan Terrace, London, NI8

AGFA-GAVAERT
125 Whittinghame Drive, GI3 1NW

AMICON LTD.
Amicon House, 2 Kingsway, Woking, Surrey, GU21 1UR

ANACHEM LTD.
20a North Street, Luton, Beds, Luton 37274

ASSAB LTD.
Bishops Walk Precinct, High Street, Tewkesbury, Glos., Tewkesbury (0684)

ASTELL HEARSON
172 Brownhill Road, Catford, London, SE6 2DL Unit 1, Brickfields, Huyton Industrial Estate, Huyton, Merseyside, L36 6HY

BDH CHEMICALS LTD.
Poole Dorset, HB12 4NN

BAIRD & TATLOCK (LONDON) LTD.
P.O. Box 1, Romford, Essex, RM1 1HA

BECKMAN-RIIC LTD. ANALYTICAL INSTRUMENTS SALES AND SERVICE OPERATION
Cressex Industrial Estate, Turnpike Road, High Wycombe, Bucks

BENNET & CO.
Brempton Common, Nr Reading, Berks.

BIO-RAD LABORATORIES LTD.
27 Homesdale Road, Bromley, Kent

BOC LTD., SPECIAL GASES
Deer Park Road, London, SW19 3UF

BOEHRINGER CORPORATION (LONDON) LTD., THE
Bell Lane, Lewes, East Sussex, BN7 1LG

BOOTS PHARMACEUTICALS
No. Higham, Nottingham

BOYDEN DATA PAPERS, LTD.
Parkhouse Street, Camberwell, London

BRITISH AMERICAN OPTICAL CO., LTD. (INSTRUMENT GROUP), REICHERT JUNG UK
820 Yeovil Road, Slough, Berks.

ALBERT BRONNE LTD.
Chancery Street, Leicester, LE1 5NA

BURGE (JOHN) (EQUIPMENT) LTD.
35 Furze Platt Road, Maidenhead, Berks.

BURKARD MANUFACTURING CO. LTD.
Woodcock Hill, Rickmansworth, Herts, WD 1PJ

BURKARD SCIENTIFIC (SALES) LTD.
P.O. Box 55, Uxbridge Middx, UB8 1LA

CALBIOCHEM LTD.
79/81 South Street, Bishops Stortford, Herts., CM23 3AL

CAMBRIDGE MEDICAL INSTRUMENTS LTD.
Rustat Road, Cambridge, CB1 3QH

CAMLAB LTD.
Nutfield Road, Cambridge CB4 1TH

CIBA
Horsham, Sussex

COMARK ELECTRONICS LTD.
Brookside Avenue, Rustington, West Sussex

CORNING LTD.
Stone, Staffordshire ST15 OBG

CRYODIFFUSION
49 Ruede Verdun, F27690 Lery, France

CRYOSERVICE LTD.
Platts Common Industrial Estate, Hawshaw Lane, Hoyland, Barnsley, Yorkshire

DAMON/IEC (U.K.) LTD.,
Unit 7, Lawrence Way, Brewers Hill Road, Dustable, Beds, LU6 1BD
DECON LABORATORIES LTD.
Conway Street, Hove, E. Sussex, BN3 3LY
DENLEY INSTRUMENTS LTD.
Daux Road, Billinghurst, Sussex, RH14 9SJ
DIFCO LABORATORIES
P.O. Box 14B, Central Avenue, East Molesey, KT8 OSE
DISTILLERS CO. (CARBON DIOXIDE) LTD., THE
Cedar House, 39 London Road, Reigate, RH2 9QE
DIVERSEY LTD.
Weston Favell Centre, Northhampton, NN3 4PD
ELGA GROUP, THE
Lane End, High Wycombe, Bucks
ENVAIR LTD.
York Ave., Haslingden, Rossendale, Lancashire, BB4 4HX
ESCO (RUBBER) LTD.
43-45 Broad Street, Teddington, Middlesex, TW11 8QZ
FISONS SCIENTIFIC APPARATUS
Bishop Meadow Road, Loughborough, LE11 ORG
FLOW LABORATORIES LTD.
P.O. Box 17, Second Avenue Industrial Estate, Irvine, Ayrshire, KA12 8NB, Scotland
FLUKA A.G.
Buchs, Switzerland CH-9470
FLUORECHEM LTD.
Dinting Vale Trading Estate, Glassop, Derbyshire
GALLENKAMP (A.) & CO. LTD.
P.O. Box 290, Technico House, Christopher Street, London EC2P 2ER
GENERAL DIAGNOSTICS, WILLIAM R. WARNER & CO. LTD.
Chestnut Avenue, Eastliegh, S05 3ZQ
GILFORD INSTRUMENTS LTD.
46-48 Church Road, Teddington, Middlesex
GLAXO LABORATORIES LTD.
Greenford, Middlesex
GONOTEC
Gesellschaft für Mess- und Regeltechnik mbH, Ringstrasse 105, D-1000 Berlin 45, Federal Republic of Germany
GORDON-KEEBLE LTD.
Petersfield House, St. Peter's St., Duxford, Cambridge CB2 4RP
GRANT INSTRUMENTS (CAMBRIDGE) LTD.
Barrington, Cambridge, CB2 5QZ
HEINICKE INSTRUMENTS
Postfach 1203, Friedrich-Ebert-Str. 10, D-8223 Trostberg/Alz, Federal Republic of Germany
HEPAIRE MANUFACTURING LTD.
Station Road, Thatcham, Berks, RG13 4JE
W.C. HERAEUS GmbH, PEW
Postfach 1220, D-3360 Osterade am Harz, Federal Republic of Germany
HOPKIN & WILLIAMS
P.O. Box 1, Romford, Essex, RM1 1HA
HORWELL (ARNOLD R.) LTD.
2 Grangeway, Kilburn High Road, London NW6 2BP
HOWORTH PARTICULATE DIVISION
Lorne Street, Farnworth, Bolton, BL4 7LZ
ILFORD LTD.
Ilford, London
JENCONS SCIENTIFIC LTD.
Mark Road, Hemel Hempstead, Herts.
JOBLING (GLASSWARE)
Stone, Staffs.
JOYCE-LOEBL LTD.
Marquisway, Team Valley, Gateshead, Tyne & Wear, NE11 OUJ
JUMO INSTRUMENT CO.
Hysol, Harlow, CM18 60Z
KODAK LTD.
P.O. Box 66, Kodak House, Station Road, Hemel Hempstead, Herts., HP1 1JU
L.I.P. (EQUIPMENT & SERVICES) LTD.
111 Dockfield Road, Shipley, West Yorkshire, BD17 7AS
LAB-EQUIP
42 Babbacombe Road, Coventry, CV3 5PD
LABORATORY IMPEX LTD.
Lion Road, Twickenham, Middx.
L'AIRE LIQUIDE
Paris, France
L'AIRE LIQUIDE (U.K.) LTD.
44 Hertford St., London W1Y 7TF
LAMB (RAYMOND A.)
6 Sunbeam Road, London NW10 6JL
LEEC
Private Road 7, Colwick, Nottingham, NG4 2AJ
LEITZ (E.) (INSTRUMENTS) LTD.
Luton, Beds.
LKB INSTRUMENTS LTD.
LKB House, 232 Addington Road, Selsdon, Croydon, CR2 8YD
MEDICAL PHARMACEUTICAL DEVELOPMENTS LTD.
Ellen Street, Portslade-by-Sea, Sussex
MERCK
Gottingen, Federal Republic of Germany
MICROFLOW LTD.
Fleet Mill, Minley Road, Fleet, Hampshire, GU13 8RD
MICROMEASUREMENTS
Shirehill Industrial Estate, Shirehill, Saffron, Walden, Essex
M.I. SCIENTIFIC
Suite 7, Second Floor, Exchange Buildings, Quayside, Newcastle-Upon-Tyne
MSE SCIENTIFIC INSTRUMENTS
Manor Royal, Crawley, West Sussex, Crawley
NEW BRUNSWICK SCIENTIFIC
26-34 Emerald Street, London WC1N 3QA
NORTHUMBRIA BIOLOGICALS
South Nelson Industrial Estate, Cranlington, Northumberland NE23 9HL
NYEGAARD & CO. A/S
Oslo, Norway
ORTHO DIAGNOSTICS (DIVISION OF ORTHO PHARMACEUTICAL LTD.)
Saunderton, High Wycombe, Bucks.
OXOID LTD.
Wade Road, Basingstoke, Hants, RG24 OPW
PACKARD INSTRUMENT LTD.
13-17 Church Road, Caversham, Berks., RG4 7AA

PALL PROCESS FILTRATION LTD.
Europa House, P.O. Box 62, Portsmouth, Hants., PO1 3PD
PHARMACIA (GREAT BRITAIN) LTD.
Prince Regent Road, Hounslow, Middx., TW3 1NE
POLAROID (U.K.) LTD.
Ashley Road, St. Albans, AL1 5PR
PORTLAND PLASTICS
Portex Ltd., The Reachfields, Hythe, Kent
PROJECTINA CO., LTD., THE
Skelmorlie, Ayrshire
RADIOCHEMICAL CENTRE LTD., THE
White Lion Road, Amersham., Bucks, HP7 9LL
REICHERT JUNG, U.K., LTD.
820 Yeovil Road, Slough, Berks.
SARTORIUS GmbH
P.O. Box 19, 3440 Gottingen, Federal Republic of Germany
SEARLE SCIENTIFIC SERVICES
Coronation Road, Cressex, High Wycombe, Bucks.
SERA LAB
Crawley Down, Sussex RH10 4LL
STERILIN LTD.
43-45 Broad Street, Teddington, TW11 8QZ
TISSUE CULTURE SERVICES LTD.
10 Henry Road, Slough, SL1 2QI
TOWNSON & MERCER LTD.
101 Beddington Lane, Croydon, Surrey, CR9 4EG
UNION CARBIDE U.K. LTD., CRYOGENICS DIVISION
Redworth Way, Aycliffe Industrial Estate, Aycliffe, Co., Durham
UNISCIENCE LTD.
Uniscience House, 8 Jesus Lane, Cambridge, CB5 8BA
VICKERS LTD., VICKERS INSTRUMENTS
Haxby Road, York, YO3 7SD
VINDON SCIENTIFIC LTD.
Ceramyl Works, Diggle, Oldham, OL3 5YJ
VOLAC
77-93 Tanner Street, Barking, Essex, IG11 8QD
WILLIAM R. WARNER & CO.
See GENERAL DIAGNOSTICS
ZEISS (CARL) (OBERKOCHEN) LTD.
Degenhardt House, 31-36 Foley Street, London W1P 8AP

North America

ADVANCED INSTRUMENTS INC.
1000 Highland Ave., Needham Heights, MA 02194
AIR PRODUCTS & CHEMICALS, INC.
P.O. Box 538, Allentown, PA 18105
AMERICAN OPTICAL CORP., SCIENTIFIC INSTRUMENT DIVISION
Sugar & Eggert Roads, Buffalo, NY 14215
AMICON CORP., SCIENTIFIC SYSTEMS DIVISION
21 Hartwell Ave., Lexington, MA 02173
ARTEK SYSTEMS CORP.
170 Finn Court, Farmingdale, NY 11735
ASEPTIC-THERMO INDICATOR COMPANY
North Hollywood, CA
BAKER CO., INC.
Sanford Airport, Sanford, ME 04073
BECKMAN INSTRUMENTS, INC., ELECTRONIC INSTRUMENTS DIVISION
3900 N. River Road, Schiller Park, IL 60176
BECTON DICKINSON & CO.
Oxnard, CA 93030
BELLCO GLASS, INC.
340 Edrudo Road, Vineland, NJ 08360
BETHESDA RESEARCH LABORATORIES
411 N. Stonestreet Ave., Rockville, MD 20850
BIO-RAD LABORATORIES
32nd & Griffin Ave., Richmond, CA 94804
BOEHRINGER MANNHEIM BIOCHEMICALS
P.O. Box 50816, Indianapolis, IN 46250
BOLEX (USA)
250 Community Drive, Great Neck, NY 11020
BRAUN, N. AMERICA
55 Cambridge Parkway, Cambridge, MA 02142
BUCK, H.J., CO.
10534 York Road, Cockeysville, MD 21030
CALBIOCHEM-BEHRING CORP.
American Hoechst Corp. P.O. Box 12087, San Diego, CA 92112
CEDANCO
P.O. Box 42, Wellesley, MA 02181
COLLABORATIVE RESEARCH, INC.
1265 Main Street, Waltham, MA 02154
COLLAGEN CORPORATION
2455 Faber Place, Palo Alto, CA 94303
CONNAUGHT MEDICAL RESEARCH LABORATORIES
Toronto, Canada
CONTAMINATION CONTROL, INC.
P.O. Box 316, Kulpsville, PA 19443
CORNING MEDICAL, CORNING GLASS WORKS
Medfield, MA 02052
COULTER ELECTRONICS, INC.
590 W. 20 Street, Hialeah, FL 33010
DAMON/IEC
300 Second Ave., Needham Heights, MA 02194
DIFCO LABORATORIES
P.O. Box 1058A, Detroit, MI 48232
DOW CORNING CORP.
P.O. Box 1767, Midland, MI 48640
DYNATECH LABORATORIES, INC., DYNATECH COR.
900 Slaters Lane, Alexandria, VA 22314
EASTMAN KODAK CO.
343 Slate Street, Rochester, NY 14650
FISHER SCIENTIFIC CO.
711 Forbes Ave., Pittsburgh, PA 15219
FLOW LABORATORIES, INC.
1710 Chapman Ave., Rockville, MD 20852
FORMA SCIENTIFIC
P.O. Box 649, Marietta, OH 45750
FUJI PHOTO FILM (U.S.A.)
350 Fifth Ave., New York, NY 10001
GELMAN SCIENCES, INC.
600 S. Wagner Road, Ann Arbor, MI 48106
GERMFREE LABORATORIES INC.
2600 S.W. 28th Lane, Miami, FL 33133

GIBCO LABORATORIES, GRAND ISLAND BIOLOGICAL CO.
3175 Staley Road, Grand Island, NY 14072
GILFORD INSTRUMENT LABORATORIES, INC.
132 Artino Street, Oberlin, OH 44074
GILSON MEDICAL ELECTRONICS, INC.
P.O. Box 27, Middleton, WI 53562
HEINICKE INSTRUMENTS CO.
3000 Taft St., Hollywood, FL 33021
HELENA LABS
P.O. Box 752, 1530 Lindberg Drive, Beaumont, TX 77704
ILFORD
West 70 Century Road, Paramus, NJ 07652
INTERNATIONAL EQUIPMENT CO., (DAMON)
300, 2nd Ave., Needham Has, MA 02194
KAHLENBERG-GLOBE EQUIPMENT CO.
Sarasota, FL
KELVINATOR COMMERCIAL INC.
Wisconsin
LAB-TEK DIVISION, MILES LABORATORIES
475 North Aurora Rd., Naperville, IL 60540
LKB INSTRUMENTS INC.
12221 Parklawn Drive, Rockville, MD 20852
LAB-LINE INSTRUMENTS, INC.
15 & Bloomingdale Ave., Melrose Park, IL 60160
LEITZ, E., INC.
Link Drive., Rockleigh, NJ 07647
LUX SCIENTIFIC CORP.
1157 Tourmaline Drive, Newbury Park, CA 91320
M.A. BIOPRODUCTS
East Coast, Building 100, Biggs Ford Rd., Walkersville, MD 21793, *West Coast,* 11841 Mississippi Ave., Los Angeles, CA 90025
MERCK CHEMICAL DIVISION, MERCK & CO., INC.
P.O. Box 2000, Rahway, NJ 07065
MICROBIOLOGICAL ASSOCIATES
5221 River Road, Bethesda, MD 20016
MILLIPORE CORP.
Ashby Road, Bedford, MA 01730
MINNESOTA VALLEY ENGINEERING
407 7th Street, N.W., New Prague, MN 56071
NAPCO
10855 S.W. Greenburg Rd., Portland, OR 97223
NEW BRUNSWICK SCIENTIFIC CO., INC.
44 Talmadge Road, Edison, NJ 08817
NEW ENGLAND NUCLEAR
549 Albany Street, Boston, MA 02118
NIKON, NIPPON KOGAKU K.K.
Fuji bldg., 2-3 Marunouchi 3-chome, Chiyoda-Ku, Tokyo, Japan
NUCLEOPORE CORP.
7035 Commerce Circle, Pleasanton, CA 94566
OLYMPUS CORPORATION OF AMERICA
4 Nevada Drive, New Hyde Park, NY 11042
ORTHO INSTRUMENTS
376 University Ave., Westwood, MA 02090
OXOID U.S.A. INC.
9017 Red Branch Rd., Columbia, MD 21045
PACKARD INSTRUMENT CO., INC.
2200 Warrenville Road, Downers Grove, IL 60515
PALL CORP.
30 Sea Cliff Ave., Glen Cove, NY 11542
PARTICLE DATA, INC.
Box 265 (111 Hahn), Elmhurst, IL 60126
PHARMACIA FINE CHEMICALS, DIVISION OF PHARMACIA, INC.
800 Centennial Ave., Piscataway, NJ 08854
POLAROID CORP.
549 Technology Square, Cambridge, MA 02139
PRECISION SCIENTIFIC CO.
3737 West Cortland St., Chicago, IL 60647
PROPPER MANUFACTURING CO., INC.
Long Island City, NY 11101
REVCO INC.
Aiken Road, Route 1, Box 275, Asheville, NC 28804
SCHERING CORRPORATION
Kenilworth, NJ 07033
SHANDON SOUTHERN INSTRUMENTS, INC.
515 Broad Street, Sewickley, PA 15143
SIGMA CHEMICAL CO.
P.O. Box 14508, St. Louis, MO 63178
TEKMAR CO.
P.O. Box 37202, Cincinnati, OH 45222
UNION CARBIDE CORP.
Linde Division, P.O. Box 372, South Plainfield, NJ 07080
VALLEY FORGE INSTRUMENT CO., INC.
55 Buckwalter Road, Phoenixville, PA 19460
VERNITRON/BETTERBUILT
S. Empire Blvd., Carlstadt, NJ 07072
VICKERS INSTRUMENTS, INC.
300 Commercial Street, Malden, MA 02148
WHEATON SCIENTIFIC
1000 N. Tenth Street, Millville, NJ 08332
ZEISS, CARL, INC.
444 Fifth Ave., New York, NY 10018

SCIENTIFIC SOCIETIES WITH INTERESTS IN TISSUE CULTURE

American Tissue Culture Association
Secretary: Robert T. Dell'Orco, Samuel Roberts Noble Foundation, Inc., Route One, Ardmore, OK 73401
European Tissue Culture Society (ETCS)
Secretary: Mme. Monique Adolphe, Institute de Pharmocologie, 21 rue de l'Ecole de Médécine, F-75006 Paris
ETCS can also supply information regarding National Tissue Culture Societies in Europe.
American Society for Cell Biology
Executive Officer: Richard S. Young, 9650 Rockville Pike, Bethesda, MD 20814
European Cell Biology Organisation
(Federation: Membership via National Cell Biology Societies in Europe)
Secretary General: Dr. Michael Balls, Department of Human Morphology, University of Nottingham

European Society for Animal Cell Culture Technology (ESACT)
Secretary: Dr. Bryan Griffiths, C.A.M.R. Porton, Porton Down, Salisbury, Wiltshire
British Society for Cell Biology
Secretary: Dr. Paul Whur, Marie Curie Institute, Oxted, Surrey, England

CELL BANKS, REPOSITORIES, AND INDICES

American Type Culture Collection
12301 Parklawn Drive, Rockville, MD
Human Genetic Mutant Cell Repository
Institute for Medical Research, Copewood and Davis Street, Camden, NJ 08103
Catalog of Human and Other Animal Cell Cultures
Naval Biosciences Laboratory, Cell Culture Department, Naval Supply Center, Oakland, CA 94625
Culture de Cellules Eucaryotes. Repertoire ces Utilisateurs.
M. Adolphe, D. Gourdji, A. Tixier-Vidal, R. Robineaux, Inserm Publications, 101 Rue de Tolbiac, 75654 Paris, Cedex 13
British Society for Cell Biology, List of Cell lines and Cell Strains
(Contact Secretary, P. Whur, Marie Curie Institute, Surrey, England.)
Commercial companies such as Flow Laboratories and GIBCO will also supply cell cultures.

Glossary*

Adaptation. Induction or repression of synthesis of a macromolecule (usually a protein) in response to a stimulus, e.g., *enzyme adaptation* — an alteration in enzyme activity brought about by an inducer or repressor and involving an altered rate of enzyme synthesis or degradation

Allograft. *See* homograft.

Amniocentesis. Prenatal sampling of the amniotic cavity

Anchorage-dependent. Requiring attachment to a solid substrate for survival or growth

Anemometer. An instrument for measuring air flow rate

Aneuploid. Not an exact multiple of the haploid chromosome number (haploid = that number present in germ cells after meiosis; i.e., each chromosome represented once)

Balanced salt solution. An isotonic solution of inorganic salts present in approximately the correct physiological concentrations. May also contain glucose but usually free of other organic nutrients

Cell culture. Growth of cells dissociated from the parent tissue by spontaneous migration or mechanical or enzymatic dispersal

Cell fusion. Formation of single cell body by fusion of two other cells; either spontaneously or, more often, by induced fusion with inactivated sendai virus or polyethylene glycol

Cell line. A propagated culture after the first subculture

Cell strain. A characterized cell line derived by selection or cloning

Chemically defined. Used of medium to imply that it is made entirely from pure defined constituents. Distinct from "serum free" where other poorly characterized constituents may be used to replace serum

Clone. A population of cells derived from one cell

Confluent. Where all the cells are in contact all round their periphery with other cells, and no available substrate is left uncovered

Contact inhibition. Inhibition of cell membrane ruffling and cell motility when cells are in complete contact with other adjacent cells, as in a confluent culture. Often precedes cessation of cell proliferation but not necessarily causally related

Continuous cell line or cell strain. One having the capacity for infinite survival. Previously known as "established" and often referred to as "immortal"

Deadaptation. Reversible loss of a specific property due to the absence of the appropriate inducer (not always defined)

*[See also Schaeffer, 1979]

Dedifferentiation. A term implying irreversible loss of the specialized properties that a cell would have expressed *in vivo.* As evidence accumulates that cultures "dedifferentiate" by a combination of selection of undifferentiated cells or stromal cells and deadaptation resulting from the absence of the appropriate inducers, this term is going out of favor. It is still correctly applied to progressive loss of differentiated morphology in histological observations of, for example, tumor tissue

Density limitation of growth. Mitotic inhibition correlated with an increase in cell density

Diploid. Each chromosome represented as a pair, identical in the autosomes and female sex chromosomes and non-identical in male sex chromosomes, and corresponding to the chromosome number and morphology of most somatic cells of the species from which the cells were derived

Ectoderm. The inner-most germ layer of the embryo giving rise to the epithelial component of organs such as the gut, liver, and lungs

Endoderm. The outer germ layer of the embryo giving rise to the epithelium of the skin

Endothelium. An epithelial-like cell layer lining spaces within mesodermally derived tissues, such as blood vessels, and derived from the mesoderm of the embryo

Epithelial. Used of a culture to imply cells derived from epithelium but often used more loosely to describe any cells of a polygonal shape with clear sharp boundaries between cells. "Pavement-like". More correctly this should be termed "epithelioid" or "epithelial-like"

Epithelium. A covering or lining of cells, as in the surface of the skin or lining of the gut, and derived from the embryonic endoderm or ectoderm

Euploid. Exact multiples of the haploid chromosome set. The correct morphology characteristic of each chromosome pair in the species from which the cells were derived is not implicit in the definition but is usually assumed to be the case. Otherwise it should be stated as "euploid but with some chromosomal aberrations"

Explant. A fragment of tissue transplanted from its original site and maintained in an artificial medium

Fibroblast. A proliferating precursor cell of the mature differentiated fibrocyte

Fibroblastic. Resembling fibroblasts, i.e., spindle shaped (bipolar) or stellate (multipolar); usually arranged in parallel arrays at confluence if contact inhibited. Often used indiscriminately for undifferentiated mesodermal cells regardless of their relationship to the fibrocyte lineage. Implies a migratory type of cell with processes exceeding the nuclear diameter by threefold or more

Finite cell line. A culture which has been propogated by subculture but is only capable of a limited number of cell generations *in vitro* before dying out

Generation number. The number of population doublings (estimated from dilution at subculture) that a culture has undergone since explantation. Necessarily contains an approximation of the number of generations in primary culture

Haploid. That chromosome number where each chromosome is represented once. In most higher animals it is the number present in the gametes and half of the number found in most somatic cells

Heterokaryon. Genetically different nuclei in a common cytoplasm, usually derived by cell fusion

Heteroploid. A term used to describe a culture (not a cell) where the cells comprising the culture have chromosome numbers other than diploid

Histotypic. A culture resembling tissue-like morphology *in vivo*. It is usually implied that this is a three dimensional culture recreated from dispersed cell culture which attempts to retain, by cell proliferation and multilayering or by reaggregation, the tissue-like structure. Organ cultures cannot be propagated whereas histotypic cultures can

Homiothermic. Able to maintain a constant body temperature in spite of environmental fluctuation

Homograft. (Allograft). A graft derived from a genetically different donor of the same species as the recipient

Homokaryon. Genetically identical nuclei in a common cytoplasm usually a product of cell fusion

Hybrid cell. Mononucleate cell which results from the fusion of two different cells, leading to the formation of a synkaryon

Ideogram. The arrangement of (in the case of genetic analysis of a cell) the chromosomes in order by size and morphology so that the karyotype may be studied

Induction. An increase in effect produced by a given stimulus. *Embryonic induction.* The interaction of cells from two different germ layers, promoting differentiation, often reciprocal. *Enzyme induction.* An increase in synthesis of an enzyme produced by, for example, hormonal stimulation

Isograft. (Syngraft). A graft derived from a genetically identical or nearly identical donor of the same species as the recipient

Karyotype. The distinctive chromosomal complement of a cell

Laminar flow. The flow of a fluid that closely follows the shape of a streamlined surface without turbulence. Used in connection with laminar air flow cabinets to imply a stable flow of air over the work area such as to minimize turbulence

Laminar flow cabinet or hood. A work station with filtered air flowing in a laminar nonturbulent flow parallel to (horizontal laminar flow) or perpendicular to (vertical laminar flow) the work surface, such as to maintain the sterility of the work

Malignant. A term to describe a tumor which has become invasive or metastatic (i.e. colonizing other tissues). Usually progressive leading to destruction of host cells and ultimately death of the host

Manometer. A "U" shaped tube containing liquid, the levels of which in each limb of the "U" reflect the pressure difference between the ends

Medium. A mixture of inorganic salts and other nutrients capable of sustaining cell survival *in vitro* for 24 hrs. *Growth medium.* That medium which is used in routine culture such that the cell number increases with time. *Maintenance medium.* A medium which will retain cell survival without growth (cell proliferation), e.g. a low serum or serum free medium used with serum dependent cells to maintain cell survival without cell proliferation

Mesenchyme. Loose, often migratory, embryonic tissue derived from the mesoderm, giving rise to connective tissue, cartilage, muscle, hemopoietic cells, etc. in the adult

Mesoderm. A germ layer in the embryo arising between the ectoderm and endoderm and giving rise to connective tissue etc. (as above for mesenchyme)

Monoclonal. Derived from a single clone of cells. *Monoclonal antibody.* Antibody produced by a clone of lymphoid cells either *in vivo* or *in vitro*. *In vitro* the clone is usually derived from a hybrid of a sensitized spleen cell and continuously growing myeloma cell

Myeloma. A tumor derived from myeloid cells. Used in monoclonal antibody production when the myeloma cell can produce immunoglobulin

Neoplastic. A new, unnecessary, proliferation of cells giving rise to a tumor

Organ culture. The maintenance or growth of organ primordia or the whole or parts of an organ *in vitro* in a way that may allow differentiation and preservation of the architecture and/or function

Passage. The transfer or subculture of cells from one culture vessel to another. Usually, but not necessarily, implies subdivision of a proliferating cell population enabling propagation of a cell line or cell strain. *Passage number* — the number of times a culture has been subcultured

Plating efficiency. The percentage of cells seeded at subculture giving rise to colonies. If each colony can be said to be derived from one cell this is synonomous with cloning efficiency. Sometimes used loosely to describe the number of cells surviving after subculture but this is better termed the "seeding efficiency"

Poikilothermic. Body temperature close to that of the environment and not regulated by metabolism

Population density. The number of monolayer cells per unit area of substrate. For cells growing in suspension this term is identical to the cell concentration

Population doubling time. The interval required for a cell population to double at the middle of the logarithmic phase of growth

Primary culture. A culture started from cells, tissue or organs taken directly from an organism and before the first subculture

Pseudodiploid. Numerically diploid chromosome number but with chromosomal aberrations

Quasidiploid. See pseudodiploid

Saturation density. Maximum number of cells attainable per cm^2 (monolayer culture) or per ml (suspension culture) under specified culture conditions

Seeding efficiency. The percentage of the inoculum which attaches to the substrate within a stated period of time (implying viability, or survival but not necessarily proliferative capacity)

Somatic cell genetics. The study of cell genetics by recombination and segregation of genes in somatic cells. Usually by cell fusion

Split Ratio. The divisor of the dilution ratio of a cell culture at subculture, e.g., one flask divided into four or 100 ml up to 400 ml would be a split ratio of 4

Sub confluent. Less than confluent. All of the available substrate is not covered

Subculture. See passage

Substrate. The matrix or solid underlay upon which a monolayer culture grows

Super confluent. When a monolayer culture progresses beyond the state where all the cells are attached to the substrate and multilayering occurs

Suspension culture. Where cells will multiply suspended in medium

Synkaryon. A hybrid cell which results from the fusion of the nuclei it carries

Tetraploid. Twice the diploid (four times the haploid) number of chromosomes

Tissue culture. Properly, the maintenance of fragments of tissue *in vitro* but now commonly applied as a generic term to include tissue explant culture, organ culture and dispersed cell culture, including the culture of propagated cell lines and cell strains

Xenograft. Transplantation of tissue to a different species from which it was derived. Often used to describe implantation of human tumours in athymic (nude), immune deprived, or immune suppressed mice

References

Aaronson, S.A., Todaro, G.J., Freeman, A.E. (1970) Human sarcoma cells in culture, identification by colony-forming ability on monolayers of normal cells. Exp. Cell Res. 61:1–5.

Abercrombie, M., Heaysman, J.E.M. (1954) Observations on the social behaviour of cells in tissue culture, II. "Monolayering" of fibroblasts. Exp. Cell Res. 6:293–306.

Adams, D.O. (1979) Macrophages. In Jakoby, W.B., Pastan, I.H. (eds): "Methods of Enzymology, Vol. LVII, Cell Culture." New York, Academic Press, pp. 494–506.

Adams, R.L.P. (1980) "Laboratory Techniques in Biochemistry and Molecular Biology. Cell Culture for Biochemists." Work, T.S., Burdon, R.H. (eds.) Amsterdam, Elsevier/North Holland Biomedical Press.

Ambrose, E.J., Dudgeon, J.A., Easty, D.M., Easty, G.C. (1961) The inhibition of tumor growth by enzymes in tissue culture. Exp. Cell Res. 24:220–227.

Ames, B.N. (1980) Identifying environmental chemicals causing mutations and cancer. Science 204:587–593.

Andersson, L.C., Nilsson, K. Gahmberg, C.G. (1979a) K562—a human erythroleukemic cell line. Int. J. Cancer 23:143–147.

Andersson, L.C., Jokinen, M., Klein, E., Klein, G., Nilsson, K. (1979b) Presence of erythrocytic components in the K562 cell line. Int. J. Cancer 24:514.

Andersson, L.C., Jokinen, M., Gahmberg, C.G. (1979c) Induction of erythroid differentiation in the human leukaemia cell line K562. Nature 278:364–365.

Antoniades, H.N., Scher, C.D., Stiles, C.D. (1979) Purification of human platelet-derived growth factor. Proc. Natl. Acad. Sci. USA 76:1809.

Arrighi, F.E., Hsu, T.C. (1974) Staining constitutive heterochromatin and Giemsa crossbands of mammalian chromosomes. In Yunis, J. (ed.): "Human Chromosome Methodology, 2nd Ed." New York, Academic Press.

Au, A.M.-J., Varon, S. (1979) Neural cell sequestration on immunoaffinity columns. Exp. Cell Res. 120:269.

Aub, J.C., Tieslau, C., Lankester, A. (1963) Reactions of normal and tumor cell surfaces to enzymes, I. wheat-germ lipase and associated mucopolysaccharides. Proc. Natl. Acad. Sci. USA 50:613–619.

Auerbach, R., Grobstein, C. (1958) Inductive interaction of embryonic tissues after dissociation and reaggregation. Exp. Cell Res. 15:384–397.

Augusti-Tocco, G., Sato, G. (1969) Establishment of functional clonal lines of neurons from mouse neuroblastoma. Proc. Natl. Acad. Sci. USA 64:311–315.

Avrameas, S. (1970) Immunoenzyme techniques: Enzymes as markers for the localization of antigens and antibodies. In Bourne, G.H., Danielli, J.F., (eds): "International Review of Cytology." New York, Academic Press, pp. 349–385.

Balin, A.K., Goodman, B.P., Rasmussen, H., Cristofalo, V.J. (1976) The effect of oxygen tension on the growth and metabolism of WI-38 cells. J. Cell. Physiol. 89:235–250.

Ballard, P.L. (1979) Glucocorticoids and differentiation. Glucocorticoid Horm Action 12:439–517.

Ballard, P.L., Tomkins, G.M. (1969) Dexamethasone and cell adhesion. Nature 224:344–345.

Bard, D.R., Dickens, M.J., Smith, Audrey U., Sarek, J.M. (1972) Isolation of living cells from mature mammalian bone. Nature 236:314–315.

Barde, Y.A., Lindsay, R.M., Monard, D., Thoenen, H. (1978) New factor released by cultured cells supporting survival and growth of sensory neurones. Nature 274:818.

Barkley, W.E. (1979) Safety considerations in the cell culture laboratory. In Jacoby, W.B., Pastan, I. (eds.): "Methods of Enzymology, LVIII." Chap. 4. New York, Academic Press, pp. 36–43.

Barnes, D., Sato, G. (1980) Methods for growth of cultured cells in serum-free medium. Anal. Biochem. 102:255–270.

Barnstable, C. (1980) Monoclonal antibodies which recognize different cell types in the rat retina. Nature 286:231–234.

Barski, G., Sorieul, S., Cornefert, F. (1960) Production dans les cultures *in vitro* de deux souches cellulaires en association de cellules de caractère "hybride". C.R. Acad. Sci. [D] Paris 251:1825.

Baserga, R. (1962) A study of nucleic acid synthesis in ascites tumor cells by two-emulsion autoradiography. J. Cell Biol. 12:633–637.

Bateman, A.E., Peckham, M.J., Steel, G.G. (1979) Assays of drug sensitivity for cells from human tumours: *In vitro* and *in vivo* tests on a xenografted tumour. Br. J. Cancer 40:81.

Bell, E., Ivarsson, B., Merrill, C. (1979) Production of a tissue-like structure by contraction of collagen lattices by human fibroblasts of different proliferative potential *in vitro*. Proc. Natl. Acad. Sci. USA 76:1274–1279.

Benda, P., Lightbody, J., Sato, G., Levine, L., Sweet, W. (1968) Differentiated rat glial cell strain in tissue culture. Science 161:370.

Benya, P.D., Padilla, S.R., Nimni, M.E. (1978) Independent regulation of collagen types by chondrocytes during the loss of differentiated function in culture. Cell 15:1313–1321.

Berger, S.L. (1979) Lymphocytes as resting cells. In Jakoby, W.B., Pastan, I.H. (eds): "Methods in Enzymology, Vol. LVII, Cell Culture." New York, Academic Press, pp. 486–494.

Bergerat, J.P., Barlogie, B., Drewinko, B. (1979) Effects of cis-dichloro-diammineplatinum (II) on human colon carcinoma cells *in vitro*. Cancer Res. 39:1334.

Berky, J.J., Sherrod, P.C., eds (1977) "Short Term *in vitro* Testing for Carcinogenesis, Mutagenesis and Toxicity." Philadelphia, The Franklin Institute Press.

Berry, M.N., Friend, D.S. (1969) High yield preparation of isolated rat liver parenchymal cells. A biochemical and fine structural study. J. Cell Biol. 43:506–520.

Berry, R.J., Laing, A.H., Wells, J. (1975) Fresh explant cultures of human tumours *in vitro* and the assessment of sensitivity to cytotoxic chemotherapy. Br. J. Cancer 31:218–227.

Biedler, J.L. (1976) Chromosome Abnormalities in human tumour cells in culture. In Fogh, J. (ed): "Human Tumour Cells in Vitro." New York, Academic Press.

Biggers, J.D., Gwatkin, R.B.C., Heyner, S. (1961) Growth of embryonic avian and mammalian tibiae on a relatively simple chemically defined medium. Exp. Cell Res. 25:41.

Birch, J.R., Pirt, S.J. (1970) Improvements in a chemically-defined medium for the growth of mouse cells (strain LS) in suspension. J. Cell Sci. 7:661–670.

Birch, J.R., Pirt, S.J. (1971) The quantitative glucose and mineral nutrient requirements of mouse LS (suspension) cells in chemically-defined medium. J. Cell Sci. 8:693–700.

Birnie, G.D., Simons, P.J. (1967) The incorporation of ^{3}H-thymidine and ^{3}H-uridine into chick and mouse embryo cells cultured on stainless steel. Exp. Cell Res. 46:355–366.

Blaker, G.J., Birch, J.R., Pirt, S.J. (1971) The glucose, insulin and glutamine requirements of suspension cultures of HeLa cells in a defined culture medium. J. Cell Sci. 9:529–537.

Bobrow, M., Madan, J., Pearson, P.L. (1972) Staining of some specific regions on human chromosomes, particularly the secondary constriction of no. 9. Nature 238:122–124.

Booyse, F.M., Sedlak, B.J., Rafelson, M.E. (1975) Culture of arterial endothelial cells. Characterization and growth of bovine aortic cells. Thromb. Diathes. Haemorrh. 34:825–839.

Bornstein, M.B., Murray, M.R. (1958) Serial observations on patterns of growth, myelin formation, maintenance and degeneration in cultures of newborn rat and kitten cerebellum. J. Biophys. Biochem. Cytol. 4:499.

Boyum, A. (1968a) Isolation of leucocytes from human blood. A two-phase system for removal of red cells with methylcellulose as erythrocyte aggregative agent. Scand. J. Clin. Lab. Invest. (Suppl. 97)21:9–29.

Boyum, A (1968b) Isolation of leucocytes from human blood. Further observations. Methylcellulose, dextran and Ficoll as erythrocyte aggregating agents. Scand. J. Clin. Lab. Invest. (Suppl. 97)31:50.

Braaten, J.T., Lee, M.J., Schewk, A., Mintz, D.H. (1974) Removal of fibroblastoid cells from primary monolayer cultures of rat neonatal endocrine pancreas by sodium ethylmercurithiosalicylate. Biochem. Biophys. Res. Comm. 61:476–482.

Bradley, T.R., Metcalf, D. (1966) The growth of mouse bone marrow cells *in vitro*. Aust. J. Biol. Med. 44:287–300.

Bradley, N.J., Bloom, H.J.G., Davies, A.J.S., Swift, S.M. (1978) Growth of human gliomas in immune-deficient mice: A possible model for pre-clinical therapy studies. Br. J. Cancer 38:263.

Breen, G.A.M., De Vellis, J. (1974) Regulation of glycerol phosphate dehydrogenase by hydrocortisone in dissociated rat cerebral cell cultures. Dev. Biol. 41:255–266.

Brockes, J.P., Fields, K.L., Raff, M.C. (1979) Studies on cultured rat Schwann cells. I. Establishment of purified populations from cultures of peripheral nerve. Brain Res. 165:105.

Brouty-Boyé, D., Gresser, I., Baldwin, C. (1979) Reversion of the transformed phenotype to the parental phenotype by subcultivation of x-ray transformed $C_3H/10T½$ at low cell density. Int. J. Cancer 2:253–260.

Brouty-Boyé, D., Tucker, R.W., Folkman, J. (1980) Transformed and neoplastic phenotype: Reversibility during culture by cell density and cell shape. Int. J. Cancer 26:501–507.

Brunk, C.F., Jones, K.C., James, T.W. (1979) Assay for nanogram quantities of DNA in cellular homogenates. Anal. Biochem. 92:497–500.

Buehring, G.C. (1972) Culture of human mammary epithelial cells. Keeping abreast of a new method. J. Natl. Cancer Inst. 49:1433–1434.

Buick, R.N., Stanisic, T.H., Fry, S.E., Salmon, S.E., Trent, J.M., Krosovich, P. (1979) Development of an Agar-methyl cellulose clonogenic assay for cells of transitional cell carcinoma of the human bladder. Cancer Res. 39:5051–5056.

Buonassisi, V., Sato, G., Cohen, A.I. (1962) Hormone-producing cultures of adrenal and pituitary tumor origin. Proc. Natl. Acad. Sci. USA 48:1184–1190.

Burgess, A.W., Metcalf, D. (1980) The nature and action of granulocyte-macrophage colony stimulating factors. Blood 56:947–958.

Burke, J.M., Ross, R. (1977) Collagen synthesis by monkey arterial smooth muscle cells during proliferation and quiescence in culture. Exp. Cell Res. 107:387–395.

Burwen, S.J., Pitelka, D.R. (1980) Secretory function of lactating moose mammary epithelial cells cultured on collagen gels. Exp. Cell Res. 126:249–262.

Cahn, R.D., Lasher, R. (1967) Simultaneous synthesis of DNA and specialized cellular products by differentiating cartilage cells *in vitro*. Proc. Natl. Acad. Sci. USA 58:1131–1138.

Cahn, R.D., Cooh, H.G., Cahn, M.B. (1967) In Wilt, F.H., Wessells, N.K. (eds): "Methods in Developmental Biology." New York, Thomas Y. Crowell, pp. 493.

Carlsson J., Gabel, D., Larsson, E., Westermark, B. (1979) Protein-coated agarose surfaces for attachment of cells. In Vitro 15:844–50.

Carney, D.N., Bunn, P.A., Gazdar, A.F., Pagan, J.A., Minna, J.D. (1981) Selective growth in serum-free hormone-supplemented medium of tumor cells obtained by biopsy from patients with small cell carcinoma of lung. Proc. Natl. Acad. Sci. USA 78:3185–3189.

Carpenter, G., Cohen, S. (1977) Epidermal growth factor. In Acton, R.T., Lynn, J.D. (eds): "Cell Culture and Its Application." New York, Academic Press, pp. 83–105.

Carrel, A. (1912) On the permanent life of tissues outside the organism. J. Exp. Med. 15:516–528.

Caspersson, T., Farber, S., Foley, G.E., Kudynowski, J., Modest, E.J., Simonsson, E., Wagh, U., Zech, L. (1968) Chemical differentiation along metaphase chromosomes. Exp. Cell Res. 49:219–222.

Catsimpoolas, N., Griffith, A.L., Skrabut, E.M., Valeri, C.R. (1978) An alternate method for the preparative velocity sedimentation of cells at unit gravity. Anal. Biochem. 87:243–248.

Center for Disease Control, Office of Biosafety, Atlanta, GA 00333, USA "Proposed Biosafety Guidelines for Microbiological and Bacteriological Laboratories." Publications Dept., DHHS, Public Health Service.

Ceriani, R.L., Taylor-Papadimitriou, J., Peterson, J.A., Brown, P. (1979) Characterization of cells cultured from early lactation milks. In Vitro 15:356–362.

Chen, T.R. (1977) In situ detection of mycoplasm contamination in cell cultures by fluorescent Hoechst 33258 stain. Exp. Cell Res. 104:255.

Choi, K.W., Bloom, A.D. (1970) Cloning human lymphocytes *in vitro*. Nature 227:171–173.

Clark, J., Hirtenstein, M., Gebb, C. (1979) Critical parameters in the microcarrier culture of animal cells. Presented at the Third General Meeting of the European Society of Animal Cell Technology, Oxford, October 2–5, 1979.

Clark, J.M., Pateman, J.A. (1978) Long-term culture of Chinese hamster Kupffer cell lines isolated by a primary cloning step. Exp. Cell Res. 112:207–217.

Cohen, J., Balazs, R., Hojos, F., Currie, D.N., Dutton, G.R. (1978) Separation of cell types from the developing cerebellum. Brain Res. 148:313–331.

Cohen, S. (1962) Isolation of a mouse submaxillary gland protein accelerating incisor eruption and eyelid opening in the new-born animal. J. Biol. Chem. 237:1555–1562.

Cole, R.J., Paul, J. (1966) The effects of erythropoietin on haem synthesis in mouse yolk sac and cultured foetal liver cells. J. Embryol. Exp. Morphol. 15:245–260.

Collins, S.J., Gallo, R.C., Gallagher, R.E. (1977) Continuous growth and differentiation of human myeloid leukaemic cells in suspension culture. Nature 270:347–349.

Committee on Standardized Genetic Nomenclature for Mice. (1972) Standard karyotype of the mouse *Mus musculis*. J. Hered. 63:69.

Conkie, D., Affara, N., Harrison, P.R., Paul, J., and Jones, K., (1974) *In situ* localization of globin messenger RNA formation. II. After treatment of Friend virus-transformed mouse cells with dimethyl sulphoxide. J. Cell Biol. 63:414–419.

Coon, H.G., Cahn, R.D. (1966) Differentiation *in vitro:* Effects of sephadex fractions of chick embryo extract. Science 153:1116–1119.

Coons, A.H., Kaplan, M.M. (1950) Localization of antigen in tissue cells. II. Improvements in a method for the detection of antigen by means of fluorescent antibody. J. Exp. Med. 91:1–13.

Cooper, G.W. (1965) Induction of somite chondrogenesis by cartilage and notochord: A correlation between inductive activity and specific stages of cytodifferentiation. Dev. Biol. 12:185–212.

Cooper, P.D., Burt, A.M., Wilson, J.N. (1958) Critical effect of oxygen tension on rate of growth of animal cells in continuous suspended culture. Nature 182:1508–1509.

Cour, I., Maxwell, G., Hay, R.J. (1979) Tests for bacterial and fungal contaminants in cell cultures as applied at the ATCC. In Evans, V.J., Perry, V.P., Vincent, M.M. (eds.): "Manual of the American Tissue Culture Association." 5:1157–1160.

Courtenay, V.D., Selby, P.J., Smith, I.E., Mills, J., Peckham, M.J. (1978) Growth of human tumor cell colonies from biopsies using two soft-agar techniques. Br. J. Cancer 38:77–81.

Cox, R.P., ed. (1974) "Cell Communication." New York, John Wiley & Sons.

Creasey, A.A., Smith, H.S., Hackett, A.J., Fukuyama, K., Epstein, W.L., Madin, S.H. (1979) Biological properties of human melanoma cells in culture. In Vitro 15:342.

Crissman, H.A., Mullaney, P.F., Steinkamp, J.A. (1975) Methods and applications of flow systems for analysis and sorting of mammalian cells. In Prescott, D.M. (ed): "Methods in Cell Biology, Vol. IX." New York, Academic Press.

Curtis, A.S.G., Seehar, G.M. (1978) The control of cell division by tension of diffusion. Nature 274:52–53.

Defendi, V. (1976) In Nelson, D.S. (ed): "Immunobiology of the Macrophage." New York, Academic Press, pp. 275–286.

Defendi, V., ed. (1964) "Retention of Functional Differentiation in Cultured Cells." Philadelphia, The Wistar Institute Press.

Dendy, P.P. (1976) Some problems in the use of short-term cultures of human tumours for *in vitro* screening of cytotoxic drugs. Chemotherapy 7:341–350.

De Ridder, L., Mareel, M. (1978) Morphology and ^{125}I-concentration of embryonic chick thyroids cultured in an atmosphere of oxygen. Cell Biol. Int. Rep. 2:189–194.

De Vonne, T.L., Mouray, H. (1978) Human α_2-macroglobulin and its antitrypsin and antithrombin activities in serum and plasma. Clin. Chim. Acta 90:83–85.

Dexter, T.M., Allen, T.D., Scott, D., Teich, N.M. (1979) Isolation and characterisation of a bipotential haematopoietic cell line. Nature 277:471–474.

Dexter, T.M., Allen, T.D., Lajtha, L.G. (1977) Conditions controlling the proliferation of haemopoietic stem cells *in vitro*. J. Cell Physiol. 91:335–345.

De Vitry, F., Camier, M., Czernichow, P., Benda, Ph., Cohen, P., Tixier-Vidal, A. (1974) Establishment of a clone of mouse hypothalamic neurosecretory cells synthesizing neurophysin and vasopressin. Proc. Natl. Acad. Sci. USA 71:3575–3579.

DiPaolo, J.A. (1965) *In vitro* test systems for cancer chemotherapy. III. Preliminary studies of spontaneous mammary tumors in mice. Cancer Chemother. Rep. 44:19–24.

Douglas, W.H.J., McAteer, J.A., Dell'Orco, R.T., Phelps, D. (1980) Visualization of cellular aggregates cultured on a three-dimensional collagen sponge matrix. In Vitro 16:306–312.

Douglas, W.H.J., Moorman, G.W., Teel, R.W. (1976) The formation of histotypic structures from monodispersed rat lung cells cultured on a three-dimensional substrate. In Vitro 12:373–381.

Duksin, D., Maoz, A., Fuchs, S. (1975) Differential cytotoxic activity of anticollagen serum on rat osteoblasts and fibroblasts in tissue culture. Cell 5:83–86.

Dulak, N.C., Temin, H.M. (1973a) A partially purified rat liver cell conditioned medium with multiplication-stimulating activity for embryo fibroblasts. J. Cell. Physiol. 81:153–160.

Dulak, N.C., Temin, H.M. (1973b) Multiplication-stimulating activity for chicken embryo fibroblasts from rat liver cell conditioned medium: A family of small peptides. J. Cell. Physiol. 81:161–170.

Dulbecco, R., Vogt, M. (1954) Plaque formation and isolation of pure cell lines with poliomyelitis viruses. J. Exp. Med. 199:167–182.

Dulbecco, R., Freeman, G. (1959) Plaque formation by the polyoma virus. Virology 8:396–397.

Dulbecco, R., Elkington, J. (1973) Conditions limiting multiplication of fibroblastic and epithelial cells in dense cultures. Nature 246:197–199.

Eagle, H., Foley, G.E., Koprowski, H., Lazarus, H., Levine, E.M., Adams, R.A. (1970) Growth characteristics of virus-transformed cells. J. Exp. Med. 131:863–879.

Eagle, H. (1955) The specific amino acid requirements of mammalian cells (strain L) in tissue culture. J. Biol. Chem. 214:839.

Eagle, H. (1959) Amino acid metabolism in mammalian cell cultures. Science 130:432.

Eagle, H. (1973) The effect of environmental pH on the growth of normal and malignant cells. J. Cell. Physiol. 82:1–8.

Earle, W.R., Schilling, E.L., Stark, T.H., Straus, N.P., Brown, M.F., Shelton, E. (1943) Production of malignancy *in vitro*. IV. The mouse fibroblast cultures and changes seen in the living cells. J. Nat. Can. Inst. 4:165–212.

Easton, T.G., Valinsky, J.E., Reich, E. (1978) M540 as a fluorescent probe of membranes: Staining of electrically excitable cells. Cell 13:476–486.

Easty, D.M., Easty, G.C. (1974) Measurement of the ability of cells to infiltrate normal tissues *in vitro*. Br. J. Cancer 29:36–49.

Easty, G.C., Easty, D.M., Ambrose, E.J. (1960) Studies of cellular adhesiveness. Exp. Cell Res. 19:539–548.

Ebendal, T., Jacobson, C.O. (1977) Tissue explants affecting extension and orientation of axons in cultured chick embryo ganglia. Exp. Cell Res. 105:379–387.

Ebendal, T. (1979) Stage-dependent stimulation of neurite outgrowth exerted by nerve growth factor and chick heart in cultured embryonic ganglia. Dev. Biol. 72:276.

Ebendal, T. (1976) The relative roles of contact inhibition and contact guidance in orientation of axons extending on aligned collagen fibrils *in vitro*. Exp. Cell Res. 98:159–169.

Edelman, G.M. (1973) Nonenzymatic dissociations. B. Specific cell fractionation on chemically derivatized surfaces. In Kruse, P.F., Jr., Patterson, M.K., Jr. (eds): "Tissue Culture Methods and Applications." New York, Academic Press, pp. 29–36.

Edwards, P.A.W., Easty, D.M., Foster, C.S. (1980) Selective culture of epithelioid cells from a human squamous carcinoma using a monoclonal antibody to kill fibroblasts. Cell Biol. Int. Rep. 4:917–922.

Eisen, H., Bach, R., Emery, R. (1977) Induction of spectrin in Friend erythroleukaemic cells. Proc. Natl. Acad. Sci. U.S.A. 74:3898–4002.

Eisinger, M., Lee, J.S., Hefton, J.M., Darzykiewicz, A., Chiao, J.W., Deharven, E. (1979) Human epidermal cell cultures—growth and differentiation in the absence of dermal components or medium supplements. Proc. Natl. Acad. Sci. U.S.A. 76:5340.

Elsdale, T., Bard, J. (1972) Collagen substrata for studies on cell behaviour. J. Cell Biol. 54:626–637.

Espmark, J.A., Ahlqvist-Roth, L. (1978) Tissue typing of cells in cultures. I. Distinction between cell lines by the various patterns produced in mixed haemabsorption with selected multiparous sera. J. Immunol. Methods 24:141–153.

Evans, V.J., Bryant, J.C. (1965) Advances in tissue culture at the National Cancer Institute in the United States of America. In: "Tissue Culture." Ramakrishnan, C.V. (ed.) The Hague, W. Junk pp. 145–167.

Evans, V.J., Bryant, J.C., Fioramonti, M.C., McQuilkin, W.T., Sanford, K.K., Earle, W.R. (1956) Studies of nutrient media for tissue C cells *in vitro*. I. A protein-free chemically defined medium for cultivation of strain L cells. Cancer Res. 16:77.

Federoff, S. (1975) In Evans, V.J., Perry, V.P., Vincent, M.M. (eds.): "Manual of the Tissue Culture Association." 1:53–57.

Fell, H.B. (1953) Recent advances in organ culture. Sci. Progr. 162:212.

Finbow, M.E., Pitts, J.D. (1981) Permeability of junctions between animal cells. Exp. Cell Res. 131:1–13.

Fisher, H.W., Puck, T.T., Sato, G. (1958) Molecular growth requirements of single mammalian cells: The action of fetuin in promoting cell attachment of glass. Proc. Natl. Acad. Sci. USA 44:4–10.

Fisher, M., Solursh, M. (1979) The influence of the substratum on mesenchyme spreading *in vitro*. Exp. Cell Res. 123:1.

Foley, J.F., Aftonomos, B. Th. (1973) Pronase. In Kruse, P.F., Jr., Patterson, M.K. Jr. (eds): "Tissue Culture Methods and Applications." New York: Academic Press, pp. 185–188.

Fogh, J. (1973) "Contamination in Tissue Culture." New York: Academic Press.

Folkman, J., Moscona, A. (1978) Role of cell shape in growth control. Nature 273:345–349.

Folkman, J., Haudenschild, C. (1980) Angiogenesis *in vitro*. Nature 288:551–556.

Folkman, J., Haudenschild, C.C., Zetter, B.R. (1979) Long-term culture of capillary endothelial cells. Proc. Natl. Acad. Sci. USA 76:5217.

Folkman, J., Tucker, R.W. (1980) Cell configuration, substrate and growth control. In Subtellny, S., Wessells, N.K. (eds): "Cell Surface, Mediator of Developmental Processes." New York, Academic Press.

Foreman, J., Pegg, D.E. (1979) Cell preservation in a programmed cooling machine: The effect of variations in supercooling. Cryobiology 16:315–321.

Frame, M., Freshney, R.I., Shaw, R., Graham, D.I. (1980) Markers of differentiation in glial cells. Cell Biol. Int. Rep. 4:732.

Fraser, C.M., Venter, J.C. (1980) The synthesis of B-adrenergic receptors in cultured human lung cells: Induction by glucocorticoids. Biochem. Biophys. Res. Commun. 94:390–398.

Fredin, B.L., Seiffert, S.C., Gelehrter, T.D. (1979) Dexamethasone-induced adhesion in hepatoma cells: The role of plasminogen activator. Nature 277:312–313.

Freedman, V.H., Shin, S. (1974) Cellular tumorigenicity in nude mice: Correlation with cell growth in semi-solid medium. Cell 3:355–359.

Freeman, A.E., Igel, H.J., Herrman, B.J., Kleinfeld, K.L. (1976) Growth and characterisation of human skin epithelial cultures. In Vitro 12:352–62.

Freshney, R.I. (1972) Tumour cells disaggregated in collagenase. Lancet 2:488–489.

Freshney, R.I. (1976a) Separation of cultured cells by isopycnic centrifugation in metrizamide gradients. In Rickwood, D. (ed): "Biological Separations." London and Washington, Information Retrieval, Ltd., pp. 123–130.

Freshney, R.I. (1976b) Some observations on assay of anticancer drugs in culture. In Dendy, P.P. (ed.): "Human Tumours in Short Term Culture." New York, Academic Press, pp. 150–158.

Freshney, R.I. (1978) Use of tissue culture in predictive testing of drug sensitivity. Cancer Topics 1:5–7.

Freshney, R.I. (1980) Culture of glioma of the brain. In Thomas, D.G.T., Graham, D.I. (eds): "Brain Tumours, Scientific Basic, Clinical Investigation and Current Therapy." London: Butterworths, pp. 21–50.

Freshney, R.I., Hart, E. (1982) Clonogenicity of human glia in suspension. Br. J. Cancer 46:463.

Freshney, R.I., Celik F., Morgan, D. (1982) Analysis of cytotoxic and cytostatic effects. Excerpta Medica (in press).

Freshney, R.I., Hart, E., Russell, J.M. (1982) Isolation and purification of cell cultures from human tumours. In Reid, E., Cook, G.M.W., Morre, D.J. (eds): "Cancer Cell Organelles. Methodological Surveys (B): Biochemistry, Vol. II." Chichester, England, Horwood, pp. 97–110.

Freshney, R.I., Morgan, D., Hassanzadah, M., Shaw, R., Frame, M. (1980a) Glucocorticoids, proliferation and the cell surface. In Richards, R.J., Rajan, K.T. (eds): "Tissue Culture in Medical Research (II)." Oxford, Pergamon Press, pp. 125–132.

Freshney, R.I., Paul, J., Kane, I.M. (1975) Assay of anti-cancer drugs in tissue culture: Conditions affecting their ability to incorporate ^{3}H-leucine after drug treatment. Br. J. Cancer 31:89–99.

Freshney, R.I., Sherry, A., Hassanzadah, M., Freshney, M., Crilly, P., Morgan, D. (1980b) Control of cell proliferation in human glioma by glucocorticoids. Br. J. Cancer 41:857–866.

Friend, C., Patuleia, M.C., Nelson, J.B. (1966) Antibiotic effect of tylosine on a mycoplasma contaminant in a tissue culture leukemia cell line. Proc. Soc. Exp. Biol. Med. 121:1009.

Friend, C., Scher, W., Holland, J.G., Sato, T. (1971) Hemoglobin synthesis in murine virus-induced leukemic cells *in vitro*. 2. Stimulation of erythroid differentiation by dimethyl sulfoxide. Proc. Natl. Acad. Sci. USA 68:378–382.

Friend, K.K., Dorman, B.P., Kucherlapati, R.S., Ruddle, F.H. (1976) Detection of interspecific translocations in mouse-human hybrids by alkaline Giemsa staining. Exp. Cell Res. 99:31–36.

Fritz, G.R., Knobil, E. (1964) Amino acid transport and protein synthesis in muscle. Action of insulin. Proc. Exp. Biol. Med. 116:873–875.

Fry, J., Bridges, J.W. (1979) The effect of phenobarbitone on adult rat liver cells and primary cell lines. Toxicol. Letters 4:295–301.

Gartler, S.M. (1967) Genetic Markers as Tracers in Cell Culture. Second Dicennial Review Conference on Cell, Tissue and Organ Culture. NCI Monographs, pp. 167–195.

Gaush, C.R., Hard, W.L., Smith, T.F. (1966) Characterization of an established line of canine kidney cells (MDCK). Proc. Soc. Exp. Biol. Med. 122:931–933.

Geppert, E.F., Williams, M.C., Mason, R.J. (1980) Primary culture of rat alveolar type II cells on floating collagen membranes. Exp. Cell Res. 128:363–374.

Gey, G.O., Coffman, W.D., Kubicek, M.T. (1952) Tissue culture studies of the proliferative capacity of cervical carcinoma and normal epithelium. Cancer Res. 12:364–365.

Gilbert, S.F., Migeon, B.R. (1975) D-valine as a selective agent for normal human and rodent epithelial cells in culture. Cell 5:11–17.

Gilbert, S.F., Migeon, B.R. (1977) Renal enzymes in kidney cells selected by D-Valine medium. J. Cell. Physiol. 92:161–168.

Gilchrest, B.A., Nemore, R.E., Maciag, T. (1980) Growth of human keratinocytes on fibronectin-coated plates. Cell Biol. Int. Repts. 4:1009–1016.

Gilden, D.H., Wroblewska, Z., Eng, L.F., Rorke, L.B. (1976) Human brain in tissue culture. Part 5. Identification of glial cells by immunofluorescence. J. Neurol. Sci. 29:177–184.

Gillis, S., Watson, J. (1981) Interleukin-2 dependent culture of cytolytic T. cell lines. Immunol. Rev. 54:81–109.

Gimbrone, M.A., Jr., Cotran, R.S., Folkman, J. (1974) Human vascular endothelial cells in culture, growth and DNA synthesis. J. Cell Biol. 60:673–684.

Giovanella, B.C., Stehlin, J.S., Williams, L.J. (1974) Heterotransplantation of human malignant tumors in "nude" mice. II. Malignant tumors induced by injection of cell cultures derived from human solid tumors. J. Natl. Can. Inst. 52:921.

Goldberg, B. (1977) Collagen synthesis as a marker for cell type in mouse 3T3 lines. Cell 11:169–172.

Golde, D.W., Cline, M.J. (1973) Cultivation of normal and neoplastic human bone marrow leucocytes in liquid suspension. In: "Proc. 7th Leucocyte Culture Conference." New York: Academic Press.

Good, N.E., Winget, G.D., Winter, W., Connolly, T.N., Izawa, S., Singh, R.M.M. (1966) Hydrogen ion buffers and biological research. Biochemistry 5:467–477.

Gorham, L.W., Waymouth, C. (1965) Differentiation in vitro of embryonic cartilage and bone in a chemically defined medium. Proc. Soc. Exp. Biol. Med. 119:287–290.

Gospodarowicz, D. (1974) Localization of fibroblast growth factor and its effect alone and with hydrocortisone on 3T3 cell growth. Nature 249:123–127.

Gospodarowicz, D., Delgado, D., Vlodavsky, I. (1980) Permissive effect of the extracellular matrix on cell proliferation *in vitro*. Proc. Natl. Acad. Sci. 77:4094–4098.

Gospodarowicz, D., Greenburg, G., Bialecki, H., Zetter, B.R. (1978) Factors involved in the modulation of cell proliferation *in vivo* and *in vitro*: The role of fibroblast and epidermal growth factors in the proliferative response of mammalian cells. In Vitro 14:85–118.

Gospodarowicz, D., Greenburg, G., Birdwell, C.R. (1978) Determination of cell shape by the extra cellular matrix and its correlation with the control of cellular growth. Cancer Res. 38:4155–4171.

Gospodarowicz, D., Mescher, A.L. (1977) A comparison of the responses of cultured myoblasts and chondrocytes to fibroblast and epidermal growth factors. J. Cell. Physiol. 93:117–128.

Gospodarowicz, D., Moran, J. (1974) Growth factors in mammalian cell cultures. Ann. Rev. Biochem. 45:531–558.

Gospodarowicz, D., Moran, J.S., Braun, D.L. (1977) Control of proliferation of bovine vascular endothelial cells. J. Cell. Physiol. 91:377–386.

Gospodarowicz, D., Moran, J., Braun, D., Birdwell, C. (1976) Clonal growth of bovine vascular endothelial cells: Fibroblast growth factor as a survival agent. Proc. Natl. Acad. Sci. USA 73:4120–4124.

Granner, D.K., Hayashi, S., Thompson, E.B., Tomkins, G.M. (1968) Stimulation of tyrosine aminotransferase synthesis by dexamethasone phosphate in cell culture. J. Mol. Biol. 35:291–301.

Green, A.E., Athreya, B., Lehr, H.B., Coriell, L.L. (1967) Viability of cell cultures following extended preservation in liquid nitrogen. Proc. Soc. Exp. Biol. Med. 124:1302–1307.

Green, H. (1977) Terminal differentiation of cultured human epidermal cells. Cell 11:405–416.

Green, H., Kehinde, O., Thomas, J. (1979) Growth of cultured human epidermal cells into multiple epithelia suitable for grafting. Proc. Natl. Acad. Sci. USA 76:5665–5668.

Green, H., Thomas, J. (1978) Pattern formation by cultured human epidermal cells: Development of curved ridges resembling dermatoglyphs. Science 200:1385–1388.

Greenleaf, R.D., Mason, R.J., Williams, M.C. (1979) Isolation of alveolar type II cells by centrifugal elutriation. In Vitro 15:673.

Griffiths, J.B., Pirt, G.J. (1967) The uptake of amino acids by mouse cells (Strain LS) during growth in batch culture and chemostat culture: The influence of cell growth rate. Proc. R. Soc. Biol. 168:421–438.

Grobstein, C. (1953) Morphogenetic interaction between embryonic mouse tissues separated by a membrane filter. Nature 4384:869–871.

Grobstein, C. (1953) Epithelio-mesenchymal specificity in the morphogenesis of mouse submandibular rudiments *in vitro*. J. Exp. Zool. 124:383.

Guilbert, L.J., Iscove, N.N. (1976) Partial replacement of serum by selenite, transferrin, albumin and lecithin in haemopoietic cell cultures. Nature 263:594–595.

Gullino, P.M., Knazak, R.A. (1979) Tissue culture on artificial capillaries. In Jakoby, W.B., Pastan, I. (eds.): "Methods in Enzymology, Vol. LVIII. Cell Culture." New York: Academic Press, pp. 178–184.

Guillouzo, A., Guguen-Guillouzo, C., Bourel, M. (1981) Hepatocytes in culture: Expression of differentiated functions and their application to the study of metabolism. Triangle (Sandoz J. Med. Sci.) 20:121–8.

Guner, M., Freshney, R.I., Morgan, D., Freshney, M.G., Thomas, D.G.T., Graham, D,I, (1977) Effects of dexamethasone and betamethasone on *in vitro* cultures from human astrocytoma. Br. J. Cancer 35:439–47.

Gwatkin, R.B.L. (1973) Pronase. In Kruse, P.F., Jr., Patterson, M.K., Jr. (eds): "Tissue Culture Methods and Applications." New York: Academic Press, pp. 3–5.

Habel, K., Salzman, N.P., eds. (1969) Fundamental Techniques in Virology. New York: Academic Press.

Ham, R.G. (1963) An improved nutrient solution for diploid Chinese hamster and human cell lines. Exp. Cell Res. 29:515.

Ham, R.G. (1965) Clonal growth of mammalian cells in a chemically defined synthetic medium. Proc. Natl. Acad. Sci. USA 53:288.

Ham, R.G., McKeehan, W.L. (1978) Development of improved media and culture conditions for clonal growth of normal diploid cells. In Vitro 14:11–22.

Ham, R.G., McKeehan, W.L. (1979) Media and growth requirements. In Jakoby, W.B., Pastan, I.H. (eds): "Methods in Enzymology, Volume LVIII, Cell Culture." Academic Press, New York, pp. 44–93.

Hamburger, A.W., Salmon, S.E. (1977) Primary bioassay of human tumor stem cells. Science 197:461–463.

Hamilton, W.G., Ham, R.G. (1977) Clonal growth of Chinese hamster cell lines in protein-free media. In Vitro 13:537–547.

Hanks, J.H., Wallace, R.E. (1949) Relation of oxygen and temperature in the preservation of tissues by refrigeration. Proc. Exp. Biol. Med. 71:196.

Harrington, W.N., Godman, G.C. (1980) A selective inhibitor of cell proliferation from normal serum. Proc. Natl. Acad. Sci. USA. Biol. Sci. 77:423–427.

Harris, H., Watkins, J.F. (1965) Hybrid cells derived from mouse and man: Artificial heterokaryons of mammalian cells from different species. Nature 205:640–646.

Harris, L.W., Griffiths, J.B. (1977) Relative effects of cooling and warming rates on mammalian cells during the freeze-thaw cycle. Cryobiology 14:662–669.

Harris, H., Hopkinson, D.A. (1976) "Handbook of Enzyme Electrophoresis in Human Genetics." New York, American Elsevier.

Harrison, P.R., Conkie, D., Affara, N., Paul, J. (1974) *In situ* localization of globin messenger RNA formation. I. During mouse foetal liver development. J. Cell Biol. 63:402–413.

Harrison, R.G. (1907) Observations on the living developing nerve fiber. Proc. Soc. Exp. Biol. Med. 4:140–143.

Hart, I.R., Fidler, I.J. (1978) An *in vitro* quantitative assay for tumor cell invasion. Cancer Res. 38:3218–3224.

Hassbroek, F.J., Neggle, J.C., Fleming, A.L. (1962) High-resolution auto-radiography without loss of water-soluble ions. Nature 195:615–616.

Hauschka, S.D., Konigsberg, I.R. (1966) The influence of collagen on the development of muscle clones. Proc. Natl. Acad. Sci. USA 55:119–126.

Hay, R.J., Kern, J., Caputo, J. (1979) Testing for the presence of viruses in cultured cell lines. In; "Manual of American Tissue Culture Association," 5:1127–1130.

Hay, R.J., Strehler, B.L. (1967) The limited growth span of cell strains isolated from the chick embryo. Exp. Gerontol. 2:123.

Hayashi, I., Sato, G.H. (1976) Replacement of serum by hormones permits growth of cells in a defined medium. Nature 259:132–134.

Hayflick, L., Moorhead, P.S. (1961) The serial cultivation of human diploid cell strains. Exp. Cell Res. 25:585–621.

Heldin, C.H., Westermark, B., Wasteson, A. (1979) Platelet-derived growth factor: purification and partial characterization. Proc. Natl. Acad. Sci. USA 76:3722–3726.

Hemstreet, G.P., Enoch, P.G., Pretlow, T.G. (1980) Tissue disaggregation of human renal cell carcinoma with further isopyknic and isokinetic gradient purification. Cancer Res 40:1043–1049.

Herzenberg, L.A., Sweet, R.G., Herzenberg, L.A. (1976) Fluorescence-activated cell sorting. Sci. Am. 234:108–117.

Higuchi, K. (1977) Cultivation of mammalian cell lines in serum-free chemically defined medium. Methods Cell Biol. 14:131.

Hilwig, I., Gropp, A. (1972) Staining of constitutive heterochromatin in mammalian chromosomes with a new fluorochrome. Exp. Cell Res. 75:122–126.

Hokin, L.E., Hokin, M.R. (1963) Biological transport. Ann Rev Biochem. 32:553–577.

Holden, H.T., Lichter, W., Sigel, M.M. (1973) Quantitative methods for measuring cell growth and death. In Kruse, P.F., Jr., Patterson, M.K., Jr (eds): "Tissue Culture Methods and Applications." New York, Academic Press, pp. 408–412.

Hollenberg, M.D., Cuatrecasas, P. (1973) Epidermal growth factor: Receptors in human fibroblasts and modulation of action by cholera toxin. Proc. Natl. Acad. Sci. 70:2964–2968.

Holley, R.W., Armour, R., Baldwin, J.H. (1978) Density-dependent regulation of growth of BSC-1 cells in cell culture: Growth inhibitors formed by the cells. Proc. Natl. Acad. Sci. USA 75:1864–1866.

Honn, K.V., Singley, J.A., Chavin, W. (1975) Fetal bovine serum: A multivariate standard (38805). Proc. Soc. Exptl. Biol. Med. 149:344–347.

Horibata, K., Harris, A.W. (1970) Mouse myelomas and lymphomas in culture. Exp. Cell Res 60:61–77.

Horita, A., Weber, L.J. (1964) Skin penetrating property of drugs dissolved in dimethylsulfoxide (DMSO) and other vehicles. Life Sci. 3:1389–1395.

House, W. (1973) Bulk culture of cell monolayers. In Kruse, P.F., Jr., Patterson, M.K., Jr. (eds): "Tissue Culture Methods and Applications." New York, Academic Press, pp. 338–344.

House, W., Shearer, M., Maroudas, N.G. (1972) Method for bulk culture of animal cells on plastic film. Exp. Cell Res. 71:293–296.

Howard, M., Kessler, S., Chused, T., Paul, W.E. (1981) Long term culture of normal mouse B lymphocytes. Proc. Natl. Acad. Sci. USA 78:5788–5792.

Howard, B.V., Macarak, E.J., Gunson, D., Kefalides, N.A. (1976) Characterization of the collagen synthesized by endothelial cells in culture. Proc. Natl. Acad. Sci. USA. 73:2361–2364.

Howie Report (1978) "Code of Practice for Prevention of Infection in Clinical Laboratories and Post-Mortem Rooms." London, H.M. Stationary Office.

Hume, D.A., Weidemann, M.J. (1980) Mitogenic lymphocyte transformation. Amsterdam, Elsevier/North Holland Biomedical Press.

Hynes, R.O. (1973) Alteration of cell-surface proteins by viral transformation and by proteolysis. Proc. Natl. Acad. Sci. USA 70:3170–3174.

Hynes, R.O. (1974) Role of cell surface alterations in cell transformation. The importance of proteases and cell surface proteins. Cell 1:147–156.

Hynes, R.O. (1976) Cell surface proteins and malignant transformation. Biochim. Biophys. Acta 458:73–107.

Ilsie, A.W., Puck, T.T. (1971) Morphological transformation of Chinese hamster cells by dibutyryl adenosine cycline 3′:5′-monophosphate and testosterone. Proc. Natl. Acad. Sci. USA. 2:358–361.

An international system for human cytogenetic nomenclature. (1978) Report of the Standing Committee on Human Cytogenetic Nomenclature. The National Foundation—March of Dimes.

Iscove, N., Melchers, F. (1978) Complete replacement of serum by albumin, tranferrin and soybean lipid in cultures of lipopolysaccharide-reactive B lymphocytes. J. Exp. Med. 147:923–933.

Iscove, N.N., Guilbert, L.W., Weyman, C. (1980) Complete replacement of serum in primary cultures of erythropoitin-dependent red cell precursors (CFU-E) by albumin, transferrin, iron, unsaturated fatty acid, lecithin and cholesterol. Exp. Cell Res. 126:121–126.

Itagaki, A., Kimura, G. (1974) TES and HEPES buffers in mammalial cell cultures and viral studies: Problems of carbon dioxide requirements. Exp. Cell Res. 83:351–360.

Jones, T.L., Haskill, J.S. (1973) Polyacrylamide: An improved surface for cloning of primary tumors containing fibroblasts. J. Natl. Cancer Inst. 51:1575–1580.

Jones, T.L., Haskill, J.S. (1976) Use of polyacrylamide for cloning of primary tumors. Methods Cell Biol. 14:195.

Kahn, P., Shin, S.-L. (1979) Cellular tumorigenicity in nude mice. Test of association among loss of cell-surface fibronectin, anchorage independence, and tumor-forming ability. J. Cell Biol. 82:1.

Kaltenbach, J.P., Kaltenbach, M.H., Lyons, W.B. (1958) Nigrosin as a dye for differentiating live and dead ascites cells. Exp. Cell Res. 15:112–117.

Kao, F.-T., Puck, T.T. (1968) Genetics of somatic mammalial cells. VII. Induction and isolation of nutritional mutants in Chinese hamster cells. Proc. Natl. Acad. Sci. USA 60:1275–1281.

Katsuta, H., (ed). (1978) "Nutritional requirements of cultured cells." Tokyo, Japan Scientific Societies Press, Baltimore, University Park Press.

Kaufmann, M., Klinga, K., Runnebaum, B, Kubli, F. (1980) *In vitro* adriamycin sensitivity test and hormonal receptors in primary breast cancer. Eur. J. Cancer 16:1609–1613.

Kawamura, A., Jr., ed. (1969) "Fluorescent Antibody Techniques and Their Applications." Tokyo, University of Tokyo Press.

Kelley, D.S., Becker, J.E., Potter, V.R. (1978) Effect of insulin, dexamethasone, and glucagon on the amino acid transport ability of four rat hepatoma cell lines and rat hepatocytes in culture. Cancer Res. 38:4591–4601.

Kempner, E.S., Miller, J.H. (1962) Autoradiographic resolution of doubly labeled compounds. Science 135:1063–1064.

Kim, Y.S., Whitehead, J.S., Perdomo, J. (1979) Glycoproteins of cultured epithelial cells from human colonic adenocarcinoma and fetal intestine. Eur. J. Cancer 15:725–735.

Kinlough, M.A., Robson, H.N. (1961) Chromosome preparations obtained directly from peripheral blood. Nature 192:684.

Kissane, J.M., Robbins, E. (1958) The fluorometric measurement of deoxyribonucleic acid in animal tissues with specific reference to the central nervous system. J. Biol. Chem. 233:184–188.

Kitos, P.A., Sinclair, R., Waymouth, C. (1962) Glutamine metabolism by animal cells growing in a synthetic medium. Exp. Cell. Res. 27:307–316.

Kleinman, H.K., McGoodwin, E.B., Rennard, S.I., Martin, G.R. (1979) Preparation of collagen substrates for cell attachment: Effect of collagen concentration and phosphate buffer. Anal. Biochem. 94:308.

Klevjer-Anderson, P., Buehring, G.C. (1980) Effect of hormones on growth rates of malignant and nonmalignant human mammary epithelia in cell culture. In Vitro 16:491–501.

Knazek, R.A. (1974) Solid tissue masses formed *in vitro* from cells cultured on artificial capillaries. Fed. Proc. 33:1978–1981.

Knazek, R.A., Gullino, P., Kohler, P.O., Dedrick, R. (1972) Cell culture on artificial capillaries. An approach to tissue growth *in vitro*. Science 178:65–67.

Knazek, R.A., Kohler, P.O., Gullino, P.M. (1974) Hormone production by cells grown *in vitro* on artificial capillaries. Exp. Cell Res. 84:251.

Knox, P., Wells, P. (1979) Cell adhesion and proteoglycans. I. The effect of exogenous proteoglycans on the attachment of chick embryo fibroblasts to tissue cultured plastic and collagen. J. Cell Sci. 40:77–88.

Kohler, G., Milstein, C. (1975) Continuous cultures of fused cells secreting antibody of predefined specificity. Nature 256:495–497.

Konigsberg, I.R. (1979) Skeletal myoblasts in culture. In Jakoby, W.B., Pastan, I.H. (eds.): "Methods in Enzymology, Vol. LVII, Cell Culture." New York, Academic Press, pp. 511–527.

Kopriwa, B.M. (1963) A Model Dark Room Unit for Radioautography. J. Histochem. Cytochem. 11:553–555.

Kosher, R.A., Church, R.L. (1975) Stimulation of *in vitro* somite chondrogenesis by procollagen and collagen. Nature 258:327–330.

Kreisberg, J.I., Sachs, G., Pretlow, T.G.E., McGuire, R.A. (1977) Separation of proximal tubule cells from suspensions of rat kidney cells by free-flow electrophoresis. J. Cell Physiol. 93:169–172.

Kreth, W., Herzenberg, L.A. (1974) Fluorescence-activated cell sorting of human T and B lymphocytes. Cell. Immunol. 12:396–406.

Kruse, P.F., Jr., Keen, L.N., Whittle, W.L. (1970) Some distinctive characteristics of high density perfusion cultures of diverse cell types. In Vitro 6:75–88.

Kruse, P.F., Jr., Miedema, E. (1965) Production and characterization of multiple-layered populations of animal cells. J. Cell Biol. 27:273.

Kuchler, R.J. ed. (1974) "Animal Cell Culture and Virology." Stroudsburg, Pa., Dowden, Hutchinson & Ross, Inc.

Kuchler, R.J. (1977) "Biochemical Methods in Cell Culture and Virology." New York, Academic Press.

Kuriharcuch, W., Green, H. (1978) Adipose conversion of 3T3 cells depends on a serum factor. Proc. Natl. Acad. Sci. 75:6107–6110.

Kurtz, J.W., Wells, W.W. (1979) Automated fluorometric analysis of DNA, protein, and enzyme activities: Application of methods in cell culture. Anal. Biochem. 94:166.

Lasfargues, E.Y. (1973) Human mammary tumors. In Kruse, P, Patterson, M.K. (eds.): "Tissue Culture Methods and Applications." New York, Academic Press, pp. 45–50.

Laug, W.E., Tokes, Z.A., Benedict, W.F., Sorgente, N. (1980) Anchorage independent growth and plasminogen activator production by bovine endothelial cells. J. Cell Biol. 84:281–293.

Lechner, J.F., Haugen, A., Autrup, H., McClendon, I.A., Trump, B.F., Harris, C.C. (1981) Clonal growth of epithelial cells from normal adult human bronchus. Cancer Res. 41:2294–2304.

Leder, A., Leder, P. (1975) Butyric acid, a potent inducer of erythroid differentiation in cultured erythroleukemic cells. Cell 5:319–322.

Leibo, S.P., Mazur, P. (1971) The role of cooling rates in low-temperature preservation. Cryobiology 8:447–452.

Leibovitz, A. (1963) The growth and maintenance of tissue cell cultures in free gas exchange with the atmosphere. Am. J. Hyg. 78:173–183.

Leighton, J., Mark, R., Rush, G. (1968) Patterns of three-dimensional growth in collagen coated cellulose sponge: Carcinomas and embryonic tissues. Cancer Res. 28:286–296.

Leighton, J. (1951) A sponge matrix method for tissue culture. Formation of organized aggregates of cells *in vitro*. J. Natl. Cancer Inst. 12:545–561.

Lesser, B., Brent, T.P. (1970) Cold storage as a method for accumulating mitotic HeLa cells without impairing subsequent synchronous growth. Exp. Cell Res. 62:470–473.

Levi-Montalcini, R. (1964) Growth control of nerve cells by a protein factor and its antiserum. Science 143:105–110.

Levi-Montalcini, R.C.P. (1979) The nerve-growth factor. Sci. Am. 240:68.

Levine, E.M., Becker, B.G. (1977) Biochemical methods for detecting mycoplasma contamination. In: McGarrity, G.T., Murphy, D.G., Nichols, W.W. (eds.): "Mycoplasma Infection of Cell Cultures." New York, Plenum Press, pp. 87–104.

Ley, K.D., Tobey, R.A. (1970) Regulation of initiation of DNA synthesis in Chinese hamster cells. II. Induction of DNA synthesis and cell division by isoleucine and glutamine in G_1-arrested cells in suspension culture. J. Cell Biol. 47:453–459.

Lieber, M., Mazzetta, J., Nelson-Rees, W., Kaplan, M., Todaro, G. (1975) establishment of a continuous tumor-cell line (PANC-1) from a human carcinoma of the exocrine pancreas. Int. J. Cancer 15:741–747.

Liebermann, D., Sachs, L. (1978) Nuclear control of neurite induction in neuroblastoma cells. Exp. Cell Res. 113:383–390.

Lillie, J.H., MacCallum, D.K., Jepsen, A. (1980) Fine structure of subcultivated stratified squamous epithelium grown on collagen rafts. Exp. Cell Res. 125:153–165.

Limburg, H., Heckmann, U. (1968) Chemotherapy in the treatment of advanced pelvic malignant disease with special reference to ovarian cancer. J. Obstet. Gynaec. Brit. Cwlth. 75:1246–1255.

Lin, C.C., Uchida, I.A. (1973) Fluorescent banding of chromosomes (Q-bands). In Kruse, P.F., Patterson, M.K. (eds.): "Tissue Culture Methods and Applications." New York, Academic Press, pp. 778–781.

Lin, M.A., Latt, S.A., Davidson, R.L. (1974) Identification of human and mouse chromosomes in human-mouse hybrids by centromere fluorescence. Exp. Cell Res. 87:429–433.

Lindgren, A., Westermark, B., Ponten, J. (1975) Serum stimulation of stationary human glia and glioma cells in culture. Exp. Cell Res. 95:311–319.

Lindsay, R.M. (1979) Adult rat brain astrocytes support survival of both NGF-dependent and NGF-insensitive neurones. Nature 282:80.

Linser, P., Moscona, A.A. (1980) Induction of glutamine synthetase in embryonic neural retina-localization in Muller fibers and dependence on cell interaction. Proc. Natl. Acad. Sci. USA 76:6476–6481.

Littauer, U.Z., Giovanni, M.Y., Glick, M.C. (1979) Differentiation of human neuroblastoma cells in culture. Biochem. Biophys. Res. Com. 88:933–939.

Littlefield, J.W. (1964) Selection of hybrids from matings of fibroblasts *in vitro* and their presumed recombinants. Science 145:709–710.

Litwin, J. (1973) Titanium disks. In Kruse, P.F., Patterson, M.K. (eds): "Tissue Culture Methods and Applications." New York, Academic Press, pp. 383–387.

Lloyd, K.O., Travassos, L.R., Takahashi, T., Old, L.J. (1979) Cell surface glycoproteins of human tumor cell lines: Unusual characteristics of malignant melanoma. J. Natl. Cancer Inst. 63:623.

Lovelock, J.E., Bishop, M.W.H. (1959) Prevention of freezing damage to living cells by dimethyl sulphoxide. Nature 183:1394–1395.

Maciag, T., Cerondolo, J., Ilsley, S., Kelley, P.R., Forand, R. (1979) Endothelial cell growth factor from bovine hypothalamus-identification and partial characterization. Proc. Natl. Acad. Sci. USA 76:5674–5678.

Macieira-Coelho, A. (1973) Cell cycle analysis. A. Mammalian cells. In Kruse, P.F., Patterson, M.K. (eds.): "Tissue Culture Methods and Applications." New York, Academic Press, pp. 412–422.

Macpherson, I. (1973) Soft agar techniques. In Kruse, P.F., Patterson, M.K. (eds.): "Tissue Culture Methods and Applications." New York, Academic Press, pp. 276–280.

Macpherson, I., Bryden, A. (1971) Mitomycin C treated cells as feeders. Exptl. Cell Res. 69:240–241.

Macpherson, I., Montagnier, L. (1964) Agar suspension culture for the selective assay of cells transformed by polyoma virus. Virology 23:291–294.

Macpherson, I., Stoker, M. (1962) Polyoma transformation of hamster cell clones—an investigation of genetic factors affecting cell

competence. Virology 16:147.
Macy, M. (1978) Identification of cell line species by isoenzyme analysis. Man. Am. Tissue Cult. Assoc. 4:833–836.
Mahdavi, V., Hynes, R.O. (1979) Proteolytic enzymes in normal and transformed cells. Biochim. Biophys. Acta 583:167–178.
Malan-Shibley, L., Iype, P.T. (1981) The influence of culture conditions on cell morphology and tyrosine aminotransferase levels in rat liver epithelial cell lines. Exp. Cell Res. 131:363–371.
Malinin, T.I., Perry, V.P. (1967) A review of tissue and organ viability assay. Cryobiology 4:104–115.
Maltese, W.A., Volpe, J.J. (1979) Induction of an oligodendroglial enzyme in C-6 glioma cells maintained at high density or in serum-free medium. J. Cell. Physiol. 101:459–470.
Maramorosch, K. (1976) "Invertebrate Tissue Culture." New York: Academic Press.
Marcus, M., Lavi, U., Nattenberg, A., Ruttem, S., Markowitz, O. (1980) Selective killing of mycoplasmas from contaminated cells in cell cultures. Nature 285:659–660.
Mardh, P.H. (1975) Elimination of mycoplasmas from cell cultures with sodium polyanethol sulphonate. Nature 254:515–516.
Mareel, M.M., Bruynell, E., Storme, G. (1980) Attachment of mouse fibrosarcoma cells to precultured fragments of embryonic chick heart. Virchows Arch B. Cell Path. 34:85–97.
Mareel, M., Kint, J., Meyvisch, C. (1979) Methods of study of the invasion of malignant C3H-mouse fibroblasts into embryonic chick heart *in vitro*. Virchows Arch. B. Cell Path. 30:95–111.
Martin, G.R. (1978) Advantages and limitations of teratocarcinoma stem cells as models of development. In Johnson, M.H. (ed.): "Development in Mammals, Vol. 3." Amsterdam, North-Holland Publishing Co., p. 225.
Martin, G.R. (1975) Teratocarcinomas as a model system for the study of embryogenesis and neoplasia. Cell 5:229–243.
Martin, G.R., Evans, M.J. (1974) The morphology and growth of a pluripotent teratocarcinoma cell line and its derivatives in tissue culture. Cell 2:163–172.
Mather, J. (1979) Testicular cells in defined medium. In Jakoby, W.B., Pastan, I.H. (eds.): "Methods in Enzymology, Vol. LVII, Cell Culture." New York, Academic Press, p. 103.
Mather, J.P., Sato, G.H. (1979a) The growth of mouse melanoma cells in hormone supplemented, serum-free medium. Exp. Cell Res. 120:191.
Mather, J.P., Sato, G.H. (1979b) The use of hormone supplemented serum free media in primary cultures. Exp. Cell Res. 124:215.
Matsamura, T., Nitta, K., Yoshikawa, M., Takaoka, T., Katsuta, H. (1975) Action of bacterial protease on the dispersion of mammalian cells in tissue culture. Jpn. J. Exp. Med. 45:383–392.
Mazur, P., Leibo, S.P., Farrant, J., Chu, E.H.Y., Hanna, M.G., Jr., Smith, C.H. (1970) Interactions of cooling rate, warming rate and protective additive on the survival of frozen mammalian cells. In Wolstenholme, G.E.W., O'Conor, M (eds): "The Frozen Cell." CIBA Foundation Symposium. London, J.A. Churchill, pp. 69–85.
McCool, D., Miller, R.J., Painter, R.H., Bruch, W.R. (1970) Erythropoietin sensitivity of rat bone marrow cells separated by velocity sedimentation. Cell Tissue Kinet. 3:55–66.
McCoy, T.A., Maxwell, M., Kruse, P.F. (1959) Amino acid requirements of the Novikoff hepatoma in vitro. Proc. Soc. Exp. Biol. Med. 100:115–118.
McGarrity, G.J. (1979) Detection of contamination. In Jakoby, W.B., Pastan, I.H. (eds.): "Methods in Enzymology." New York, Academic Press, pp. 18–27.
McKeehan, W.L. (1977) The effect of temperature during trypsin treatment on viability and multiplication potential of single normal human and chicken fibroblasts. Cell Biol. Int. Rep. 1:335–343.
McKeehan, W.L., Ham, R.G. (1976) Stimulation of clonal growth of normal fibroblasts with substrata coated with basic polymers. J. Cell Biol. 71:727–734.
McKeehan, W.L., Hamilton, W.G., Ham, R.G. (1976) Selenium is an essential trace nutrient for growth of WI-38 diploid human fibroblasts. Proc. Natl. Acad. Sci. USA 73:2023–2027.
McKeehan, W.L., McKeehan, K.A. (1979) Oxocarboxylic acids, pyridine nucleotide-linked oxidoreductases and serum factors in regulation of cell proliferation. J. Cell Physiol. 101:9–16.
McKeehan, W.L., McKeehan, K.A., Hammond, S.L., Ham, R.G. (1977) Improved medium for clonal growth of human diploid cells at low concentrations of serum protein. In Vitro 13:399–416.
McLimans, W.F. (1979) Mass culture of mammalian cells. In Jakoby, W.B., Pastan, I.H. (eds.): "Methods in Enzymology, Vol. LVII, Cell Culture." New York, Academic Press, pp. 194–211.
Meera Khan, P. (1971) Enzyme electrophoresis on cellulose acetate gel: Zymogram patterns in man-mouse and man-Chinese hamster somatic cell hybrids. Arch. Biochem. Biophys. 145:470–483.
Meier, S., Hay, E.D. (1974) Control of corneal differentiation by extracellular materials. Collagen as a promoter and stabilizer of epithelial stroma production. Dev. Biol. 38:249–270.
Meier, S., Hay, E.D. (1975) Stimulation of corneal differentiation by interaction between cell surface and extracellular matrix. J. Cell Biol. 66:275–291.
Meistrich, M.L., Meyn, R.E., Barlogie, B. (1977a) Synchronization of mouse L-P59 cells by centrifugal elutriation separation. Exp. Cell Res. 105:169.
Meistrich, M.L., Grdina, D.J. Meyn, R.E., Barlogie, B. (1977b) Separation of cells from mouse solid tumors by centrifugal elutriation. Cancer Res. 37:4291–4296.
Melamad, M.R., Mullaney, P.F., Mendelsohn, M.L. (1979) "Flow Cytometry and Sorting." New York, John Wiley & Sons.
Metcalf, D. (1970) Studies on colony formation *in vitro* by mouse bone marrow cells. J. Cell. Physiol. 76:89–100.
Michalopoulos, G., Pitot, H.C. (1975) Primary culture of parenchymal liver cells on collagen membranes. Fed. Proc. 34:826.
Miller, D.R., Hamby, K.M., Allison, D.P., Fischer, S.M., Slaga, T.J. (1980) The maintenance of a differentiated state in cultured mouse epidermal cells. Exp. Cell Res. 129:63–71.
Miller, G.G., Walker, G.W.R., Giblack, R.E. (1972) A rapid method to determine the mammalian cell cycle. Exp. Cell Res. 72:533–538.
Miller, R.G., Phillips, R.A. (1969) Separation of cells by velocity sedimentation. J. Cell. Physiol. 73:191–201.
Milo, G.E., Ackerman, G.A., Noyes, I. (1980) Growth and ultrastructural characterization of proliferating human keratinocytes *in vitro* without added extrinsic factors. In Vitro 16:20–30.
Milstein, C., Galfre, G., Secher, D.S., Springer, T. (1979) Mini review, monoclonal antibodies and cell surface antigens. Cell Biol. Internat. Rep. 3:1–16.
Minna, J., Glazer, D., Nirenberg, M. (1972) Genetic dissection of neural properties using somatic cell hybrids. Nature New Biol. 235:225–231.
Minna, J., Gilman, A. (1973) Genetic analysis of the nervous system using somatic cell hybrids. In Davidson, R.L., de la Cruz, F.F. (eds): "Somatic Cell Hybridization." New York, Raven Press, pp. 191–196.
Montagnier, L. (1968) Corrélation entre la transformation des cellule BHK21 et leur résistance aux polysaccharised acides en milieu gélifié. C.R. Acad. Sci., D. 267:921–924.
Moore, G.E., Gerner, R.E., Franklin, H.A. (1967) Culture of normal human leukocytes. J. Amer. Med. Assoc. 199:519–524.
Morgan, J.F., Morton, H.J., Parker, R.C. (1950) Nutrition of animal cells in tissue culture. I. Initial studies on a synthetic medium. Proc. Soc. Exp. Biol. Med. 73:1.
Morton, H.J. (1970) A survey of commercially available tissue cul-

ture media. In Vitro 6:89–108.

Moscona, A.A. (1952) Cell suspension from organ rudiments of chick embryos. Exp. Cell Res. 3:535.

Moscona, A.A., Piddington, R. (1966) Stimulation by hydrocortisone of premature changes in the developmental pattern of glutamine synthetase in embryonic retina. Biochim. Biophys. Acta 121:409–411.

Munthe-Kaas, A.C., Seglen, P.O. (1974) The use of metrizamide as a gradient medium for isopycnic separation of rat liver cells. FEBS Lett. 43:252–256.

Nagy, B., Ban, K., Brdar, B. (1977) Fibrinolysis associated with human neoplasia: Production of plasminogen activator by human tumors. Int. J. Cancer 19:614–620.

Nardone, R.M., Todd, G., Gonzalex, P., Gaffney, E.V. (1965) Nucleoside incorporation into strain L cells: Inhibition by pleuropneumonia-like organisms. Science 149:1100–1101.

Nelson, P.G., Lieberman, M. (1981) "Excitable Cells in Tissue Culture." New York: Plenum.

Nelson-Rees, W., Flandermeyer, R.R. (1977) Inter- and intraspecies contamination of human breast tumor cell lines HBC and BrCa5 and other cell cultures. Science 195:1343–1344.

Newton, A.A., Wildy, P. (1959) Parasynchronous division of HeLa cells. Exp. Cell Res. 16:624–635.

Neugut, A.I., Weinstein, I.B. (1979) Use of agarose in the determination of anchorage-independent growth. In Vitro 15:351.

Nichols, E.A., Ruddle, F.H. (1973) A review of enzyme polymorphism, linkage and electrophoretic conditions for mouse and somatic cell hybrids in starch gels. J. Histochem. Cytochem. 21:1066–1081.

Nicolson, G.L. (1976) Trans-membrane control of the receptors on normal and tumor cells. II. Surface changes associated with transformation and malignancy. Biochim. Biophys. Acta 458:1–72.

Nicosia, R.F., Leighton, J. (1981) Angiogenesis *in vitro*: Light microscopic, radioautographic and ultrastructural studies of rat aorta in histophysiological gradient culture. In Vitro 17:204.

Nilos, R.M., Makarski, J.S. (1978) Control of melanogenesis in mouse melanoma cells of varying metastatic potential. J. Natl. Cancer Inst. 61:523–526.

Noguchi, P., Wallace, R., Johnson, J., Earley, E.M., O'Brien, S., Ferrone, S., Pellegrino, M.A., Milstein, J., Needy, C., Browne, W., Petricciani, J. (1979) Characterization of WiDr: A human colon carcinoma cell line. In Vitro 15:401.

Novak, J. (1962) A high-resolution autoradiographic apposition method for water-soluble tracers and tissue constituents. Int. J. Appl. Radiat. Isot. 13:187–190.

O'Brien, S.J., Kleiner, G. (1977) Enzyme polymorphisms as genetic signatures in human cell cultures. Science 195:1345–1348.

O'Farrell, P.H. (1975) High resolution two-dimensional electrophoresis of proteins. J. Biol. Chem. 250:4007–4021.

O'Hare, M.J., Ellison, M.L., Neville, A.M. (1978) Tissue culture in endocrine research: Perspectives, pitfalls, and potentials. Curr. Top. Exp. Endocrinol. 3:1–56.

Okazaki, K., Holtzer, H. (1966) Myogenesis: Fusion, myosin synthesis and the mitotic cycle. Proc. Natl. Acad. Sci. USA 56:1484–1489.

Olmsted, C.A. (1967) A physico-chemical study of fetal calf sera used as tissue culture nutrient correlated with biological tests for toxicity. Exp. Cell Res. 48:283–299.

Orly, J., Sato, G., Erickson, G.F. (1980) Serum suppresses the expression of hormonally induced functions in cultured granulosa cells. Cell 20:817–827.

Olsson, I., Ologsson, T. (1981) Induction of differentiation in a human promyelocytic leukemic cell line (HL-60). Exp. Cell Res. 131:225–230.

Osborne, C.K., Hamilton, B., Tisus, G., Livingston, R.B. (1980) Epidermal growth factor stimulation of human breast cancer cells in culture. Cancer Res. 40:2361–2366.

Owens, R.B., Smith, H.S., Hackett, A.J. (1974) Epithelial cell culture from normal glandular tissue of mice. Mouse epithelial cultures enriched by selective trypsinisation. J. Natl. Cancer Inst. 53:261–269.

Pahlman, S., Ljungstedt-Pahlman, A., Sanderson, P.J., Ward, G.A., Hermon-Taylor, J. (1979) Isolation of plasma-membrane components from cultured human pancreatic cancer cells by immuno-affinity chromatography of anti-β3M Sepharose 6MB. Br. J. Cancer 40:701.

Parenjpe, M.S., Boone, C.W., Ande Eaton, S.del. (1975) Selective growth of malignant cells by *in vitro* incubation on Teflon. Exp. Cell Res. 93:508–512.

Paris Conference (1971), Supplement (1975): Standardization in human cytogenetics. Cytogenet. Cell Genet. 15:201–238.

Pastan, I. (1979) Cell transformation. In Jakoby, W.B., Pastan, I.H. (eds): "Methods in Enzymology, Vol. LVII, Cell Culture." New York, Academic Press, pp. 368–370.

Patueleia, M.C., Friend, C. (1967) Tissue culture studies on murine virus-induced leukemia cells: Isolation of single cells in agar-liquid medium. Cancer Res. 27:726–730.

Paul, J., Fottrell, P.F. (1961) Molecular variation in similar enzymes from different species. Ann. NY Acad. Sci. 94:668–677.

Paul, J., Conkie, D., Freshney, R.I. (1969) Erythropoietic cell population changes during the hepatic phase of erythropoiesis in the foetal mouse. Cell Tissue Kinet. 2:283–294.

Paul, J. (1975) "Cell and Tissue Culture." Edinburgh, Churchill Livingstone, pp. 172–184.

Paul, W.E., Sredni, B., Schwartz, R.H., (1981) Long-term growth and cloning of non-transformed lymphocytes. Nature 294:697–699.

Pearse, A.G.E. (1968) "Histochemistry, Theoretical and Applied." Boston, Little, Brown & Co., pp. 255–264.

Peehl, D.M., Ham, R.G. (1980) Clonal growth of human keratinocytes with small amounts of dialysed serum. In Vitro 16:526–540.

Pereira, M.E.A., Kabat, E.A. (1979) A versatile immunoadsorbent capable of binding lectins of various specificities and its use for the separation of cell populations. J. Cell Bio. 82:185–194.

Perper, R.J., Zee, T.W., Mickelson, M.M. (1968) Purification of lymphocytes and platelets by gradient centrifugation. J. Lab. Clin. Med 72:842–868.

Pertoft, H., Laurent, T.C. (1977) Isopycnic separation of cells and cell organelles by centrifugation in modified colloidal silica gradients. In Catsimpoolas, N. (ed): "Methods of Cell Separation." New York, Plenum Press.

Petersen, D.F., Anderson, E.C., Tobey, R.A. (1968) Mitotic cells as a source of synchronized cultures. In Prescott, D.M. (ed): "Methods in Cell Physiology." New York, Academic Press, pp. 347–370.

Peterson, E.A., Evans, W.H. (1967) Separation of bone marrow cells by sedimentation at unit gravity. Nature 214:824–825.

Petricciani, J.C., Hopps, H.E., Chapple, P.J., eds. (1979) "Cell Substrates: Their Use in the Production of Vaccines and Other Biologicals." New York: Plenum Press.

Phillips, P., Kumar, P., Kumar, S., Waghe, M. (1979) Isolation and characterization of endothelial cells from adult rat brain white matter. J. Anat. 129:261.

Phillips, P., Steward, J.K., Kumar, S. (1976) Tumor angiogenesis factor (TAF) in human and animal tumors. Int. J. Cancer 17:549–558.

Pike, B.L., Robinson, W.A. (1970) Human bone marrow colony growth in agar-gel. J. Cell. Physiol. 76:77–84.

Pitot, H., Periano, C., Morse, P., Potter, V.R. (1964) Hepatomas in tissue culture compared with adapting liver *in vivo*. Natl. Cancer

Inst. Monogr. 13:229–245.
Platsoucas, C.D., Good, R.A., Gupta, S. (1979) Separation of human lymphocyte-T subpopulations (T-mu, T-gamma) by density gradient electrophoresis. Proc. Natl. Acad. Sci. USA 76:1972.
Pluznik, D.H., Sachs, L. (1965) The cloning of normal "mast" cells in tissue culture. J. Cell. Comp. Physiol. 66:319–324.
Polinger, I.S. (1970) Separation of cell types in embryonic heart cell cultures. Exp. Cell Res. 63:78–82.
Pollock, M.F., Kenny, G.E. (1963) Mammalian cell cultures contaminated with pleuro-pneumonia-like organisms. III. Elimination of pleuro-pneumonia-like organisms with specific antiserum. Proc. Soc. Exp. Biol. Med. 112:176–181.
Pollack, M.S., Heagney, S.D., Livingston, P.O., Fogh, J. (1981) HLA-A, B, C & DR alloantigen expression on forty-six cultured human tumor cell lines. J. Natl. Cancer Inst. 66:1003–1012.
Pontecorvo, G. (1975) Production of mammalian somatic cell hybrids by means of polyethylene glycol treatment. Somat. Cell Genet. 1:397–400.
Pontén, J., Macintyre, E. (1968) Interaction between normal and transformed bovine fibroblasts in culture. II. Cells transformed by polyoma virus. J. Cell Sci. 3:603–668.
Pontén, J. (1975) Neoplastic human glia cells in culture. In Fogh, J. (ed): "Human Tumor Cells *in vitro*." New York, Plenum Publishing, pp. 175–206.
Pontén, J., Westermark, B (1980) Cell generation and aging of nontransformed glial cells from adult humans. In Fedoroff, S., Hertz, L. (eds): "Advances in Cellular Neurobiology, Vol. 1." New York, Academic Press, pp. 209–227.
Pretlow, T.P., Stinson, A.J., Pretlow, T.G., Glover, G.L. (1978) Cytologic appearance of cells dissociated from rat colon and their separation by isokinetic and isopyknic sedimentation in gradients of Ficoll. J. Natl. Cancer Inst. 61:1431–1437.
Pretlow, T.G. (1971) Estimation of experimental conditions that permit cell separations by velocity sedimentation on isokinetic gradients of Ficoll in tissue culture medium. Anal. Biochem. 41:248–255.
Prince, G.A., Jenson, A.B., Billups, L.C., Notkins, A.L. (1978) Infection of human pancreatic beta cell cultures with mumps virus. Nature 271:158–161.
Prop, F.J.A., Wiepjes, G.J. (1973) Sequential enzyme treatment of mouse mammary gland. In Kruse, P.F., Patterson, M.K. (eds): "Tissue Culture Methods and Applications." New York, Academic Press, pp. 21–24.
Puck, T.T., Marcus, P.I. (1955) A rapid method for viable cell titration and clone production with HeLa cells in tissue culture: The use of X-irradiated cells to supply conditioning factors. Proc. Natl. Acad. Sci. USA 41:432–437.
Quarles, J.M., Morris, N.G., Leibovitz, A. (1980) Carcinoembryonic antigen production by human colorectal adenocarcinoma cells in matrix-perfusion culture. In Vitro 16:113–118.
Quastler, H. (1963) The analysis of cell population kinetics. In Lamerton, L.F., Fry, R.J.M. (eds): "Cell Proliferation." Philadelphia, Davis, pp. 18–34.
Rabito, C.A., Tchao, R., Valentich, J., Leighton, J. (1980) Effect of cell substratum interaction of hemicyst formation by MDCK cells. In Vitro 16:461–468.
Raff, M.C., Fields, K.L., Hakomori, S.L., Minsky, R., Pruss, R.M., Winter, J. (1979) Cell-type-specific markers for distinguishing and studying neurons and the major classes of glial cells in culture. Brain Res. 174:283–309.
Raff, M.C., Abney, E., Brockes, J.P., Hornby-Smith, A. (1978) Schwann cell growth factors. Cell 15:813–822.
Reddy, J.K., Rao, M.S., Warren, J.R., Minnick, O.T. (1979) Concanavalin A agglutinability and surface microvilli of dissociated normal and neoplastic pancreatic acinar cells of the rat. Exp. Cell Res. 120:55–61.
Reel, J.R., Kenney, F.T. (1968) "Superinduction" of tyrosine transaminase in hepatoma cell cultures: Differential inhibition of synthesis and turnover by actinomycin D. Proc. Natl. Acad. Sci. USA 61:200–206.
Reid, L.M., Rojkind, M. (1979) New techniques for culturing differentiated cells: Reconstituted basement membrane rafts. In Jakoby, W.B., Pastan, I.H. (eds.): "Methods in Enzymology, Vol. LVII, Cell Culture." New York, Academic Press, pp. 263–278.
Reitzer, L.J., Wice, B.M., Kennel, D. (1979) Evidence that glutamine, not sugar, is the major energy source for cultured HeLa cells. J. Biol. Chem. 254:2669–2677.
Rheinwald, J.G., Beckett, M.A. (1981) Tumorigenic keratinocyte lines requiring anchorage and fibroblast support cultured from human squamous cell carcinomas. Cancer Res. 41:1657.
Rheinwald, J.G., Green, H. (1975) Formation of a keratinizing epithelium in culture by a cloned cell line derived from a teratoma. Cell 6:317–330.
Rheinwald, J.G., Green, H. (1975) Serial cultivation of strains of human epidermal keratinocytes: The formation of keratinizing colonies from single cells. Cell 6:331–344.
Rheinwald, J.G., Green, H. (1977) Epidermal growth factor and the multiplication of cultured human epidermal keratinocytes. Nature 265:421–424.
Richler, C., Yaffe, D. (1970) The *in vitro* cultivation and differentiation capacities of myogenic cell lines. Dev. Biol. 23:1–22.
Rickwood, D., Birnie, G.D. (1975) Metrizamide, a new density-gradient medium. FEBS Lett. 50:102–110.
Riddle, P.N. (1979) Time-lapse cinemicroscopy. In Treherne, J.E., Rubery, P.H. (eds.): "Biological Techniques Series." Academic Press.
Rifkin, D.B., Loeb, J.N., Moore, G., Reich, E. (1974) Properties of plasminogen activators formed by neoplastic human cell cultures. J. Exp. Med. 139:1317–1328.
Rindler, M.J., Chuman, L.M., Shaffer, L., Saier, M.H., Jr. (1979) Retention of differentiated properties in an established dog kidney epithelial cell line (MDCK). J. Cell Biol. 81:635–648.
Rockwell, G.A., Sato, G.H., McClure, D.B. (1980) The growth requirements of SV40 virus transformed Balb/c-3T3 cells in serum-free monolayer culture. J. Cell. Physiol. 103:323–331.
Rogers, A.W. (1979) "Techniques of Autoradiography (3rd Edition)." The Netherlands, Elsevier/North-Holland Biomedical Press.
Rojkind, et al. (1980) Connective tissue biomatrix: Its isolation and utilization for long term cultures of normal rat hepatocytes. J. Cell Biol. 87:255–263.
Rosenberg, M.D. (1965) The culture of cells and tissues at the saline-fluorocarbon interface. In Ramakrishnan, C.V. (ed.): "Tissue Culture." The Hague, W. Junk, pp. 93–107.
Ross, R. (1971) The smooth muscle cell. II. Growth of smooth muscle in culture and formation of elastic fibers. J. Cell Biol. 50:172–186.
Rossi, G.B., Friend, C. (1967) Erythrocytic maturation of (Friend) virus-induced leukemic cells in spleen clones. Proc. Natl. Acad. Sci. USA 58:1373–1380.
Rothfels, K.H., Siminovitch, L. (1958) An air drying technique for flattening chromosomes in mammalian cells grown *in vitro*. Stain Technol. 33:73–77.
Rotman, B., Papermaster, B.W. (1966) Membrane properties of living mammalian cells as studies by enzymatic hydrolysis of fluorogenic esters. Proc. Natl. Acad. Sci. USA 55:134–141.
Rutzky, L.P., Tomita, J.T., Calenoff, M.A., Kahan, B.D. (1979) Human colon adenocarcinoma cells. III. *In vitro* organoid expression and carcino-embryonic antigen kinetics in hollow fiber culture. J. Natl. Cancer Inst. 63:893.

Salpeter, M.M. (1974) Electron microscope autoradiography: A personal assessment. In Wisse, E., Daems, W.Th., Molenaar, I., van Duijn, P. (eds.): "Electron Microscopy and Cytochemistry." Amsterdam, North-Holland Publishing Company, pp. 315–326.

Sandberg, A.A. (1980) "The Chromosomes in Human Cancer and Leukaemia." Elsevier/North-Holland Press.

Sandström, B. (1965) Studies on cells from liver tissue cultivated *in vitro*. I. Influence of the culture method on cell morphology and growth pattern. Exp. Cell Res. 37:552–568.

Sanford, K.K., Handieman, S.L., Jones, G.M. (1977) Morphology and serum dependence of cloned cell lines undergoing spontaneous malignant transformation in culture. Cancer Res. 37:821–830.

Sanford, K.K., Earle, W.R., Likely, G.D. (1948) The growth *in vitro* of single isolated tissue cells. J. Natl. Cancer Inst. 9:229.

Sanford, K.K., Earle, W.R., Evans, V.J., Waltz, H.K., Shannon, J.E. (1951) The measurement of proliferation in tissue cultures by enumeration of cell nuclei. Natl. Cancer Inst. 11:773.

Sato, G., ed. (1981) "Functionally Differentiated Cell Lines." New York, Alan R. Liss, Inc.

Sato, G. (1979) The growth of cells in serum-free hormone-supplemented medium. In Jakoby, W.B., Pastan, I.H. (eds.): "Methods in Enzymology." New York, Academic Press, pp. 94–109.

Sato, G.H., Yasumura, Y. (1966) Retention of differentiated function in dispersed cell culture. Trans. NY Acad. Sci. 28:1063–1079.

Sato, G., Reid, L. (1978) Biochemical mode of action of hormones II. In Richenburg, H.V., (ed.): "International Review of Biochemistry." Baltimore, University Park Press, pp. 219–251.

Sato, G.H., Pardee, A.B., Sirbasku, D.A. (eds.) (1982) "Growth of Cells in Hormonally Defined Media." Cole Spring Harbor Conference on Cell Proliferation, 9. Cold Spring Harbor, Maine, Cold Spring Harbor Laboratory.

Sattler, C.A., Michalopoulos, G., Sattler, G.L., Pitot, H.C. (1978) Ultrastructure of adult rat hepatocytes cultured on floating collagen membranes. Cancer Res. 38:1539–1549.

Savage, C.R., Jr., Bonney, R.J. (1978) Extended expression of differentiated function in primary cultures of adult liver parenchymal cells maintained on nitrocellulose filters. I. Induction of phosphoenol pyruvate carboxykinase and tryosine aminotransferase. Exp. Cell Res. 114:307–315.

Schaeffer, W.I. (1979) Proposed usage of animal tissue culture terms. Usage of vertebrate cell, tissue and organ culture terminology. In Vitro 15:649–653.

Schengrund, C.L., Repman, M.A. (1979) Differential enrichment of cells from embryonic rat cerebra by centrifugal elutriation. J. Neurochem. 33:283.

Scher, W., Holland, J.G., Friend, C. (1971) Hemoglobin synthesis in murine virus-induced leukemic cells *in vitro*. I. Partial purification and identification of hemoglobins. Blood 37:428–437.

Schimmelpfeng, L., Langenberg, U., Peters, J.M. (1968) Macrophages overcome mycoplasma infections of cells *in vitro*. Nature 285:661.

Schleicher, J.B. (1973) Multisurface stacked plate propagators. In Kruse, P.F., Patterson, M.K. (eds.): "Tissue Culture Methods and Applications." New York, Academic Press, pp. 333–338.

Schneider, E.L., Stanbridge, E.J. (1975) A simple biochemical technique for the detection of mycoplasma contamination of cultured cells. Methods Cell Biol. 10:278–290.

Schneider, H., Muirhead, F.E., Zydeck, F.A. (1963) Some unusual observations of organoid tissues and blood elements in monolayer cultures. Exp. Cell Res. 30:449–459.

Schnook, L.B., Otz, U., Lazary, S., De Week, A.L., Minowada, J., Odavic, R., Kniep, E.M., Edy, V. (1981) Lymphokine and monokine activities in supernatants from human lymphoid and myeloid cell lines. Lymphokines 2:1–19.

Schousboe, A., Thorbek, P., Hertz, L., Krogsgaard-Larsen, P. (1979) Effects of GABA analogues of restricted conformation on GABA transport in astrocytes and brain cortex slices and on GABA receptor binding. J. Neurochem. 33:181.

Schulman, H.M. (1968) The fractionation of rabbit reticulocytes in dextran density gradients. Biochim. Biophys. Acta 148:251–255.

Schwartz, S.M. (1978) Selection and characterization of bovine aortic endothelial cells. In Vitro 14:966.

Segal, S. (1964) Hormones, amino-acid transport and protein synthesis. Nature 203:17–19.

Selby, P.J., Thomas, J.M., Monaghan, P., Sloane, J., Peckham, M.J. (1980) Human tumor xenografts established and serially transplanted in mice immunologically deprived by thymectomy, cytosine arabinoside and whole-body irradiation. Br. J. Cancer 41:52.

Shall, S. (1973) Sedimentation in sucrose and Ficoll gradients of cells grown in suspension culture. In Kruse, P.F., Patterson, M.K., (eds.): "Tissue Culture Methods and Applications." New York, Academic Press, pp. 198–204.

Shall, S., McClelland, A.J. (1971) Synchronization of mouse fibroblast LS cells grown in suspension culture. Nat. New Biol. 229:59–61.

Shows, T.B., Sakaguchi, A.Y. (1980) Gene transfer and gene mapping in mammalian cells in culture. In Vitro 16:55–76.

Sinclair, R., Morton, R.A. (1963) Variations in X-ray response during the division cycle of partially synchronized Chinese hamster cells in culture. Nature 199:1158–60.

Sirica, A.E., Hwand, C.G., Sattler, G.L., Pitot, H.C. (1980) Use of primary cultures of adult rat hepatocytes on collagen gel-nylon mesh to evaluate carcinogen-induced unscheduled DNA synthesis. Cancer Res. 40:3259–3267.

Sladek, N.E. (1973) Bioassay and relative cytotoxic potency of cyclophosphamide metabolites generated *in vitro* and *in vivo*. Cancer Res. 33:1150–1158.

Smith, H.S., Owens, R.B., Hiller, A.J., Nelson-Rees, W.A., Johnston, J.O. (1976) The biology of human cells in tissue culture. I. Characterization of cells derived from osteogenic sarcomas. Int. J. Cancer 17:219–234.

Smith, H.S., Lan, S., Ceriani, R., Hackett, A.J., Stampfer, M.R. (1981) Clonal proliferation of cultured non-malignant and malignant human breast epithelia. Cancer Res. 41:4637–4643.

Sorieul, S., Ephrussi, B. (1961) Karylogical demonstration of hybridization of mammalian cells *in vitro*. Nature 190:653–654.

Splinter, T.A.W., Beudeker, M., Beek, A.V. (1978) Changes in cell density induced by isopaque. Exp. Cell Res. 111:245–251.

Sredni, B., Sieckmann, D.G., Kumagai, S.H., Green, I., Paul, W.E. (1981) Long term culture and cloning of non-transformed human B-lymphocytes. J. Exp. Med. 154:1500–1516.

Stampfer, M., Halcones, R.G., Hackett, A.J. (1980) Growth of normal human mammary cells in culture. In Vitro 16:415–425.

Stanbridge, E.J., Doersen, C.-J. (1978) Some effects that mycoplasmas have upon their injected host. In McGarrity, G.J., Murphy, D.G., Nichols, W.W. (eds.): "Mycoplasma Infection of Cell Cultures." New York, Plenum Press, pp. 119–134.

Stanley, M.A., Parkinson, E.,. (1979) Growth requirements of human cervical epithelial cells in culture. Int. J. Cancer 24:407–414.

Stanners, C.P., Eliceri, G.L., Green, H. (1971) Two types of ribosome in mouse-hamster hybrid cells. Nat. New Biol. 230:52–54.

Steck, P.A., Voss, P.G., Wang, J.L. (1979) Growth control in cultured 3T3 fibroblasts. J. Cell Biol. 83:562–575.

Stein, G.H. (1979) T98G: An anchorage-independent human tumor cell line that exhibits stationary phase G1 arrest *in vitro*. J. Cell Physiol. 99:43–54.

Stein, H.G., Yanishevsky, R. (1979) Autoradiography. In Jakoby, W.B., Pastan, I.H. (eds.): "Methods in Enzymology, Vol. LVII. Cell Culture." New York, Academic Press, pp. 279–292.

Stephenson, J.R., Axelrad, A.A., McLeod, D.I., Schreeve, M.M.

(1971) Induction of colonies of hemoglobin-synthesizing cells by erythropoietin *in vitro*. Proc. Natl. Acad. Sci. USA 68:1542–1546.

Stockdale, F.E., Topper, Y.J. (1966) The role of DNA synthesis and mitosis in hormone dependent differentiation. Proc. Natl. Acad. Sci. USA 56:1283–1289.

Stoker, M., O'Neill, C., Berryman, S., Waxman, B. (1968) Anchorage and growth regulation in normal and virus transformed cells. Int. J. Cancer 3:683–693.

Stoker, M.G.P. (1973) Role of diffusion boundary layer in contact inhibition of growth. Nature 246:200–203.

Stoker, M.G.P., Rubin, H. (1967) Density dependent inhibition of cell growth in culture. Nature 215:171–172.

Stoner, G.D., Harris, C.C., Myers, G.A., Trump, B.F., Connor, R.D. (1980) Putrescine stimulates growth of human bronchial epithelial cells in primary culture. In Vitro 16:399–406.

Stoner, G.D., Katoh, Y., Foidart, J-M., Trump, B.F., Steinert, P., Harris, C.C. (1981) Cultured human bronchial epithelial cells: Blood group antigens, keratin, collagens and fibronectin. In Vitro 17:577–587.

Strickland, S., Beers, W.H., (1976) Studies on the role of plasminogen activator in ovulation. *In vitro* response of granulosa cells to gonadotropins, Cyclic nucleotides, and prostaglandins. J. Biol. Chem. 251:5694–5702.

Stubblefield, E. (1968) Synchronization methods for mammalian cell cultures. In Prescott, D.M., (ed.): "Methods in Cell Physiology." New York, Academic Press, pp. 25–43.

Styles, J.A. (1977) A method for detecting carcinogenic organic chemicals using mammalian cells in culture. Br. J. Cancer 36:558.

Sykes, J.A., Whitescarver, J., Briggs, L., Anson, J.H. (1970) Separation of tumor cells from fibroblasts with use of discontinuous density gradients. J. Natl. Cancer Inst. 44:855–864.

Takahashi, K., Okada, T.S. (1970) Analysis of the effect of "conditioned medium" upon the cell culture at low density. Dev. Growth Differ. 12:65–77.

Tashjian, A.H., Jr. (1979) Clonal strains of hormone-producing pituitary cells. In Jakoby, W.B., Pastan, I.H. (eds.): "Methods in Enzymology, Vol. LVII, Cell Culture." New York, Academic Press, pp. 527–535.

Tashjian, A.H., Jr., Yasumura, Y., Levine, L., Sato, G.H., Parker, M.C. (1968) Establishment of clonal strains of rat pituitary tumor cells that secrete growth hormone. Endocrinology 82:342–368.

Taub, M., Saier, M.H., Jr. (1979) An established but differentiated kidney epithelial cell line (MDCK). In Jakoby, W.B., Pastan, I.H. (eds.): "Methods in Enzymology, Vol. LVII, Cell Culture." New York, Academic Press, pp. 552–560.

Taylor, C.R. (1978) Immunoperoxidase techniques. Arch. Pathol. Lab. Med. 102:113–121.

Taylor-Papadimitriou, J., Purkiss, P., Fentiman, I.S. (1980) Choleratoxin and analogues of cyclic AMP stimulate the growth of cultured human epithelial cells. J. Cell Physiol. 102:317–322.

Taylor-Robinson, D. (1978) Cultural and serologic procedures for mycoplasmas in tissue culture. In McGarrity, G., Murphy, D.G., Nichols, W.W. (eds.): "Mycoplasma Infection of Cell Cultures." New York, Plenum Press, pp. 47–56.

Temin, H.M. (1966) Studies on carcinogenesis by avian sarcoma viruses. III. The differential effect of serum and polyanions on multiplication of uninfected and converted cells. J. Natl. Cancer Inst. 37:167–175.

Temin, H.M., Rubin, H. (1958) Characteristics of an assay for Rous sarcoma virus and Rous sarcoma cells in tissue culture. Virology 6:669–688.

Thilly, W.G., Levine, D.W. (1979) Microcarrier culture: A homogenous environment for studies of cellular biochemistry. In Jakoby, W.B., Pastan, I.H., (eds.): "Methods in Enzymology, Vol. LVII, Cell Culture." New York, Academic Press, pp. 184–194.

Thomas, J.A., ed. (1970) "Organ Culture." New York, Academic Press.

Thurston, J.M., Joftes, D.L. (1963) Stain compatible with dipping radioautography. Stain Technol. 38:231–235.

Tobey, R.A., Anderson, E.C., Petersen, D.F. (1967) Effect of thymidine on duration of G1 in chinese hamster cells. J. Cell Biol. 35:53–67.

Tobner, J., Watts, M.T., Fu, J.J.L. (1980) An *in vitro* and *in vivo* investigation on three surface active agents as modulators of cell proliferation. Cancer Res. 40:1173–1180.

Todaro, G.J., DeLarco, J.E. (1978) Growth factors produced by sarcoma virus-transformed cells. Cancer Res. 38:4147–4154.

Todaro, G.J., Green, H. (1963) Quantitative studies of the growth of mouse embryo cells in culture and their development into established lines. J. Cell Biol. 17:299–313.

Tom, B.H., Rutzky, L.P., Jakstys, M.M., Oyasu, R., Kaye, C.I., Kahan, B.D. (1976) Human colonic adenocarcinoma cells. I. Establishment and description of a new line. In Vitro 12:180.

Toshiharu, M., Keiko, N., Masaaki, Y., Toshiko, T., Hajim, K. (1975) Action of bacterial neutral protease on the dispersion of mammalian cells in tissue culture. Japan J. Exp. Med. 45:383–392.

Tozer, B.T., Pirt, S.J. (1964) Suspension culture of mammalian cells and macromolecular growth promoting fractions of calf serum. Nature 201:375–378.

Traganos, F., Darzynkiewicz, Z., Sharpless, T., Melamed, M.R. (1977) Nucleic acid content and cell cycle distribution of five human bladder cell lines analyzed by flow cytofluorometry. Int. J. Cancer 20:30–36.

Trowell, O.A. (1959) The culture of mature organs in a synthetic medium. Exp. Cell Res. 16:118–147.

Tsao, M.C., Walthall, B.J., Ham, R.G. (1982) Clonal growth of normal human epidermal keratinocytes in a defined medium. J. Cell. Physiol. 110:219.

Turner, R.W.A., Siminovitch, L., McCulloch, E.A., Till, J.E. (1967) Density gradient centrifugation of hemopoietic colony-forming cells. J. Cell. Physiol. 69:73–81.

Twentyman, P.R. (1980) Response to chemotherapy of EMT6 spheroids as measured by growth delay and cell survival. Eur. J. Cancer 42:297–304.

Uchida, I.A., Lin, C.C. (1974) Quinacrine fluorescent patterns. In Yunis, J. (ed): "Human Chromosome Methodology, 2nd Ed." New York, Academic Press, pp. 47–58.

Unkless, J., Dano, K., Kellerman, G., Reich, E. (1974) Fibrinolysis associated with oncogenic transformation. Partial purificaition and characterization of cell factor, a plasminogen activator. J. Biol. Chem. 249:4295–4305.

Ursprung, H., ed. (1968) "The Stability of the Differentiated State." New York, Springer-Verlag.

Vago, C., ed. (1971) "Invertebrate Tissue Culture, Vol. 1." New York, Academic Press.

Vago, C., ed. (1972) "Invertebrate Tissue Culture, Vol. 2." New York, Academic Press.

Vaheri, A., Ruoslahti, E., Westermark, B., Ponten, J. (1976) A common cell-type specific surface antigen in cultured human glial cells and fibroblasts: Loss in malignant cells. J. Exp. Med. 143:64–72.

Valenti, C. (1973) Diagnostic use of cell cultures initiated from amniocentesis. In Kruse, P.F., Patterson, M.K. (eds.): "Tissue Culture Methods and Applications." New York, Academic Press, pp. 617–622.

Van Beek, W.P., Glimelius, B., Nilson, K., Emmelot, P. (1978) Changed cell surface glycoproteins in human glioma and osteosarcoma cells. Cancer Lett. 5:311–317.

Van der Bosch, J., Masui, H., Sato, G. (1981) Growth characteristics of primary tissue cultures from heterotransplanted human colorectal carcinomas in serum-free medium. Cancer Res. 41:611–618.

VanDiggelen, O., Shin, S., Phillips, D. (1977) Reduction in cellular tumorigenicity after mycoplasma infection and elimination of mycoplasma from infected cultures by passage in nude mice. Cancer Res. 37:2680–2687.

Van Someren, H., Van Hemegowyen, H.B., Los, W., Wurzer-Figurelli, E., Doppert, B., Yerylolt, M., Meera Khan, P. (1974) Enzyme electrophoresis on cellulose acetate gel II zymogram patterns in man-Chinese hamster cell hybrids. Humangenetik 25:189–201.

Van't Hof, J. (1968) In D.M. Prescott (ed): "Methods in Cell Physiology." New York, Academic Press, p. 95.

Van't Hof, J. (1973) Cell cycle analysis B. In Kruse, P., Patterson, M.K. (eds.): "Tissue Culture, Techniques and Application." New York, Academic Press, pp. 423–428.

Varon, S., Manthorpe, M. (1980) Separation of neurons and glial cells by affinity methods. In Fedoroff, S., Hertz, L. (eds.): "Advances in Cellular Neurobiology, Vol. I." New York, Academic Press, pp. 405–442.

Vlodavsky, I., Lui, G.M., Gospodarowicz, D. (1980) Morphological appearance, growth behavior and migratory activity of human tumor cells maintained on extracellular matrix versus plastic. Cell 19:607–617.

Vries, J.E., Benthem, M., Rumke, P. (1973) Separation of viable from nonviable tumor cells by flotation on a Ficoll-triosil mixture. Transplantation 15:409–410.

Wallace, D.H., Hegre, O.D. (1979) Development *in vitro* of epithelial-cell monolayers derived from fetal rat pancreas. In Vitro 15:270.

Walter, H. (1977) Partition of cells in two-polymer aqueous phases: A surface affinity method for cell separation. In Catsimpoolas, N. (ed.): "Methods of Cell Separation." New York, Plenum Press, pp. 307–354.

Walter, H. (1975) Partition of cells in two-polymer aqueous phases: A method for separating cells and for obtaining information on their surface properties. In Prescott, D.M. (ed.): "Methods in Cell Biology." New York, Academic Press, pp. 25–50.

Wang, H.C., Fedoroff, S. (1973) Karyology of cells in culture E. Trypsin technique to reveal G-bands. In Kruse, P.F., Patterson, M.J. (eds.): "Tissue Culture Methods and Applications." New York, Academic Press, pp. 782–787.

Wang, H.C., Fedoroff, S. (1972) Banding in human chromosomes treated with trypsin. Nature New Biol. 235:52–53.

Wang, R.J. (1976) Effect of room fluorescent light on the deterioration of tissue culture medium. In Vitro 12:19–22.

Warren, L., Buck, C.A., Tuszynski, G.P. (1978) Glycopeptide changes and malignant transformation. A possible role for carbohydrate in malignant behavior. Biochim. Biophys. Acta. 516:97.

Waymouth, C. (1977) In Evans, V.J., Perry, V.P., Vincent, M.M. (eds): "Manual of American Tissue Culture Association." 3:521.

Waymouth, C. (1970) Osmolality of mammalian blood and of media for culture of mammalian cells. In Vitro 6:109–127.

Waymouth, C. (1974) To disaggregate or not to disaggregate. Injury and cell disaggregation, transient or permanent? In Vitro 10:97–111.

Waymouth, C. (1959) Rapid proliferation of sublines of NCTC clone 929 (Strain L) mouse cells in a simple chemically defined medium (MB752/1). J. Natl. Cancer Inst. 22:1003.

Waymouth, C. (1979) Autoclavable medium AM 77B. J. Cell. Physiol. 100:548–550.

Weibel, E.R., Palade, G.E. (1961) New cytoplasmic components in arterial endothelia. J. Cell Biol. 23:101–102.

Weichselbaum, R., Epstein, J., Little, J.B. (1976) A technique for developing established cell lines from human osteosarcomas. In Vitro 12:833–836.

Weiss, M.C., Green, H. (1967) Human-mouse hybrid cell lines containing partial complements of human chromosomes and functioning human genes. Proc. Natl. Acad. Sci. USA 58:1104–1111.

Westermark, B., Ponten, J., Hugosson, R. (1973) Determinants for the establishment of permanent tissue culture lines from human gliomas. Acta Pathol. Microbiol. Scand. Section A 81:791–805.

Westermark, B. (1974) The deficient density-dependent growth control of human malignant glioma cells and virus-transformed glia-like cells in culture. Int. J. Cancer 12:438–451.

Westermark, B., Wasteson, A. (1975) The response of cultured human normal glial cells to growth factors. In Luft & Hall (eds.): "Advances in Metabolic Disorders, Vol. 8." New York, Academic Press, pp. 85–100.

Westermark, B. (1978) Growth control in miniclones of human glial cells. Exp. Cell Res. 111:295–299.

Whei-Yang, K.W., Prockop, D.J. (1977) Proline analogue removes fibroblasts from cultured mixed cell populations. Nature 266:63–64.

Whittle, W.L., Kruse, P.F. (1973) Replicate roller bottles. In Kruse, P.F., Patterson, M.K. (eds.): "Tissue Culture Methods and Applications." New York, Academic Press, pp. 327–331.

Whur, P., Magudia, M., Boston, J., Lockwood, J., Williams, D.C. (1980) Plasminogen activator in cultured Lewis lung carcinoma cells measured by chromogenic substrate assay. Br. J. Cancer 42:305–312.

Wiepjes, G.J., Prop, F.J.A. (1970) Improved method for preparation of single-cell suspensions from mammary glands of adult virgin mouse. Exp. Cell Res. 61:451–454.

Wigler, M., Sweet, R., Sim, G.K., Wold, B., Pellicer, A., Lacy, E., Maniatis, T., Silverstein, S., Axel, R. (1979) Transformation of mammalian cells with genes from procaryotes and eucaryotes. Cell 16:777–785.

Wilkes, P.R., Birnie, G.D., Old, R.W. (1978) Histone gene expression during the cell cycle studied by *in situ* hybridization. Exp. Cell Res. 115:441–444.

Willecke, K., Klomfass, M., Mierau, R., Dohner, J. (1979) Intraspecies transfer via total cellular DNA of the gene for hypoxanthine phosphoribosyltransferase into cultured mouse cells. Mol. Gen. Genet. 170:179–185.

Willingham, M.C., Pastan, I. (1975) Cyclic AMP modulates microvillus formation and agglutinability in transformed and normal mouse fibroblasts. Proc. Natl. Acad. Sci. USA 72:1263–1267.

Willmer, E.N., ed. (1965) "Cells and Tissues in Culture." London, Academic Press.

Wolff, E., Wolff, E. (1952) La determination de la differentiation sexuelle de la syrinx du canard cultinee *in vitro*. Bull. Biol. 86:325.

Wolff, D.A., Pertoft, H. (1972) Separation of HeLa cells by colloidal silica density gradient centrifugation. J. Cell Biol. 55:579.

Worton, R.G., Duff, C. (1979) Karyotyping. In Jakoby, W.B., Pastan, I.H. (eds.): "Methods in Enzymology, Vol. LVII, Cell Culture." New York, Academic Press, pp. 322–244.

Wright, J.E., Dendy, P.P. (1976) Identification of abnormal cells in short-term monolayer cultures of human tumor specimens. Acta Cytol. (Baltimore) 20:328–334.

Wu, A.M., Siminovitch, L., Till, J.E., McCulloch, E.A. (1968) Evidence for a relationship between mouse hemopoietic stem cells and cells forming colonies in culture. Proc. Natl. Acad. Sci. USA 59:1209–1215.

Yaffe, D. (1971) Developmental changes preceding cell fusion during muscle cell differentiation in vitro. Exp. Cell Res. 66:33–48.

Yaffe, D. (1968) Retention of differentiation potentialities during prolonged cultivation of myogenic cells. Proc. Natl. Acad. Sci.

USA 61:477–483.
Yang, J., Richards, J., Bowman, P., Guzman, R., Enami, J., McCormick, K., Hamamoto, S., Pitelka, D., Nandi, S. (1979) Sustained growth and 3-dimentional organization of primary mammary tumor epithelial cells embedded in collagen gels. Proc. Natl. Acad. Sci. USA 76:3401.
Yang, J., Elias, J.J., Petrakis, N.L., Wellings, S.R., Nandi, S. (1981) Effects of hormones and growth factors on human mammary epithelial cells in collagen gel culture. Cancer Res. 41:1021–1027.
Yang, J., Richards, J., Guzman, R., Imagawa, W., Nandi, S. (1980) Sustained growth in primary cultures of normal mammary epithelial cells embedded in collagen gels. Proc. Natl. Acad. Sci. USA 77:2088–2092.
Yavin, Z., Yavin, E. (1980) Survival and maturation of cerebral neurons on poly(L-lysine) surfaces in the absence of serum. Dev. Biol. 75:454–460.
Yerganian, G., Leonard, M.J. (1961) Maintenance of normal in situ chromosomal features in long-term tissue cultures. Science 133:1600–1601.
Yoshida, Y., Hilborn, V., Hassett, C., Mezfi, P., Byers, M.J., Freeman, A.G. (1980) Characterization of mouse fetal lung cells cultured on a pigskin substrate. In Vitro 16:433–445.
Yunis, J. (1974) "Human Chromosome Methodology, 2nd Edition." New York, Academic Press.
Zaroff, L., Sato, G.H., Mills, S.E. (1961) Single-cell platings from freshly isolated mammalian tissue. Exp. Cell Res. 23:565–575.
Zawydiwski, R., Duncan, G.R. (1978) Spontaneous ^{51}Cr release by isolated rat hepatocytes: An indicator of membrane damage. In Vitro 14:707–714.

GENERAL TEXTBOOKS FOR FURTHER READING

Paul, J. (1975) "Cell and Tissue Culture." Edinburgh, Scotland, Livingstone. Good basic textbook.
Kruse, P., Patterson, M.K. (1973) "Tissue Culture, Techniques and Applications." New York, Academic Press. Good for specialized techniques but some duplication.
Jakoby, W.B., Pastan, I.H. (eds.) (1979) "Cell Culture." In Colowick, S.P., Kaplan N.D. (series eds.) "Methods in Enzymology," Vol. LVIII. New York, Academic Press. Very good for specialized techniques, matrix, serum free media.
Pollack, R. (ed.) (1975) "Reading in Mammalian Cell Culture, 1st Edition." Cold Spring Harbor, New York, Cold Spring Harbor Laboratory. Very good compilation of key papers in the field. Good tutorial and general interest. Good for teaching.
Pollack, R. (ed.) (1981) "Reading in Mammalian Cell Culture, 2nd Edition." Cold Spring Harbor, New York, Cold Spring Harbor Laboratory. (Extensively revised from first edition, more like a second volume; you may want to have both editions.)
Crowe, R., Ozer, H., Rifkin, D. (1978) "Experiments with Normal and Transformed Cells." Cold Spring Harbor, New York, Cold Spring Harbor Laboratory. Laboratory exercises for senior undergraduate and graduate students.
Hall, D., Hawkins, S. (1975) "Laboratory Manual of Cell Biology." London, English Universities Press. Laboratory exercises for undergraduates.
Maramorosch, K. (1976) "Invertebrate Tissue Culture." New York, Academic Press. Quite useful, but restricted in species by the extent of work done.
Vago, E. (ed.) (1971, 1972) "Invertebrate Tissue Culture, Vols. 1 and 2." New York, Academic Press.
Dendy, P.P. (ed.) (1973) "Human Tumours in Short Term Culture." London, Academic Press. Characterization, Drug and Radiosensitivity. Good techniques review and some useful articles.
Federoff, S., Herz, L. (eds.) (1977) "Cell, Tissue and Organ Cultures in Neurobiology." New York, Academic Press. A good introduction to culturing neural cells.
Fogh, J. (1975) "Human Tumor Cells In Vitro." New York, Plenum. Some useful listings of human tumor types in culture.
Sato, G., Pardee, A.B., Sirbasku, D.A. (1982) "Growth of Cells in Hormonally Defined Media." Cold Spring Harbor Conferences on Cell Proliferation, Vol. 9. Cold Spring Harbor, New York, Cold Spring Harbor Laboratories. Excellent up to date review of serum free culture.
Reinert, J., Yeoman, M.M. (1982) "Plant Cell and Tissue Culture. A Laboratory Manual." Berlin, Heidelberg, New York, Springer-Verlag.
Harris, C.C., Trump, B.F., Stoner, G.D. (1981) "Normal Human Tissue and Cell Culture." In Prescott, D.M. (series ed.) "Methods in Cell Biology." New York, Academic Press.

USEFUL JOURNALS

In Vitro. Journal of American Tissue Culture Association
Journal of Cellular Physiology
Experimental Cell Research
Cell
Cell Biology, International Reports
Journal of Cell Biology
Cancer Research
International Journal of Cancer
British Journal of Cancer
Journal of the National Cancer Institute
Journal of Cell Science
Cellular Biology

Index

NOTES

NOTES

NOTES